Composites Science and Technology

Series Editor

Mohammad Jawaid, Laboratory of Biocomposite Technology, Universiti Putra Malaysia, INTROP, Serdang, Malaysia

Composites Science and Technology (CST) book series publishes cutting edge research monographs (both edited and authored volumes) comprehensively covering topics shown below:

- Composites from agricultural biomass/natural fibres include conventional composites-Plywood/MDF/Fiberboard
- Fabrication of Composites/conventional composites from biomass and natural fibers
- Wood, and Wood based materials
- Chemistry and biology of Composites and Biocomposites
- Modelling of damage of Composites and Biocomposites
- Failure Analysis of Composites and Biocomposites
- Structural Health Monitoring of Composites and Biocomposites
- Durability of Composites and Biocomposites
- Thermal properties of Composites and Biocomposites
- Flammability of Composites and Biocomposites
- Tribology of Composites and Biocomposites
- Bionanocomposites and Nanocomposites
- Applications of Composites, and Biocomposites

To submit a proposal for a research monograph or have further inquries, please contact springer editor, Ramesh Premnath (ramesh.premnath@springer.com).

More information about this series at https://link.springer.com/bookseries/16333

Jyotishkumar Parameswaranpillai · Nisa V. Salim ·
Harikrishnan Pulikkalparambil ·
Sanjay Mavinkere Rangappa ·
Ing. habil Suchart Siengchin
Editors

Micro- and Nano-containers for Smart Applications

Springer

Editors
Jyotishkumar Parameswaranpillai
Alliance University, Department of Science, Faculty of Science & Technology
Bengaluru, Karnataka, India

Harikrishnan Pulikkalparambil
King Mongkut's University of Technology North Bangkok
Bangkok, Thailand

Ing. habil Suchart Siengchin
King Mongkut's University of Technology North Bangkok
Bangkok, Thailand

Nisa V. Salim
Swinburne University of Technology
Hawthorn, VIC, Australia

Sanjay Mavinkere Rangappa
King Mongkut's University of Technology North Bangkok
Bangkok, Thailand

ISSN 2662-1819 ISSN 2662-1827 (electronic)
Composites Science and Technology
ISBN 978-981-16-8145-5 ISBN 978-981-16-8146-2 (eBook)
https://doi.org/10.1007/978-981-16-8146-2

This Springer imprint is published by the registered company Springer Nature Singapore Pte Ltd.
The registered company address is: 152 Beach Road, #21-01/04 Gateway East, Singapore 189721, Singapore

Contents

About the Editors

Dr. Jyotishkumar Parameswaranpillai is an Associate Professor at Alliance University, Bengaluru, India. He is a prolific editor and researcher who has published more than 15 edited books, 120 high-quality international research articles, and 40 book chapters. He is a frequent invited speaker and a consultant for many international organizations. He has received numerous prestigious awards including the Kerala State Young Scientist Award 2016 (Government of Kerala), INSPIRE Faculty Award 2011 (Government of India), and the best researcher award from KMUTNB 2019, Thailand. He is named in the world's Top 2% of the most-cited scientists in Single Year Citation Impact 2020, by Stanford University. His research interest includes polymer coatings, shape memory polymers, antimicrobial polymer films, green composites, nanostructured materials, water purification, polymer blends, and high-performance composites.

Dr. Nisa V. Salim is a Vice Chancellors Initiative Research Fellow at Swinburne University of Technology. She received her Ph.D. from Deakin University in 2013 on nanostructured polymer materials and joined Carbon Nexus as a Research Fellow in 2014. Her research has been focused on next generation carbon fibres; porous carbon materials; and functional fibres. Nisa won many awards as she advances her research career including AINSE Gold Medal, Smart Geelong Early Researcher award and SPE-ANZ Award. She has been awarded a number of prestigious fellowships including Victoria Fellowship and Endeavour Fellowship.

Dr. Harikrishnan Pulikkalparambil is a senior research fellow at King Mongkut's University of Technology North Bangkok, Bangkok, Thailand. He completed his B.Tech. in Polymer engineering from Mahatma Gandhi University, Kerala, India and M.Tech. in Polymer Science and Rubber Technology from Cochin University of Science & Technology, Kerala, India. His research work mainly focuses on preparation and characterization of smart polymeric materials and automotive lightweight composites. During his Ph.D., he worked on the development of self-healing epoxy coatings. He has published many papers and book chapters in high quality peer reviewed international journals, and co-edited one book.

Dr. Sanjay Mavinkere Rangappa is currently working as a Senior Research Scientist and also 'Advisor within the office of the President for University Promotion and Development towards International goals' at King Mongkut's University of Technology North Bangkok, Bangkok, Thailand. He has received the B.E. (Mechanical Engineering) in the year 2010, M.Tech. (Computational Analysis in Mechanical Sciences) in the year 2013, Ph.D. (Faculty of Mechanical Engineering Science) from Visvesvaraya Technological University, Belagavi, India in the year 2018 and Post Doctorate from King Mongkut's University of Technology North Bangkok, Thailand, in the year 2019. He is a Life Member of Indian Society for Technical Education (ISTE) and an Associate Member of Institute of Engineers (India). Also acting as a Board Member of various international journals in the fields of materials science and composites. He is a reviewer for more than 100 international Journals (for Nature, Elsevier, Springer, Sage, Taylor & Francis, Wiley, American Society for Testing and Materials, American Society of Agricultural and Biological Engineers, IOP, Hindawi, NC State University USA, ASM International, Emerald Group, Bentham Science Publishers, Universiti Putra, Malaysia), also a reviewer for book proposals, and international conferences. In addition, he has published more than 150 articles in high-quality international peer-reviewed journals indexed by SCI/Scopus, 6 editorial corners, 60 book chapters, one book, 18 books as an Editor (Published by lead publishers such as Elsevier, Springer, Taylor & Francis, Wiley), and also presented research papers at national/international conferences. In 2021, his 17 articles have got top-cited article status in various top journals (Journal of Cleaner Production, Carbohydrate Polymers, International Journal of Biological Macromolecules, Journal of Natural Fibers, Journal of Industrial Textiles). He is a lead editor of Several special issues. He has delivered many keynote and invited talks in various international conferences and workshops. His current research areas include Natural fiber composites, Polymer Composites, and Advanced Material Technology. He is a recipient of the DAAD Academic exchange- PPP Programme between Thailand and Germany to Institute of Composite Materials, University of Kaiserslautern, Germany. He has received a 'Top Peer Reviewer 2019' award, Global Peer Review Awards, Powered by Publons, Web of Science Group. The KMUTNB selected him for the 'Outstanding Young Researcher' Award 2020. He is recognized by Stanford University's list of the world's Top 2% of the Most- Cited Scientists in Single Year Citation Impact 2019 and also for the year 2020.

Prof. Dr.-Ing. habil Suchart Siengchin is President of King Mongkut's University of Technology North Bangkok. He has received his Dipl.-Ing. in Mechanical Engineering from University of Applied Sciences Giessen/Friedberg, Hessen, Germany in 1999, M.Sc. in Polymer Technology from University of Applied Sciences Aalen, Baden-Wuerttemberg, Germany in 2002, M.Sc. in Material Science at the Erlangen-Nürnberg University, Bayern, Germany in 2004, Doctor of Philosophy in Engineering (Dr.-Ing.) from Institute for Composite Materials, University of Kaiserslautern, Rheinland-Pfalz, Germany in 2008 and Postdoctoral Research from Kaiserslautern University and School of Materials Engineering, Purdue University, USA. In 2016 he received the habilitation at the Chemnitz University in Sachen, Germany. He worked as a Lecturer for Production and Material Engineering Department at The Sirindhorn International Thai-German Graduate School of Engineering (TGGS), KMUTNB. He has been full Professor at KMUTNB and became the President of KMUTNB. He won the Outstanding Researcher Award in 2010, 2012 and 2013 at KMUTNB. His research interests in Polymer Processing and Composite Material. He is Editor-in-Chief: KMUTNB International Journal of Applied Science and Technology and the author of more than 250 peer-reviewed Journal Articles, 8 editorial corners, 50 book chapters, one book, and 20 books as an Editor. He has participated with presentations in more than 39 International and National Conferences with respect to Materials Science and Engineering topics. He has recognized and ranked among the world's top 2% scientists listed by prestigious Stanford University.

Polymer Based Microcapsules for Encapsulation

Siddhant Bhutkar and Kadhiravan Shanmuganathan

1 Introduction

The concept of encapsulating delicate material inside a protective shell can be observed in living cells, where the semi-permeable cell walls control the flow of materials and protect the core [1]. The first widespread use of the technology was demonstrated for manufacturing carbon-less paper for copying in the 1950s [2, 3]. Complex coacervates using gelatin and gum arabic were used to prepare these capsules. Since then, microencapsulation has evolved and is used effectively for protecting active ingredients in applications ranging from pharmaceuticals, personal care, agriculture, perfumery, food additives and functional coatings. In most of these applications, polymers (natural or synthetic) are used as the shell material. This chapter highlights methods used for synthesis and characterization of microcapsules prepared using polymeric materials and key areas where such formulations are used regularly.

2 Microencapsulation

By definition, microencapsulation is the process of entrapping small solid particles, liquid droplets or gas within an outer coating/shell. This is usually carried out for protecting the core materials against external factors present in the environment where these formulations are applied, thereby increasing their life and enhancing

S. Bhutkar · K. Shanmuganathan (✉)
Polymer Science and Engineering Division, CSIR-National Chemical Laboratory, Dr. Homi Bhabha Road, Pune 411008, Maharashtra, India
e-mail: k.shanmuganathan@ncl.res.in

K. Shanmuganathan
Academy of Scientific and Innovative Research, (AcSIR), Ghaziabad 201002, India

J. Parameswaranpillai et al. (eds.), *Micro- and Nano-containers for Smart Applications*, Composites Science and Technology,
https://doi.org/10.1007/978-981-16-8146-2_1

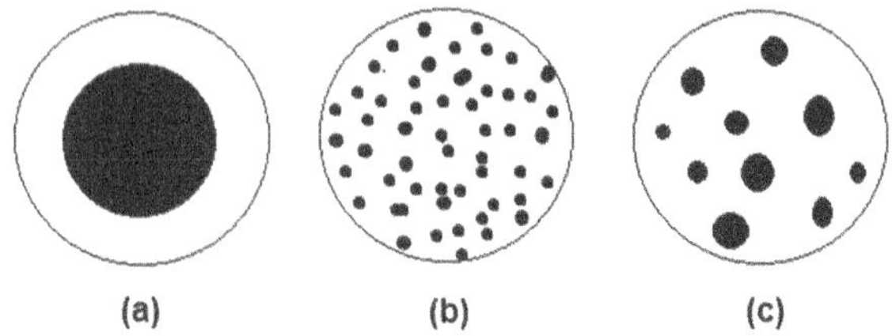

Fig. 1 Types of microcapsules, **a** core–shell, **b** matrix, **c** multi-core

performance. The coating can be engineered specifically for controlling the release of the core material. It also serves as a tool to transform liquids into solids thereby making them easier to handle. During the formation of these microcapsules, the inertness of the shell with the core is critical for stability. The *microcapsules* formed are in the size range of 1–1000 μm. These capsules can have a variety of morphologies and geometries depending on the materials used and the encapsulation process. Broadly they can be classified as shown in Fig. 1:

- Core–shell microcapsules with a single core and a shell around it.
- Matrix-type encapsulation where the core is entrapped homogenously in a matrix of the shell
- Multi-core microcapsules with multiple pockets of the core material inside a single shell

3 Microencapsulation Techniques

The choice of shell material depends on the final application. The most commonly used polymers are highlighted in Table 1: Polymers used for microencapsulation [4–6]. Apart from these, inorganic materials like silica, titanium dioxide and sodium silicate are also used as shell materials for specific applications [7]. The entire process of encapsulation can be divided into three different steps [8] involving formation of shell wall around the core materials, storing the core materials inside the shell without release and finally controlled release at the desired rate and at a particular time. The final application usually dictates the selection of materials (core and shell) and a compatible process needs to be tabbed for synthesis. Various schemes have been used to classify the types of microencapsulation processes. Most commonly, these processes are classified as physical, chemical, and physicochemical. Theis [9] simply divided these techniques into two types—Type A and B.

Type A processes are those where the capsules are formed in a vessel filled with a liquid and Type B processes include those where a coating is deposited onto the core (solid or liquid) in a gaseous medium or vacuum. On similar lines, Oakley compiled a list of various microencapsulation techniques into two broad heads of physical and emulsion/suspension type of processes [10]. Table 2 depicts this classification. The physical processes have been further classified into atomization, spray coating or extrusion, based on the mechanism used to form the matrix or shell. The emulsion

Table 1 Polymers used for microencapsulation [4–6]

	Types	Materials
Natural polymers	Carbohydrates	Agarose
		Alginate
		Chitosan Polydextran
		Polystarch
		Starch Cellulose derivatives
	Proteins	Albumin
		Collagen
		Gelatin
	Others	Calcium carbonate
		Lipids
		Tricalcium phosphate
Synthetic Polymers	Non-biodegradable	Acrolein
		Glycidyl methacrylate
		Lactides
		Polyanhydride
		Polymethylmetharylate
		Polyiminocarbonates
		Urea/melamine–formaldehyde
	Biodegradable	Glycolides
		Epoxy polymers
		Hydrogels
		Paraffin Polyvinyl alcohol
		Pegylated poly(lactide)
		Poly(lactide-co-glycolide)
		Polyacrylates
		Polyacrylonitrile
		Polyamide
		Polyamino acids
		Polycaprolactones
		Polyelectrolytes
		Polyester
		Polyethylene Glycol
		Polyphosphazenes
		Polyurea
		Polyurethane

Table 2 Microencapsulation processes [10, 20]

	Process	Morph.[a]	Payload[b]	Size[c] (μm)	General process steps
Physical processes	*Atomization*				• Dissolve/disperse active and coating in solution • Atomize/spray the solution • Heat/cool and collect the microcapsules
	Spray drying	M	L–H	5–500	
	Spray chilling	M	L–H	50–1000	
	Spray congealing	M	L	50–1000	
	Spinning disk	M/CS	L–H	5–1000	
	Jet cutting	M	L–H	100–3000	
	Electrospray	M	L–H	1–100	
	Spray coating				• Fluidize/disperse/spread the active • Spray the coating/shell material • Cool and collect the capsules
	Fluid bed/Wurster coating	CS	M-H	>75	
	Pan coating	CS	M-H	>250	
	Granulation	M	M-H	>5	
	Coextrusion				• Prepare separate solutions for active and shell • Co-extrude through syringe/nozzle/extruder • Cool/crosslink/harden microparticles
	Stationary nozzle	M/CS	M-H	500–5000	
	Vibrating nozzle	M/CS	M-H	150–5000	
	Centrifugal nozzle	M/CS	M-H	150–1500	
	Submerged nozzle	M/CS	M-H	500–5000	
	Flow focusing	M/CS	L–H	10–500	
	Microfluidics	M/CS	L–H	1–1000	
	Extrusion	M	L–H	10–5000	
Emulsion/suspensions	Coacervation	M/CS	H	5–1000	• Emulsify the active/ active solution in a continuous medium • Initiate polymerization/self-assembly/evaporation to generate microcapsules • Separate microcapsules from the continuous medium
	In situ polymerization	CS	H	1–100	

(continued)

Table 2 (continued)

Process	Morph.[a]	Payload[b]	Size[c] (μm)	General process steps
Interfacial polymerization	CS	H	1–500	
Solvent evaporation	M	L–H	<1–100	
Liposomes	CS/M	L-M	<1–10	
Sol–gel	CS/M	L–H	<1–100	
Layer-by-layer	CS	H	<1–500	
Molecular Complexation	M	L	n/a	

[a] M, microsphere; CS, core–shell

[b] $L \leq 30\%$, $M = 30\%$-60%, $H \geq 60\%$

[c] General size range, subject to variation based on formation and process conditions

type of processes involves emulsification followed by the shell/matrix formation due to a chemical reaction or phase separation.

While selecting an appropriate process, factors (listed in Table 2) like size, type of core and shell and the final size and dispersity of microcapsules are crucial. Apart from this, while selecting the polymer, the final application plays a major role. For non-biological applications (such as paints, pesticides, detergents), polymers like poly (methyl methacrylate), polyurethanes, poly(ethylene), poly (vinyl pyrrolidone) were used for their physical properties like structural strength, elasticity, resistance to swelling and hydrophilicity/hydrophobicity [11]. However, for applications like drug delivery and food technology, polymers have to be completely free of impurities and benign. Polymers like polylactide and poly (lactide-co-glycolides) are increasingly used for drug delivery due to their biodegradability. The physical properties of these polymers in-turn, affect the processing conditions. Thus, all these factors are interlinked and affect the process selection and careful literature review is essential before choosing an appropriate technique.

3.1 Physical Methods

These methods usually involve techniques where a chemical reaction does not play a significant role in encapsulation. The core and the shell are brought in physical contact which consequently leads to encapsulation. Most commonly, these techniques are used for applications where the use of chemicals and solvents is restricted.

3.1.1 Spray Drying

In general terms, the process of transforming a suspension (solution, paste, dispersion) into dry particulate powder by spraying and evaporation is termed as spray drying [12]. The feed is first atomized and sprayed into the drying medium (usually hot air), wherein solvent evaporates to yield dry particulates. A detailed description of the steps and the components of a spray drying system is given by Drusch and Diekmann [13]. The apparatus is usually coupled with a cyclone separator for collection of fine particles Figure 2. Schematic representation of a spray dryer shows a schematic representation of a single-stage spray dryer.

For microencapsulation, the shell or the polymer used is dissolved in a solvent and the active (usually insoluble in the solvent) is added to the polymer solution. A dispersion or emulsion is formed, depending on whether the active is solid or liquid. This is fed into the atomizer and subsequent steps described above lead to the formation of capsules. Since the active is homogenously mixed with the solution, the resultant capsules are usually of the matrix type. The size and monodispersity of these capsules depend on the surface tension and viscosity of the liquid, pressure drop across the atomizer and the spraying rate.

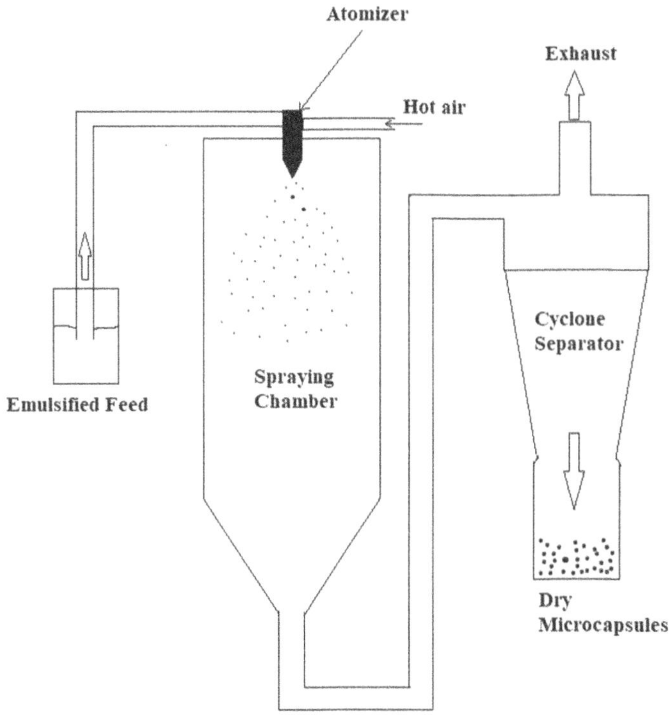

Fig. 2 Schematic representation of a single-stage spray dryer

By altering the surfactant, polymer solution (carrier), and operating parameters of the system, the properties of the dry powder obtained can be optimized as per the final application. Figure 3 illustrates the type of structures that can be obtained, based on the feed [14].

Spray drying offers a relatively simple alternative for encapsulating sensitive materials. Even the challenges involved in encapsulating volatile materials have been addressed often for applications in flavors and aromas [15–18]. Due to the continuous nature of the process, spray drying is relatively easy to scale-up and has been in use since the 1950s in the food industry using carriers like mono and disaccharides, hydrolyzed and chemically modified starches and gum Arabic or acacia [19].

In the pharmaceutical industries, cellulosic polymers like sodium carboxymethylcellulose NaCMC and hydroxypropyl methyl cellulose HPMC are used for improving the solubility of hydrophobic drugs by spray drying at low concentrations (0.5%) [21]. Other polymers used for drug delivery include cellulose acetate phthalate [22], methacrylates (like Eudragit) [23] polyacrylic acids [24] and natural polymers like chitosan [25]. Polylactide (PLA) and polylactide glycolic acid (PLGA) have often been employed for preparing biodegradable formulations [26]. All these polymers can form films over the surface of the core (either solid or liquid) during drying, which is essential for forming microcapsules.

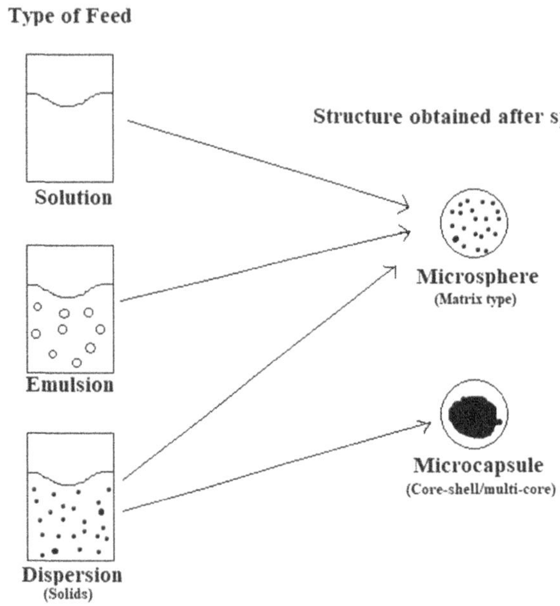

Fig. 3 Types of microparticles obtained using spray drying

3.1.2 Fluidized Bed Coating

The fluidized bed coating technology involves fluidizing the core (usually solid particulates) in a stream of air and simultaneous spraying of the shell coating material using spray nozzles. The coating material is usually dissolved in a solvent to facilitate spraying. During the process, this carrier solvent evaporates leaving behind the coating on the surface of the particulates, which upon cooling form a hard shell. The formed microcapsules are usually in the range of 10–100 μm. Dr. Dale E. Wurster, at the Univesity of Wisconsin, first used this technology in the 1950s and his design, which involves a bottom spray is commonly termed as Wurster bed coating and was patented in the 1960s for small particle coatings [27–29]. These patents form the basis for most of the bottom spray designs today.

The processes involved in coating vary slightly depending on the type of fluidized bed used. However, the basic principle of fluidization remains constant throughout. The concept was explained in detail by Wurster [30]. It involves injecting air (or the fluidizing medium) through a column containing the particulates. At low velocities, the air passes through the voids in particle bed without any "lift" or particle movement. As velocity increases, at a certain point, the particles begin to suspend with the air flow and are free to move (like a fluid). This point is called the minimum or incipient fluidization and is affected by the physical properties of the particles. Further increase in velocity leads to turbulent conditions and the particles begin to slug until eventually they are transported with the air flow. Figure 4 shows the

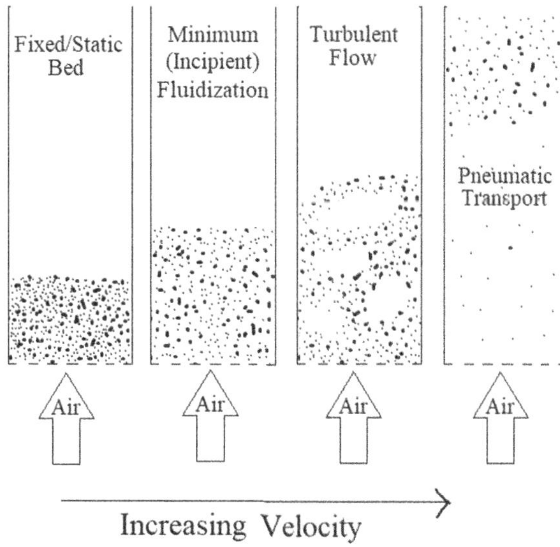

Fig. 4 Modes of fluidization observed as a function of increasing jet velocity

schematic diagram.

A detailed description of the various configurations of the fluidized bed technology and aspects involved during scale-up can be found in resources focusing only on this method [31, 32]. The most common types of fluid bed systems for encapsulation include the top spray and bottom spray (Wurster unit). Top spray units are usually preferred for coating materials like waxes which are sprayed in the form of melts and not for coatings in solution which use volatile solvents. The counter-current nature of the operation facilitates rapid evaporation. Consequently, this leads to uneven deposition of the shell material on the fluidized particulates.

In the Wurster unit, the coating solution is sprayed from the bottom of the fluidized bed, co-currently with the air flow. As shown in Fig. 5, the column is divided into annular sections by a partition. The differential air flow is designed to carry the fluidized particles in a circular motion. The particles in the central chamber move upwards with the air flow, where they are coated with the atomized material. In the expansion zone, the solvent evaporates, leaving behind core–shell type of capsules which fall back. This process can be repeated until a desired thickness of the shell is achieved.

Materials that can be used for coating are innumerable since a wide variety of additives can be dissolved or dispersed in the carrier solvent. However, the ability to form films is crucial for an effective coating over the suspended particles Essentially, most of these ingredients are polymers, however, materials like salts, sugars, waxes and clays have also been used. Frey has listed some of these materials [33] out of which Table 3 depicts the materials used.

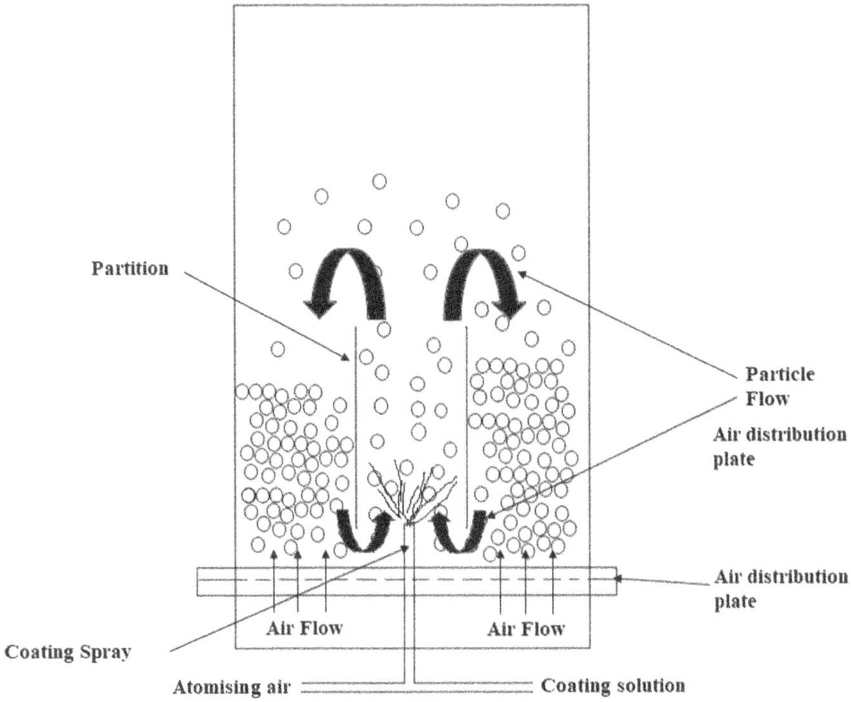

Fig. 5 Schematic representation of Wurster fluidized bed coating system

Table 3 Examples of polymers used for fluidized bed coating [32, 35, 36]

Poly(vinyl pyrrolidone)	Poly(amino acids)	Ethyl cellulose
Poly(vinyl alcohol)	Poly(ethylene)	Methyl cellulose
Poly(vinyl chloride)	Poly(methacrylates)	Ethylene vinyl acetate
Poly(vinylidene chloride)	Poly(ethylene glycol)	Gelatin
Starches	Poly(vinyl acetate)	Hydroxy propyl cellulose
Nylon	Cellulose Acetate	Gum Arabic
Poly(lactides)	Cellulose Acetate butyrate	Waxes (Carnauba, beeswax, paraffin)
Soyabean oil	Dextrins	

3.1.3 Pan Coating

Patented in the 19th Century [34], pan coating is one of the oldest methods used for coating tablets in the pharmaceutical industry. Apart from controlled release of the active, improving appearance or taste, hiding unplesant odors, protection against contamination, ease of printing tablets etc. can be some of the advantages of pan coating. It is preferred over fluidized bed coating when the materials used need

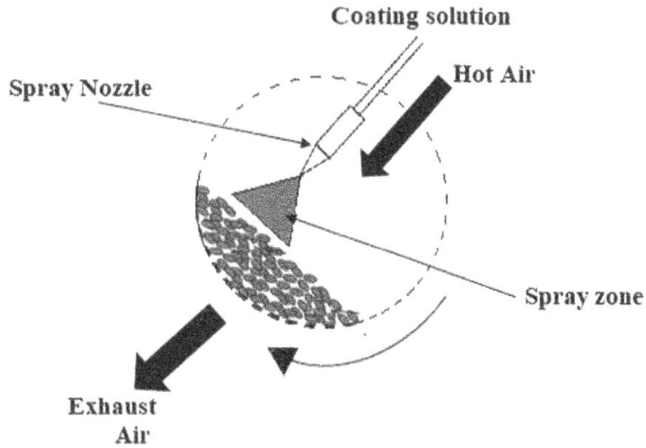

Fig. 6 Schematic representation of the Pan coating process

to be protected from excessive mechanical attrition. The process usually involves placing the tablet (or intermediate granules) to be coated in a rotating pan. The coating materials (usually with a solvent or water), are atomized and sprayed over these tablets using a spray gun. As the pan rotates, the tablets on the surface move towards the bottom and expose a fresh layer of tablets. This layer gets coated and dries out before the tablets fall back and the entire cycle is repeated [37]. Figure 6 shows the schematic representation. Analysis of various types of pan coaters, process parameters and considerations during scale-up is given by Porter et al. [38]. The materials used for coating can in principle, be the same as those used for fluidized bed coating (Table 1.3) where the film forming ability is a necessity.

3.1.4 Extrusion

Extrusion, as applied to melts is an encapsulation process most commonly used for entrapping flavors. In the extrusion process, a carbohydrate matrix which is used for encapsulation, is heated to form a melt. A plasticizer like glycerol or water can be added to reduce the brittleness of the final product. The temperature is normally in the range of 110 °C [39], to protect the flavor against thermal degradation. The flavor (core) or essential oil is added with stirring to create a fine dispersion. This dispersion (the melt + flavor) is extruded through a die into a cold organic fluid (like isopropanol). The melt droplets harden on quenching trapping the flavor. This type of extrusion is also termed as screw-less extrusion. The assembly is normally vertical using gravity as a driving force.

The hot-melt extrusion process uses a horizontal extruder for handling materials that are too viscous to be processed using processes like spray drying or fluidized bed coatings. The two most commonly used extruder configurations are shown in Fig. 7. During the process, the ingredients—actives, encapsulating matrix and additives like plasticizers—are fed to extruder using the hopper. This feed is carried forward into the melt section, which can be maintained at a specific temperature by external heating. Throughout the length of the screw, several processes like melting, mixing,

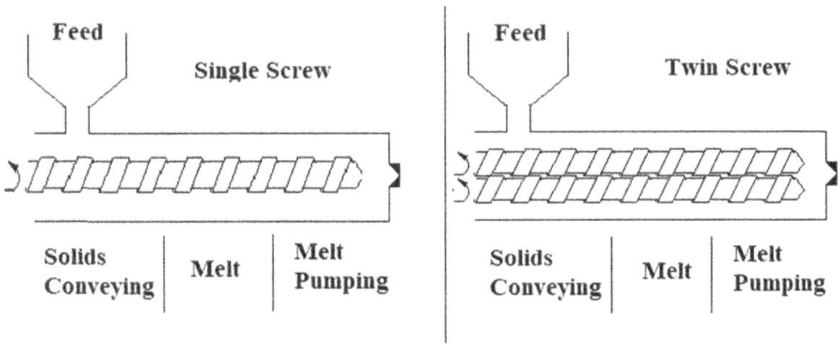

Fig. 7 Schematic representation of the single and twin-screw extruders

attrition, particle size reduction, kneading etc. [40] occur and the characteristics of the materials fed like feed rate continuity, type of feed, flowability, and degree of fill all affect the homogeneity of the final product [41]. Towards the end of the screw barrel, the material is extruded through a die and cooled to a glassy state by various methods, including air cooling or quenching in a bath.

Thermoplastic polymers are frequently used as the encapsulating (carrier) materials along with waxes and plasticizers. These ingredients bind the flavors or drugs and act as controlled release vessels. Table 4 shows a list of materials used for hot-melt extrusions. These polymers selected should have low processing temperatures, to avoid thermal degradation of the sensitive flavors or drugs [43].

3.1.5 Vibrating Nozzle

The process of forming equal-sized droplets from a laminar jet of liquid, by applying controlled vibrations, is another commonly used extrusion technology for microencapsulation. Lord Rayleigh analyzed the instabilities in laminar jets and showed that the frequency of maximum instability was dependent on the velocity of the jet and the diameter of the opening [44]. Spray drying, like the vibrating nozzle, also involves the formation of a droplet and subsequent conversion into capsules, where the central principle employed is of droplet formation at an orifice.

The mechanisms followed at the orifice are affected by friction, gravity, surface tension and jet outflow velocity [45] as shown in Fig. 8. At low jet velocities (A), individual droplets form at the outlet whose diameter is given by

$$D = \sqrt[3]{\frac{6d\sigma}{g\Delta\rho}}$$

where D is the droplet diameter, d is the orifice diameter, σ is the surface tension and ρ is the density (all in SI units). As the jet velocity increases, an uninterrupted stream flows out before breaking due to surface tension and axial vibrations (D).

Table 4 Polymers used for hot melt extrusion [42, 43]

Chemical name	Trade name	T_g (°C)	Tm (°C)
Poly(ethylene oxide)	Polyox® WSR	−67	65–80
Poly(ethylene glycol)	Carbowax®	−20	37–63
Poly(vinyl pyrrolidone)	Kollidon®	168	–
Poly(vinyl acetate)	Sentry® plus	35–40	–
Poly(dimethylaminoethylmethacrylate-comethacrylic esters)	Eudragit® E	50	–
Poly(methyl acrylate-co-methyl methacrylate-co-methacrylic acid) 7:3:1	Eudragit® 4135F	48	–
Poly(methacrylic acid-co-methyl methacrylate) 1:2	Eudragit® S	160	–
Methacrylic acid copolymer type C	Eudragit® L100-55	–	104.4
Ammonio methacrylate copolymer	Eudragit ® RS/RL	64	–
Poly(ethylene-co-vinyl acetate)	Elvax® 40 W	−36	45
Polyethylene	–	−125	140
Poly(vinyl acetate-co-methacrylic acid)	CIBA-I	–	84–145
Poly(lactide-co-glycolide)	PLGA	–	–
Polyvinyl alcohol	–	–	–
Hydroxypropyl cellulose	Klucel®	130	–
Ethyl cellulose	Ethocel®	133	–
Cellulose acetate butyrate	CAB 381–0.5	125	157
Hydroxypropyl methylcellulose phthalate	–	137	150
Polyvinylpyrrolidone-co-vinyl acetate	Kollidon® VA64	101	–
Hydroxypropyl methylcellulose	Methocel®	175	–
Hydroxypropyl methylcellulose Acetate succinate	Aqoat-AS®	–	–
Hypromellose	Methocel® E5	170–180	
Chitosan lactate	Sea-Cure®	–	–
Pectin	Obipektin®	–	–
Polycarbophil	Noveon® AA-1	–	–
Carnauba wax	–	–	82–85
Potato starch	–	–	–
Citric acid	–	–	153
Chitosan	–	203	–
Microcrystalline cellulose	Avicel® PH 101	–	–

Further increase in velocity causes a statistical distribution of the droplet size either due to symmetrical vibrations or due to the frictional forces prevalent while spraying (E and F) [46].

Out of these mechanisms, only two (D and E) are industrially viable and only D is commonly used. A sinusoidal force with predetermined frequency can be applied

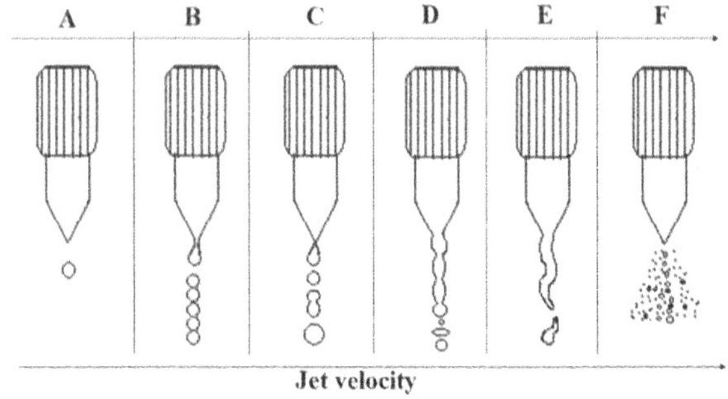

Fig. 8 Mechanisms for droplet formation as a function of increasing jet velocity

to D to get droplets of the required size. Unlike the previously described droplet diameter which depends only on intrinsic characteristics, droplet diameter in this process can be controlled by changing the frequency of vibrations employed. This vibration can be applied using a vibrating nozzle, varying the diameter of the orifice or by pulsating the polymer cell suspension [41]. Alginate microcapsules are most commonly prepared by this method, where Na^{+}alginates are dropped into a solution of calcium chloride. Other polymeric materials used for preparing capsules using this vibrating technology include chitosan, cellulose sulfate [47], and gelatin [48].

3.2 Emulsion Processes

Apart from the absence of a chemical reaction, the techniques described previously do not involve an emulsion/dispersion stabilization around which the encapsulation process occurs. The methods that are described in this section, may or may not involve a chemical reaction, but a stable emulsion of phases containing the core and shell materials is a key factor for encapsulation.

3.2.1 Microfluidics

For encapsulation processes involving emulsification, controlling the size of the droplets/particles formed is essential. Conventional techniques like high-speed homogenization, ultrasonication or static mixing rarely provides control over particle size distribution. Monodisperse formulations are crucial for maintaining repeatability among batches and making methods scalable.

Microfluidics involves manipulation and precise control of fluids at small (sub-milliliters) scale. When applied to the processes used for emulsification, droplet-based microfluidics leads to the generation of droplets by the interaction of the phases flowing through microchannels. Advantages like improved control over the size distribution, minimal variations among batches and scalability have led to increased use of such systems [49].

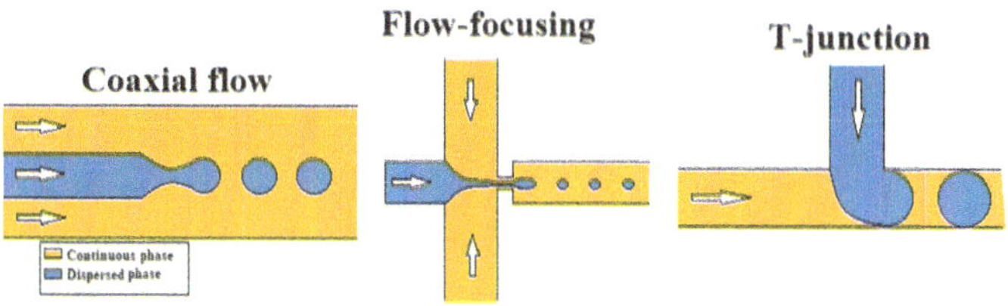

Fig. 9 Three major techniques used for microfluidic droplet formation

Various configurations of microchannels have been used for structuring capsules [50, 51]. Broadly, these configurations can be classified into two categories [52]: two dimensional devices, where the microchannels are engraved onto a substrate, and three-dimensional devices, formed by nesting capillary tubes. Figure 9 shows the schematic of the three most common droplet formation techniques: coaxial flow, flow focusing and a T-junction [53]. The basic principle involves interaction of two co-flowing immiscible phases, where one isolates into distinct droplets forming the dispersed phase, while the other surrounds it to form the continuous phase [54].

Modifications of these techniques have been used for structuring complex microcapsules involving multiple core–shell combinations [46], however, the fundamental principle remains unchanged. Moreover, these techniques, by themselves, cannot be termed as encapsulation processes. However, a combination of microfluidics with a physical or chemical encapsulation process has been used effectively for synthesizing polymeric nanospheres [55], complex coacervates [56], magnetic microcapsules [57], and multi-compartment or multi-emulsion globules [46].

Polymeric materials were introduced in the 2000s for fabricating microfluidic devices. The most common material used is polydimethylsiloxane due to its, cost effectiveness, ease of replacement and optical transparency [58]. Other polymers include polymethylmethacrylate [59], polystyrene [60] and polycarbonate [61]. These are more preferred over the glass and silica based devices, since they are cheaper and optically transparent.

3.2.2 Complex Coacervation

The term coacervates originated from the Latin word ‘coacervus,’ implying a process of association. The phenomenon, coacervation, was first described in 1911 [62] and further demonstrated using gelatin and gum Arabic in the1920s and 1940s [63]. Coacervation can be of two types: simple and complex. In simple coacervation, single polymer coacervates are formed around an active in the reaction medium by adding a coacervating agent. For example, gelatin coacervates are formed by adding ethanol. In complex coacervation, oppositely charged polyelectrolytes associate to form the coacervates by virtue of ionic interactions. The phase containing the assemblies is called the coacervate phase, and the other, which has very little coacervates is called the equilibrium phase.

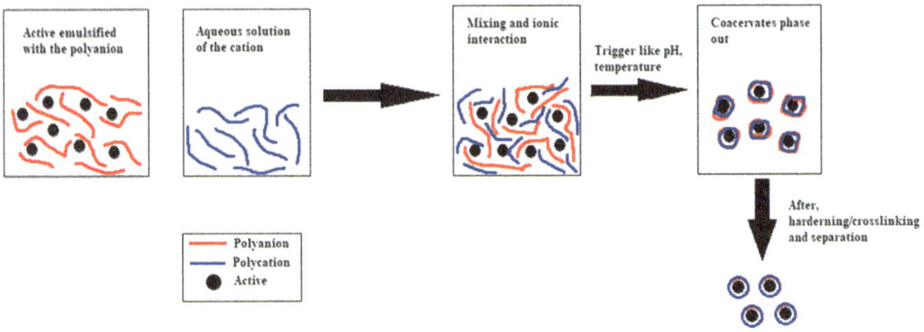

Fig. 10 Schematic representation of the complex coacervation process

The reaction mixture for complex coacervation involves three phases: the active ingredient, the polymers (shell material) and the solvent. An aqueous solution of the polymers is prepared above the isoelectric point and the hydrophobic active is emulsified with one of the polymer solutions. The second polymer is added to this emulsion and phase separation and coacervation is triggered by external stimuli like changing temperature or pH or adding a coacervating agent (like a salt) (Fig. 10). The mixture is maintained at elevated temperatures for hardening the polymer shell. Initially, very fine coacervates start appearing making the solution turbid, which is termed as micro-coacervation, if this process is prolonged, these particles coalesce to form larger particles which settle down at the bottom. This is termed as macro-coacervation and two distinct phases are clearly visible [64].

The most commonly used materials for complex coacervation are gelatin and gum arabic. Gelatin (type A) has an isoelectric point in the range of pH 7 to 9. Below this pH gelatin is positively charged and has the ability to form coacervates with gum arabic (negatively charged) provided the temperature, concentration and ratio of the two polymers are appropriate [65]. Chitosan [66], whey protein [67], and soy protein [68] are some of the polycations being researched as a replacement for gelatin. Materials like Chia seed protein [69] and flaxseed protein isolate (FPI) [70] are being investigated to increase the compatibility of microcapsules for specific applications.

3.2.3 Solvent Evaporation

In this widely used encapsulation technology, a volatile solvent is evaporated from an emulsion. Most commonly, the polymer (shell) is dissolved in a volatile solvent like dichloromethane and the active ingredient is dissolved or dispersed in this solution. This solution is emulsified into an aqueous phase using surfactants and the volatile solvent is evaporated by heating. This gives rise to spherical capsules encapsulating the active in the polymer matrix. Figure 11 shows the schematic representation of the process.

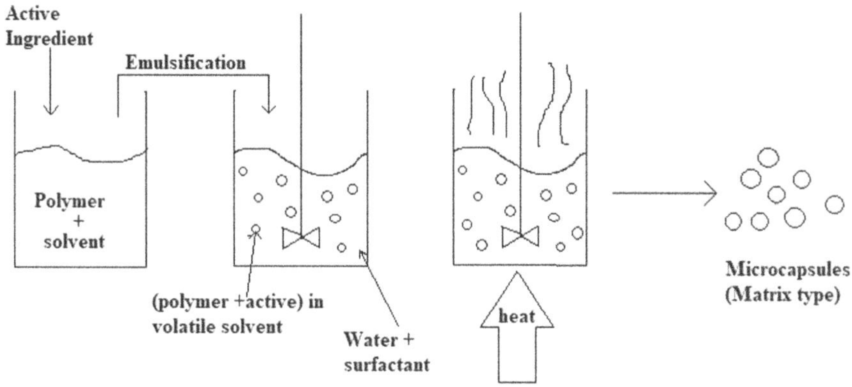

Fig. 11 Schematic representation of the solvent evaporation process

This type of oil in water emulsion described hitherto is not suitable for actives that are highly hydrophilic because they do not dissolve in the organic solvent and may escape into the continuous phase during emulsification [71]. For such systems, modification to the process is required for effective encapsulation.

In the w/o/w double emulsion method, an aqueous solution of the active is emulsified into an organic solvent (w/o) which in turn is further emulsified in an aqueous solution forming a double emulsion (w/o/w). Here, the shell material is dissolved in the organic phase (middle) which is evaporated to form the capsules [72]. The o/w co-solvent method is used when the active is not soluble in the organic solvent. Here, the active is pre-dissolved in a co-solvent (like an alcohol) and then in the polymer solution. The entire organic phase is then emulsified with the aqueous phase to form the o/w emulsion [73]. Other methods include o/o emulsions using a non-volatile organic solvent as the continuous phase and o/w dispersion method where a solid active is dispersed in the organic polymer solution.

The solvents used need to dissolve the polymer and the active, be relatively insoluble in water and have high volatility (low b.p). The most common solvents include chloroform and dichloromethane. If the final application of the capsules is biological (drug delivery, food etc.) the polymer has to be biodegradable—like, poly (lactide-co-glycolide), polylactide, poly-3-hydroxybutyrate, ethyl cellulose to list a few [67]. Apart from these polystyrene can also be used as a shell, where dichloromethane can be used as a solvent.

3.2.4 Sol–Gel Microencapsulation

In general, sol–gel chemistry refers to the formation of inorganic polymers or ceramics from liquid precursors. The sol–gel method uses an assembly of chemical reactions which most commonly yields ceramic oxides [74]. During microparticle synthesis, an interfacial polymerization reaction takes place where the dispersed droplets of an emulsion act as microreactors for the hydrolysis of alkoxides. The resultant colloidal solution formed after hydrolysis and polymerization is termed as "sol." Fig. 12 shows the schematic of routes which this "sol" can take to form various macrostructures or "gels." The structure and strength of the final gel depends on the type of interactions between the "sol." Strong interactions involve crosslinked covalent bonds, whereas weak interactions include Van der Waal's forces, hydrogen bonding or weakly interconnected metal complexes [75].

Specifically, for microparticle formation and encapsulation, the O/W or W/O emulsion techniques are the two main procedures used [76]. Figure 13 shows the flow charts for both these techniques. For the W/O process, the active ingredient is mixed with aqueous metal alkoxide solution along with a suitable catalyst. This entire solution is emulsified with a non-polar solvent using suitable surfactants. As the alkoxides undergo hydrolysis and subsequent polymerization, an oxide shell is built around the droplet. Typically, the active gets homogeneously dispersed in the polymer droplet.

forming matrix type of microparticles. The size of microparticles can be controlled by controlling the size of the droplets within a wide range of 10 nm–100 μm [76]. In the case of the O/W systems, the active ingredient is mixed with a water-insoluble alkoxide solution and is dispersed in an aqueous surfactant solution. The microparticles are formed by hydrolytic condensation using an acid or a base as a catalyst

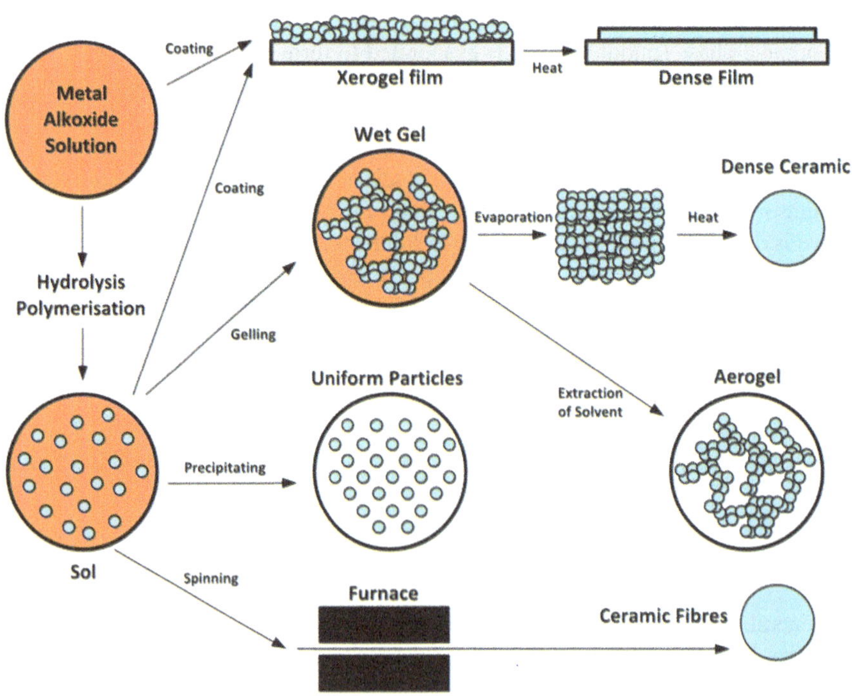

Fig. 12 Schematic representation for various routes of sol–gel technology [77]

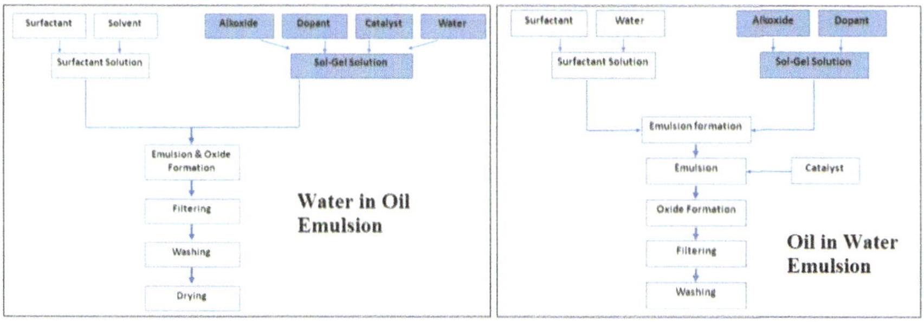

Fig. 13 Two major techniques for microparticle synthesis by sol–gel method. Reprinted from Ciriminna et al. [76] with permission from American Chemical Society, Copyright (2011)

[78]. Silica is the chief material used for encapsulation using sol–gel techniques from precursors like tetramethyl orthosilicate (TMOS) and tetraethyl orthosilicate. Natural polymers like alginate and pectin can also be dissolved in water and crosslinked to form gels entrapping active ingredients.

3.2.5 Encapsulation Using Supercritical CO_2

Supercritical fluids are those having properties between gases and liquids, at temperatures and pressures beyond the critical point. With high compressibility like gases along with densities similar to liquids (0.1–1 g/mL), these fluids have been used as solvents during many processes. Unique properties of supercritical fluids include [79]:

(1) Solvent strength of these fluids is a function of their densities, which can be controlled by varying the temperature and pressure. This property can be effectively utilized for applications such as extraction
(2) Considerably higher rates of mass transfer than liquid solvents, facilitate faster extraction. In chromatographic techniques, higher diffusion rates lead to higher efficiencies (at a particular mobile phase velocity) than liquids.
(3) Low viscosities improve flowability.
(4) Ease of separation, since most of the supercritical fluids are gases at atmospheric pressure. A simple decompression can be used for recovering the solutes dissolved.
(5) Most of the substances used as supercritical fluids are inert, cheaply available with high purity and non-toxic.

During the encapsulation process, the active and the polymer are dissolved in the supercritical fluid maintained at temperatures and pressure above the critical point. The pressure is then released by passing the solution through a tiny nozzle. The supercritical fluid gets discharged into the atmosphere, and due to the desolvation, the polymer gets coated on the active forming microcapsules. Carbon dioxide is

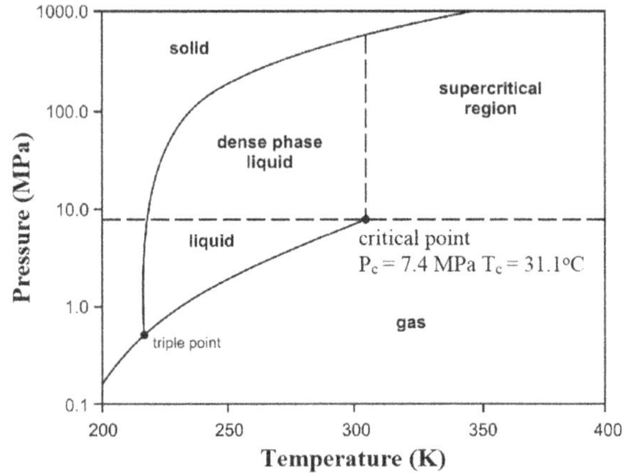

Fig. 14 Phase diagram for CO_2

the most commonly used fluid for this process, having temperature and pressure at 7.4 MPa and 31.1 °C. Figure 14 shows the phase diagram for carbon dioxide.

The supercritical CO_2 assisted microencapsulation processes can broadly be divided into three categories based on the function of CO_2 in each [80]. In the rapid expansion of supercritical solution (RESS) technique, it acts as a solvent (method discussed in the previous paragraph). During the gas anti-solvent method (GAS), the CO_2 is added to an organic solvent, which leads to the precipitation of microparticles. Polymers like hydroxyl-propyl methyl cellulose [81], polyesters and polyesteramides [82], poly-lactic-co-glycolic acid (PLGA) [83] have effectively been used for encapsulating actives using this method. The particles from gas saturated solutions (PGSS) method uses the high solubility of supercritical CO_2 in fats, lipids and polymers at moderate pressures. Here, the coating material is first saturated with CO_2 and the active is emulsified in this solution. Subsequently, gas-saturated solution is expanded through a nozzle which causes a rapid reduction in temperature. The coating material becomes solid during this expansion encapsulating the active [84]. A detailed list of various carrier materials used in all these processes has been given by Martin et al. [76].

3.2.6 Interfacial Polymerization

Interfacial polymerization is one of the most important processes for microencapsulation. It is commonly employed for large scale industrial manufacture of microcapsules. Using polymerization chemistry, solids, liquid droplets or sometimes even gas bubbles are covered in a polymer shell (core–shell capsules) or embedded in a matrix using this process. Theis systematically classified the process into five major types [9]. Figure 15 shows a schematic of these types.

Figure 15a shows the encapsulation mechanism where one of the monomers(A) is dissolved in the core and this solution is dispersed in a continuous phase. The other

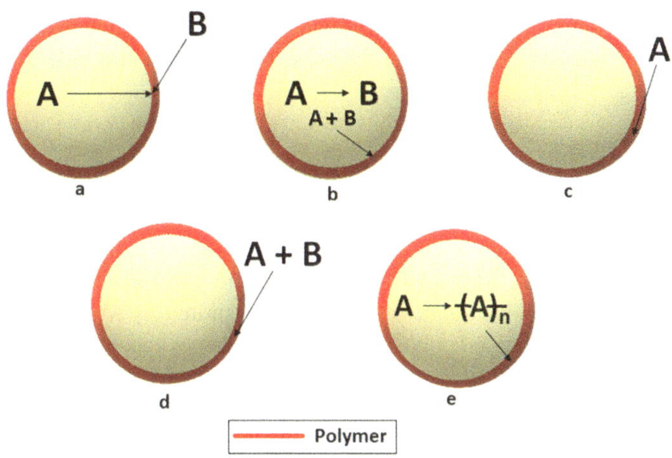

Fig. 15 Types of interfacial polymerization reactions, A and B represent reactants/monomers

monomer(B) is dissolved in the continuous phase; The reaction takes place at the interface, where both the monomers diffuse and react together.

Figure 15b depicts a condition when a monomer in the dispersed phase reacts with the continuous phase to form an intermediate, which in turn reacts with the unreacted monomer in the dispersed phase to form the shell. Figure 15c shows the case where a single monomer is transferred from the continuous phase to the interface where it polymerizes to form the shell.

Figure 15d represents the case of in-situ polymerization. Microcapsules prepared using systems like urea–formaldehyde or melamine–formaldehyde are prepared using this method. An organic core material is dispersed in the aqueous phase. A pre-polymer of urea–formaldehyde is added into the continuous phase and the pH of the system is reduced (to 3–4.5) and the entire mixture is maintained at elevated temperatures (40–60 °C) for several hours.

Figure 15e represents a less commonly used method, where a vinyl monomer along with an initiator and the core material is dispersed in the aqueous phase. Upon heating, the polymerization leads to shell deposition at the interface of the two-phases.

Materials most commonly used for encapsulation using interfacial polymerization include polyurea, polyurethanes, polyamides, polycarbonates and polyesters. In each of these, one of the monomers used is water-soluble while the other is oil-soluble. Apart from these materials, the urea/melamine–formaldehyde continues to be a favorite for industrial use. Duan has enlisted all of these polymers along with major fields where each of them has been employed [85]. The porosity of the shell formed plays a crucial role in determining the efficacy of these capsules. Hence, to improve the strength of the shell and to tune the release of active from the capsules, nanomaterials like carbon nanotubes, nano-clay etc. have been incorporated during the interfacial polymerization process [86, 87]. Rajagopalan et al. have used a double encapsulation method, wherein urea–formaldehyde microcapsules were entrapped again in a starch matrix to control the release rates of a solid pesticide core [88].

3.2.7 Layer-By-Layer Assembly

The Layer-By-Layer assembly method involves formation of microcapsules where layers of oppositely charged polyelectrolytes or complementary polymers are deposited onto a colloid. Multi-walled microcapsules can be effectively prepared using this method. This technique was first demonstrated by Möhwald's group in 1998 [89]. Oppositely charged polyelectrolytes were deposited onto colloids or "sacrificial templates" successively to prepare hollow microcapsules using electrostatic interactions. Figure 16 shows a schematic diagram for the process.

Hollow microcapsules prepared by alternative deposition of oppositely charged polyelectrolytes have been used for drug delivery, biosensors, micro-cages and micro-reactors [48]. By choosing suitable combinations of polymers and templates, the characteristics of such capsules can be fine-tuned. Reaction parameters like pH, temperature and ionic strength can be used for controlling shell permeability.

In recent times, the interactions between materials used for layer-by-layer assemblies have encompassed hydrogen bonds, covalent bonding, metal coordination, charge transfer, biospecific recognition etc. This has led to the use of materials other than oppositely charged polyelectrolytes like proteins, polysaccharides, inorganic particles, dendrimers, vesicles, micelles and virus particles [90]. Due to this variety of combinations available, microcapsules prepared using this method have been used for many applications [91–93] especially in the biomedical field.

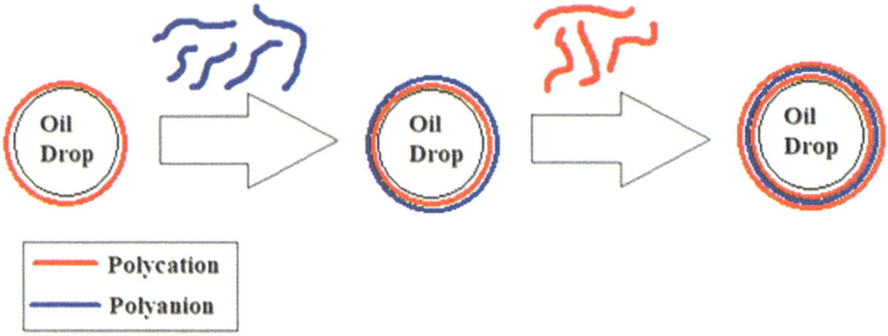

Fig. 16 Schematic representation of the layer-by-layer deposition method

4 Characterization Techniques

The previous section highlighted some of the typical techniques used for synthesizing capsules. Once formed, these capsules are characterized for attributes like payload of the active, size, shape, morphology, strength and release patterns of the active. This section contains a description of the techniques used for characterizing microcapsules. Although there may be specific techniques developed for niche applications, the discussion in this section is limited only to those that are most commonly used universally.

Size and morphology are characteristics that are studied for all microcapsules, irrespective of the application. Optical microscopy and Scanning Electron Microscopy (SEM) are two techniques that are used for determining the size and morphology. Optical microscopy is often, the most commonly used preliminary technique to determine the size of microcapsules. Especially, in the case of emulsion techniques, optical microscopy is also used for observing the stability of emulsion formed. The images obtained from the optical microscope can also be used for determining the particle size distribution using imaging software like ImageJ. Figure 17 shows one such image of polyurethane microspheres. The pixelated version is produced using ImageJ and used for measuring the size of the droplets. The technique can be modified by the inclusion of colored dyes in the emulsions and use of fluorescence microscopy for better imaging.

Microcapsules over a micrometer in size can be effectively imaged with typical microscopes (with up to 60× magnification capacity). Oil immersion techniques can be used for improving the resolution (up to 400×) for smaller sizes. For observing the morphology of capsules, SEM can be employed. With most of the controlled release applications, the porosity of the polymeric shell plays a crucial role in determining the release rates of the core from the capsule. SEM images can be used for observing the nature of the polymeric shell. Irregularities present in the capsules (for example, broken shell) can also be viewed using SEM. Figure 18 shows polyurethane micro-

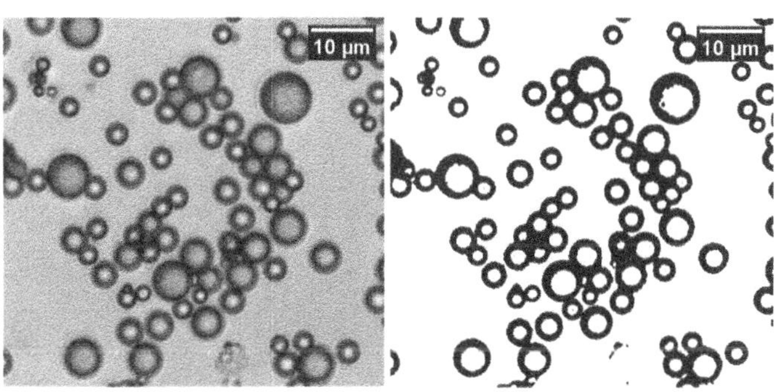

Fig. 17 Optical image of polyurethane microspheres and pixelated version using ImageJ for analysis

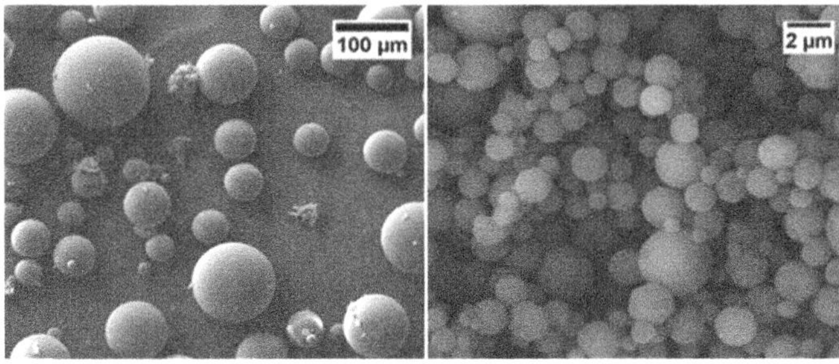

Fig. 18 SEM images showing polyurethane microcapsules with an insect repellent core

capsules with an insect repellent core. Such images can be combined with BET study to quantify the porosity of microcapsules.

When nanomaterials are incorporated into the capsule wall, Transmission Electron Microscopy (TEM) can be used to view these materials in the shell. Functionalized carbon nanotubes were used for improving the strength of microcapsules by Jagtap et al. [83]. Figure 19 shows the TEM image of fragments from the microcapsule wall. Figure 19a shows a grey cloud where no nanotubes were present. In the rest of the three figures (b, c and d), strands of nanotubes can be clearly observed.

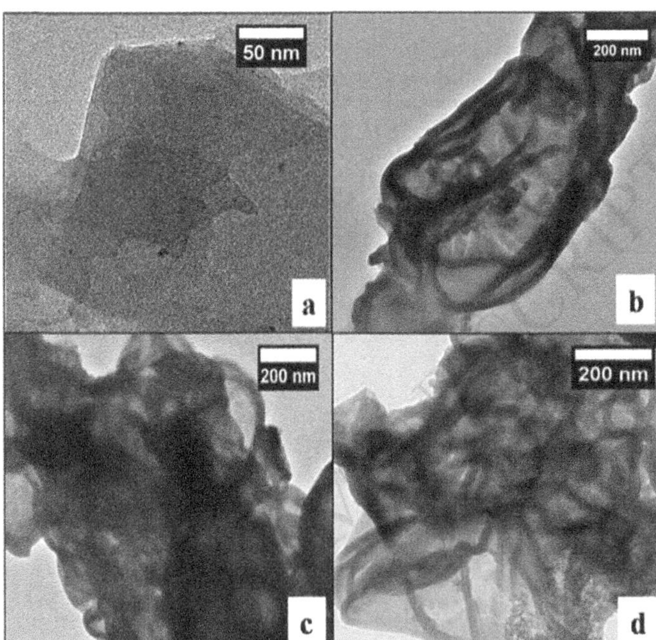

Fig. 19 TEM images for polyurea capsules incorporated with carbon nanotubes. (see text for details). Reprinted from Jagtap et al. [85], with permission from Elsevier, Copyright (2018)

Payload or the amount of core present in the capsules is another important criterion to be considered while characterizing microcapsules. Encapsulation efficiencies are also calculated using this metric. Normally, the core material is extracted into a solvent in which the polymeric shell is insoluble. Subsequently, common analytical techniques like UV–Visible spectrophotometry, gas or liquid chromatography can be used for determining the active content extracted into the solvent.

Furthermore, spectral techniques like Fourier Transform Infrared spectroscopy (FTIR) and Nuclear Magnetic Resonance (NMR) can be used for analyzing the core material to ensure that it does not undergo any chemical changes during the encapsulation process [94]. Similarly, FTIR and NMR can also be used to confirm functionalization and modification made to the polymeric shell for improving the properties of microcapsules.

Improving the thermal stability of the core is another important function that encapsulation serves. Thermogravimetric analysis (TGA) and differential scanning calorimetry (DSC) are used for measuring this attribute. By comparing the thermograms for the unencapsulated core, the polymer used as the shell and the capsules produced after encapsulation, the effect of encapsulation on the thermal stability of the core can be determined. Figure 20 shows one such thermogram comparing tung oil formulations where tung oil was encapsulated using polysulfones [92]. It can be observed that the curve for microcapsules lies in between the curve for polysulfones and the tung oil which indicates an increase in thermal stability and confirms encapsulation.

A key aspect in terms of the performance of the capsules, is the release rates of the core material. In-vitro release studies are conducted using the capsules wherein, they are immersed in a suitable dissolution media and the amount of core material released into this medium is monitored over a certain period of time. Normally, measured aliquots are withdrawn from the medium after specific time intervals and the concentration of the core material is measured. Figure 21 shows release profiles

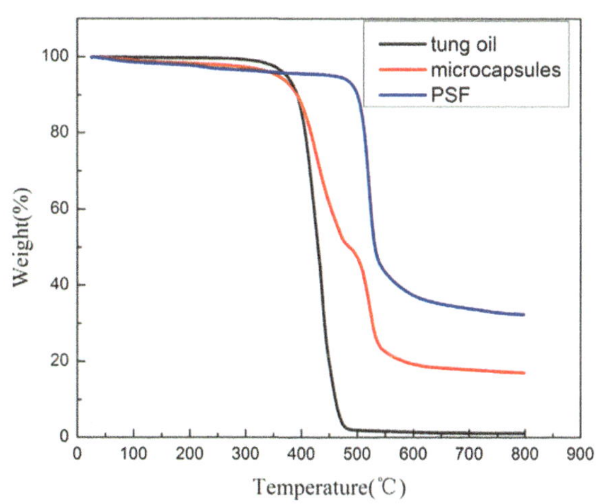

Fig. 20 TGA thermogram for tung oil, polysulfone and polysulfone microcapsules containing tung oil. Reprinted from Li et al. [95] with permission from Elsevier, Copyright (2017)

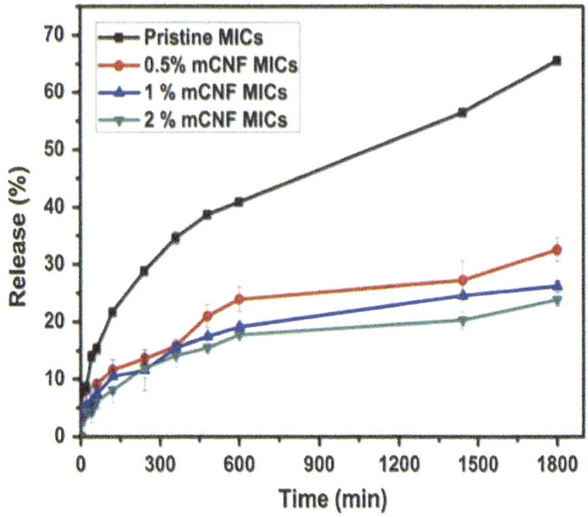

Fig. 21 Release profiles for DEET from polyurethane microcapsules incorporated with varying amounts of cellulose nanofibrils. Reprinted from Kadam et al. [100] with permission from Elsevier, Copyright (2019)

for *N*, *N*-diethyl-3-methylbenzamide (DEET) encapsulated in polyurethane microcapsules [97]. Modified cellulose nanofibrils were incorporated into the polymer shell to control the release rates of DEET. Distilled water was used as the dissolution medium for the study. It was observed that the release rates for DEET reduced significantly upon the incorporation of just 0.5% of cellulose nanofibrils. Several mathematical equations have been developed empirically to describe these release profiles [96–98]. Most notably, the equation given by Peppas [99] is frequently used for modeling the release profile in terms of an exponential expression. The experimental release profiles are mathematically modeled using these expressions.

There can be various ways in which the active ingredient or the core releases from the capsule. The shell can break or rupture due to mechanical forces. Sometimes, the polymeric shell wall disintegrates when exposed to certain external environmental conditions (like pH etc.). The encapsulated core can also diffuse through the polymeric shell wall. While using encapsulated formulations, one of these methods or a combination is put to use depending on the application. The techniques described hitherto, represent methods used for characterizing almost any capsule, irrespective of the application. Additional experiments and techniques can be designed depending upon final use.

5 Applications

Encapsulation technology has shown widespread promise in a variety of applications ranging from pharmaceuticals, agriculture, food, coatings, cosmetics and health care. New areas like catalysis, energy storage and defense have started utilizing microencapsulation [101]. This section highlights the use of polymeric encapsulation technologies in some of these areas.

5.1 Food Technology

Today, consumers worldwide are demanding healthier foods. Functional components used for enhancing the efficacy of foods include vitamins, essential oils, flavors and minerals. However, incorporation of these compounds into food products is challenging. Some of them degrade when exposed to the environment (for example, by oxidation). Some are essentially hydrophobic in nature and adding them to aqueous food formulations is difficult. Compounds that are considered healthy occasionally have a bad odor.

Most importantly, all of these are consumed orally and therefore, must not degrade in the digestive tract. Encapsulation has provided food corporations, a viable solution to tackle these complications. Furthermore, encapsulation also gives rise to dry powders, which are easier to handle. In 1960s, attempts were made to encapsulate essential oils to prevent lipid oxidation, volatile compound losses and aroma-controlled release. Since then, numerous studies have been published [102].

Aditya et al. used water in oil in water emulsions to encapsulate curcumin and catechin for use in beverages. Both these compounds are proven to be effective against ailments like cancer, cardiovascular diseases and obesity [103]. Like double emulsions, modified structures like multi-layered microcapsules have been used for encapsulating essential oils. Zhang et al. used cyclodextrin for encapsulating lavender essential oil [104]. Furthermore, a second coating layer of chitosan and alginate was applied using complex coacervation to improve the long term sustained release properties of the microcapsules. Studies have also shown co-encapsulation of probiotic bacteria with fatty acids using complex coacervation [105]. Apart from these common food ingredients, ethylene gas was also encapsulated in cyclodextrin for use in plant growth and fruit ripening [106]. Shell materials employed for food applications have to fall under the 'generally recognized as safe (GRAS)' category. Common materials include starch and cellulose derivatives, plant exudates like gum Arabic and chia seed gum, polysaccharides, alginate, chitosan, gelatin, waxes, and phospholipids [107]. Timilsena et al. have provided a detailed overview of encapsulation in the food industry, which covers most of these materials and their applications [108].

5.2 Perfumes and Fragrances

Perfumes and fragrances are volatile compounds that are used in applications like detergents, textiles, cosmetics, deodorants, and air-fresheners. These compounds are frequently encapsulated to prolong their effect at the application site. When used in detergents, these compounds frequently get lost during washing. Upon encapsulation, the microcapsules tend to adhere to the fabric and release the fragrance during drying or while the fabric is in use. New techniques to analyze the behavior of such systems quantitatively are being developed. Zhao et al. studied the broken and sustained release behavior of a softening essence oil encapsulated in melamine–formaldehyde capsules [109] using a solid-phase micro-extraction technique coupled with gas chromatography and mass spectrometer. Most of these non-topical applications are still dominated industrially by urea/melamine–formaldehyde based polymers. However, studies with polyurea/polyurethane are increasing, where mainly interfacial polymerization technique is being employed for encapsulation [110–112]. In the case of topical applications, biopolymers like chitosan-glycolic acid, cellulose derivatives and polyvinyl alcohol have been used for encapsulating fragrances using techniques like coacervation and solvent evaporation [113–115].

5.3 Agriculture

Farmers often use excessive fertilizers and pesticides with a misconception that it would lead to better yields. However, this excess usage is counterproductive and leads to soil contamination, which decreases soil fertility in the long run. These chemicals can also pollute groundwater by seeping through the layers of soil, or make their way into nearby water bodies due to surface run-off. Encapsulation can be used for the precise application of the compounds and help mitigate these problems. Covering the fertilizer or pesticide with a coating also ensures safe handling. These advantages have been demonstrated for decades. Sinclair reported the use of biocompatible polymers of lactic acid to encapsulate urea using a simple compression molding method in 1973 [116]. Scher et al. used isocyanate chemistry to encapsulate herbicides and insecticides [117]. The study discussed that encapsulation could lead to reduced mammalian toxicity, reduced evaporation, protect the pesticide from degradation and to extend the activity. Pesticides like Quinolphos and Chlorpyrifos were encapsulated using urea–formaldehyde and showed a drastic reduction in release rates even with payloads as high as 80% [118, 119]. Recently, pheromones have become a biocompatible alternative to chemical pesticides. These compounds, specifically sex pheromones, disrupt the mating cycle of pests and help reduce their population. Polyurea microcapsules [120] and complex coacervates of gelatin and gum Arabic [121] were used for encapsulating such pheromones and prolonging their release. Recently, the "Attract and Kill" approach has gained importance for controlling agricultural pests. A pheromone coupled with a killing agent and trapping device is sprayed to lure the pests in and kill them [122]. Patel et al. have published multiple studies where such attractants were co-encapsulated with bio-pesticides using calcium alginate beads [123, 124].

5.4 Surface Coatings

Ever since Scott White's work with self-healing microcapsules [125], the concept of encapsulating a healing agent in a capsule and embedding these capsules on surfaces has gained tremendous importance. In this system, the microcapsule containing a healing agent is embedded in a polymer matrix along with a catalyst that induces crosslinking. When a crack develops in such a matrix, the capsule breaks open and releases the healing agent to fill up the cracks. The catalyst induces crosslinking of the agent which consequently repairs the crack. Commonly urea/melamine formaldehyde systems have been used for such applications. Recently this concept has also been applied to healing micro-cracks in cement. Capsules prepared using complex coacervation of gelatin and acacia gum were used for entrapping sodium silicate [126]. On-site, the sodium silicate gets released into the concrete cracks and hydrolyses in the presence of moisture to yield products that aid in healing. Coatings used for aerospace applications have also employed this technology. Using titanium oxide or silica as the inorganic shell material, UV-responsive microcapsules were synthesized [127]. Once a crack develops, the microcapsules in its immediate vicinity degrade due to UV radiation and release the healing agent, which repairs the crack.

Anti-fouling paints and coatings also make use of encapsulation technology. A serious challenge encountered by the shipping industry is the attack of marine organisms on the surfaces in contact with water. The biofouling on the underside of ships can promote deterioration, make the hull heavier and increasing frictional resistance, which leads to an increase in fuel consumption. Polyurethane-polyurea microcapsules were successfully used for encapsulating an anti-fouling agent [128]. Here, encapsulation serves two purposes: it maintains a minimal effective concentration of the biocide on the surface for long durations and it prevents the excess release of biocide into the water body. A similar technique was used for preparing anti-fouling fishing nets and cages. Ethyl cellulose microcapsules encapsulating the anti-fouling agent were prepared using spray drying and embedded as a coating onto fishing nets [129]. These biocidal coatings can also be used as exterior paints to avoid attacks by algae/mold. Polyurethane microcapsules were used for encapsulating biocide zinc pyrithione were prepared using interfacial polymerization [130].

5.5 Drug Delivery

Controlled release of drugs is another major area where encapsulation technology has found widespread use. Protecting the encapsulated drug, slow sustained release to prolong its effect and targeted release at a specific location in the body are some of the ways in which the drug release mechanisms have been applied. Eudragit is a popular example of engineered polymer used for targeted release. It is the trade name given to polymethacrylates and the functional groups determine their physicochemical properties. These polymers have a unique characteristic of being insoluble in acidic condition (<7 pH) and soluble in alkaline conditions. They can be used as enteric formulations for targeting release to various parts of the gastrointestinal tract. Different combinations of various grades of these polymers can be used for obtaining a desired release profile.

The capsules are generally prepared using direct compression, granulation or melt extrusion [131]. Cell encapsulation is being used prominently for delivery of therapeutics and cell/tissue transplantation. The polymeric shell acts as a semi-permeable membrane and protects the cell contents from the immune system of the host. Common cell encapsulation materials include biocompatible polymers like alginate, poly (ethylene glycol), chitosan, and poly (lactic-co-glycolic acid) [132]. Enzyme immobilization is another technology prominently used in biomedical applications where polymeric materials play a crucial role. Enzymes can be bound to polymeric materials via adsorption, ionic binding, covalent binding, encapsulation or entrapment [133]. Specifically, encapsulated enzymes have been used, among other applications, for enzyme replacement therapy [134], degradable capsules for delivery of bioactive compounds [135], and biosensors [136]. Microcapsules prepared using microfluidics have recently been used for drug delivery applications [137, 138]. These can be used for overcoming the heterogeneity observed in conventional drug delivery systems. Also, the ability to accurately synthesize multi-compartment droplets makes microfluidics, a viable technique to produce artificial cells.

6 Conclusions

Polymers dominate the field of microencapsulation. Almost all the prominent techniques used for encapsulating active ingredients make use of polymeric shell materials. As the world shifts towards green technology, the need for efficient and sustainable formulations will increase. With widespread applications across a variety of fields, polymeric microcapsules will play a crucial role in this shift. This chapter aims to acquaint the reader with techniques routinely used for microencapsulation. Individually, the chapter provides an overview of microencapsulation technology and points towards reviews and reports that can prove to be crucial for an in-depth understanding of the subject. Hence, for an audience that is new to microencapsulation, it can serve as a starting point for their research.

References

1. Fanger GO (1974) Microencapsulation a brief history and introduction. In: Vandegaer JE (ed) Microencapsulation. Springer, Boston, MA, pp 1–20
2. Green BK (1960) U.S. Reissue 24,899
3. Green BK (1957) Schleicher L, U.S. 2800457

4. Lin CY, Lin SJ, Yang YC, Wang DY, Cheng HF, Yeh MK (2015) Biodegradable polymeric microsphere-based vaccines and their applications in infectious diseases. Human Vaccines Immunotherapeutics 11(3):650–656
5. Dragostin I, Dragostin O, Pelin AM, Grigore C, Zamfir CL (2017) The importance of polymers for encapsulation process and for enhanced cellular functions. J Macromol Sci Part A 54(7):489–493
6. Abdelkader H, Hussain SA, Abdullah N, Kmaruddin S (2018) Review on micro-encapsulation with Chitosan for pharmaceuticals applications. MOJ Current Res Rev 1(2):77–84
7. Sevault A, Kauko H, Bugge M, Banasiak K, Haugen N, Skreiberg O (2017) Phase change materials for thermal energy storage in low- and high-temperature applications: a state-of-the-art. Report, SINTEF Energy Research
8. Mishra MK (2016) Overview of encapsulation and controlled release. In: Mishra MK (ed) Handbook of encapsulation and controlled release. CRC Press, pp 3–22
9. Theis C (2005) Microencapsulation, Encyclopedia of polymer science and technology. Wiley
10. Oakley J (2016) Process selection criteria. In: Mishra MK (ed) Handbook of encapsulation and controlled release. CRC Press, pp 23–33
11. Sri Vajra Priya V, Roy HK, Jyothi N, Lakshmi Prasanthi N (2016) Polymers in Drug delivery technology, types of polymers and applications. Scholars Acad J Pharmacy 5(7):305–308
12. Masters K (1985) Spray drying handbook-, 4th edn. Wiley & Sons Inc., Publishers. New York, Halsted Press, p 696
13. Drusch S, Diekmann S (2016) Microencapsulation by spray drying. In: Mishra MK (ed) Handbook of encapsulation and controlled release. CRC Press, pp 35–44
14. Re MI (2006) Formulating drug delivery systems by spray drying. Drying Technol 24:433–446
15. Rosenberg M, Kopelman IJ, Talmon Y (1990) Factors affecting retention in spray-drying microencapsulation of volatile materials. J Agric Food Chem 38:1288–1294
16. Re MI (1998) Microencapsulation by spray drying. Drying Technol 16(6):1195–1236
17. Soottitantawat A, Yoshii H, Furuta T, Ohkawara M, Linko P (2003) Microencapsulation by spray drying: influence of emulsion size on the retention of volatile compounds. J Food Sci 68(7):2256–2262
18. Popplewell LM, Hans KT, Henson L, Lavallee CT, Wolff EJ, Wright M (2013) Spray-dried compositions capable of retaining volatile compounds and methods of producing the same. US 20130022728
19. Reineccius GA (2004) The spray drying of food flavors. Drying Technol 22(6):1289–1324
20. Zuidam NJ, Shimoni E (2010) Overview of microencapsulates for use in food products or processes and methods to make them. In: Zuidam N, Nedovic V (eds) Encapsulation technologies for active food ingredients and food processing. Springer, New York, pp 3–29
21. Ermis D, Yuksel A (1999) Preparation of spray-dried microspheres of indomethacin and examination of the effects of coating on dissolution rates. J Microencapsul 16(3):315–324
22. Singh A, Van den Mooter G (2016) Spray drying formulations of amorphous solid dispersions. Adv Drug Deliv Rev 100:27–50
23. Pradhan R, Kim SY, Yong CS, Kim JO (2016) Preparation and characterization of spray-dried valsartan-loaded Eudragit® E PO solid dispersion microparticles. Asian J Pharm Sci 11(6):744–750
24. Vidgren P, Vidgren M, Arppe J, Hakuli T, Laine EE, Paronen P (1992) In vitro evaluation of spray-dried mucoadhesive microspheres for nasal administration. Drug Dev Ind Pharm 18(5):581–597
25. Acosta N, Sánchez E, Calderón L, Cordoba-Diaz M, Cordoba-Diaz D, Dom S, Heras A (2015) Physical stability studies of semi-solid formulations from natural compounds loaded with chitosan microspheres. Marine Drugs 13:5901–5919
26. Arpagaus C, Schafroth N (2009) Laboratory scale spray drying of biodegradable polymers. Respir Drug Deliv Eur 2:269–274
27. Wurster DE (1963) Granulating and coating process for uniform granules. US Patent No 3,089,824

28. Wurster DE (1965) Apparatus for the encapsulation of discrete particles. US Patent No 3,196,827
29. Wurster DE (1966) Particle coating apparatus. US Patent No 3,241,520
30. Wurster DE (1990), Particle-coating methods. In: Lieberman HA (ed) Pharmaceutical dosage forms: tablets, vol 3, rev 90. Marcel Dekker, New York, pp 161–197
31. Guignon B, Duquenoy A, Dumoulin ED (2002) Fluid bed encapsulation of particles: principles and practice. Drying Technol 20(2):419–447
32. Jones DM, Rajabi-Siahboomi AR (2017) Fluid bed technology, process robustness, and scale-up. In: Rajabi-Siahboomi A (eds) Multiparticulate drug delivery. advances in delivery science and technology. Springer, New York, NY
33. Frey CR (2016) Encapsulation via fluidized bed coating technology. In: Mishra MK (ed) Handbook of encapsulation and controlled release. CRC Press, pp 35–44
34. Upjohn WE (1885) Process of making pills. US Patent No. 312041
35. Barrier P, Rousseau JY (1997) New fluid powder containing micro-encapsulated fish oil. French Patent No. 2758055A1
36. Gautam A, Patrick M, Dagerath ML (2007) Nutrition bar or other food product and process of making. Worldwide Patent No. 2006058634A1
37. Agrawal AM, Pandey P (2015) Scale up of pan coating process using quality by design principles. J Pharm Sci 104:3589–3611
38. Porter S, Sackett G, Liu L (2009) Development, optimization, and scale-up of process parameters: pan coating. In: Qiu Y, Chen Y, Zhang GGZ, Liu L, Porter WR (eds) Developing solid oral dosage forms. Academic Press, San Diego, California, pp 761–805
39. Ubbink J (2013) Flavour delivery systems. In: Kirk-Othmer (ed) Encyclopedia of chemical technology. Wiley and Sons, New York
40. Leister MD, Geilen T, Geissler T (2012) Twin-screw extruders for pharmaceutical hot-melt extrusion: Technology, techniques and practices. In: Douroumis D (ed) Hot-melt extrusion: pharmaceutical applications, 1st edn. Wiley, Chichester, U.K., pp 23–42
41. Patil H, Tiwari RV, Repka MA (2016) Encapsulation via hot-melt extrusion. In: Mishra MK (ed) Handbook of encapsulation and controlled release. CRC Press, pp 213–233
42. Lu M, Guo Z, Li Y, Pang H, Lin L, Liu X, Pan X, Wu C (2014) Application of hot melt extrusion for poorly water-soluble drugs: limitations. Adv Future Prospects 20(3):369–387
43. Crowle MM, Zhang F, Repka MA, Thumma S, Upadhye SB, Battu SK, McGinity JW, Martin C (2007) Pharmaceutical applications of hot-melt extrusion: part I. Drug Dev Ind Pharm 33:909–926
44. Rayleigh L (1878) On the stability of jets. Proc Lond Math Soc 10:4–13
45. Heinzen C, Berger A, Marison IW (2004) Use of vibration technology for jet break-up for encapsulation of cells and liquids in monodisperse microcapsules. In: Nedovic V, Willaert R (eds) Fundamentals of cell immobilisation technology. Kluwer Academic Publishers, Dordrecht, pp 257–275
46. Whelan M, Marison IW (2011) Microencapsulation using vibrating technology. J Microencapsul 28(8):669–688
47. Gugerli R (2003) Polyelectrolyte-complex and covalent-complex microcapsules for encapsulation of mammalian cells: potential and limitations. Chemical Engineering, Lausanne, Ecole Polytechnique Federale de Lausanne
48. Marison IW, Peters A, Heinzen C (2004) Liquid core capsules for applications in biotechnology. In: Nedovic V, Willaert R (eds) Fundamentals of cell immobilisation biotechnology. Kluwer Academic Publishers, Dordrecht, pp 185–204
49. Liu Z, Fontana F, Python A, Hirvonen JT, Santos HA (2020) Microfluidics for production of particles: mechanism, methodology, and applications. Small 16(9): 1904673
50. Wang JT, Wang J, Han JJ (2011) Fabrication of advanced particles and particle-based materials assisted by droplet based microfluidics. Small 7(13):1728–1754
51. Chen JM, Kuo MC, Liu CP (2011) Control of droplet generation in flow-focusing microfluidic device with a converging-diverging nozzle-shaped section. Japanese J Appl Phys 50(10R): 107301

52. Bah MG, Bilal HM, Wang J (2019) Fabrication and application of complex microcapsules: a review. Soft Matter 16:570–590
53. Zhao CX, Middelberg AP (2011) Two phase microfluidic flows. Chem Eng Sci 66:1394–1411
54. Collins DJ, Nield A, deMello A, Qun Liu A, Ai Y (2015) The Poisson distribution and beyond: methods for microfluidic droplet production and single cell encapsulation. Lab Chip 15:3439–3459
55. Kucuk I, Edirisinghe M (2014) Microfluidic preparation of polymer nanospheres. J Nanopart Res 16:2626
56. Deng NN, Huck WTS (2017) Microfluidic formation of monodisperse coacervate organelles in liposomes. Angew Chem Int Ed 56(33):9736–9740
57. Hou L, Ren Y, Liu W, Deng X, Chen X, Jiang T, Wu G, Jiang H (2020) Eccentric magnetic microcapsule for on-demand transportation, release, and evacuation in microfabrication fluidic networks. Colloids Surfaces A Physicochem Eng Aspects 599:124905
58. Lei KF (2018) Introduction: the origin, current status, and future of microfluidics. microfluidics: fundamental, devices and applications. In: Song Y, Cheng D, Zhao L (eds) Microfluidics: fundamental, devices and applications. Wiley-VCH Verlag GmbH & Co. KGaA, pp 1–18
59. Klank H, Kutter JP, Geschke O (2002) CO_2-laser micromachining and back-end processing for rapid production of PMMA-based microfluidic systems. Lab Chip 4:242–246
60. Chen CS, Breslauer DN, Luna JI (2008) Shrinky-dink microfluidics: 3D polystyrene chips. Lab Chip 8:622–624
61. Wabuyele MB, Ford SM, Stryjewski W (2001) Single molecule detection of double-stranded DNA in poly(methylmethacrylate) and polycarbonate microfluidic devices. Electrophoresis 22:3939–3948
62. Lemetter C, Meeuse F, Zuidam N (2009) Control of the morphology and the size of complex coacervate microcapsules during scale-up. AIChE J 55(6):1487–1496
63. de Kruif CG, Weinbreck F, de Vries R (2004) Complex coacervation of proteins and anionic polysaccharides. Curr Opin Colloid Interface Sci 9(5):340–349
64. Timilsena YP, Akanbi TO, Khalid N, Adhikari B, Barrow CJ (2019) Complex coacervation: principles, mechanisms and applications in microencapsulation. Int J Biol Macromol 121:1276–1286
65. Yan M (2016) Microencapsulation with coacervation. In: Mishra MK (ed) Handbook of encapsulation and controlled release. CRC Press, pp 235–243
66. Deveci SS, Basal G (2009) Preparation of PCM microcapsules by complex coacervation of silk fibroin and chitosan. Colloid Polym Sci 287(12):1455–1467
67. Yan C, Zhang W, Jin Y, Webber LA, Barrow C (2008) Vegetarian microcapsules. WO/2008085997
68. Mendanha DV, Ortiz SEM, Favaro-Trindade CS, Mauri A, Monterrey-Quintero ES, Thomazini M (2009) Microencapsulation of casein hydrolysate by complex coacervation with SPI/pectin. Food Res Int 42:1099–1104
69. Timilsena YP, Vongsvivut J, Tobin MJ, Adhikari R, Barrow C, Adhikari B (2019) Investigation of oil distribution in spray-dried chia seed oil microcapsules using synchrotron-FTIR microspectroscopy. Food Chem 275:457–466
70. Pham BL, Wang B, Zisu B, Truong T, Adhikari B (2020) Microencapsulation of flaxseed oil using polyphenol-adducted flaxseed protein isolate-flaxseed gum complex coacervates. Food Hydrocolloids 107:105944
71. Li M, Rouaud O, Poncelet D (2008) Microencapsulation by solvent evaporation: State of the art for process engineering approaches. Int J Pharm 363:26–39
72. Amasya G, Badilli U, Aksu B, Tarimci N (2016) Quality by design case study 1: design of 5-fluorouracil loaded lipid nanoparticles by the W/O/W double emulsion—solvent evaporation method. Eur J Pharm Sci 84:92–102
73. Lai MK, Tsiang RCC (2005) Microencapsulation of acetaminophen into poly(L-lactide) by three different emulsion solvent-evaporation methods. J Microencapsul 22(3):261–274
74. Allain LR, Sorasaenee K, Xue Z (1997) Doped thin-film sensors via a sol-gel process for high-acidity determination. Anal Chem 69(15):3076–3080

75. Danks AE, Hall SR, Schnepp Z (2016) The evolution of 'sol–gel' chemistry as a technique for materials synthesis. Mater Horizon 3:91–112
76. Ciriminna R, Sciortino M, Alonzo G, Schrijver AD, Pagliaro M (2010) From molecules to systems: sol–gel microencapsulation in silica-based materials. Chem Rev 111(2):765–789
77. Sol-Gel Scheme.svg [Online]. Available: https://commons.wikimedia.org/wiki/File:Sol-Gel_Scheme.svg. Accessed 11.08.2020
78. Magdassi S, Avnir D, Seri-levy A, Lapidot N, Rottman C, Sorek Y, Gans O (2001) Method for the preparation of oxide microcapsules loaded with functional molecules and the products obtained thereof. US 6303149
79. Hitchen SM, Dean JR (1993) Properties of supercritical fluids. In: Dean JR (ed) Applications of supercritical fluids in industrial analysis. Springer, Dordrecht, pp 1–11
80. Martín A, Fraile M, Rodríguez-Rojo S, José Cocero M (2016) Supercritical fluid technology for encapsulation. In: Mishra MK (ed) Handbook of encapsulation and controlled release. CRC Press, pp 447–468
81. Karim FT, Ghafoor K, Ferdosh S, Al-Juhaimi F, Ali E, Yunus KB, Hamed MH, Islam A, Asif M, Zaidul M, Sarker I (2017) Microencapsulation of fish oil using supercritical antisolvent process. J Food Drug Anal 25(3):654–666
82. Suttiruengwong S, Rolker J, Smirnova I, Arlt W, Seiler M, Luederitz L, Perez de Diego Y, Jansens PJ (2006) Hyperbranched polymers as drug carriers: microencapsulation and release kinetics. Pharm Dev Technol 11(1):55–70
83. Zabihi F, Yang M, Leng Y, Zhao Y (2015) PLGA–HPMC nanoparticles prepared by a modified supercritical antisolvent technique for the controlled release of insulin. J Supercrit Fluids 99:15–22
84. Martın A, Varona S, Navarrete A, Cocero MJ (2010) Encapsulation and co-precipitation processes with supercritical fluids: applications with essential oils. Open Chem Eng J 4(1):31–41
85. Duan B (2015) Microencapsulation via interfacial polymerization. In: Mishra MK (eds) Handbook of encapsulation and controlled release. CRC Press, pp 297–305
86. Jagtap SB, Patil VD, Suresh K, Ram F, Mohan MS, Rajput SS, Patil S, Shukla PG, Shanmuganathan K (2018) Functionalized carbon nanotube reinforced polymer nanocomposite microcapsules with enhanced stiffness. Colloids Surf, A 550:82–89
87. Shukla PG (2018) Microcapsule modified with nanomaterial for controlled release of active agent and preparation process thereof. US 20180161746
88. Rajagopalan N, Bhaskar C, Bankar VS, Pokharkar VB, Shukla PG, Regupathy A, Khilar KC (1995) Starch urea formaldehyde matrix encapsulation of solid agrochemicals: III. Studies and bio efficacy trials on double encapsulation. Pesticide Sci 45:123–131
89. Donath E, Sukhorukov GB, Caruso F, Davis SA, Möhwald H (1998) Novel hollow polymer shells by colloid-templated assembly of polyelectrolytes. Angew Chem Int Ed 37(16): 2201–2205
90. Jia Y, Li J (2019) Molecular assemblies of biomimetic microcapsules. Langmuir 35:8557–8564
91. Ariga K, Yamauchi Y, Rydzek G, Ji Q, Yonamine Y, Wu KCW, Hill JP (2014) Layer-by-layer nanoarchitectonics: invention, innovation, and evolution. Chem Lett 43(1):36–68
92. He Q, Cui Y, Li J (2009) Molecular assembly and application of biomimetic microcapsules. Chem Soc Rev 38(8):2292–2303
93. Matsusaki M, Ajiro H, Kida T, Serizawa T, Akashi M (2012) Layer-by-layer assembly through weak interactions and their biomedical applications. Adv Mater 24(4):454–474
94. Shukla PG (2006) Microencapsulation of liquid active agents. In: Ghosh SK (eds) Functional coatings. WILEY-VCH Verlag GmbH & Co. KGaA, Weinheim, pp 153–186
95. Li H, Cui Y, Wang H, Zhu Y, Wang B (2017) Preparation and application of polysulfone microcapsules containing tung oil in self-healing and self-lubricating epoxy coating. Colloids Surf, A 518:181–187
96. Higuchi T (1961) Rate of release of medicaments from ointments bases containing drugs in suspension. J Pharm Sci 50:874–875

97. Weilbull W (1951) A statistical distribution function of wide applicability. J Appl Mech 18:293–297
98. Korsmeyer RW, Gurny R, Doelker EM, Buri P, Peppas NA (1983) Mechanism of solute release from porous hydrophilic polymers. Int J Pharm 15:25–35
99. Ritger PL, Peppas NA (1987) A simple equation for describing of solute release. I. Fickian and non-Fickian release from non-swellable devices in the form of slabs, spheres, cylinders or discs. J Control Release 5:23–36
100. Kadam SL, Yadav P, Bhutkar S, Patil VD, Shukla PG, Shanmuganathan K (2019) Sustained release insect repellent microcapsules using modified cellulose nanofibers (mCNF) as pickering emulsifier. Colloids Surf A Physicochem Eng Aspects 582: 123883
101. Dubey R, Shami TC, Bhasker Rao KU (2009) Microencapsulation technology and applications. Defence Sci J 59(1):82–95
102. Rodrigues do Amaral PH, Andrade PL, Costa de Conto L (2019) Microencapsulation and its uses in food science and technology: a review. In: Microencapsulation—processes, technologies and industrial applications. https://doi.org/10.5772/intechopen.81997
103. Aditya N, Aditya S, Yang HJ, Kim HW, Park SO, Lee J, Ko S (2015) Curcumin and catechin co-loaded water-in-oil-in-water emulsion and its beverage application. J Funct Foods 15:35–43
104. Zhang T, Luo Y, Wang M, Chen F, Liu J, Meng K, Zhao H (2020) Double-layered microcapsules significantly improve the long-term effectiveness of essential oil. Polymers (Basel) 12(8):1651
105. Eratte D, McKnight S, Gengenbach TR, Dowling K, Barrow CJ, Adhikari BP (2015) Co-encapsulation and characterisation of omega-3 fatty acids and probiotic bacteria in whey protein isolate–gum Arabic complex coacervates. J Funct Foods 19:882–892
106. Ho BT, Joyce DC, Bhandari BR (2011) Encapsulation of ethylene gas into α-cyclodextrin and characterisation of the inclusion complexes. Food Chem 127(2):572–580
107. Wandrey C, Bartkowiak A, Harding SE (2010) Materials for encapsulation. In: Zuidam NJ, Nedović VA (eds) Encapsulation technologies for active food ingredients and food processing. Springer, New York, pp 31–100
108. Timilsena YP, Haque MA, Adhikari BR (2020) Encapsulation in the food industry: a brief historical overview to recent developments. Food Nutr Sci 11:481–508
109. Zhao H, Fei X, Cao L, Zhang B, Liu X (2019) The fabrication of fragrance microcapsules and their sustained and broken release behavior. Materials (Basel, Switzerland) 12(3): 393
110. Tekin R, Bac N, Erdogmus H (2013) Microencapsulation of fragrance and natural volatile oils for application in cosmetics, and household cleaning products. Macromol Symposia 333:35–40
111. Rodrigues Teixeira CS (2010) Microencapsulation of perfumes for application in textile industry. https://repositorio-aberto.up.pt/bitstream/10216/57560/1/000143206.pdf. Accessed 25 Aug 2020
112. Rodrigues SN, Martins IM, Fernandes IP, Gomes PB, Mata VG, Barreiro MF, Rodrigues AE (2009) Scentfashion®: microencapsulated perfumes for textile application. Chem Eng J 149:463–472
113. Casanova F, Santos L (2015) Encapsulation of cosmetic active ingredients for topical application—a review. J Microencapsul 33(1):1–17
114. Sansukcharearnpon A, Wanichwecharungruang S, Leepipatpaiboon N, Kerdcharoen T, Arayachukeat S (2010) High loading fragrance encapsulation based on a polymer-blend: preparation and release behavior. Int J Pharm 391(1–2):267–273
115. Anchisi C, Meloni MC, Maccioni AM (2007) Chitosan beads loaded with essential oils in cosmetic formulations. Int J Cosmet Sci 29(6):485–485
116. Sinclair RG (1973) Slow-release pesticide system. Polymers of lactic and glycolic acids as ecologically beneficial, cost-effective encapsulating materials. Environ Sci Technol 7(10): 955–956
117. Scher HB, Rodson M, Lee KS (1998) Microencapsulation of pesticides by interfacial polymerization utilizing isocyanate or aminoplast chemistry. Pestic Sci 54(4):394–400

118. Rajagopalan N, Bhaskar C, Shukla PG, Amarnath N (1994) Novel controlled-release (CR) agrochemical formulations: development and evaluations. Res Develop Controlled Release Formul Pesticides, Vienna, Austria I:91–110
119. Bhaskar C, Shukla PG, Rajagopalan N (2000) An improved process for the preparation of micorcapsular formulations of agrochemicals. Indian Patent IN184975
120. Mihou AP, Michaelakis A, Krokos FD, Majomenos BE, Couladouros EA (2007) Prolonged slow release of (Z)-11-hexadecenyl acetate employing polyurea microcapsules. J Appl Entomol 131(2):128–133
121. Zengliang C, Yuling F, Zhongning Z (2007) Synthesis and assessment of attractiveness and mating disruption efficacy of sex pheromone microcapsules for the diamondback moth, *Plutella xylostella* (L). Chin Sci Bull 57(10):1365–1371
122. Gregg PC, Del Socorro AP, Landolt PJ (2018) Advances in attract-and-kill for agricultural pests: beyond pheromones. Annu Rev Entomol 63(1):453–470
123. Humbert P, Vemmer M, Mävers F, Schumann M, Vidal S, Patel AV (2018) Development of an attract-and-kill co-formulation containing Saccharomyces cerevisiae and neem extract attractive towards wireworms. Pest Manag Sci 74(7):1575–1585
124. Humbert P, Vemmer M, Giampà M, Bednarz H, Niehaus K, Patel AV (2017) Co-encapsulation of amyloglucosidase with starch and Saccharomyces cerevisiae as basis for a long-lasting CO_2 release. World J Microbiol Biotechnol 33(4):71
125. White SR, Sottos NR, Geubelle PH, Moore JS, Kessler MR, Sriram SR, Brown EN, Viswanathan S (2001) Autonomic healing of polymer composites. Nature 409:794–797
126. Kanellopoulos A, Giannaros P, Palmer D, Kerr A, Al-Tabbaa A (2017) Polymeric microcapsules with switchable mechanical properties for self-healing concrete: synthesis, characterization and proof of concept. Smart Mater Struct 26: 045025
127. Zhu Y, Cao K, Chen M, Wu L (2019) Synthesis of UV-responsive self-healing microcapsules and their potential application in aerospace coatings. ACS Appl Mater Interfaces 11(36):33314–33322
128. Feio R, Ferreira O, Bordado JC, Marques AC, Silva ER (2015) Microencapsulation of biocides: a new strategy for biofouling control. https://fenix.tecnico.ulisboa.pt/downloadFile/1126295043835061/Extended%20Abstract.pdf. Accessed 26 Aug 2020
129. Kartal GE, Sarıışık AM (2020) Providing antifouling properties to fishing nets with encapsulated econea. J Ind Text. https://doi.org/10.1177/1528083720920568
130. Shukla PG, Sivaram S (2006) Polyurethane microcapsules containing biocide and process for the preparation thereof. US 2006/0251688 A1
131. Joshi M (2013) Role of Eudragit in targeted drug delivery. Int J Current Pharmaceutical Res 5(2):58–62
132. Blanchette J (2016) Cell encapsulation. In: Mishra MK (ed) Handbook of encapsulation and controlled release. CRC Press, pp 917–931
133. Brena BM, Batista-Viera F (2006) Immobilization of enzymes. In: Guisan JM (eds) Immobilization of enzymes and cells. Methods in biotechnology™, vol 22. Humana Press, pp 15–30
134. Genta I, Perugini P, Pavanetto F, Maculotti K, Modena T, Casado B, Lupi A, Iadarola P, Conti B (2001) Enzyme loaded biodegradable microspheres in vitro: ex vivo evaluation. J Control Release 77(3):287–295
135. Borodina T, Markvicheva E, Kunizhev S, Möhwald H, Sukhorukov GB, Kreft O (2007) Controlled release of DNA from self-degrading microcapsules. Macromol Rapid Commun 28:1894–1899
136. Khattab TA, Fouda MMG, Abdelrahman MS, Othman SI, Bin-Jumah M, Alqaraawi MA, Al Fassam H, Allam AA (2019) Co-encapsulation of enzyme and tricyanofuran hydrazone into alginate microcapsules incorporated onto cotton fabric as a biosensor for colorimetric recognition of urea. React Funct Polym 142:199–206

137. Tong W, Gao C (2016) Multilayer microcapsules with tailored structures and properties as delivery carriers for drugs and growth factors. In: Gao C (eds) Polymeric biomaterials for tissue regeneration. Springer, Singapore
138. Fontana F, Ferreira MPA, Correia A, Hirvonen J, Santos HA (2016) Microfluidics as a cutting-edge technique for drug delivery applications. J Drug Delivery Sci Technol 34:76–87

Electrospinning Techniques for Encapsulation

Nalan Oya San Keskin and Sena Kardelen Dinç

1 Introduction

Nanotechnology remains one of the most consequential areas utilized in the manipulation of molecular-sized materials and in the creation and processing of new products. For this reason, in the recent years, nanotechnology has gained strategic importance for development of systems, structures and networks which has benefits from the outstanding features on the atomic or molecular levels of nano substances. Nanotechnology involves contributions from multidiscipline fields like physics, chemistry, biology and engineering. Nanomaterials at least one dimension should be on the nanoscale. Among different type of materials, nanofibers with their unique structure are exceedingly encouraging matters that have potential usage in many fields. Lately, nanofibers and electrospinning method enable a plausible way for encapsulation compounds [100].

While microencapsulation generates particles larger than 100 nm, nanoencapsulation provides a protecting barrier at nanometric scale and generates particles of smaller than 100 nm [65]. The nanoencapsulation method is still one of the most promising technologies among bioactive substances capturing technologies today. Nanoencapsulation of bioactive substances ensures to multi-directional approach in target-specific delivery and effective absorption along cells [30]. Capsulation septum protects molecules like food and flavors in a nanosize levels [68]. This method uses several techniques like nanostructuration, nanocomposite, nanoemulsification and electrospinning so, it provides functional output that includes controlled release property. Nanoencapsulation techniques have the makings of solving challenges in food industry concerning delivering functional substance effectively in health aspects and

N. O. San Keskin (✉) · S. K. Dinç
Polatlı Science and Literature Faculty, Biology Department, Nanosan Laboratory, Ankara Hacı Bayram Veli University, Polatlı, Ankara, Turkey
e-mail: nalan.san@hbv.edu.tr

J. Parameswaranpillai et al. (eds.), *Micro- and Nano-containers for Smart Applications*, Composites Science and Technology,
https://doi.org/10.1007/978-981-16-8146-2_2

releasing flavour substances in a controlled manner. Under favor of electrospinning, nanocomposite, nanostructuration and nanoemulsification techniques, in nanoencapsulation important compounds wrap in a small scale. Nanoencapsulation is thought as a service vehicle for the transporting substance to the necessary region for activity. They act as a protector against degradation chemically or biologically in the course of production, storing and beneficial occupancy. They are component at controlled release of functional compounds. Hereby, the resulting delivery system is consonant with attributes of obtained product in many aspects [2].

2 Electrospun Nanofibers

The size range of 1–100 nm defines the nanoscale to which structured materials refer like nanorods, nanotubes or other fibrous materials, electrospun fibers are a representative of one-dimensional-nanostructure. Electrospinning process known since the 1930s for ongoing fiber manufacturing from different materials like polymers, ceramics and composites. Electrospinning process was first patented in the 1900s. The cone-shaped structure formed at the tip of fluid droplet during the process was modeled mathematically by Taylor in 1960s. He researched the form of the polymer droplet generated on the needle tip at the time of performed an electrical field and has shown that the cone-shaped structure and the bubbles are expelled from the cone edges. In the literatures, this cone shape is known "Taylor cone". In the 1990s, several research groups indicated that many polymers could be electrospun into nano-micro fibers. Reneker produced fibers which have dimension of smaller than 5 μm through using high voltage to charge the polymer dispersion [51]. After these progress electrospinning has attracted notice from researchers across the world [61].

Figure 1 show the summary of various electrospinning types and polymers that used in the techniques. There are three electrospinning types and these are melt electrospinning, solution electrospinning and suspension electrospinning. The most common type is solution electrospinning that enable to study various kind of polymers. In the polymer types which are used in melt-electrospinning generally have high molecular weight and have solubility problem. In this method the nanofiber with diameter less than 500 nm cannot be produced and this is a limitation also, during the process constant heat supplier should be used and higher voltage should be applied compared to other electrospinning type for the fiber formation. In suspension electrospinning, it is being used just hydrous dispersion of water insoluble polymers (lattices). However, this technique is definitely a rewarding choice to attain fresh, also confidential electrospinning so it is thought that suspension electrospinning is promising for future in applications in biomedical fields [28, 39].

Electrospinning is one of the well-known economical and easy technologies to obtain nanofiber or nanofibrous webs from particular polymers [31, 39] or non-polymers [19]. In addition to polymers, supramolecular structures, polymer mixtures, inorganic materials and composite materials are also used for electrospinning technology [20]. Furthermore, different forms like porous, core–shell, hollow, aligned

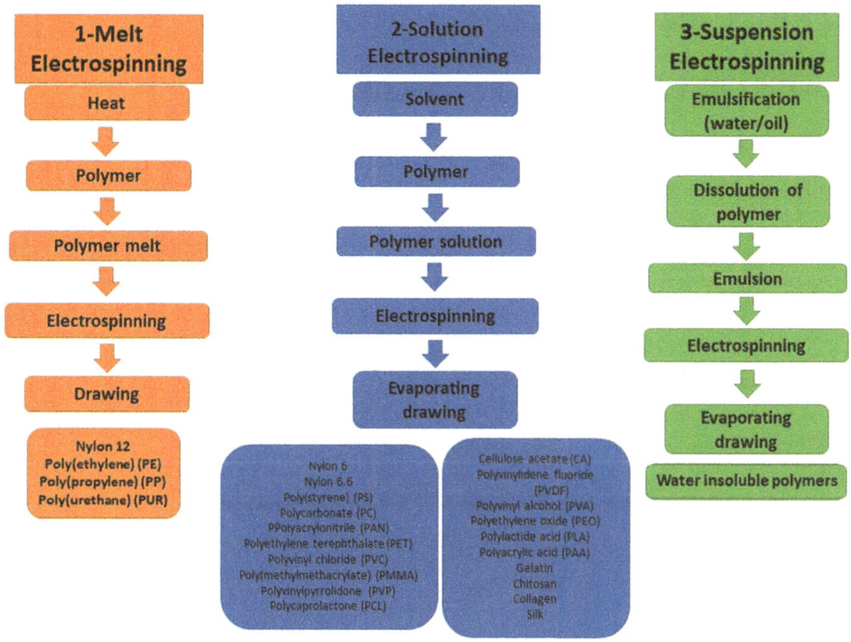

Fig. 1 Various type of electrospinning and polymers that used in these systems [28]

and ribbon-like nanofibers can be produced. Many applications in tissue engineering makes use of parallel-aligned nanofibers which is produced via the orientation of the nanofibers because these nanofibers ensure the desired growth of living cells. In the coaxial electrospinning technique, functionalities can be maintained (mostly core part) by allowing the active substances (drug, antibacterial, antioxidant, protein, enzyme, virus and bacteria) used for the functionalization of nanofibers to pass through a separate liquid medium. As a result, electrospinning is modern and effective fiber production method and enables to fabricate fibers with dimensions in the range of micrometers to nanometers [72, 93].

Electrospun nanofibers have the most amazing properties like high surface area-to-volume ratio, being feather-light, nano-porous nature, adaptability in surface functionalities, good mechanical properties and high permeability. Simultaneously, fibers produced by electrospinning method and have a wide and specific surface area characteristically have a diameter of 30–200 m^2/g. The produced nanofibers by electrospinning have a porous form and by making the smooth surface, the porosity can be reduced to smaller dimensions and the ratio can be increased. Due to these properties, 1D-nanofibers use for different applications in different areas, such as biomaterials, medical textiles, tissue engineering, electronics, information technology, defense industry, aerospace technologies, composites, filters, skin care applications, biomedical applications [93].

2.1 Electrospinning Fundamentals

Electrospinning technique is a well-established polymer fiber production method. There is no doubt that electrospinning is the most suitable and controllable technique that used to make ultra-thin fibers. For electrospinning, two configuration types, vertical and horizontal, are available (Fig. 2). In vertical setup; with the help of physical forces caused by gravity, there is no need for an extra syringe pump, as the polymer solution can easily exit the capillaries. However, without a syringe pump the controllability of the system decreases. Therefore, horizontal setups or vertical setup with syringe pumps are commonly used [53].

Figure 3 shows the schematic set up of the technique and the SEM micrographs of cellulose acetate (CA) nanofibrous.

As shown in Fig. 2, electrospinning basic set up contains a syringe pump with a capillary tube or a needle, a high-power voltage supply (1–30 kV) and a grounded collector. Firstly, suitable polymer solution in a compatible solvent of moderate to high viscosity prepared for the production of nanofibers by electrospinning technique. Then, capillary tube, which includes a polymeric or non-polymeric solution connect the high-power supply with electrical wires as well as to the target. Different targets such as aluminum foil or plates, copper plates, rotating drums, and a metal surface have been utilized [82]. The electrospinning technique depends on the principle that the powerful reciprocal electrical repulsive forces of booster overcome surface

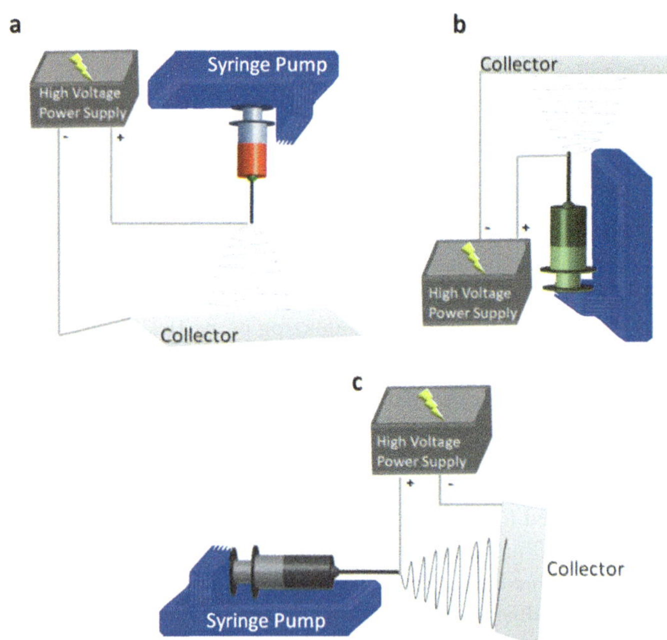

Fig. 2 Vertical (**a**, **b**) and horizontal (**c**) electrospinning set ups

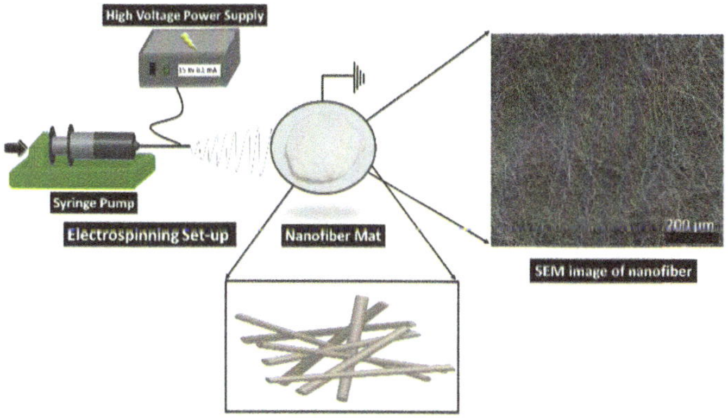

Fig. 3 Schematic set up of electrospinning, the micrograph of cellulose acetate (CA) nanofibrous

tension weaker forces. In the process, polymer solution is pumped with a constant flow-rate and high voltage source produce a charge of a certain polarity into a polymer which is then moved towards collector. When the electrostatic pushing is bigger than the surface tension, the liquid starts to deform into a conically form structure. Due to vaporization of the solvent between the spinneret tip and collector area, the jet travels through the atmosphere and leaving the dry fibers. Finally, a nonwoven nanofiber mat is collected on the collector or between the collectors. Type of a collector affect the fiber morphology of electrospun fibers by using different types of collectors. Figure 4 shows two type of collector: flat and rotating drum collector. For using rotating drum disk (due to drum speed), the diameter of the fiber becomes controllable and this collector shows an advantage like numerous fibers are residue on the margin of the disc and the aligned fibers are collected as nanofibers [33].

It is difficult to work in this field as the production of fibers by electro-spinning is affected by several parameters. In Fig. 5, we have discussed many factors and their impacts on fiber features.

Solution factors are important like viscosity, molecular weight, conductivity and surface. Molecular weight is related with the chain entanglement and important to forming bead or fibers. Polymer which has a low molecular weight used in electrospinning process resulted bead form. After increment of concentration, long and bead free nanofibers finally produced. However, high molecular mass polymers change the formation of fibers into ribbons forms [67]. Polymer viscosity is also a main parameter affecting the fiber morphology. Due to higher viscosity of polymer against to Coulombic repulsion forces, hard ejection of the jet or formation of larger fibers is occurred. When the lower viscosity and lower viscoelastic forces occur, beaded fibers produce [38]. The other factor is surface tension which affect the fibers morphology directly. Beadless fibers produce from using low surface tension solutions. Solvent volatility is also affecting the surface micro texture like pore or rugosity by enhancement of volatility [101].

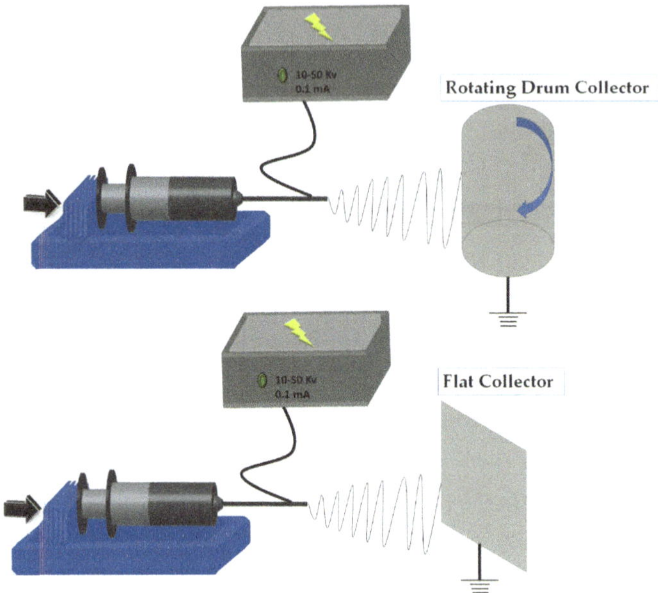

Fig. 4 Type of collectors: rotating drum and flat collector

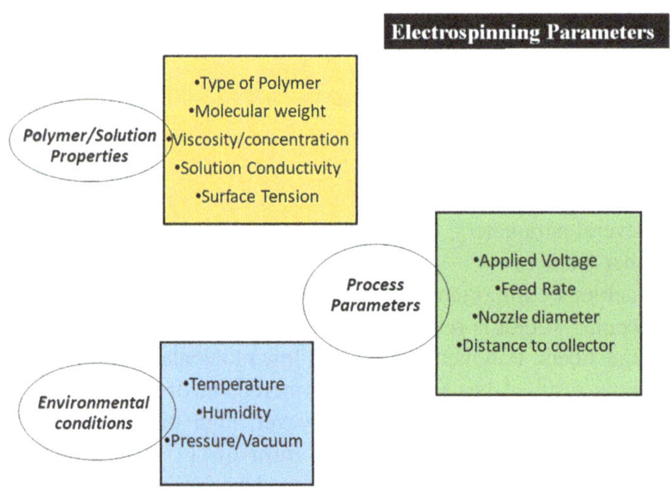

Fig. 5 The process parameters of electrospinning [13]

The other is process parameters include applied voltage, gap (distance between the needle tip and the collector), and flow velocity. Among these, applied voltage is a prominent parameter in the electrospinning procedure. Generally, the voltage increase leads to fiber length increases and fiber size decreases [24]. Also, the flow rate affects the flow rate of polymer solution and the jet velocity. The diameter of the fiber is increased by increasing the flow rate, and beaded morphology is observed [59]. Another parameter is the needle tip-to-collector distance and it is a significant factor to obtain uniform fibers owing to the sufficient flight time of the solution jet. Larger tip to collector distance causes the thinner diameter of nanofibers. Stronger electrostatic field which is important for the formation of beaded form is related with short distance. When the tip to collector distance increased less beaded electrospun fibers were produced due to higher stretching and elongation time before the fiber deposition on the collector. Finally, ambient parameters such as humidity, temperature and air speed in the electrospinning cabinet is important. During electrospinning process humidity variation influences the morphology. For instance, the diameter of polyethylene oxide (PEO) [103] decreases, the diameter of cellulose acetate (CA) [17] increases with increasing relative humidity. In a study done by Li and Xia, it is recommend that high humidity lends to the discharge of fibers throughout electrospinning [56]. As a result, while low humidity could raise the velocity of the solvent evaporation and so the fibers completely dry. However, high humidity will form thicker fiber diameter because the charges on the jet can be neutralized and the stretching forces will decrease. Temperature is another ambient parameter that affect the fiber morphology. In some cases increase in ambient temperature from 25 °C to higher temperatures like 75 °C electrospun silk nanofibers circular shape changes to flat shape [104]. Furthermore, studies report that biological substances such as proteins and enzymes which are sensitive to temperature can lose their functionality during electrospinning [37].

2.2 *Polymer Usage in Production of Nanofibers*

Natural, synthetic, ceramic, polymeric, ceramic or even biomolecules like nucleic acid, proteins and polysaccharide matters can be used in the electrospinning process. Chitosan [4], collagen [55], hyaluronic acid [106], gelatin [22], and silk protein [22] are natural polymers. However, synthetic polymers have advantageous like desired degradation rate and the required mechanical properties than natural polymers. For this reason, synthetic polymers like polylactide (PLA), poly(ethylene oxide) (PEO), polyvinyl alcohol (PVA), polysulfone (PSU) and poly (ἑ-caprolactone) (PCL) [70, 76, 105, 109] usually used in electrospinning process.

3 Applications Used Electrospun Nanofibers

In company with the combination of biotechnology, nanotechnology and information technology, a new opportunity has advanced. In this content, nanofibers used in many applications because the main reasons are nanofibers have special characteristics like the volume to weight ratio, high strength and smooth handling in the structures. These properties are turned to account for a wide range of applications from consumer to medical products and industrial that has technological applications for transistors, aerospace, drug delivery systems, condensers, power storage, information technology, battery separators, and fuel cells [108].

3.1 Medical Applications

Electrospinning has aroused interests on producing nanofibers because electrospun nanofibers have high porosity, large surface-area to volume ratio and superior mechanical characteristics due to nanoscale of biological systems, nanostructured materials (nanofibers, hollow nanofibers, nanotubes, nanorods or core–shell nanofibers) have a multitude of possible applications in medical and pharmaceutical applications. Among these application areas, nanomaterials usually use in wound healing, drug delivery, tissue engineering or other medical devices [63, 66].

3.2 Filtration Applications

Because of the population growth and environment contamination, providing safe and adequate water is becoming challenging. Especially, uncontrolled wastewater discharge has become a main concern by reason of the adverse effect to community health and the environment [15]. Micron size polymeric fibers have been used in water applications but researches have shown that nano size filtration materials are advantageous as it could offer matchless properties like high surface area to volume ratio along with porousness at nanoscale which makes them promising as filter and membrane materials. For this reason, nanofibers have begun to be used by the filtration industry and in some cases have produced products that are commercially available [11, 74, 79].

3.3 Composite Applications

Because of their optical, thermal, mechanical, electrochemical and catalytic properties, composite materials can be apply in specific areas such as filtration and catalyzes, protective textiles, tissue scaffolds, nanofibrous structures, drug delivery systems [49, 50]. Products which produced from nanofibers used composites sector has advantageous such as much lower and lighter in weight at the compare to same or improved mechanical norms traditional composites [1].

3.4 Food Applications

In science related food, nanotechnology applications include altering the texture of food components, extending the shelf-life of food may be added antimicrobial agent for use as food packaging material, encapsulating different food components, additives or flavors and controlling the release of these molecules and/or increasing the bioavailability of nutritional compounds and developing new tastes. Nowadays, materials which source is generally agricultural like zein [7], soy protein [83], and whey protein [23, 89] has been electrospun. Electrospinning allows for encapsulation of insoluble particles within a nanofiber mat. For this aim, antimicrobial agents incorporated into packaging material to prevent microorganism growth. Electrospun nanofibers have many advantages over films. Firstly, polymeric film can produce at high temperature as nanofibers processed at room temperature. For this reason, encapsulation of essential oils and flavors which are prone to temperature preferred electrospun nanofibers. Secondly, nanofibers show response to solution changes like temperature, pH or relative humidity so in this content, it is possible to release of the active agents [99].

3.5 Military and Defense Applications

Due to small pore size that enables resistance against chemicals and the greater surface area, fabrics produced from nanofibers are appropriate for clothing protection in military and defense applications. Nanofibers fabric enables breathing of the clothes. Nanofibers with high pores with tiny pore sizes enable good resistance against chemicals diffusion to the fabrics [75].

3.6 Agricultural Applications

One of the applications used nanofiber is agricultural applications. One of the grand issues in agri-food system is ineffective delivery of fertilizers and pesticides. This situation effects the food safety, security, safety, and quality. For this reason, it is explicit that nanotechnology, can and will play an important role regard to the inadequacies in agrichemical delivery and food safety. Recently in research and applications, plant covered with nanofiber mat or webs contain pesticide, insecticide or fertilizer for improve the productivity [75, 99].

3.7 Electrical and Optical Applications

The nanofibers are basically used in the manufacture of small electronic instruments and the certain machines production. These fibers capable of transmitting electricity. Furthermore, because of high surface area, electrospun nanofibers used to improve high-performance batteries [21, 58].

3.8 Delivery Systems and Controlled Release

Because of the undesirable condition of processing and storage, masking undesirable taste, masking unpleasant tastes, and low-solubility problems, nanoencapsulation become favorable technique for protection of ingredients like bioactive compounds, antioxidants, flavors, pharmaceutical compounds and antimicrobials.

One of the promising approaches for the encapsulation process is electrospun polymer nanofiber. The nanoencapsulation usually use in food science. 'Nanofood' is term relevant to the advancement of food color, prolongation of shelf life and preservation, detection of microorganism, and antibacterial characteristics, and intelligent packaging matters. In addition, food packaging is included in nanofood [52]. In food packaging, conventional active packaging have involved the dispersion of active agents in in carriers but the surface area is limited and during production and storage negligible losses occur [40]. Recently, electrospinning has drawn attention to be used in systems created for packaging functional and active foods. Furthermore, encapsulation living microorganism while preserving its activity is important for many applications.

4 Encapsulation Process Using Electrospun Nanofibers

In the last period, awakening about wellness problems builds up in parallel with growing cases chronicle and age-associated and diseases, such as diabetes, diseases related to progressive neuronal loss in brain and cancer, affecting quality of life [25, 29]. Besides, due to increased average life span and costs required to maintain and improve health, functional nourishment has been attracted attention and has become a new and growing market [25]. Medical progress and protection from illness with nutrients have prompted researchers about studying bioactive compounds like probiotics, vitamins, minerals, omega-3-fatty acids, polyphenols and peptides in pharmaceutical and nutraceutical industries [25, 100].

Many bioactive compounds cannot be included to commercial nutrition and pharmaceuticals in their pure status because of physicochemical and biotic constraints such as instability, low solubility, poor bioavailability, low permeation, susceptibility and readily impairment towards environmental stresses. Due to their low dissolubility in oil or water and physical, chemical and enzymatic susceptibility against degradation while they are processing, storage and preparation requires extra protection. Because some nutrients leave a bad taste in the mouth in first place, they are not accepted into food industry and it is necessary to suppress this bad taste with aroma. One of the challenges about nutrient substances is reducing of different nutrients each other's activities and stability in some cases [48, 84]. For this reason, these nutrients should be separated from each other or not come into contact with each other. In these and similar situations, the need to encapsulation for production and proper release of bioactive nutrients arose while producing [26].

Nowadays, considering the structure or number of the nozzles, four different electrospinning approaches have been established for encapsulation process.

(1) Single nozzle which is most common and easy technique to produce nanofibers from single polymer solutions [32].
(2) Emulsion electrospinning that manufacturing core—sheath structured nanofibers using single nozzles [36]
(3) Side-by-side nozzle; feasible for diverse solutions through distinct capillaries
(4) Coaxial electrospinning which producing core—shell structured fibers offers unequaled properties for developed delivery applications [45].

4.1 *The Encapsulation of Essential Oils in Nanofibers*

Figure 6 shows the summary of different encapsulated materials, encapsulation process using electrospinning and applications of the produced nanofibers.

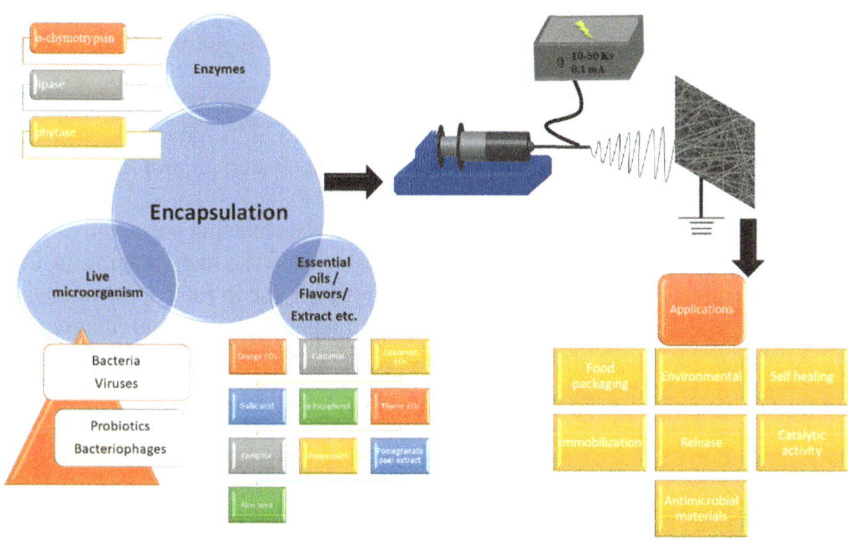

Fig. 6 Summary of the encapsulated substance, electrospinning process and applications

In addition, Table 1 shows the materials and polymers are used for the nanoencapsulation by electrospinning. For this content, essential oils (EOs) are detailed. For this content, EOs are often used by extracting the organs such as buds, flowers, fruits, leaves, twigs, rhizomes, seeds, roots of flowering and aromatic plants in tropical climates such as Mediterranean countries with mild seasons clove, jasmine, eucalyptus, anise, ginger, rosemary, lavender, cardamom and rose can be given as examples of organs that are known to produce essential oil and have antioxidant properties [6]. They have complex structures consisting of mixtures of isoprenoids such as monoterpenes and sesquiterpenes in different proportions [16].

Since they are obtained from nature, they have chemical and biological properties that do not harm the environment. Also, numerous studies have shown antioxidant and biocide activities of EOs. On account of their biocidal activities, they are used in medical fields [5]. It is known that there are 3000 essential oils that have been discovered so far. About 300 of them have gained an important place in industries and are commonly used in both food, dental therapy and other medical applications, as well as in the cosmetic and agricultural industries, and provide healing from nature. They have applications in many areas from soap to perfume, which are used for household cleaning and personal hygiene, to flavor foods. They have applications in many areas such as soap for household cleaning and personal hygiene, from soap to perfume, flavoring to foods, bath products for use in aromatherapy. In addition, it is known that they exhibit medical properties and help treatments in the functional disorders that occur in the organs and diseases related to the general system [10, 43, 69, 85]. Another example of its use in medical fields is using as a liquid or aerosol in equipment and surface disinfection in places such as operating rooms, hospitals, and in cleaning the air of the environment [14, 98]. Its rich ingredients have enabled them to inhibit the growth of pathogenic microorganisms, since they contain dozens of terpenoid compounds and rich content. In drug industries, it is used in different forms such as capsules, ointment, syrup, spray. Their uses and applications are increasing day by day [5].

Since they are suitable for use in food packaging, they can be added directly to foods or can be gradually combined at particular phases of the process. The release of essential oils in processes such as transportation and storage prolongs the shelf life and increases the quality of the products [6].

Although EOs are seen as an alternative to chemical-containing preservative materials, there are restrictions that need to be resolved for use in packaging in food industries. The most important features that limit their use in food applications are high incidence of volatilization, low solubility in aqueous media, and intense odor [6].

Since they have intense odor and taste properties, they should be used in the smallest doses that they can function in foods; this is also one of its disadvantages. In studies for conducting to prevent all these disadvantages, it was observed that when encapsulated with appropriate substances, their resistance to external factors that cause their reduction increased and the rate of encountering unwanted changes decreased [42].

Table 1 Nanoencapsulation of materials by electrospinning

Encapsulated materials	Polymers/Non-polymers	Applications	Reference
Pomegranate peel extract (PE)	Chitosan/Poly(ethylene oxide) (CS/PEO)	Food packaging material	[90]
Aloe Vera skin extract (AVE)	Poly(ethylene oxide) (PEO)	Food packaging material	[87]
Thiabendazole (TBZ)	Hydroxypropyl-β-cyclodextrin (HPβCD)	Environmental	[35]
Thyme	Potato starch	Food packaging material	[34]
Epoxy, mercaptan	Polyacrylonitrile (PAN)	Self healing	[62]
Probiotics; *Lactobacillus paracasei KS-199*	Poly(vinyl alcohol) (PVA) and sodium alginate (SA)	Functional foods	[107]
Probiotics; *Bifidobacterium animalis* Bb12	CS-PVA	Functional foods	[64]
Spores of *Volvariella volvacea*	PVA and polyvinyl pyrrolidone (PVP)	Immobilization	[9]
1,8 cineole-rich EOs	Non-polymers–Zein Powder	Edible Coating	[42]
Phytase	Polyvinyl-alcohol-sodium alginate (PVA-SA)	Catalytic activity	[27]
Peppermint EOs and chamomile EOs	Gelatin	Food packaging material	[91]
Pomegranate seed oil, Sea-buckthorn oil	Polylactide and Poly(vinyl-pyrrolidone) (PVP-based fibers)	Cosmetic Products	[60]
Thyme EOs	Gelatin/Chitosan	Food Processing	[96]
Camphor	Cyclodextrins [hydroxypropyl-b-cyclodextrin (HPβCD) and hydroxypropyl-γ-cyclodextrin (HPγCD)]	Release	[18]
Live bacterium	HPβCD	Environmental	[80]
Orange EOs	Gelatin (G) and Gelatin-cross-linked tannic acid (GT) nanofibers	Controlled Release	[92]
Curcumin	CS-phospholipid	Controlled release	[100]
Cinnamon EOs	PV (polyvinyl alcohol), β-CD (β-cyclodextrin)	Food Packaging	[100]
Gallic acid	HPβCD & polylactic acid (PLA)	Food packaging material	[8]

(continued)

Table 1 (continued)

Encapsulated materials	Polymers/Non-polymers	Applications	Reference
Lipase	Polylactic acid- chitosan (PLA-CS)	Immobilization	[86]
α-tocopherol	Zein-polyethyleneoxide-chitosan (ZN-PEO-CN)	Release	[102]
Cinnamaldehyde (CA)	Chitosan/ poly(ethylene oxide)	Erading infections	[73]
Lipase	Polycaprolactone (PCL)	Catalytic activity	[88]
Lipase	Polystyrene	Catalytic activity	[77]
Two bacteria (*Escherichia coli, Staphylococcus albus*) and bacterial viruses (T7, T4, λ)	PVA	Immobilization	[78]
α-chymotrypsin	Polystyrene	Biocatalysts	[47]

Polymeric micro and nanofibers were produced by methods like phase separation, template synthesis, bicomponent extrusion, drawing, and melt blowing [3, 41, 54, 94]. To the contrary, for applications about food, bigger parts of these methods remain incapable due to various deficiencies such as being difficult to control, impossible to apply to all kinds of desired substances and affection on food substance and health concern related to toxicity. Also, they demonstrate compatibility and degradability, as well as the potency of bioactive compounds in achieving desired and proper release [6, 71]. Increasing the effectiveness of the encapsulated substance and the production process brings this method to an important place [46].

In the literature, there are many studies of essential oils encapsulated by electrospinning towards the cosmetics, health and food industries. In a study done by Miletić et al. sea-buckthorn and pomegranate seed oil were encapsulated into PVP-based fibers electrospun nanofibers to avoid from losses of valuable active compounds and was investigated antioxidant activity. They stated that these materials have a high potency for use in the cosmetics industry [60]. In other research done by Tang and al., gelatin nanofibers incorporated with chamomile essential oil (CO) and peppermint essential oil (PO) were produced by electrospinning for edible packaging process. They confirmed the characteristics of the gelatin/EO nanofibers have potency for edible packaging [91]. In 2020, Fonseca et al. nanoencapsulate thyme essential oil (TEO) in electrospun nanofibers made by using starch and formic acid. They state that electrospun starch/TEO nanofibers can be carried out in food packaging technologies and food products [34].

4.2 The Encapsulation of Living Organism in Nanofibers

For many biotechnological applications such as bioremediation, phage therapy, regenerative medicine transplant cells and microbial fuel cell (MFC) systems, encapsulated cells are useful [110]. Encapsulated microorganisms bring advantages compare to free cells like reusability potential, lower space and growth medium necessities. In addition, encapsulation provide higher resistance in extreme environments. But the real problem is it is not easy to preserve activity of livings. For this reason, the encapsulation of complete organism such as bacteria, viruses, microalgae, bacteriophage and spores with active and dry form is important due to using them in alternative applications. In the literatures, there has latterly been an extensively raised interest in usage of encapsulated microorganism in nanofiber. For example, Salalha et al. encapsulated two bacteria, *Escherichia coli*, *Staphylococcus albus* and bacterial viruses (T7, T4, λ) in poly(vinyl alcohol) (PVA). In this work, electrostatic field used as 1.1 kV cm^{-1}. The diameter of the produced nanofibers ranging between 250 and 400 nm with bead formation due to embed microorganism. After electrospinning process, when compare the viability, *Staphylococcus albus* maintains viability completely while *Escherichia coli* show a reduction in colony forming ability to 19%. However, when bacteria suspended in 5% glycerol before electrospinning, viability of bacteria increased about 50%. In addition, nanofibers contain microorganisms stored at different temperatures like −55, −20, 4 °C, room temperature and periodically examined. Results showed that at the room temperature and 4 °C, microorganism in nanofiber lost their viability completely but two lower temperatures were suitable for the storage for several weeks. As a results microorganism endured the electrospinning process while maintaining their viability at fairly high levels [78]. In few studies, scientist encapsulated probiotics which are live microorganisms that are intended to have health benefits when consumed or applied to the body in nanofiber. Heunis et al. used polyethylene oxide (PEO) nanofiber to encapsulation of probiotics: *Lactobacillus plantarum* 423 and its bacteriocin. In the electrospinning process 10 kV constant electric field used. After encapsulation process, only 2% of the bacteria cells remained viable. The damage is from high voltage and whipping or shearing action of electrospinning [44]. Another application of PVA for encapsulation of probiotics by Lopez-Rubio et al. For encapsulating material, poly(vinyl alcohol) (PVOH) was used. *Bifidobacterium animalis* Bb12 encapsulated within the electrospun PVOH fibers and stored for 40 days at room temperature and for 130 days at refrigeration temperature. However, viability of control bacterium that was not encapsulated decreased [57]. In another study, San et al. encapsulated bacteria into nanofiber from cyclodextrin (CD) which is a nontoxic cyclic oligosaccharide produced from the enzymatic degradation of starch for wastewater treatment application. The encapsulated bacteria successfully stand viable at least 7 days at 4 °C temperature condition. Besides, encapsulated bacteria nanofiber still had a the bioremediation capability for the treatment of heavy metals like Nickel(II) and Chromium(VI) and dye like Reactive Black 5. As a result, this research shows that CD nanofibers suitable platforms for the encapsulation of living bacterial cells and the novel biocomposites can used for bioremediation [80]. In another study, polyvinyl alcohol (PVA) and polyethylene oxide (PEO) which are two water based and biocompatible polymers used for the production of nanofiber to encapsulate bacteria cells [81]. Researchers used these materials for biodegradation of wastewater contains reactive dye. They found that encapsulated bacterial cells can survive by being stored at 4 °C for three months and so this method has potency for storage.

4.3 The Encapsulation of Enzyme in Nanofibers

Enzymes are proteins that catalyzes extensive stereo-, chemo- and regiospesific chemical reactions and accelerates the reaction under mild conditions [12, 95]. These specificity properties of enzymes with a supernal impact for applications in high quality of chemical and biological production, food processing, proteomic analyses about proteins, biosensors fabrication and bioremediation [97].

The biggest reasons that restricts using enzymes in many industrial applications are their sensitivity to external factors, their instability, costly production and reusability properties. Due to these limitations, enzymes need to be preserved from damaging circumstances until time of use and they should be recycled for subsequent usage in order to bring down expenses [12, 97].

Researches have shown that the mesoporous silica material limits the enzyme–substrate binding on the surfaces of the enzymes and this reduces the enzyme activity. Nanoparticles and nanotubes, on the other hand, have impressive properties such as reducing the limitation of mass transfer, but they have the disadvantage that diffusion and recycling are difficult. Except for these materials, nanofibers obtained by electrospinning method offer a solution to these limitations and show promise in encapsulating enzymes [97].

The broad surface area of nanofibers enables them to be used for the protection of enzymes and their rapid release when it's necessary. It also has the reusability feature in which the enzymes are integrated with the nanofiber. With unexampled characteristics of nanofibers like high surface region, adjustable surface morphology, ability of pursue electrostatic charges, permeability, nanofibers have the criteria sought in the encapsulation and immobilization of enzymes [12]. Properties such as the variety of polymers that can be used in nanofiber production, adjustable surface as desired and ability to prevent mass transfer at a low rate make the electrospinning technique stand out in encapsulation processes [97].

It has been confirmed by numerous studies that electrospinning technique is an effective and sought-after method in encapsulating enzymes [95].

The first study about enzyme encapsulation with nanofibers was done by Jia et al. in 2002. In this research, bioactive polystyrene (PS) nanofibers were produced by electrospinning method after a chemical attachment of α-chymotrypsin which is a model enzyme. They achieved to cover 27.4% surface of nanofiber with enzyme. They stated that the encapsulated enzyme showed 65% more specific activity than the enzyme found in its natural state. As a result, nanofibrous enzymes showed high catalytic efficiency, improved stability and prolonged half-life by comparison with native enzyme [47]. In another study about enzyme encapsulation done by Sakai et al., lipase was encapsulated in electrospun polystyrene fibers and fabricated polystyrene non-women fabric membrane encapsulating lipase. These PS/lipase fibers showed excellent recyclability and faster initial reaction [77]. More recently, in the study carried out by Kamaci et al., phytase enzyme obtained from cowpea seeds by various processes was encapsulated in polyvinyl (PVA)—sodium alginate (SA) nanofibers. As a result of study, it has been shown that the catalytic activity of the immobilized phytase enzyme is higher in a wider range of pH and temperature in comparison with free phytase enzyme [27].

5 Conclusions

For improvement of health and environment, nanotechnology has gained major potential in improving the effectiveness of bioactive compounds delivery. When considering encapsulation systems, electrospun polymer/non polymer nanofibers offer promising platform by providing advantages. While there are different methods such as phase separation, template synthesis, drawing and self-assembly methods to produce nanofibers one unique method, electrospinning, has received much attention lately. Because of the cost effective, easy and repeatable properties, the electrospinning process remains attractive. Owing to unequalled characteristics like high surface area and engineered different forms, electrospun nanofiber applications become wider.

The bioactive components like live cells, proteins, nucleic acids, enzymes, essential oils and flavors have low stability and viability properties. For this reason, encapsulation process is needed. In this regard, electrospinning is an easy and multiple encapsulation technique convenient for the manufacturing of nano-fibers including bioactive compounds. This process presents advantages such as the electrospinning process doesn't require heat which is suitable for the bioactive compounds, size of the encapsulates decreased that allows successful incorporation. Furthermore, nanofiber provides protection in the extreme environments.

Future studies should enable the use of bioactive compounds encapsulated nanofibers as products in industrial sectors.

References

1. . Colfen (2003) Precipitation of carbonates. Curr Opin Colloid Interface Sci 8:145–155. https://doi.org/10.1016/S1359-0294
2. Ahmad MU (2006) Nanotechnology: emerging interest, opportunities, and challenges, lipids in nanotechnology. AOCS Press. https://doi.org/10.1016/B978-0-9818936-7-9.50004-4
3. Almetwally AA et al (2017) Technology of nano-fibers: production techniques and properties—critical review. J Text Assoc 78(1):5–14

4. Amiri N et al (2018) Optimization of Chitosan-Gelatin nanofibers production: investigating the effect of solution properties and working parameters on fibers diameter. Bionanoscience 8(3):778–789. https://doi.org/10.1007/s12668-018-0540-5
5. Asbahani A El et al (2015) Essential oils: from extraction to encapsulation. Int J Pharm. Elsevier B.V. 483(1–2):220–243. https://doi.org/10.1016/j.ijpharm.2014.12.069
6. Ataei S et al (2020) Essential oils-loaded electrospun biopolymers: a future perspective for active food packaging. Adv Polym Technol 2020:1–21. https://doi.org/10.1155/2020/9040535
7. Aytac Z et al (2017) 'Antibacterial electrospun zein nanofibrous web encapsulating thymol/cyclodextrin-inclusion complex for food packaging. Food Chem Elsevier Ltd. 233:117–124. https://doi.org/10.1016/j.foodchem.2017.04.095
8. Aytac Z et al (2016) Encapsulation of gallic acid/cyclodextrin inclusion complex in electrospun polylactic acid nanofibers: release behavior and antioxidant activity of gallic acid. Mater Sci Eng C. Elsevier B.V. 63:231–239. https://doi.org/10.1016/j.msec.2016.02.063
9. Azaddin AF et al (2020) Entrapment of Volvariella volvacea spores in electrospun nanofibers. 1:15–20
10. Bakkali F et al (2008) Biological effects of essential oils—a review. Food Chem Toxicol 46(2):446–475. https://doi.org/10.1016/j.fct.2007.09.106
11. Balamurugan R, Sundarrajan S, Ramakrishna S (2011) Recent trends in nanofibrous membranes and their suitability for air and water filtrations. Membranes 1(3):232–248. https://doi.org/10.3390/membranes1030232
12. Balogh-Weiser D et al (2018) Electrospun nanofibers for entrapment of biomolecules. IntechOpen, pp 135–147. https://doi.org/10.1016/j.colsurfa.2011.12.014
13. Bhardwaj N, Kundu SC (2010) Electrospinning: a fascinating fiber fabrication technique. Biotechnol Adv 28(3):325–347. https://doi.org/10.1016/j.biotechadv.2010.01.004
14. De Billerbeck VG (2007) Huiles essentielles et bactéries résistantes aux antibiotiques. Phytotherapie 5(5):249–253. https://doi.org/10.1007/s10298-007-0265-z
15. Bui X-T et al (2019) In: Agarwal AK, Pandey A (eds) Water and wastewater treatment technologies energy, environment, and sustainability. http://www.springer.com/series/15901
16. Can Başer KH, Buchbauer G (2015) Handbook of essential oils: science, technology, and applications, 2nd edn. https://doi.org/10.1201/b19393
17. Casper CL et al (2004) Controlling surface morphology of electrospun polystyrene fibers: effect of humidity and molecular weight in the electrospinning process. Macromolecules 37(2):573–578. https://doi.org/10.1021/ma0351975
18. Celebioglu A et al (2018) Encapsulation of camphor in cyclodextrin inclusion complex nanofibers via polymer-free electrospinning: enhanced water solubility, high temperature stability, and slow release of camphor. J Mater Sci Springer US 53(7):5436–5449. https://doi.org/10.1007/s10853-017-1918-4
19. Celebioglu A, Uyar T (2011) Electrospinning of polymer-free nanofibers from cyclodextrin inclusion complexes. Langmuir 27(10):6218–6226. https://doi.org/10.1021/la1050223
20. Celebioglu A, Uyar T (2011b) Electrospun porous cellulose acetate fibers from volatile solvent mixture. Mater Lett Elsevier B.V. 65(14):2291–2294. https://doi.org/10.1016/j.matlet.2011.04.039
21. Cho DK et al (2009) A UDP-based protocol for mobile robot control over wireless internet. In: 2009 2nd international conference on robot communication and coordination, RoboComm 2009. https://doi.org/10.4108/icst.robocomm2009.6076
22. Chomachayi MD, Solouk A, Mirzadeh H (2016) Electrospun silk-based nanofibrous scaffolds: fiber diameter and oxygen transfer. Prog Biomater Springer Berlin Heidelberg 5(1):71–80. https://doi.org/10.1007/s40204-016-0046-6
23. Colín-Orozco J et al (2015) Properties of poly (ethylene oxide)/whey protein isolate nanofibers prepared by electrospinning. Food Biophys 10(2):134–144. https://doi.org/10.1007/s11483-014-9372-1
24. Deitzel JM et al (2001) The effect of processing variables on the morphology of electrospun. Polymer 42:261–272

25. Dias MI, Ferreira ICFR, Barreiro MF (2015) Microencapsulation of bioactives for food applications. Food Funct R Soc Chem 6(4):1035–1052. https://doi.org/10.1039/c4fo01175a
26. Donsì F et al (2011) Encapsulation of bioactive compounds in nanoemulsion—based delivery systems. Procedia Food Sci Elsevier Srl 1:1666–1671. https://doi.org/10.1016/j.profoo.2011.09.246
27. Duru Kamaci U, Peksel A (2020) Enhanced catalytic activity of immobilized phytase into polyvinyl alcohol-sodium alginate based electrospun nanofibers. Catal Lett Springer US (0123456789). https://doi.org/10.1007/s10562-020-03339-0
28. Esfahani MZ et al (2011) Polypropylene nanofibre. Nanofibers and nanotechnology research advances
29. Espín JC, García-Conesa MT, Tomás-Barberán FA (2007) Nutraceuticals: facts and fiction. Phytochemistry 68(22–24):2986–3008. https://doi.org/10.1016/j.phytochem.2007.09.014
30. Ezhilarasi PN et al (2013) Nanoencapsulation techniques for food bioactive components: a review. Food Bioprocess Technol 6(3):628–647. https://doi.org/10.1007/s11947-012-0944-0
31. El Fawal G (2019) Polymer nanofibers electrospinning: a review. Egypt J Chem. https://doi.org/10.21608/ejchem.2019.14837.1898
32. Feng X et al (2019) Electrospun polymer micro/nanofibers as pharmaceutical repositories for healthcare. J Control Release Elsevier 302(January):19–41. https://doi.org/10.1016/j.jconrel.2019.03.020
33. Fennessey SF, Farris RJ (2004) Fabrication of aligned and molecularly oriented electrospun polyacrylonitrile nanofibers and the mechanical behavior of their twisted yarns. Polymer 45(12):4217–4225. https://doi.org/10.1016/j.polymer.2004.04.001
34. Fonseca LM et al (2020) Electrospun potato starch nanofibers for thyme essential oil encapsulation: antioxidant activity and thermal resistance. J Sci Food Agric 100(11):4263–4271. https://doi.org/10.1002/jsfa.10468
35. Gao S et al (2020) Encapsulation of thiabendazole in hydroxypropyl-β-cyclodextrin nanofibers via polymer-free electrospinning and its characterization. Pest Manag Sci (February). https://doi.org/10.1002/ps.5885
36. García-Moreno PJ et al (2016) 'Encapsulation of fish oil in nanofibers by emulsion electrospinning: physical characterization and oxidative stability. J Food Eng Elsevier Ltd 183:39–49. https://doi.org/10.1016/j.jfoodeng.2016.03.015
37. Garg K, Bowlin GL (2011) Electrospinning jets and nanofibrous structures. Biomicrofluidics 5(1):1–19. https://doi.org/10.1063/1.3567097
38. Ghorani B, Tucker N (2015) Fundamentals of electrospinning as a novel delivery vehicle for bioactive compounds in food nanotechnology. Food Hydrocoll Elsevier Ltd 51:227–240. https://doi.org/10.1016/j.foodhyd.2015.05.024
39. Greiner A, Wendorff JH (2007) Electrospinning: a fascinating method for the preparation of ultrathin fibers. Angew Chem Int Ed 46(30):5670–5703. https://doi.org/10.1002/anie.200604646
40. Guillard V et al (2009) Food preservative content reduction by controlling sorbic acid release from a superficial coating. Innov Food Sci Emerg Technol Elsevier Ltd 10(1):108–115. https://doi.org/10.1016/j.ifset.2008.07.001
41. Gómez-Mascaraque LG et al (2017) Impact of microencapsulation within electrosprayed proteins on the formulation of green tea extract-enriched biscuits. LWT Food Sci Technol 81:77–86. https://doi.org/10.1016/j.lwt.2017.03.041
42. Göksen G et al (2020) Phytochemical-loaded electrospun nanofibers as novel active edible films: characterization and antibacterial efficiency in cheese slices. Food Control 112: 107133. https://doi.org/10.1016/j.foodcont.2020.107133
43. Hajhashemi V, Ghannadi A, Sharif B (2003) Anti-inflammatory and analgesic properties of the leaf extracts and essential oil of Lavandula angustifolia Mill. J Ethnopharmacol 89(1):67–71. https://doi.org/10.1016/S0378-8741(03)00234-4
44. Heunis TDJ, Botes M, Dicks LMT (2010) Encapsulation of Lactobacillus plantarum 423 and its bacteriocin in nanofibers. Probiotics Antimicrob Proteins 2(1):46–51. https://doi.org/10.1007/s12602-009-9024-9

45. Hu Q, Wu C, Zhang H (2020) Preparation and optimization of a biomimetic triple-layered vascular scaffold based on coaxial electrospinning. Appl Biochem Biotechnol 190(3):1106–1123. https://doi.org/10.1007/s12010-019-03147-2
46. Jafari SM (ed) (2017) Nanoencapsulation technologies for the food and nutraceutical industries. Academic Press
47. Jia H et al (2002) Enzyme-carrying polymeric nanofibers prepared via electrospinning for use as unique biocatalysts. Biotechnol Prog 18(5):1027–1032. https://doi.org/10.1021/bp020042m
48. Joye IJ, Davidov-Pardo G, McClements DJ (2014) Nanotechnology for increased micronutrient bioavailability. Trends Food Sci Technol Elsevier Ltd 40(2):168–182. https://doi.org/10.1016/j.tifs.2014.08.006
49. Kamel BM et al (2016) Rheology and thermal conductivity of calcium grease containing multi-walled carbon nanotube. Fuller Nanotub Carbon Nanostructures Taylor & Francis 24(4):260–265. https://doi.org/10.1080/1536383X.2016.1143462
50. Karim SA et al (2018) Mechanical properties and the characterization of polyacrylonitrile/carbon nanotube composite nanofiber. Arab J Sci Eng Springer, Berlin Heidelberg 43(9):4697–4702. https://doi.org/10.1007/s13369-018-3065-x
51. Kim JS, Reneker DH (1999) Polybenzimidazole nanofiber produced by electrospinning. Polym Eng Sci 39(5):849–854. https://doi.org/10.1002/pen.11473
52. Kour H et al (2015) Nanotechnology –new lifeline for food industry. Crit Rev Food Sci Nutr 8398:00–00. https://doi.org/10.1080/10408398.2013.802662
53. Kriegel C, Kit KM, McClements DJ, Weiss J (2009) Influence of surfactant type and concentration on electrospinning of chitosan–poly(ethylene oxide) blend nanofibers. Food Biophys 4(3):213–228. https://doi.org/10.1007/s11483-009-9119-6
54. Lassalle V, Ferreira ML (2007) PLA nano- and microparticles for drug delivery: an overview of the methods of preparation. Macromol Biosci 7(6):767–783. https://doi.org/10.1002/mabi.200700022
55. Law JX et al (2017) Electrospun collagen nanofibers and their applications in skin tissue engineering. Tissue Eng Regen Med Korean Tissue Engineering and Regenerative Medicine Society 14(6):699–718. https://doi.org/10.1007/s13770-017-0075-9
56. Li D, Xia Y (2004) Direct fabrication of composite and ceramic hollow nanofibers by electrospinning. Nano Lett 4(5):933–938. https://doi.org/10.1021/nl049590f
57. López-Rubio A et al (2009) Encapsulation of living bifidobacteria in ultrathin PVOH electrospun fibers. Biomacromol 10(10):2823–2829. https://doi.org/10.1021/bm900660b
58. Matsumoto H, Tanioka A, Materials P (2021) Encyclopedia of polymeric nanomaterials, pp 9–12. https://doi.org/10.1007/978-3-642-36199-9
59. Megelski S et al (2002) Micro- and nanostructured surface morphology on electrospun polymer fibers. Macromolecules 35(22):8456–8466. https://doi.org/10.1021/ma020444a
60. Miletić A et al (2019) Encapsulation of fatty oils into electrospun nanofibers for cosmetic products with antioxidant activity. Appl Sci (Switzerland) 9(15). https://doi.org/10.3390/app9152955
61. Mirjalili M, Zohoori S (2016) Review for application of electrospinning and electrospun nanofibers technology in textile industry. J Nanostructure Chem Springer, Berlin Heidelberg 6(3):207–213. https://doi.org/10.1007/s40097-016-0189-y
62. Mirmohammad Sadeghi SA et al (2020) Single nozzle electrospinning of encapsulated epoxy and mercaptan in PAN for self-healing application. Polymer Elsevier Ltd 186(November 2019):122007. https://doi.org/10.1016/j.polymer.2019.122007
63. Mo X, Chen Z, Weber HJ (2007) Electrospun nanofibers of collagen-chitosan and P(LLA-CL) for tissue engineering. Front Mater Sci Chin 1(1):20–23. https://doi.org/10.1007/s11706-007-0004-2
64. Mojaveri SJ, Hosseini SF, Gharsallaoui A (2020) Viability improvement of Bifidobacterium animalis Bb12 by encapsulation in chitosan/poly(vinyl alcohol) hybrid electrospun fiber mats. Carbohydr Polym Elsevier 241(February):116278. https://doi.org/10.1016/j.carbpol.2020.116278

65. Murthy KNC et al (2018) Nanoencapsulation : an advanced nanotechnological nanoencapsulation : an advanced nanotechnological approach to enhance the biological efficacy of curcumin. (October). https://doi.org/10.1021/bk-2018-1286.ch021
66. Nemati S et al (2019) Current progress in application of polymeric nanofibers to tissue engineering. Nano Converg Springer Singapore 6(1). https://doi.org/10.1186/s40580-019-0209-y
67. Neo YP, Ray S, Perera CO (2018) Nanostructures for food applications, role of materials science in food bioengineering. Elsevier Inc. https://doi.org/10.1016/B978-0-12-811448-3/00004-8
68. Paredes AJ et al (2016) Nanoencapsulation in the food industry: manufacture, applications and characterization. 1(1):56–79
69. Perry NSL et al (2003) Salvia for dementia therapy: review of pharmacological activity and pilot tolerability clinical trial. Pharmacol Biochem Behav 75(3):651–659. https://doi.org/10.1016/S0091-3057(03)00108-4
70. Pourhojat F et al (2017) Evaluation of poly ε-caprolactone electrospun nanofibers loaded with Hypericum perforatum extract as a wound dressing. Res Chem Intermed 297–320. https://doi.org/10.1007/s11164-016-2623-7
71. Pérez-Masiá R, Lagaron JM, Lopez-Rubio A (2015) Morphology and stability of edible lycopene-containing micro- and nanocapsules produced through electrospraying and spray drying. Food Bioprocess Technol 8(2):459–470. https://doi.org/10.1007/s11947-014-1422-7
72. Ramakrishna S et al (2006) Electrospun nanofibers: solving global issues. Mater Today Elsevier Ltd 9(3):40–50. https://doi.org/10.1016/S1369-7021(06)71389-X
73. Katrina A, Rieger Jessica D, Schiffman (2014) Electrospinning an essential oil: cinnamaldehyde enhances the antimicrobial efficacy of chitosan/poly(ethylene oxide) nanofibers. Carbohydr Polym 113:561–568. https://doi.org/10.1016/j.carbpol.2014.06.075
74. Roslan NSA et al (2018) Nylon electrospun nanofibre water filtration media for wastewater treatment. Mater Res Express 5(10). https://doi.org/10.1088/2053-1591/aada94
75. Rostamabadi H et al (2020) Electrospinning approach for nanoencapsulation of bioactive compounds; recent advances and innovations. Trends Food Sci Technol Elsevier 100(March):190–209. https://doi.org/10.1016/j.tifs.2020.04.012
76. Safdari F et al (2017) Enhanced properties of poly(ethylene oxide)/cellulose nanofiber biocomposites. Cellulose Springer, Netherlands 24(2):755–767. https://doi.org/10.1007/s10570-016-1137-1
77. Sakai S et al (2010) Enhanced catalytic activity of lipase in situ encapsulated in electrospun polystyrene fibers by subsequent water supply. Catal Commun Elsevier B.V. 11(6):576–580. https://doi.org/10.1016/j.catcom.2009.12.023
78. Salalha W et al (2006) Encapsulation of bacteria and viruses in electrospun nanofibres. Nanotechnology 17(18):4675–4681. https://doi.org/10.1088/0957-4484/17/18/025
79. San NO et al (2014) Reusable bacteria immobilized electrospun nanofibrous webs for decolorization of methylene blue dye in wastewater treatment. RSC Adv 4(61):32249–32255. https://doi.org/10.1039/c4ra04250f
80. San Keskin NO et al (2018) Encapsulation of living bacteria in electrospun cyclodextrin ultrathin fibers for bioremediation of heavy metals and reactive dye from wastewater. Colloids Surf B Biointerfaces Elsevier B.V. 161:169–176. https://doi.org/10.1016/j.colsurfb.2017.10.047
81. Sarioglu OF et al (2017) Bacteria encapsulated electrospun nanofibrous webs for remediation of methylene blue dye in water. Colloids Surf B Biointerfaces Elsevier B.V. 152:245–251. https://doi.org/10.1016/j.colsurfb.2017.01.034
82. Schiffman JD, Schauer CL (2007) One-step electrospinning of cross-linked Chitosan fibers. Biomacromol 8(9):2665–2667. https://doi.org/10.1021/bm7006983
83. Shankar A, Seyam AFM, Hudson SM (2013) Electrospinning of soy protein fibers and their compatibility with synthetic polymers. J Text Appar Technol Manag 8(1)
84. Shishir MRI et al (2018) 'Advances in micro and nano-encapsulation of bioactive compounds using biopolymer and lipid-based transporters. Trends Food Sci Technol Elsevier Ltd 78:34–60. https://doi.org/10.1016/j.tifs.2018.05.018

85. Silva J et al (2003) Analgesic and anti-inflammatory effects of essential oils of Eucalyptus. J Ethnopharmacol 89(2–3):277–283. https://doi.org/10.1016/j.jep.2003.09.007
86. Siqueira NM et al (2015) Poly (lactic acid)/chitosan fiber mats: Investigation of effects of the support on lipase immobilization. Int J Biol Macromol. Elsevier B.V. 72:998–1004. https://doi.org/10.1016/j.ijbiomac.2014.08.048
87. Solaberrieta I et al (2020) Encapsulation of bioactive compounds from aloe vera agrowastes in electrospun poly (ethylene oxide) nanofibers. Polymers 12(6). https://doi.org/10.3390/polym12061323
88. Song J et al (2012) Enhanced catalytic activity of lipase encapsulated in PCL nanofibers. Langmuir 28(14):6157–6162. https://doi.org/10.1021/la300469s
89. Sullivan ST et al (2014) Electrospinning and heat treatment of whey protein nanofibers. Food Hydrocoll Elsevier Ltd 35:36–50. https://doi.org/10.1016/j.foodhyd.2013.07.023
90. Surendhiran D et al (2020) Fabrication of high stability active nanofibers encapsulated with pomegranate peel extract using chitosan/PEO for meat preservation. Food Packag Shelf Life Elsevier 23(March 2019):100439. https://doi.org/10.1016/j.fpsl.2019.100439
91. Tang Y et al (2019) Electrospun gelatin nanofibers encapsulated with peppermint and chamomile essential oils as potential edible packaging. J Agric Food Chem 67(8):2227–2234. https://doi.org/10.1021/acs.jafc.8b06226
92. Tavassoli-Kafrani E, Goli SAH, Fathi M (2018) Encapsulation of orange essential oil using cross-linked electrospun gelatin nanofibers. Food Bioprocess Technol 11(2):427–434. https://doi.org/10.1007/s11947-017-2026-9
93. Teo WE, Ramakrishna S (2006) A review on electrospinning design and nanofibre assemblies. Nanotechnology 17(14). https://doi.org/10.1088/0957-4484/17/14/R01
94. Torres-Giner S et al (2010) Stabilization of a nutraceutical omega-3 fatty acid by encapsulation in ultrathin electrosprayed zein prolamine. J Food Sci 75(6). https://doi.org/10.1111/j.1750-3841.2010.01678.x
95. Tran DN, Balkus KJ (2012) Enzyme immobilization via electrospinning. Top Catal 55(16–18):1057–1069. https://doi.org/10.1007/s11244-012-9901-4
96. Vafania B, Fathi M, Soleimanian-Zad S (2019) 'Nanoencapsulation of thyme essential oil in chitosan-gelatin nanofibers by nozzle-less electrospinning and their application to reduce nitrite in sausages. Food Bioprod Process Inst Chem Eng 116:240–248. https://doi.org/10.1016/j.fbp.2019.06.001
97. Wang ZG et al (2009) Enzyme immobilization on electrospun polymer nanofibers: an overview. J Mol Catal B Enzym 56(4):189–195. https://doi.org/10.1016/j.molcatb.2008.05.005
98. Warnke PH et al (2009) The battle against multi-resistant strains: renaissance of antimicrobial essential oils as a promising force to fight hospital-acquired infections. J Cranio-Maxillofac Surg European Association for Cranio-Maxillofacial Surgery 37(7):392–397. https://doi.org/10.1016/j.jcms.2009.03.017
99. Wen P et al (2016) Fabrication of electrospun polylactic acid nanofilm incorporating cinnamon essential oil/β-cyclodextrin inclusion complex for antimicrobial packaging. Food Chem Elsevier Ltd 196:996–1004. https://doi.org/10.1016/j.foodchem.2015.10.043
100. Wen P et al (2017) Electrospinning: a novel nano-encapsulation approach for bioactive compounds. Trends Food Sci Technol Elsevier 70(May):56–68. https://doi.org/10.1016/j.tifs.2017.10.009
101. Wills KM et al (2016) 2016 reciprocal meat conference—muscle and lipid biology and biochemistry meat and muscle biology ™ effects of pomegranate rind extract on ground beef color, p 2016
102. Wongsasulak S, Pathumban S, Yoovidhya T (2014) Effect of entrapped α-tocopherol on mucoadhesivity and evaluation of the release, degradation, and swelling characteristics of zein-chitosan composite electrospun fibers. J Food Eng Elsevier Ltd 120(1):110–117. https://doi.org/10.1016/j.jfoodeng.2013.07.028
103. Yang Y et al (2006) Experimental investigation of the governing parameters in the electrospinning of polyethylene oxide solution. IEEE Trans Dielectr Electr Insul 13(3):580–584. https://doi.org/10.1109/TDEI.2006.1657971

104. Yang Z et al (2010) Crystallization behavior of poly(ε-caprolactone)/layered double hydroxide nanocomposites. J Appl Polym Sci 116(5):2658–2667. https://doi.org/10.1002/app
105. Yang Z et al (2019) Morphological, Mechanical and thermal properties of poly(lactic acid) (PLA)/cellulose nanofibrils (CNF) composites nanofiber for tissue engineering. J Wuhan Univ Technol Mater Sci Ed 34(1):207–215. https://doi.org/10.1007/s11595-019-2037-7
106. Yasakci V et al (2018) Hyaluronic acid-modified [19F]FDG-conjugated magnetite nanoparticles: in vitro bioaffinities and HPLC analyses in organs. J Radioanal Nucl Chem Springer International Publishing 318(3):1973–1989. https://doi.org/10.1007/s10967-018-6282-6
107. Yilmaz MT et al (2020) An alternative way to encapsulate probiotics within electrospun alginate nanofibers as monitored under simulated gastrointestinal conditions and in kefir. Carbohydrate Polymers Elsevier 244(February):116447. https://doi.org/10.1016/j.carbpol.2020.116447
108. YukiKogyo Co.Ltd, D. integrity issue (2019) No TitleEΛENH. Aγαη 8(5):55
109. Zahmatkeshan M et al (2019) Polymer-based nanofibers: preparation, fabrication, and applications. Handbook of nanofibers. https://doi.org/10.1007/978-3-319-53655-2_29
110. Zussman E (2011) Encapsulation of cells within electrospun fibers. Polym Adv Technol 22(3):366–371. https://doi.org/10.1002/pat.1812

Fibers as Containers for Encapsulation

Subrata Mondal

1 Introduction

Encapsulation of functional substances is an area of massive interest in both industry and academia. Functional molecules like drugs, micro-nutrients are susceptible to degradation or denaturation, therefore, design of effective encapsulation, protection, targeting and substantial release for these functional molecules are crucial [1]. Encapsulation process consists of the entrapment of functional substances by a protective material [2]. The aims of encapsulation is to provide a protective coating on active substances by thin wall of encapsulating material in order to protect the active substance from aggressive environment and to control the rate of release of active substances [3]. Depending on types of core functional materials, various types of synthetic organic or biopolymers can be used as encapsulating materials [2]. Micro and nano encapsulations are widely used for the protection and controlled release of various drugs, micro nutrients or other functional substances [4–8]. Due to its interesting properties like high aspect ratio and specific surface area etc. fibers can be used as container for the encapsulation of various useful compounds [9–15].

Encapsulated fibers can be manufactured mainly by melt extrusion and electrospinning techniques in order to encapsulate wide verities of functional substances [1, 16, 17]. Extrusion technique is a simple method to manufacture micro fibers for the container of encapsulation. On the other hand electrospinning is a convenient and versatile technique to prepare nanofibers. However, difficulties may arise during the encapsulation of functional compounds by using conventional electrospinning method. Modified electrospinning viz. suspension, emulsion and coaxial electrospinning method can be used for the encapsulation of useful compounds

S. Mondal (✉)
Department of Mechanical Engineering, National Institute of Technical Teachers' Training and Research (NITTTR) Kolkata, FC Block, Sector III, Salt Lake City, Kolkata 700106, West Bengal, India

J. Parameswaranpillai et al. (eds.), *Micro- and Nano-containers for Smart Applications*, Composites Science and Technology,
https://doi.org/10.1007/978-981-16-8146-2_3

into the fiber matrix. Suspension and emulsion electrospinning methods produced micro/nanosphere embedded encapsulated fibers, while, coaxial electrospinning method produced core–shell structure consist of functional molecules core which is being encapsulated by a shell material. The shell material and the core liquids are spun together to prepare core–shell nanofiber structure. Properties of fibers are influenced by various factors such as miscibility of pair solution, viscosity ratio, conductivity ratio, flow-rate ratio, interfacial tension, applied electric filed, distance between nozzle and target plate etc. [18].

This chapter presents on fibers as a container for the encapsulation of functional substances. Various types of materials used for the encapsulation are included. Several techniques for manufacturing of encapsulated fibers such as melt extrusion, coaxial spinning, and electrospinning are reviewed. Major applications of encapsulated fibers in medicine, food industry, textiles and some of the other key areas are covered. Finally, the chapter has been concluded with few challenges of fibers as container for encapsulation.

2 Materials Used for Encapsulations

The capsule consists of core and shell structure. Depending on size, capsules can be termed as microcapsule and nanocapsule. Microcapsules has size range from 2 to 2000 μm approximately [19]. Whereas, nanocapsule has size up to 100 nm [20, 21]. In a core and shell structure of micro/nanocapsule has two primary components viz. protective i.e. coating material and functional i.e. core material. Selection of coating (wall) material and thickness of coating determined the physical and chemical properties of micro/nano capsules. Wall material should be able to form nonreactive chemically compatible cohesive membrane on active core substance, and provide desired properties such as strength, flexibility, impermeability etc. Various types of polymers such as hydrophilic and hydrophobic or combination of hydrophilic and hydrophobic polymers are used as coating materials of encapsulation. Core material is a functional material on which shell material has been applied to serve the specific purposes. Core material in core/sheath encapsulated structure may exist in the form of solid, droplets of dispersion, and dispersion [19, 22–25]. Apart from the polymers, others materials such as carbon nanotube, boron nitride tube, silica, graphene nest etc. are used as encapsulating materials [26–30]. Due to the toxicity, drugs can be encapsulated and deliver to the targeted cells to enhance the efficiency and reduce the toxicity to normal cells [26, 31]. Ghahremani et al. reported encapsulation of anticancer drug in a boron nitride tube, as the boron nitride is non-toxic and biocompatible. Boron nitride tube can be used as vehicle for fine drug delivery system [32]. Dolinina et al. presents encapsulation of protein drug with improved pharmacological and consumer properties by using silica as container. Protein encapsulation in silica can be carried out by one-step and multi-step methods. Encapsulation of bovine serum albumin (BSA) in one step method and synthesis of silica containers for the multi steps encapsulation of BSA can be performed by sol–gel emulsion techniques [27].

3 Techniques for Manufacturing of Encapsulated Fibers

Encapsulation is a group of techniques to manufacture small particles which is termed as capsules of core and shell structure. Core substance from the encapsulated structure can be released at specific rate under certain surrounding condition. Further, functional substance can also be encapsulated by embedded technique. Melt extrusion, emulsion and suspension electrospinning can produced encapsulated functional substance embedded structure, while, coaxial extrusion technique, e.g. coaxial extrusion electrospinning method can produce core substance encapsulated by sheath structure. Core-sheath and embedded encapsulated structures of fiber cross sections are schematically shown in Fig. 1.

3.1 Melt Extrusion

Matrix material used for the extrusion of melt spinning should have good process ability and barrier property. Some of the materials used for the melt extrusion of encapsulated fibers include carbohydrate and polysaccharide, such as starch, maltodextrin, modified starch, carbohydrate mixture, proteins such as sodium caseinate, soy and pea protein isolates etc. [33]. Poly-L-lactic acid (PLA) hollow fibers as container for liquid drug or pesticide can be manufactured by high speed melt spinning. Outer and inner diameter of hollow fibers depends on polymer throughput rate, winding speed and extrusion temperature. Outer and inner diameter of hollow fibers increases with increasing polymer throughput rate, while, decreases with

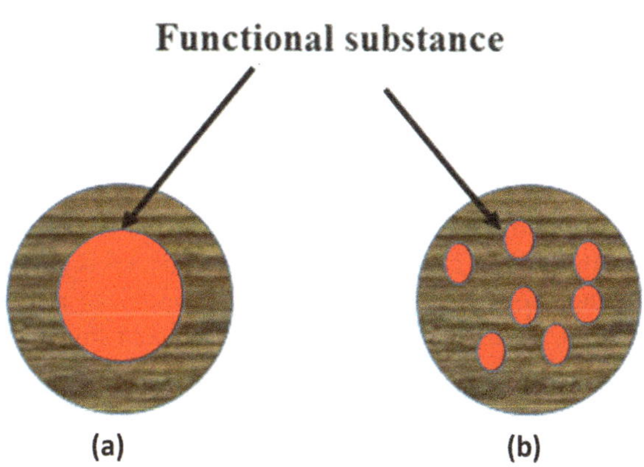

Fig. 1 Schematic for the cross section of fibers, showing: **a** core-sheath structure of fiber as container for encapsulation; **b** functional substance embedded structure of fiber as container for encapsulation

winding speed. Inner diameter of hollow fiber decreases with increasing extrusion temperature, however, extrusion temperature has no influence on outer diameter of hollow fiber [34]. Cherif et al. reported bi-component fibers prepared by melt extrusion. Physical properties of fibers can be improved by drawing of prepared fibers at different draw ratios. Tensile properties of polypropylene sheath and polypropylene and wax blend fibers has been improved with increasing draw ratio, while, elongation at break decreases [35]. Viscosity of individual components influenced the morphology of bi component fibers [36]. Prior to the melt extrusion of polymer fibers, microencapsulated functional materials e.g. micro phase change materials (MicroPCM) can be dispersed to the polymer matrix. Mineral oil can be used during mechanical mixing process of polymer granule and MicroPCM. Multicomponent yarn can be obtained by extrusion through specially designed spinneret and subsequent quenching in water bath [37].

3.2 Electrospinning

In electrospinning technique, high voltage is applied between positive and negative poles. Positive pole is being attached with needle of the syringe. Electrospun nanofibers mat are collected onto an aluminum foil which is attached to the negative pole of high voltage. Polymers are dissolved in a conductive solvent and pumped at a constant rate by using a syringe pump. Major factors which influences the properties of electrospun fibers are solvent properties, feed rate, applied voltage, distance between positive and negative pole etc. [3]. In order to encapsulate, functional substances are dispersed in the polymer solution prior to the electrospinning. Lopez-Rubio et al. reported encapsulation of bifidobacteria in poly (vinyl alcohol) fiber having diameter of 200 nm by using electrospinning technique. Microscopic images revealed homogeneous distribution of fibers and optical microscopic image using a florescence revealed uniform distribution of bacteria (Fig. 2) [38]. Conventional electrospinning is a convenient and versatile technique for the preparation of nanofibers. However, difficulties may arise in case of encapsulation of useful substances into the electrospun fiber. Modified techniques such as emulsion, suspension and coaxial electrospinning have been developed to prepare micro or nanosphere embedded or core sheath structure of encapsulated fibers, respectively [39–42].

Core shell nanofibers encapsulated with functional substance can be prepared by coaxial electrospinning technique. Co-extrusion consists of extrusion of two immiscible liquid core and sheath materials. Core and sheath materials are pumped separately through concentric feed tubes and subsequently core-sheath fibers are exit through concentric orifice of nozzle [33]. Two syringes feed, inner separated and coaxial are used for inner fluid and coaxial fluids, respectively. Similar like electrospinning technique, the Taylor core shell compound structure is formed during the stretching of conductive fluids between positive and negative poles. Finally core sheath structure of functional core material encapsulated fibers can be deposited on the collector plate. McCann et al. reported coaxial electrospinning to prepared

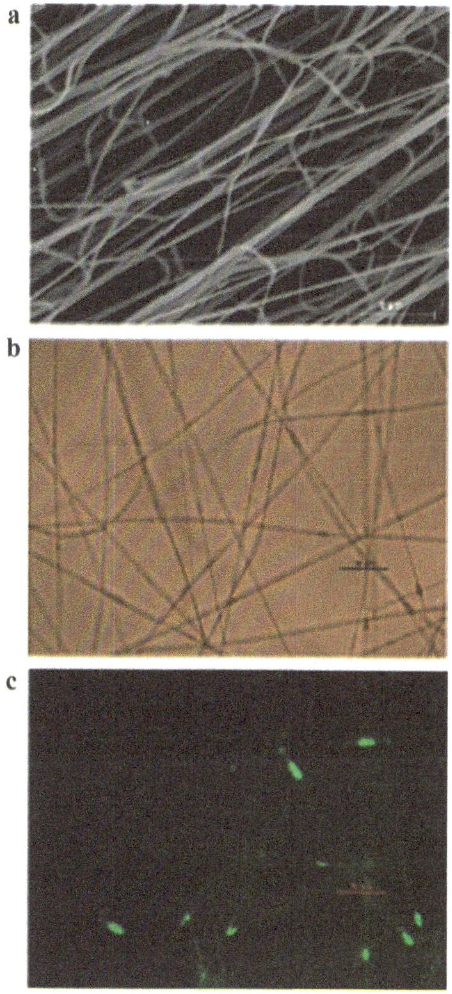

Fig. 2 *Bifidobacterium animalis* subsp. *lactis* Bb12 encapsulated hybrid poly (vinyl alcohol) fibers prepared by electrospinning technique. **a** Scanning electron micrograph; **b** optical micrograph using polarized light and **c** optical micrograph using flurocence light (Reproduced with permission from Lopez-Rubio et al. *Biomacromolecules*, 10 (10), 2823–2829, 2009 © 2009 American Chemical Society [38].)

functional substance encapsulated nanofibers. Coaxial electrospinning set up is schematically shown in Fig. 3 [43].

Lancuski et al. reported core sheath fiber manufactured by using coaxial electrospinning technique. The core structure is made of glycerol for the suspension medium of cells and this can be encapsulated by sheath of starch-formate material. Figure 4 shows the core sheath structure of electrospun fiber prepared by coaxial electrospinning. Glycerol core fluid has been successfully entrapped in starch sheath material (Fig. 4a, b). Cryogenic fractured cross section (Fig. 4c) showed fully loaded with glycerol after encapsulation. Average diameter of the fiber was 4.13 μm [1].

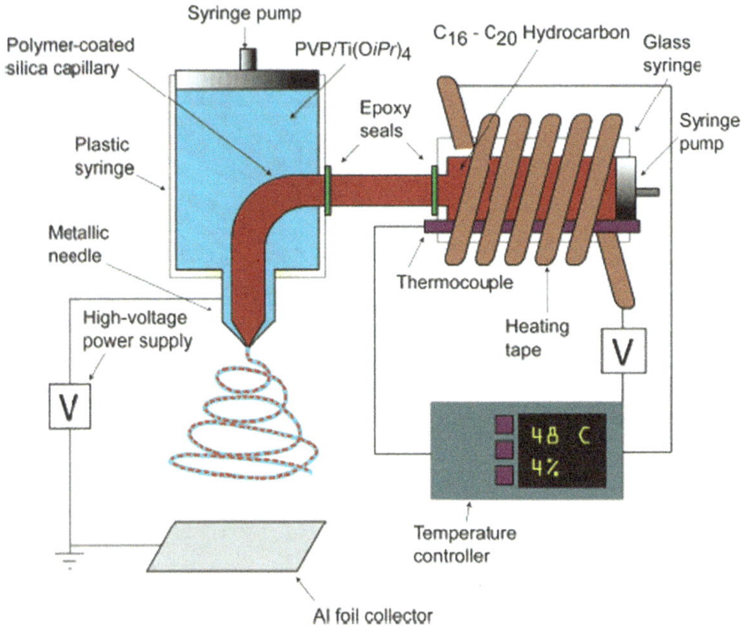

Fig. 3 Schematic showing melt coaxial electrospinning set up to prepare phase change materials encapsulated nanofibers (Reproduced with permission from McCann et al. *Nano Lett.*, 6 (12), 2868–2872, 2006 © 2006 American Chemical Society [43].)

4 Applications of Encapsulated Fibers

4.1 Application in Medicine

In medicine, micro or nano encapsulation of drug is required to protect drugs for controlled and targeted delivery to specific body sites [19, 44, 45]. Fibers can be used as container for the encapsulation of drugs and cells for biomedical applications [46–51]. Encapsulated electro spun fibers are widely used for controlled drug delivery system due to the several interesting properties of nanofibers such as higher surface area, ability to encapsulate wide range of drugs, capacity to delivery drugs at specific target site etc. [52]. Souza et al. reported nanofiber of encapsulated diclofenac sodium drug by using poly (3-hydroxybutyrate-*co*-3-hydroxyvalerate) (PHBV) prepared by electrospinning and solution blow spinning techniques. Both techniques efficiently and effectively encapsulated drugs by using PHBV polymer fibers, and the average fiber diameter was 500 nm. In vitro drug release of encapsulated fibers depends on drug concentration and temperature. Higher rate of drug release was observed with higher concentration drug encapsulated fiber at higher temperature. The encapsulated fiber has potential applications in controlled-release devices of human and veterinary medicine [52]. PLA hollow fibers can be used as container for the encapsulation of

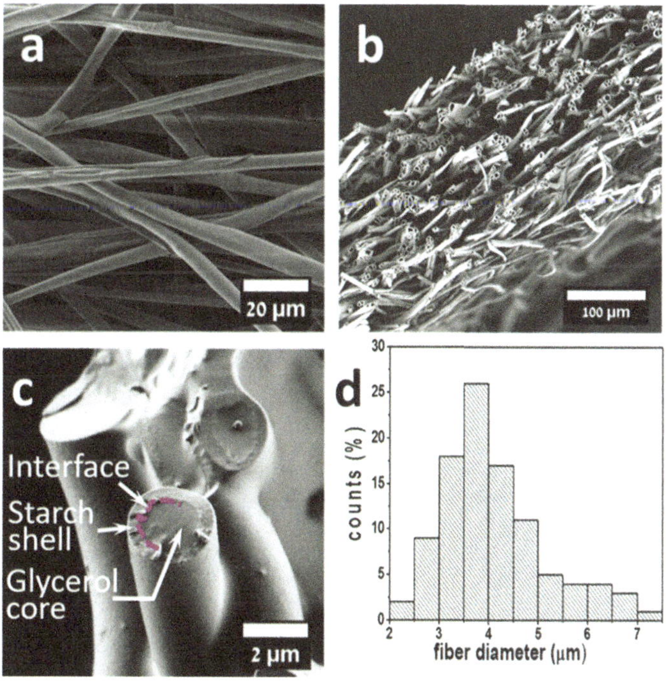

Fig. 4 **a** High resolution (HR) scanning electron microscopic (SEM) image of electro spun starch-formate/glycerol fiber; **b** cross section view image by conventional SEM; **c** cross section view captured by cryogenic HR-SEM and **d** fiber diameter distribution (Reproduced with permission from Lancuski et al. *Carbohydr. Polym.*, 158, 68–76, 2017 © 2016 Elsevier Ltd. [1].)

liquid drug. PLA is biodegradable and formed byproducts of CO_2 and H_2O. Liquid drug encapsulated fibers can be used to prepare bandage for drug delivery on human skin to treat dermal patient [34]. Beaded nano fibers can be prepared by coaxial electrospinning of hydrophobic core and hydrophilic shell polymer. Hydrophobic functional fluid can be efficiently encapsulated into the beads and the concept can be used for the encapsulation of drug in beaded structure for drug delivery [53]. Wang and Windbergs compared drug delivery for two types of encapsulated fibers viz. blend electrospun and coaxial electrospun fiber. Morphology of fibers and drug distribution varied depending on the selected spinning techniques, and these all influences subsequent release of encapsulated fibers. Encapsulated fiber prepared by coaxial spinning has average diameter of around 220 nm, while, blend spinning produced fibers with an average diameter of around 244 nm. Encapsulated fibers produced by blend electrospinning, showed extended release of drug up to 15 days (96.2%). However, coaxial electrospun fibers showed rapid release of encapsulated drug and around 83% of encapsulated released after 1 day, thereafter, extended release till 10 days (~98%) (Fig. 5) [54]. In the blend electrospun fibers, drug molecules should be homogeneously distributed throughout the whole matrix, while, in case of coaxial

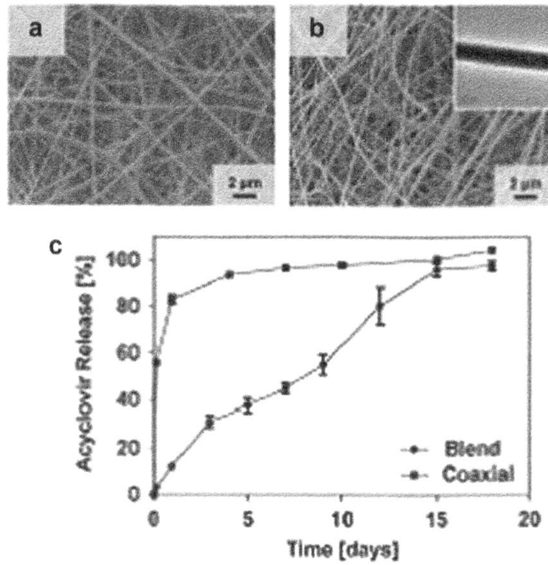

Fig. 5 **a** Scanning electron microscopic (SEM) image of fibre morphology prepared by blend spinning; **b** SEM image of morphology of fibers prepared by coaxial spinning; **c** Acyclovir drug release profile from fibers prepared by blend or coaxial electrospinning (Reproduced with permission from Wang and Windbergs, *International Journal of Pharmaceutics*, vol. 556, pp. 363–371, 2019 © 2018 Elsevier B. V. [54].)

electrospun fibers, core sheath structure of encapsulated drug in core of the fiber and biodegradable polymer in the sheath of the core-sheath has been formed [28].

Jindal et al. studied encapsulation and release of Zafirlukast, an anti-inflammatory drug from electrospun polyisobutylene fiber mat. Their result revealed complete encapsulation of drugs with the use of coaxial spinning. The process ensured higher loading of drug and slower release of drug for better treatment [55]. Intestinal microbial play a major role in human health by inhabiting pathogen colonization, boosting immune system and metabolizing nutrients. Lopez-Rubio et al. reported encapsulation and stabilization of bifidobacterial strains in poly (vinyl alcohol) (PVA) fiber by using electrospinning. PVA has been selected due to its high oxygen barrier in dry state and water solubility. Encapsulated bacteria within the PVA fiber remain viable for 40 and 130 days at room temperature and refrigeration temperature, respectively. In contrast, significant viability reduction was observed for bacteria without encapsulation [38].

4.2 Applications in Food Industry

Flavor, fragrance and bioactive compounds are often required to supply in accurate dosing in final product for effective absorption by our body. Further, these are highly sensitive volatile bioorganic substance needs to protect from aggressive environment such as light, moisture, oxidation, evaporation etc. to extend the self-life and control release rate [33, 56, 57]. Apart from nutritional role, bioactive components are physiologically active to provide health benefits. Some of the common bioactive components include vitamins, minerals, antioxidants, lipids etc. Bioactive components can be destroyed by influence of heat, light and other environmental conditions. Therefore, these useful ingredients must be protected from the environment before it is absorbed by our body. Micro or nano encapsulation are commonly used techniques to stabilized sensitive compounds [58, 59]. Torkamani et al. reported encapsulation of polyphenolic antioxidant into zein/gelatin shell core fibers by coaxial electrospinning. Average antioxidant encapsulated fiber diameter ranges from 311 to 380 nm, and there is no chemical reaction between encapsulated core and shell wall materials at various loading rate. Encapsulation efficiency of antioxidant is over 80% with 5–15% loading of antioxidant. The process resulted in robust fibers with high encapsulation efficiency and good shelf stability. Further, two polymers, gelatin and zein with different solubility ensured controlled release feature within the body [60]. Fernandez et al. reported encapsulation of β-carotene in ultra-thin fiber of zein prolamine by using electrospinning technique. β-carotene is a light sensitive colorant and antioxidant molecules extensively used in the food industry. The process provided stable and widely dispersed antioxidant molecules in the zein matrix. Further, encapsulated in fiber provided significant improvement in light stability when exposed to UV–vis radiation [3].

4.3 Applications in Textiles

Thermo-regulating textiles can be prepared by phase change material (PCM) incorporated fibers. Hu and Yu reported bio PCM based on natural soy wax encapsulated polyurethane fibers prepared by coaxial electrospinning technique. Experimental result demonstrated homogeneous surface morphology of fibers with uniform distribution of core PCM material in the matrix. Enthalpy of fiber increases with increasing content of PCM material [61]. Cherif et al. reported melt spinning of PCM incorporated bi-components fibers consisting blend of paraffin wax and polypropylene core, and polypropylene sheath material. PCM incorporated fibers has latent heat up to 22 g/J which is three times higher than the controlled fibers. Thermo-regulated bi-component fibers can be used for smart textiles or other technical applications [35]. PCM encapsulated nanofibers can be manufactured by coaxial electrospinning. Due its higher specific surface area, electrospun PCM encapsulated nanofibers can absorb, retain and release large amount of thermal energy in a certain temperature range, and composite fibers can be used for multi-functional assemblies [43]. In general, to prepare thermo-regulated textiles, PCM are encapsulated into a core shell structure of PCM core and thin layer of polymer shell, and then encapsulated PCMs are applied on textile substrate by using binder. Some of the disadvantages for the conventional PCM treated textiles are leakage of PCM due to poor encapsulation, ineffective heat transfer from PCM to the textiles substrate due to the presence of shell polymer material, poor wash/rubbing fastness of conventional PCM treated textiles etc. and these disadvantages can be overcome by incorporation of PCM in the fiber matrix instead of fiber surface, and fabrication of PCM core and polymer fiber sheath structure. In order to prevent the leakage of PCM, Lu et al. reported paraffin wax (PW) core and polyacrylonitrile (PAN) sheath fiber by coaxial electrospinning technique. Further, hexagonal cesium tungsten bronze ($Cs_{0.32}WO_3$) has been incorporate to utilize solar energy. Prepared fiber showed around 54% encapsulation efficiency and with latent heat of 63.1 J/g with no significant change of latent heat after 500 cycles. Figure 6 shows the core sheath fiber structure with different core flow rate. Diameter of fiber has been increased from 188 $\pm$ 29 nm to 221 $\pm$ 25 nm when core speed increased from 0.1 ml/h to 0.5 ml/h, and these can be attributed to the successful encapsulation of paraffin wax core in the PAN sheath fiber. Heat of fusion of encapsulated fibers are 26.81, 49.78, 59.26 J/g for fibers with 0.1, 0.3 and 0.5 ml/h core flow rate, respectively [62].

Iqbal et al. reported multi-component micro PCM encapsulated polypropylene yarn by melt extrusion technique. PCM e.g. n-octadecane with melting point temperature close to human body temperature can be used to fabricate thermo-regulated

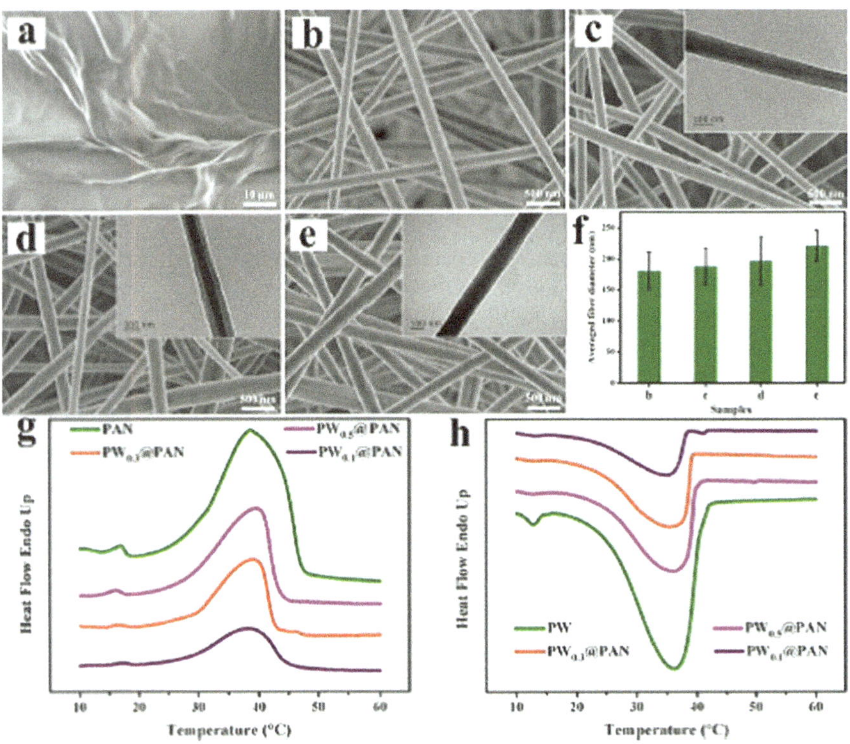

Fig. 6 Filed emission electron microscopic images: **a** pure paraffin wax; **b** pure polyacrylonitrile (PAN). Core sheath PW/PAN with core flow rate: **c** 0.1 ml/h; **d** 0.3 ml/h; **e** 0.5 ml/h; **f** average fiber diameter of fibres (**b**–**e**). Differential scanning calorimetry curves of pure PAN and encapsulated PAN fibers: **g** heating cycle; **h** cooling cycle. Inset in (**c**–**e**) shows the corresponding transmission electron microscopic images of PAN fiber (Reproduced with permission from Y. Lu et al., *Chem. Eng. J.*, 355, 532–539, 2019 © 2018 Elsevier B. V. [62].)

textiles, such as garment and upholstery to enhance thermal comfort [37]. Few specific applications of PCM incorporated fibers include sportswear, hospital apparel, bedding and accessories, space suits, gloves etc. [63]. In order to reduce the thermal stress, active wear should have thermal balance between heat generated by the body and subsequent heat released into the environment during physical activities. PCM encapsulated fibers can be used to manufacture garments for sports persons to dynamically control the body temperature of individual. During the sports activities, the body temperature will increase and the excess heat would be absorbed by PCM in the encapsulated fibers. During the resting time, absorbed heat will release to keep the wearer warm [64]. Encapsulated PCM in the composite fibers can interact with micro climate changes across the human body, and subsequently could respond to the fluctuation in temperature changes during patient activity level and external environmental temperature changes. Therefore, PCM incorporated fibers can be used for the manufacturing of surgical apparel, bandages, bedding accessories for patient, and other medical textile products. Other applications of PCM encapsulated fibers include quilt, bed sheet, pillow cover, mattress etc. in the bedding and accessories category. Moreover, PCM encapsulated fibers can be used in decoration, curtains, blanket, home furnishing, fire fighter garment. Further, PCM encapsulated fibers can be used for the preparation of structures for car interior and seat cover in order to provide superior thermal comfort to the passengers [63].

4.4 Other Applications

Phase change material (PCM) encapsulated fibers can be used for the electrical application for thermal management of electrical devices. Thermal management is an important aspect as the size of electronic device decreases whereas its speed increases. In an ineffective thermal management of electronic device can reduces its performance and in extreme case it can also damage the electronic devices [65, 66]. Sinha-Ray et al. presents co electrospinning of dicyclopentadiene (DCPD) (a self-healing material) encapsulation into polyacrylonitrile (PAN) sheath. Healing material encapsulated fibers can be used for the development of smart composites with embedded self-healing agent [67]. PCM encapsulated fibers can be used in shoes and accessories. Some of the other applications of PCM encapsulated fibers include pipe wrap and building insulation in civil engineering, temperature regulated containers for foods and drinks, thermal insulation of household appliances etc. [68, 69]. Keskin et al. reported living bacteria encapsulation in electrospun cyclodextrin fiber (CD-F) for the removal of heavy metal and reactive dye. Figure 7a shows that *Lysinibacillus* sp. NOSK bacterial cells which emitting green fluorescence owing to their viability nature and Fig. 7b confirmed successful encapsulation of bacteria by CD electrospun fiber (CD-F). CD-F acts as both carrier matrix and feeding source for the encapsulated bacteria, due to the non-toxic property of CD. Removal efficiency of Ni(II), Cr(IV) and RB5 by using encapsulated fiber are around 70, 58 and 82%, respectively. Pollutant removal capability of encapsulated fiber is higher than that of

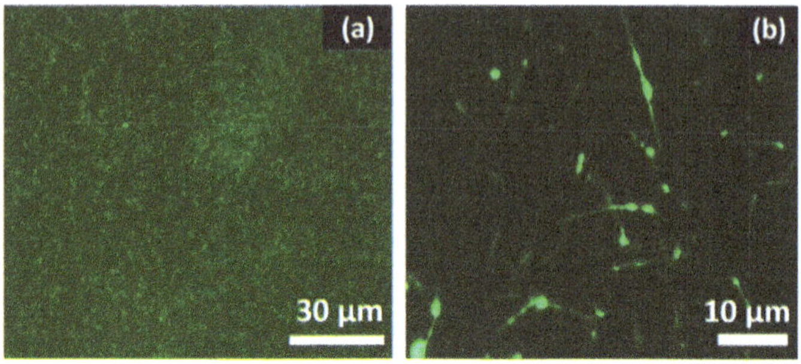

Fig. 7 **a** Fluorescence microscopic images of bacteria, before encapsulation and **b** after encapsulation in CD-F (Reproduced with permission from Keskin et al. *Colloid Surf. B-Biointerfaces,* 161, 169–176, 2018 © 2017 Elsevier B. V. [70].)

the free bacteria, as bacteria can use CD as an extra carbon source for their growth rate [70].

Hybrid organic–inorganic perovskites (HOIP) are used as potential light harvesting materials for solar cells. However, HOIP possesses inherent material instability due to the moisture driven degradation. Murphy et al. presents HOIP encapsulated core in polystyrene shell fiber by using coaxial electrospinning. Protection of HOIP core in the polystyrene shell is due to the two primary reasons such as significant increase of contact angle, secondly, polystyrene shell material provided a barrier for the moisture sensitive HOIP [71].

5 Conclusions: Challenges and Future Prospects

Encapsulated functional substance loaded fibers plays an important role in food technology, medicine, textiles and many other areas in order to protect the functional materials from aggressive environment during processing and usage. This chapter presents an overview of fibers as a container for the encapsulation of various functional substances. Some of the challenges of fiber as container for encapsulation are effective loading of useful component, encapsulation efficiency and release rate from the encapsulated fibers. Further, for medicine application, some of the other challenges of fibers as container for encapsulation are residual solvent removal from the encapsulated fiber, influence of initial burst drug release, uniform distribution of drugs etc. [39]. Electrospinning techniques are widely used for the preparation of encapsulated fibers. Major challenges of electrospinning is low production rate, hence, high production cost and this hindered the possibility of scale up in bulk industrial applications, especially for food, agro industry, textiles etc. More researches in these areas are encouraging for further development of fibers as container for the encapsulation of various functional substances.

References

1. Lancuski A, Abu Ammar A, Avrahami R, Vilensky R, Vasilyev G, Zussman E (2017) Design of starch-formate compound fibers as encapsulation platform for biotherapeutics. Carbohydr Polym 158:68–76. https://doi.org/10.1016/j.carbpol.2016.12.003
2. Gomes MHG, Kurozawa LE (2020) Improvement of the functional and antioxidant properties of rice protein by enzymatic hydrolysis for the microencapsulation of linseed oil. J Food Eng 267, Art no 109761. https://doi.org/10.1016/j.jfoodeng.2019.109761
3. Fernandez A, Torres-Giner S, Lagaron JM (2009) Novel route to stabilization of bioactive antioxidants by encapsulation in electrospun fibers of zein prolamine. Food Hydrocolloids 23(5):1427–1432. https://doi.org/10.1016/j.foodhyd.2008.10.011
4. Khan WA, Butt MS, Pasha I, Jamil A (2020) Microencapsulation of vitamin D in protein matrices: in vitro release and storage stability. J Food Meas Charact 14:1172–1182. https://doi.org/10.1007/s11694-019-00366-3
5. Lavanya MN, Kathiravan T, Moses JA, Anandharamakrishnan C (2020) Influence of spray-drying conditions on microencapsulation of fish oil and chia oil. Drying Technol 38(3):279–292. https://doi.org/10.1080/07373937.2018.1553181
6. Lozinska N, Glowacz-Rozynska A, Artichowicz W, Lu YQ, Jungnickel C (2020) Microencapsulation of fish oil—determination of optimal wall material and encapsulation methodology. J Food Eng 268, Art no 109730. https://doi.org/10.1016/j.jfoodeng.2019.109730
7. Kurd F, Fathi M, Shekarchizadeh H (2019) Nanoencapsulation of hesperetin using basil seed mucilage nanofibers: characterization and release modeling. Food Biosc 32, Art no 100475. https://doi.org/10.1016/j.fbio.2019.100475
8. Walia N, Dasgupta N, Ranjan S, Ramalingam C, Gandhi M (2019) Methods for nanoemulsion and nanoencapsulation of food bioactives. Environ Chem Lett 17(4):1471–1483. https://doi.org/10.1007/s10311-019-00886-w
9. Zompero RHD, Lopez-Rubio A, de Pinho SC, Lagaron JM, de la Torre LG (2015) Hybrid encapsulation structures based on beta-carotene-loaded nanoliposomes within electrospun fibers. Colloid Surf B-Biointerfaces 134:475–482. https://doi.org/10.1016/j.colsurfb.2015.03.015
10. Cheng J, Park D, Jun Y, Lee J, Hyun J, Lee SH (2016) Biomimetic spinning of silk fibers and in situ cell encapsulation. Lab Chip 16(14):2654–2661. https://doi.org/10.1039/c6lc00488a
11. Li XQ, Su Y, Liu SP, Tan LJ, Mo XM, Ramakrishna S (2010) Encapsulation of proteins in poly(L-lactide-co-caprolactone) fibers by emulsion electrospinning. Colloid Surf B-Biointerfaces 75(2):418–424. https://doi.org/10.1016/j.colsurfb.2009.09.014
12. Triches M, Brusch A, Hald J (2015) Portable optical frequency standard based on sealed gas-filled hollow-core fiber using a novel encapsulation technique. Appl Phys B-Lasers Opt 121(3):251–258. https://doi.org/10.1007/s00340-015-6224-8
13. Bauer AJP, Zeng TY, Liu JZ, Uthaisar C, Li BB (2014) The enhanced encapsulation capacity of polyhedral oligomeric silsesquioxane-based copolymers for the fabrication of electrospun core/shell fibers. Macromol Rapid Commun 35(7):715–720. https://doi.org/10.1002/marc.201400032
14. Ghitescu RE, Popa AM, Popa VI, Rossi RM, Fortunato G (2015) Encapsulation of polyphenols into pHEMA e-spun fibers and determination of their antioxidant activities. Int J Pharm 494(1):278–287. https://doi.org/10.1016/j.ijpharm.2015.08.020
15. Korehei R, Kadla JF (2014) Encapsulation of T4 bacteriophage in electrospun poly(ethylene oxide)/cellulose diacetate fibers. Carbohydr Polym 100:150–157. https://doi.org/10.1016/j.carbpol.2013.03.079
16. Aceituno-Medina M, Mendoza S, Rodriguez BA, Lagaron JM, Lopez-Rubio A (2015) Improved antioxidant capacity of quercetin and ferulic acid during in-vitro digestion through encapsulation within food-grade electrospun fibers. J Funct Food 12:332–341. https://doi.org/10.1016/j.jff.2014.11.028.
17. Chung WJ, Merzlyak A, Lee SW (2010) Fabrication of engineered M13 bacteriophages into liquid crystalline films and fibers for directional growth and encapsulation of fibroblasts. Soft Matter 6(18):4454–4459. https://doi.org/10.1039/c0sm00199f

18. Zussman E (2011) Encapsulation of cells within electrospun fibers. Polym Adv Technol Rev 22(3):366–371. https://doi.org/10.1002/pat.1812
19. Singh MN, Hemant KSY, Ram M, Shivakumar HG (2010) Microencapsulation: a promising technique for controlled drug delivery. Res Pharm Sci 5(2):65–77
20. Ordanini S, Cellesi F (2018) Complex polymeric architectures self-assembling in unimolecular micelles: preparation, characterization and drug nanoencapsulation. Pharm Rev 10(4):19, Art no 209. https://doi.org/10.3390/pharmaceutics10040209
21. Kumari A, Singla R, Guliani A, Yadav SK (2014) Nanoencapsulation for drug delivery. Excli J Rev 13:265–286 [Online]. https://www.excli.de/index.php/excli/article/view/700
22. Zhu F (2017) Encapsulation and delivery of food ingredients using starch based systems. Food Chem Rev 229:542–552. https://doi.org/10.1016/j.foodchem.2017.02.101
23. Aditya NP, Espinosa YG, Norton IT (2017) Encapsulation systems for the delivery of hydrophilic nutraceuticals: food application. Biotechnol Adv Rev 35(4):450–457. https://doi.org/10.1016/j.biotechadv.2017.03.012
24. Ruiz JCR, Vazquez EDO, Campos MRS (2017) Encapsulation of vegetable oils as source of omega-3 fatty acids for enriched functional foods. Crit Rev Food Sci Nutr Rev 57(7):1423–1434. https://doi.org/10.1080/10408398.2014.1002906
25. Fatih IM, Ibrahim C (2016) Encapsulation of probiotic bacteria with alginate-starch and evaluation of viability in storage conditions and in food. Res J Biotechnol 11(12):31–37 [Online]. https://worldresearchersassociations.com/Archives/RJBT/Vol(11)2016/December2016.aspx
26. Yoosefian M, Sabaei S, Etminan N (2019) Encapsulation efficiency of single-walled carbon nanotube for Ifosfamide anti-cancer drug. Comput Biol Med 114:8, Art no 103433. https://doi.org/10.1016/j.compbiomed.2019.103433
27. Dolinina ES, Akimsheva EY, Parfenyuk EV (2019) Silica microcapsules as containers for protein drugs: direct and indirect encapsulation. J Mol Liq Proc Paper 287: Art no 110938. https://doi.org/10.1016/j.molliq.2019.110938
28. Chen JL et al (2019) Encapsulation and release of drug molecule pregabalin based on ultrashort single-walled carbon nanotubes. J Phys Chem C 123(14):9567–9574. https://doi.org/10.1021/acs.jpcc.9b00675
29. Mlaouah M, Tangour B, El Khalifi M, Gharbi T, Picaud F (2018) The encapsulation of the gemcitabine anticancer drug into grapheme nest: a theoretical study. J Mol Model 24(4):9, Art no 102. https://doi.org/10.1007/s00894-018-3627-6
30. Su YC, Zhao H, Wu JR, Xu JH (2016) One-step fabrication of silica colloidosomes with in situ drug encapsulation. RSC Adv 6(113):112292–112299. https://doi.org/10.1039/c6ra19048k
31. Peng PC, Hong RL, Tsai T, Chen CT (2019) Co-Encapsulation of chlorin e6 and chemotherapeutic drugs in a pegylated liposome enhance the efficacy of tumor treatment: pharmacokinetics and therapeutic efficacy. Pharmaceutics 11(11):17, Art no 617. https://doi.org/10.3390/pharmaceutics11110617
32. Ghahremani S, Samadizadeh M, Khaleghian M, Shiraz NZ (2020). Theoretical study of encapsulation of Floxuridine anticancer drug into BN (9, 9–7) nanotube for medical application. Phosphorus Sulfur Silicon Relat Elem 195(4): 293-306. https://doi.org/10.1080/10426507.2019.1687479
33. Castro N, Durrieu V, Raynaud C, Rouilly A, Rigal L, Quellet C (2016) Melt extrusion encapsulation of flavors: a review. Polym Rev Rev 56(1):137–186. https://doi.org/10.1080/15583724.2015.1091776
34. Naeimirad M, Zadhoush A, Neisiany RE, Salimian S, Kotek R (2019) Melt-spun PLA liquid-filled fibers: physical, morphological, and thermal properties. J Text Inst 110(1):89–99. https://doi.org/10.1080/00405000.2018.1465336
35. Cherif C, Tran NHA, Kirsten M, Brunig H, Vogel R (2018) Environmentally friendly and highly productive bi-component melt spinning of thermoregulated smart polymer fibres with high latent heat capacity. Express Polym Lett 12(3):203–214. https://doi.org/10.3144/expresspolymlett.2018.19
36. Ayad E, Cayla A, Rault F, Gonthier A, Campagne C, Devaux E (2018) Effect of viscosity ratio of two immiscible polymers on morphology in bicomponent melt spinning fibers. Adv Polym Technol 37(4):1134–1141. https://doi.org/10.1002/adv.21772

37. Iqbal K, Sun DM (2015) Development of thermal stable multifilament yarn containing micro-encapsulated phase change materials. Fibers Polym 16(5):1156–1162. https://doi.org/10.1007/s12221-015-1156-9
38. Lopez-Rubio A, Sanchez E, Sanz Y, Lagaron JM (2009) Encapsulation of living bifidobacteria in ultrathin PVOH electrospun fibers. Biomacromol 10(10):2823–2829. https://doi.org/10.1021/bm900660b
39. Wen P, Wen Y, Zong MH, Linhardt RJ, Wu H (2017) Encapsulation of bioactive compound in electrospun fibers and its potential application. J Agric Food Chem Rev 65(42):9161–9179. https://doi.org/10.1021/acs.jafc.7b02956
40. Han FX, Zhang H, Zhao J, Zhao YH, Yuan XY (2012) In situ encapsulation of hydrogel in ultrafine fibers by suspension electrospinning. Polym Eng Sci 52(12):2695–2704. https://doi.org/10.1002/pen.23227
41. Dong B, Smith ME, Wnek GE (2009) Encapsulation of multiple biological compounds within a single electrospun fiber. Small 5(13):1508–1512. https://doi.org/10.1002/smll.200801750
42. Qi HX, Hu P, Xu J, Wang AJ (2006) Encapsulation of drug reservoirs in fibers by emulsion electrospinning: morphology characterization and preliminary release assessment. Biomacromol 7(8):2327–2330. https://doi.org/10.1021/bm060264z
43. McCann JT, Marquez M, Xia YN (2006) Melt coaxial electrospinning: a versatile method for the encapsulation of solid materials and fabrication of phase change nanofibers. Nano Lett 6(12):2868–2872. https://doi.org/10.1021/nl0620839
44. Nada AA, Abdelazeem RA, Elghandour AH, Abou-Zeid NY (2016) Ricinoleic acid encapsulation into ethyl cellulose electrospun fibers. J Nat Fibers 13(6):670–681. https://doi.org/10.1080/15440478.2015.1102792
45. Li XR, Zhang H, Li H, Yuan XY (2010) Encapsulation of proteinase K in PELA ultrafine fibers by emulsion electrospinning: preparation and in vitro evaluation. Colloid Polym Sci 288(10–11):1113–1119. https://doi.org/10.1007/s00396-010-2235-5
46. Granicka LH et al (2000) Encapsulation of parathyroid cells in hollow fibers: a preliminary report. Folia Histochem Cytobiol 38(3):129–131 [Online]. https://pubmed.ncbi.nlm.nih.gov/10970072/
47. Tammaro L, Russo G, Vittoria V (2009) Encapsulation of diclofenac molecules into poly(epsilon-caprolactone) electrospun fibers for delivery protection. J Nanomater 8, Art no 238206. https://doi.org/10.1155/2009/238206
48. Granicka LH, Kawiak J, Snochowski M, Wojcicki JM, Sabalinska S, Werynski A (2003) Polypropylene hollow fiber for cells isolation: Methods for evaluation of diffusive transport and quality of cells encapsulation. Artif Cells Blood Substit Immobil Biotechnol 31(3):249–262. https://doi.org/10.1081/bio-120023156
49. Wan ACA, Yim EKF, Liao IC, Le Visage C, Leong KW (2004) Encapsulation of biologics in self-assembled fibers as biostructural units for tissue engineering. J Biomed Mater Res Part A 71A (4):586–595. https://doi.org/10.1002/jbm.a.30158
50. Huang ZM, Yang AH (2006) Encapsulation of pure drugs ento the central part of polycaprolactone ultrafene fibers (in Chinese). Acta Polym Sin (1):48–52. https://doi.org/10.3724/SP.J.1105.2006.00048
51. Patel AC, Li SX, Yuan JM, Wei Y (2006) In situ encapsulation of horseradish peroxidase in electrospun porous silica fibers for potential biosensor applications. Nano Lett 6(5):1042–1046. https://doi.org/10.1021/nl0604560
52. Souza MA, Sakamoto KY, Mattoso LHC (2014) Release of the diclofenac sodium by nanofibers of poly(3-hydroxybutyrate-co-3-hydroxyvalerate) obtained from electrospinning and solution blow spinning. J Nanomater, Art no 129035. https://doi.org/10.1155/2014/129035
53. Diaz JE, Barrero A, Marquez M, Loscertales IG (2006) Controlled encapsulation of hydrophobic liquids in hydrophilic polymer nanofibers by co-electrospinning. Adv Func Mater 16(16):2110–2116. https://doi.org/10.1002/adfm.200600204
54. Wang J, Windbergs M (2019) Controlled dual drug release by coaxial electrospun fibers—impact of the core fluid on drug encapsulation and release. Int J Pharm 556:363–371. https://doi.org/10.1016/j.ijpharm.2018.12.026

55. Jindal A et al (2018) Encapsulation and release of Zafirlukast from electrospun polyisobutylene-based thermoplastic elastomeric fiber mat. Eur Polym J 98:254–261. https://doi.org/10.1016/j.eurpolymj.2017.11.012
56. Aceituno-Medina M, Mendoza S, Lagaron JM, Lopez-Rubio A (2015) Photoprotection of folic acid upon encapsulation in food-grade amaranth (Amaranthus hypochondriacus L.) protein isolate—pullulan electrospun fibers. LWT-Food Sci Technol 62(2):970–975. https://doi.org/10.1016/j.lwt.2015.02.025
57. Alborzi S, Lim LT, Kakuda Y (2013) Encapsulation of folic acid and its stability in sodium alginate-pectin-poly(ethylene oxide) electrospunfibres. J Microencapsul 30(1):64–71. https://doi.org/10.3109/02652048.2012.696153
58. Abbas S, Da Wei C, Hayat K, Zhang XM (2012) Ascorbic acid: microencapsulation techniques and trends-a review. Food Rev Int Rev 28(4):343–374. https://doi.org/10.1080/87559129.2011.635390
59. Shekarforoush E, Mendes AC, Baj V, Beeren SR, Chronakis IS (2017) Electrospun phospholipid fibers as micro-encapsulation and antioxidant matrices. Molecules 22(10):16, Art no 1708. https://doi.org/10.3390/molecules22101708
60. Torkamani AE, Syahariza ZA, Norziah MH, Wan AKM, Juliano P (2018) Encapsulation of polyphenolic antioxidants obtained from Momordica charantia fruit within zein/gelatin shell core fibers via coaxial electrospinning. Food Biosci 21:60–71. https://doi.org/10.1016/j.fbio.2017.12.001
61. Hu W, Yu X (2012) Encapsulation of bio-based PCM with coaxial electrospun ultrafine fibers. RSC Adv 2(13):5580–5584. https://doi.org/10.1039/c2ra20532g
62. Lu Y et al (2019) Novel smart textile with phase change materials encapsulated core-sheath structure fabricated by coaxial electrospinning. Chem Eng J 355:532–539. https://doi.org/10.1016/j.cej.2018.08.189
63. Mondal S (2008) Phase change materials for smart textiles—an overview. Appl Therm Eng 28(11–12):1536–1550. https://doi.org/10.1016/j.applthermaleng.2007.08.009
64. Shin Y, Yoo DI, Son K (2005) Development of thermoregulating textile materials with microencapsulated phase change materials (PCM). II. Preparation and application of PCM microcapsules. J Appl Polym Sci 96(6):2005–2010. https://doi.org/10.1002/app.21438
65. Aftab W, Huang X, Wu W, Liang Z, Mahmood A, Zou R (2018) Nanoconfined phase change materials for thermal energy applications. Energy Environ Sci 11(6):1392–1424. https://doi.org/10.1039/C7EE03587J
66. Liu CC et al (2020) Phase change materials application in battery thermal management system: a review. Materials 13(20), Art no 4622. https://doi.org/10.3390/ma13204622.
67. Sinha-Ray S, Pelot DD, Zhou ZP, Rahman A, Wu XF, Yarin AL (2012) Encapsulation of self-healing materials by coelectrospinning, emulsion electrospinning, solution blowing and intercalation. J Mater Chem 22(18):9138–9146. https://doi.org/10.1039/c2jm15696b
68. Dugan J and Kuckhoff E (2007) Multicomponent fiber comprising a phase change material. USA Patent, p WO 2007035483 A1
69. Chalco-Sandoval W, Fabra MJ, Lopez-Rubio A, Lagaron JM (2017) Use of phase change materials to develop electrospun coatings of interest in food packaging applications. J Food Eng 192:122–128. https://doi.org/10.1016/j.jfoodeng.2015.01.019
70. Keskin NOS, Celebioglu A, Sarioglu OF, Uyar T, Tekinay T (2018) Encapsulation of living bacteria in electrospun cyclodextrin ultrathin fibers for bioremediation of heavy metals and reactive dye from wastewater. Colloid Surf B-Biointerfaces 161:169–176. https://doi.org/10.1016/j.colsurfb.2017.10.047
71. Murphy JP, Andriolo JM, Sutton NJ, Brockway MC, Skinner JL (2017) Coaxial hybrid perovskite fibers: synthesis and encapsulation in situ via electrospinning. J Vac Sci Technol B 35(6):6, Art no 06g402. https://doi.org/10.1116/1.4991724

Bio-Based/Biodegradable Containers for Encapsulation

Ignacio Rivero Berti and Guillermo R. Castro

Abbreviations

Alg	Alginate
BC	Bacterial cellulose
BCS	Biopharmaceutics Classification System
Car	Carrageenan
Chi	Chitosan
COL	Collagen
ECM	Extracellular matrix
EDTA	Ethylenediaminetetraacetic
FDA	Food and Drug Administration (USA)
FTIR	Fourier Transform Infrared
HA	Hyaluronic acid
MP	Microparticle
MW	Molecular weight
NLC	Nanostructured lipid carriers
NP	Nanoparticle
PCL	Polycaprolactone
PEG	Polyethylene glycol
PGA	Polyglycolic acid

I. R. Berti
Laboratorio de Nanobiomateriales, CINDEFI, Departamento de Química, Facultad de Ciencias Exactas, Universidad Nacional de La Plata (UNLP) -CONICET (CCT La Plata), Calle 47 y 115, (B1900AJI), La Plata, Buenos Aires, Argentina

G. R. Castro (✉)
Max Planck Laboratory for Structural Biology, Chemistry and Molecular Biophysics of Rosario (MPLbioR, UNR-MPIbpC), Partner Laboratory of the Max Planck Institute for Biophysical Chemistry (MPIbpC, MPG), Centro de Estudios Interdisciplinarios (CEI), Universidad Nacional de Rosario, Maipú 1065, S2000, Rosario, Santa Fe, Argentina

J. Parameswaranpillai et al. (eds.), *Micro- and Nano-containers for Smart Applications*, Composites Science and Technology,
https://doi.org/10.1007/978-981-16-8146-2_4

PLA Polylactic acid
SLN Solid lipid nanoparticles

1 Introduction

The development of the new era of Biotechnology in the 70's is the result of the emergence of novel tools and concepts established in the areas of molecular biology, genetic engineering and later metabolic engineering and novel and sensitive instrumentation with the consequently discovery of new processes and molecules. Under this frame, systems for the encapsulation and delivery of molecules were synthesized at experimental level and some them reach the global market. These novel "molecular containers" were used at large scale in the food industry, cosmetics ánd later in biomedicine for drug delivery purposes. Many novel structures from biopolymers to lipids and hybrid systems having different cargoes were built during the last 10 years. Some molecular containers were developed as smart devices since they can trigger the release of molecules by specific physicochemical environmental conditions and/or diverse biological effectors.

Particularly synthetic and natural polymers, lipid structures, and many hybrid systems have playing new roles in the encapsulation and/or entrapment of many molecules from antioxidants, vitamins to antibiotics, enzymes, growth factors and many drugs. The design molecular carriers involve many factors such route of administration, target location and organ, amount of cargo, biophysical properties of the cargo, structure and composition of the carrier, environmental conditions of molecular release, kinetic of release, biodegradability of matrix, biocompatibility, potential toxicity secondary effect of carrier and its degradation products. However, synthetic polymer synthesis in general requires high temperatures and organic solvents for synthesis, expensive and time-consuming purification processes required for approval of regulatory agencies (i.e. FDA, EMA, etc.), could show unspecific interaction with the cargo and hydrolysis can trigger an inflammatory reaction produced by toxic byproducts and/or drastic changes in local physiological pH [22].

In the case of foods, molecular administration is oral, but for therapeutic purposes there are many alternatives for the design of molecular delivery carriers which depends on tissue target and drug properties (Table 1). However, many marketed drugs show problems of low solubility under physiological conditions, low permeability, rapid metabolism, and fast elimination and/or biotransformation in the body and sometimes associated with low tolerability and undesirable side effects [46].

In general, conventional oral administration systems of drugs available in the market involves the physical entrapment of drug in excipients such as flour, starch, cellulose, lactose, gums, gelatin and its mixtures and/or derivatives made by physical methods (i.e. pressure, spray dry, etc.). The main problem in such case is when the carrier is dissolved and all drug content released and spread in the body including non-targeted organs which requires high drug concentrations to reach the therapeutic

Table 1 Administration routes of therapeutic molecules in mammals

Administration routes			
Gastrointestinal	Parenteral (injectable)	Transepithelial	Inhalatory
Oral	Intra-arterial	Transdermal	Transnasal
Rectal	Intramuscular	Transmucosal	Pulmonary
	Intravenous		
	Subcutaneous		

window but could cause toxic effects in several organs. Additionally, the traditional administration systems lack the ability of drug-controlled release because of the interaction between the excipients components of the matrix and the drug in most the cases is unknown or is only physical. Consequently, the amount of released drug from the carrier must be keep inside the therapeutic window for a period of time is only determined by the stability of the carrier under specific physiological environment. In this case, two main problems could be considered: first of all, the bioavailability of the drug could be seriously compromised since high number of marketed drugs are poor soluble in aqueous physiological environment. The major obstacle of low aqueous soluble drugs is produced during the dissolution of the carrier with the release of the drug to the physiological environment in where the organic/aqueous partition coefficient reduce the interaction by passive diffusion with the cell membranes. This effect could reduce the therapy effectivity and by increasing drug concentration could induce unwanted deleterious effects in the patients. Second et all, when the free drug concentration in the body is below the minimal concentration required by the therapeutic window could induce resistance mechanisms of pathogens such is the case of the emergence of microorganisms multidrug resistant (MDR), reported as the top third worldwide public health threat by the World Health Organization [10].

Low aqueous solubility affects between 40 to 50% of 150,000 novel synthetized potential therapeutic molecules but also only 10% were successfully introduced to the market [28, 44]. For example, the 69% of the top orally administered drugs in the US during 2006 showed problems of solubility and/or permeability in the digestive tract using the proposed BCS (Table 2) [90]. In general, about of 40% of marketed drugs and about 90% of new molecular entities at development stage display poor aqueous solubility which fits on classes II and IV of BCS [59].

Table 2 Classification of top 200 drugs orally administered in the USA in 2006 based on the biopharmaceutics classification system [90]

Class	Properties		Percentage (%)
	Solubility	Permeability	
I	High	High	31
II	Low	High	32
III	High	Low	23
IV	Low	Low	6
Unclassified	–	–	11

In addition to the main drug problems of solubility, the full synthesis and development of a new drug from a research laboratory to the market requires between 10 to 15 years correlated with U$ 500 to 800 million dollars costs. Also, it was estimated than 1 molecular candidate in 1,000 pass the regulatory process and enter at the production stage [28].

All factors together represent severe limitations for the pharmaceutical industry and serious problems for the treatment of diverse and complex pathologies. The negative impact at industrial level in USA can be seen in the FDA-approved antibiotics along different decades, the number of antibiotics approved in the period of 1983 to 1992 were 15, meanwhile between 2003 to 2009 were only 5 [10]. In this sense, one the main strategies are the repositioning old drugs besides, it requires novel approaches to be effective for therapeutic purposes.

Since Biotechnology brings a new scenario by massive use of biological living organisms to developed and produce goods and novel products, the landscape of many fields had changed including the pharmaceutical industry. Also, Green Chemistry technology is a trend for sustainable processes based on eco-friendly reagents and biomaterials and minimizing synthetic chemical compounds and decreasing toxic subproducts, reducing the purification process and wastes, and consuming less energy. The growth of Biotechnology combined with Green Chemistry is also fueled by the incorporation of novel and more sensitive biophysical and analytical tools with high resolution spectroscopies provides by cyclotrons that allow to obtain very sensitive UV, Raman and FTIR spectroscopies, different X-rays techniques now available in many countries; plus novel and very sensitive microscopies such as 3D confocal, environmental and laser scanning microscopies, near-field scanning optical microscopy, atomic force microscopy, etc. Also, novel technologies such as microfluidics, 3D-printing, electrospinning, etc. presents new ways to produce novel devices, drug delivery matrices with unprecedent properties.

The convergence of these concepts, tools, technologies and the theoretical bases of quantum mechanics provides an extraordinary new platform for the emergence of micro- and nano-technologies applied to the pharmaceutical industry, this brand-new area now is called nanopharmaceuticals [48]. In particular, studies of the interactions of drugs-matrices to develop tailored carriers for specific targets, combined the *in-situ* observation and analysis of the drug delivery systems with cell, organs and living organisms can provide novel approach to solve the aforementioned problems and improve the therapeutic efficiency of drugs concomitantly with the reduction of undesirable side effects [34]. The main advantages of nanopharmaceuticals can be summarized by the advantage of specific drug targeting based on the pathology and location of disease; combined molecules with therapeutic activity can be simultaneously administered in only one nanodevice; high drug concentration can be administered locally on-site but with low systemic drug concentration; enhanced vascular permeability and reduced lymphatic drainage of inflamed tissues through improved drug spread and holding; capacity of surpassing biological barriers (i.e. brain delivery); kinetic release control of the administered drugs; keeping the homeostatic physiological state of the patients by reducing the drug toxicity and avoid damage of untargeted organs [34, 36]. Particularly, nanoparticles receive high attention because of substantial advantages compared with conventional drug delivery systems because of subcellular size, high stability, kinetic control release, high specificity, high cargo content, and targetability.

There are many types of nanosystems for encapsulations of a myriad of molecules with therapeutic activities. However, based on the new trend of bio-based/biodegradable materials polymers, lipid structures and hybrid ones are more considered to develop novel carriers.

2 Biopolymeric Carriers

Biopolymers are produced form natural resources and present high diversity in their chemical structures and function. Some of them also are byproducts of other industrial processes and represent a waste material that can be easily recycled to produce goods with low costs [11]. Among the advantageous properties of biopolymers over chemically synthesized can be considered the readily and abundantly available, high diversity in the chemical and structural composition, stereoselectivity and enantioselectivity, bioinertia, most of them are non-toxic as well as they degradation products, restricted molecular weight, bioactivity (i.e. tailored residues), wide spectrum of chemical and mechanical properties. Biopolymers can be classified by different properties such as source (i.e. natural or synthetic), structure (i.e. linear, branched chains and three-dimensional network), synthesis (i.e. addition or condensation), molecular forces (i.e. swelling, viscosity, Young modulus, etc.). Besides, the biopolymers most used for pharmaceutical applications are generally grouped as polysaccharides, proteins, and polyesters (Fig. 1).

Among the advantages of biopolymer, the degradation under physiological environments in one of the main problems. Biopolymer degradation can affect the drug release and consequently the therapeutic value of the administration by releasing the

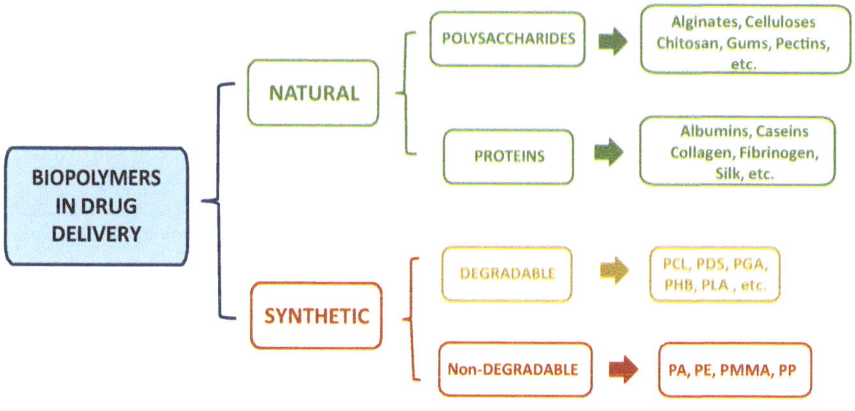

Fig. 1 Classification of drug delivery systems based on biopolymers properties

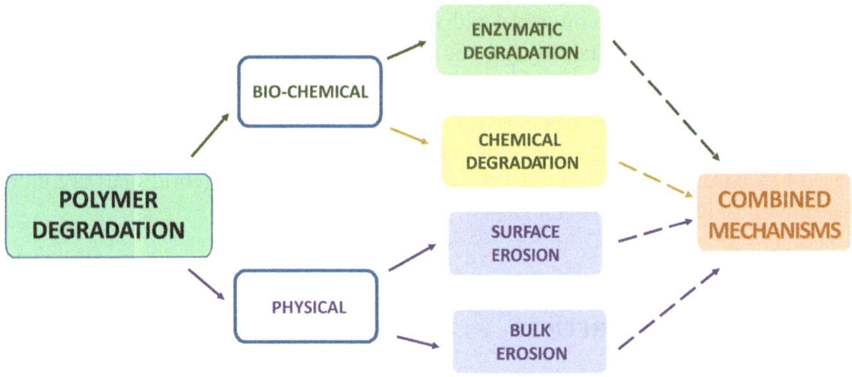

Fig. 2 Scheme of biopolymers degradation

drug out of the target and also it could be inactivated by local environmental conditions (e.g. enzymes, extreme pHs, etc.). The main mechanism of polymer degradation is the bioerosion which can be described as the transition of a water-insoluble or colloidal polymer to a water-soluble molecular structure. The mechanisms of bioerosion can be physical and/or biochemical and involves the incorporation of water into the polymer network associated with loss of tridimensional structure followed or not by endo- and/or exo-hydrolytic bond cleavage, chemically or enzymatically of polymer chains and pendant residual groups. The consequences of these processes are the reduction of mechanical properties such as strength and Young modulus, reduction of molar mass and uncontrolled release of the cargo from the polymer matrix. Besides, the increase of polymer crosslinking, molecular weight, hydrophobic properties, glass transition and crystalline domains to reduce bioerosion mechanisms (Fig. 2).

Besides, there are several strategies to reduce the biopolymer degradation including chemical or enzymatic crosslinking and/or addition of reinforcement materials.

There are many biopolymers employed to develop drug delivery carriers, and it will be difficult to enumerate and describe the main properties of all of them. However, the main properties of biopolymers more described as a biodegradable carrier are described below.

2.1 Natural Biopolymers

2.1.1 Albumins

The most common are Human serum (HSA) and bovine serum albumins (BSA) with MW $\approx$ 66.5 kDa. The physiological role of these globular proteins is the transport of hydrophobic molecules mainly hormones, steroids, and fatty acids in the blood plasm. The main secondary structure of HSA and BSA are similar and comprise 50–67% of alpha-helix under physiological conditions. The main advantage of HSA is that contain six main binding sites for small organic molecules (2), long fatty acids (2), ligands with free sulfhydryl group (1) and the last one for metallic ions. The availability of the binding sites depends on the protein conformation and molecular binding could induce structural changes in the globular protein [97].

2.1.2 Alginate (Alg)

Alginate (Alg) is a polymer mainly produced by seaweeds but also by some bacteria (i.e. *P. aeruginosa*). The polymer is a linear carbohydrate composed of 1,4 linked of poly-α-L-guluronic (G) and poly-ß-D-mannuronic (M) acids. Alginate is naturally synthesized in blocks of M or G monomers (i.e. M or G blocks) but also in mixed blocks of MG. Changes in the M/G ratios as well MG content can produce polymer of high viscosity in aqueous solutions to pseudoplastic polymer. Alginate can form hydrogels in presence of multivalent cations (i.e. calcium, magnesium, etc.). Alginate was approved the FDA for therapeutic purposes and used to develop drug delivery devices and tissue engineering [63, 83]. However, alginate gels are sensitive to breakdown by the presence of high monovalent cations, chelating compounds (i.e. EDTA, lactate, phosphate) and harsh chemical conditions such as acid pH which decrease the mechanical strength and gel structural stability [75].

2.1.3 2.1.3. Carrageenans (Car)

Carrageenans (Car) are a group of linear sulfated polygalactans extracted from the class *Rhodophyceas* algae, particularly from *Chondrus crispus*. The main chemical composition is based on D-galactose and 3,6-anhydrogalactose linked by α-1,3 and ß-1,4 glycosidic bonds and sulfated pendant groups up to 41% (w/w). Cars have different amounts of sulfate groups that brings gels with different properties and classified as kappa (κ), iota (ι) and lambda (λ) which mainly contain one, two and three sulfate ester pendant groups per two galactose groups with an median of 20%, 33% and 38% (w/w), respectively [20]. Cars are recognized by FDA as GRAS (generally regarded as safe) and commonly used as thickeners and stabilizing agents in foods and in the pharmaceutical industry. Kappa (κ) and iota (ι) Cars are able form helical curvy strict and elastic gel structures in presence of monovalent (i.e. K^+, Na^+, Li^+) or divalent (Ca^{+2}, Mg^{+2}) ions, respectively; meanwhile λ-Car does not. Car's biological activities were described as inductive of inflammatory process in mammals but with advantageous anti-tumor, immunomodulatory, antiviral, and anticoagulant activities [101].

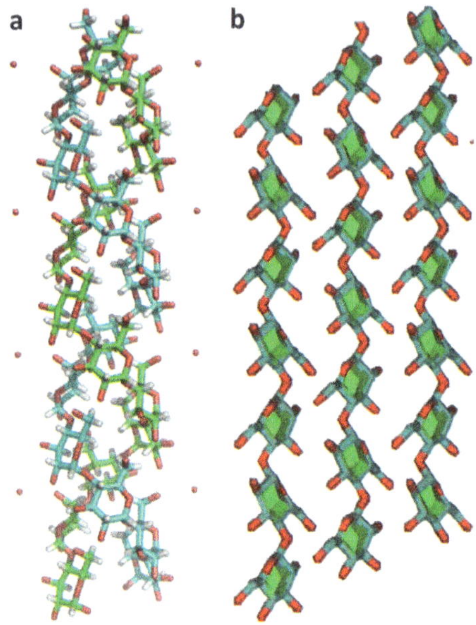

Fig. 3 Computer representation of three-dimensional structures of starch with Representation of the left-handed single chains that are parallel stranded in A-starch double helix (**a**), and structural orientation of cellulose chains of native cellulose 1β (**b**). Extracted from [80]

2.1.4 2.1.4. Cellulose

Cellulose is the most abundant natural polymer on earth obtained from the cell wall of plants and produced by several microorganisms. Cellulose is composed of a linear ß-1,4 glucose polymerized by hydrogen bonds in a 3D network (Fig. 3). Members of acetobacter family are the most abundant producer of BC with higher crystalline nanodomains than plant cellulose [18]. BC is used in a wide range of applications because of its nanofibrillar structure similar to ECM that make biopolymer useful for wound dressing, implants, and tissue engineering (*i.e.* dura mater, cartilage, dental, bones, etc.), contact lens, drug delivery, molecular immobilization, biocomposites, etc. [12, 72].

2.1.5 2.1.5. Chitosan (Chi)

Chitosan (Chi) is a product of partial deacetylation of chitin, a natural linear polymer found in many biological kingdoms. Chi is composed of two monomers of *N*-acetyl-2-amino-2-deoxy-D-glucose which are *N*-acetylated groups, and 2-amino-2-deoxy-D-glucose aminated residues [7]. Chi are polymers of different degree de deacetylation and present mucoadhesive properties, relevant for medical applications and often considered GRAS. Chi exerts antimicrobial, antidiabetic and anti-inflammatory activities, is bioadsorvable without toxic effects, and hydrolysis product are non-toxic. Biomedical applications of Chi are absorbable sutures, constituent of scaffold for tissue engineering (i.e. cartilage repair, wound dressing), and drug delivery devices, immobilization of molecules. However, Chi films and microdevices are easily breakables with low mechanical resistance and reason why is generally used in micro- and nanocomposites [2, 13].

2.1.6 Collagens

Collagens are a large family of structural proteins of the extracellular matrix with triple helix configuration extracted from mostly from connective tissue, skin and bones of many animals. Fourteen collagen types from different animal sources were so far identified that can make homo- and heterodimers. The most common collagens are fibrils and sheets depending on collagen types, globular domains, and triple helix. Collagen is biocompatible and biodegradable widely used in the development of many biomaterials such as beads, nanoparticles, films, fibers and sponges for cosmetics, health care and pharmaceutics [4, 79].

2.1.7 Gelatin

Gelatin is the product of partial hydrolysis of collagen and composed of a mixture of peptides and proteins. Gelatin is a transparent hydrocolloid with thermoreversible gelling properties commonly used in foods but also in many composites for biomedical applications. The chemical composition of gelatin depends on the collagen natural source and on the process of degradation [14, 45].

2.1.8 Hyaluronic acid (HA)

Hyaluronic acid (HA) is one main components of the extracellular matrix and very abundant in cartilage and vitreous. HA is a glycosaminoglycan composed of disaccharide repetitive units of ß-1,4-D-glucuronic acid and ß-1,3-*N*-acetyl-D-glucosamine linked with beta 1,4-inter-glycosidic linkage. HA is commonly used in cosmetics, tissue engineering and controlled release of drugs and DNA. Many HA structural modifications and composites were developed for biomedical applications [16, 52, 93].

2.1.9 2.1.9. Pectin

Pectin is complex heteropolysaccharide found in plant cell walls giving firmness and structure to the tissue. Pectins are commonly used in food as and stabilizing thickening, and gelling agent in several foods. Pectins are composed mainly by galacturonic acids linked by α-(1,4) glucosidic bonds and partially methylated at C6. The classification of pectins are based on the esterification degree into low and medium methoxylated (LMP and MMP) when the amount of methoxylated residues are lower than to 40% and 40% to 60% respectively. LMP and MMP are able to develop reversible gels in presence of multivalent cations (i.e. Ca, Mg, etc.). Pectins with esterification degree higher than 60% are considered high methoxylated pectins (HMP) are capable of make irreversible gels in presence of osmotically active molecules (i.e. glucose) and low pH. Several pectin gels and composites were proposed to be used in biomedical applications for the development of drug delivery devices and scaffolds [60, 66].

2.1.10 Silk

Silk is a natural protein fiber obtained from some lepidopteran larvae to form cocoons (i.e. *Bombyx mori*) or other arthropods (i.e. spider). *B. mori* heavy chain silk consists of alternate ß-sheet crystallizable domains and amorphous spacers with an average molecular weight of 350 kDa. and residual anionic charge ($pI_{fibroin} \approx 4$) [102, 108]. Natural silk is composed of silk fibroin and sericin. Purification of silk is required to eliminate sericin responsible of toxicity in mammalian cells. Studies of silk-based structures were applied in the development of scaffold, films, nano- and microparticles and composites for biomedicine [70, 71, 99].

2.1.11 Starch

Starch is the second abundant biopolymer in nature and GRAS polymer commonly used in the food and pharmaceutical industries. Starch is composed of two polymers amylose (20–30%) and amylopectin. Amylose is a linear polymer of α-1,4-D-glucose units with molecular weight lower than 1 million Da., generally soluble in aqueous media (Fig. 3). Amylopectin possess also α-1,4-D-glucose units and ramifications with α-1,6-D-glucose units with a molecular weight higher than millions. The starch granules size is in the range of 1 μm to 100 μm which depends on the biological source. The granules contain an amorphous core, mainly of amylose, surrounded by a helical pattern of semi-crystalline and amorphous rings made of both polymers. Several starch-based extended release devices have successfully retarded drug releases were recently reviewed in the literature [15, 23, 67].

2.2 Synthetic Biodegradable Polymers

2.2.1 Polylactic acid (PLA)

Polylactic acid (PLA) is a thermoplastic biocompatible aliphatic polyester obtained from the condensation of lactic acid with loss of water. In general, PLA is used to reinforced material in composites as reinforced polymer as well as copolymerized with other polymer to develop scaffolds, fibers, and nanoparticles [17, 21, 39, 87].

2.2.2 Polyglycolic acid (PGA)

Polyglycolic acid (PGA) was one of first polymers developed for biomedical applications. PGA is linear aliphatic biodegradable polyester with 45–55% crystallinity, melting point in the range of 220–230 °C depending on molecular weight, and glass transition temperature in the range of 35 °C to 40 °C. PGA fibers showed high tensile strength (890 MPa., higher than silk) and 8.40 GPa elasticity modulus. Despite these relevant properties, PGA has some drawback such as rapid hydrolysis in physiological environments because of the ester bonds, that could cause inflammatory response [95]. However, because of PGA high compatibility and biodegradability was approved by FDA for use in tissue engineering (i.e. neural regeneration, tendon repair bone surgery), degradable sutures and wound healing devices (Anderson et al. 2015), [9, 73].

2.2.3 Polycaprolactone (PCL)

Polycaprolactone (PCL) is a semicrystalline (up to 69%) polyester made of hexanoate. PCL is very soluble in organic solvents with melting point around 62 °C, and glass transition temperature of −55 °C. The molecular weight determines the main properties of PCL, at low MW the polymer is brittle but by increasing MW become soft and semicrystalline. PCL is commonly used as an additive in blends and composites with other polymers. PCL is generally used in esthetic surgery implants a filler for induction of collagen, but also as drug delivery devices and tissue engineering of long-term residence because of low degradation under physiological conditions [24, 56]. PCL was approved by FDA and EU for drug delivery and tissue engineering [43].

2.2.4 Polyethylene glycol (PEG)

Polyethylene glycol is a synthetic polyether with chain of variable length. PEG can be synthesized in types of structures such as comb-like, branched and star-type. The main application of PEG in biomedicine is named PEGylation in which PEGs molecules are covalently attached to a macromolecule and/or device's surface to extend the residence time in the body and reducing the corona effect of nanodevices [5]. However, there is some controversy regarding the induction immunogenic response in mammalians by pegylated macromolecules [35, 84, 96].

3 Lipid Carriers

The very hydrophobic or amphiphilic nature that lipids possess is what makes them a study material in the pharmaceutical industry to solve the aforementioned drawbacks regarding the solubility of new drugs. The first generation of this type of drug delivery systems, liposomes, were developed in the first decades of the twentieth century. Liposomes consist of an aqueous vesicle surrounded by a lipid bilayer, because of their biocompatibility, low toxicity, biodegradability and ability to encapsulate both lipophilic and hydrophilic substances liposomes are used in pharmaceutical, cosmetic, food and even farming industries extensively [3, 6]. While liposomes are currently both use and development, concerns about its stability, poor release kinetics, and the frequent need for the use of non-aqueous solvents in their preparation lead to the development of solid lipid nanoparticles (SLN).

SLN consist in an aqueous dispersion of solid lipids or a mixture of solid lipids stabilized by a surfactant. Solid nature of SLN makes them physically stable, gives them controlled release, protects encapsulated drug from degradation while conserve biocompatibility and low cytotoxicity found in liposomes. SLN can be prepared without any organic solvents and can be easily escalated from laboratory bench to pilot plant by e.g. a high-pressure homogenizer. However, over time, lipid matrix in SLN crystallize and expulse its load, this shortens shelf life as it makes it difficult to determine the free drug in the formulation at any given time.

Nanostructured lipid carriers (NLC) overcome this last issue by replacing part of the solid lipids with liquid lipids (oils) introducing entropy into the matrix and preventing crystallization. NLC also improve encapsulation efficiency and in some cases drug release profiles [88] (Fig. 4).

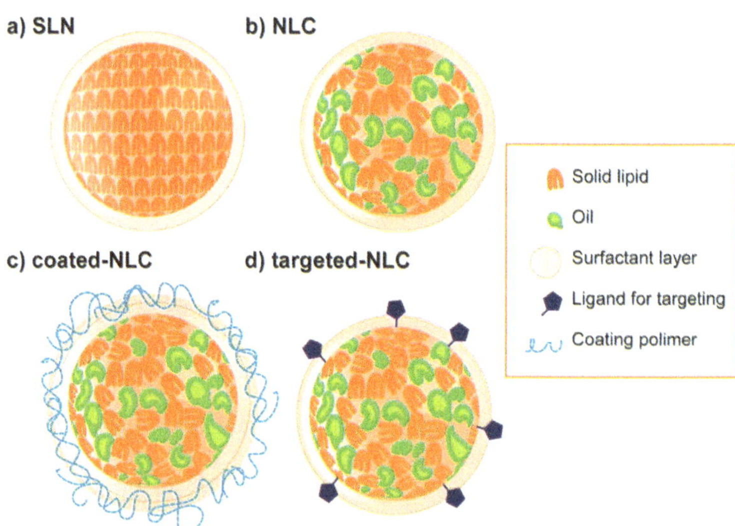

Fig. 4 Models for different solid lipid carriers. **a** SNL (Solid Lipid Nanoparticles) solid lipids in its core are highly ordered. **b** NLC (Nanostructured Lipid Carriers) Oil in its core disorganizes solid lipid molecules. **c** Coated-NLC, a polymer is used to cover NLC surface. **d** Targeted-NLC, ligand attached to carrier surface to ensure targeting

3.1 Lipids and Surfactants

Both lipids, solid or otherwise, and surfactants play the mayor role in drug loading, particle size, polydispersity index, toxicity, and of course, biodegradation of SLN and NLC.

Solid lipids can be free fatty acids, glycerol esters and waxes with stearic acid, tripalmitin and cetyl palmitate among the most commonly found in literature [41]. *Oils*, on the other hand can be free fatty acids like oleic acid, synthetic triglycerides like caprylic acid triglycerides or natural oils, like calendula, sunflower and canola oils.

In addition to integrating the carrier, lipids can play an active role in the system, e.g. stearic acid can help drugs penetrate cancer cells, and oils like linalool that can be used to nanostructure, possess therapeutic uses of their own [62, 78].

Surfactants on the other side, can be non-ionic, anionic, or cationic. Surface charge in SLN and NLC surface can in one hand decrease particle aggregation, by stablish repulsion between particles, but on the other hand, highly charged particles, specially positively, tend to be more toxic because of cell membrane destabilization [30]. However, non-ionic surfactants are by far the most utilized in SLN and NLC formulations, polysorbates and poloxamers in particular are used in more than half of the formulations in the literature. Poloxamers in addition to stabilize nanoparticles structure also provide SLN and NLC with a PEGylated surface that extend circulation lifetimes on the blood and reduce plasma proteins adsorption [40, 86].

3.2 Administration Routes

Due to the flexibility that exists when choosing the formulation components of SLN and NLC, added to the already discussed properties of biocompatibility, increased bioavailability of poorly water soluble drugs, and biodegradability, they possess a wide field of application in pharmaceutics and cosmetics. NLC and SLN have been proved to enhance *oral* bioavailability of drugs, the phenomena involved in this property include direct uptake of the particle by enterocytes, increased permeability by surfactants and decreased clearance [51]. Moreover, *topical delivery* of drug with lipid carriers has been tested for both local and transdermal delivery. While, some cosmetic NLC-based products for transdermal delivery of anti-aging products are currently in the market, therapeutic applications in this case, although very promising, have not reached the market due to the complexity of regulations that a pharmaceutical product must address before reach patients [37, 69]. *Pulmonary drug delivery* by lipid carriers has been reported, both as lipid dispersion and dry powder, it was demonstrated that carriers are retained in the lung with low cytotoxicity, and release therapeutic agents in a prolonged manner [98]. *Parenteral administration* of nanoparticles usually has the drawback that they are rapidly taken up by the mononuclear phagocytic system. As we detail this effect can be decreased to a certain extent by surface PEGylation, achieved in lipid carriers by stabilization with poloxamers. Beyond the extended circulation lifetime and prolonged release SLN and NLC present some targeting advantages.

Lipid carriers can cross blood–brain barrier by different mechanisms, because of their lipophilic nature they can diffuse through paracellular and transcellular routes, and in this way deliver drugs that normally would not affect patients brain [1]. Moreover, depending on size, SLN and NLC can experience enhanced permeability and retention effect (EPR effect), by this concept, tortuosity, and large fenestrations in cancerous tumors vessels, improve retention of nanoparticles and therefore direct carried drug into the tumor. However, EPR effect showed by itself clinical level efficacy only in some tumors [65].

3.3 Coating and Active Targeting

targeting consists in decorate particle surface with ligands to receptors that are highly expressed in the target tissue and thus increase the affinity and adhesion of the delivery systems to the target cells. Folate, has been used extensively in this type of experimental systems since in folate receptors (FOLR1) are over expressed in certain tumors, additionally when folate attaches FOLR1 a endocytosis response its initiated internalizing both folate and the conjugated nanoparticle [54, 68, 74, 77]. However, several targeting agents can be found in literature e.g. cannabidiol to target glioma cells [8], biotin and transferrin to target lung cancer [85, 92], among many others. This strategy can be improved further by "double targeting", that is, tailor carrier surface with two ligands [58, 105] or add surface modifications with active peptides to enhance cell penetration [55].

3.4 Loading

While lipid carriers possess a great encapsulation efficiency (EE) for lipophilic drugs, other small molecules can be encapsulated at acceptable ratios depending on drug structure and carrier nanostructuration. The exact mechanism by which this happens is not elucidated, but it is assumed that drug molecules could be concentrated at the lipid/surfactant interface or even on the carrier surface [41].

However, lipid carriers are not limited to small molecules drugs. Peptides and proteins have been encapsulated; grafted o even adsorbed. While not all lipid carrier preparation methods can be utilized in these cases, because of the fragility of the peptides to mechanical forces and temperature, this can be overcome, and encapsulation can also protect this drugs from external agents like pH or proteases [26, 33, 47, 61].

3.5 Commercial Prospects

All this together, presents lipid carriers, as highly flexible systems with easily modifiable characteristics, and great prospects of reaching clinical practice in next years. The low toxicity of SLN and NLC made them attractive to cosmetic industries, and there is a wide range of cosmetic products currently in the market, like coenzyme Q and vitamin A carriers and sunscreens. However, while pharmaceutical applications take longer to reach market, interest in pharmaceutical commercialization is already greater than in cosmetic applications among the shareholders. Among this interest, the most popular research topic its cancer drugs and therapies [109].

4 Biocarriers for Molecular Encapsulation and Delivery

The advantage of molecular encapsulation is based on preservation of the drug biological activity until the site of release is reached, reduce the amount of drug in the body because of molecular release on target, minimal circulating drug in the body, decrease undesirable side effects, and by biobased containers were degraded without toxic effects and some of them could help the drug effect. However, the vastest majority of bio-based and biodegradable polymer carriers employed for encapsulation shows at least one or more than one drawbacks such as do not have the properties required to keep the proper amount of the cargo molecule, unproper kinetic release, carrier instability, lack of cargo protection, some adverse reactions in the body (i.e. trigger immune response, tissue irritation, etc.). In this sense, the trend for encapsulation of active molecules to be used in tissue engineering and drug delivery is based on development of blends, coacervates, coatings, composites, and hybrid structures.

The main containers based on biopolymers can be classified based on their tridimensional structures in macro-systems such as films, microspheres and microtubes and on the other side nano-systems such as nanoparticles and nanotubes.

4.1 Biopolymeric Particles

Both nanoparticles and microparticles can be produced from biopolymers. Not only, polymer o polymers election but size, size distribution and shape of the particles, will influence their final properties. Depending on these properties, particles can be used to deliver active compounds by different administration routes, and release by erosion, diffusion, or cellular uptake.

Cellulose nanocrystals (CNC's) can be obtained from bulk cellulose by mechanical, chemical, or enzymatic treatment or by a combination of these treatments. CNC's showed encapsulation of significant amounts of doxorubicin and tetracycline even unmodified, however, with surface modification to increase hydrophobicity CNC's could load insoluble drugs as paclitaxel and docetaxel [49]. On this line, there are reports of CNC's modified with cationic porphyrin for photodynamic antibacterial activity, curcumin conjugated CNC's with antitumoral activity, pyrene conjugated CNC's as with applicability as chemosensor, surface oxidation with TEMPO to increase drug loading, folic acid conjugation to promote targeting and uptake in in cells that overexpress the folate receptor, CNC's surface tailoring with sulfate groups that demonstrated anticoagulant activity and functionalization to produce smart colon targeted drug delivery systems, recently CNC's were complexed with siRNA resulting in a nanocomposite with antitumoral properties [53, 82, 91]. While CNC's toxicity depends on its grafting, tailoring and modification, most systems show low cytotoxicity, taking this altogether, CNC's shows a flexible, biocompatible platform for a wide range of drug delivery and biomedical applications.

Chitosan is the most common polysaccharide in nanoparticle construction. Chitosan crosslinking can be achieved with polyanions like citric acid or tripolyphosphate. And chitosan nanoparticles showed to cross biological barriers and because of this, improve bioavailability of drugs and protect macromolecular drugs from degrading in biological media [29]. Particles of chitosan/alginate blends were used to carry DNA, these particles provided reduced toxicity, and comparable transfection efficiency with the commercial liposome formulation [31]. Moreover, amino groups in chitosan can be easily tailored, thiolated chitosan nanoparticles loading insulin were tested in diabetic rats, improving glucose and insulin levels compared to control treatments [89].

Pectin its commonly used in oral colon targeted systems. The rationale of its utilization it is based some of these properties. Its resistance to acids, which protects the active component from the acids of the upper digestive tract. Its responsiveness to pH, which makes the matrix hydration, and therefore, faster drug release at higher pH such as the pH of the colon. And finally, the fact that human digestive enzymes cannot degrade pectin, but it is degraded in the colon by the action of the microbiota that resides there, as a result of this, the drug in the drug delivery systems is site-specific released. However, pectin its normally used in blends, to tailor release performance. Drugs like vancomycin, 5-fluorouracil and methotrexate were encapsulated in pectin/chitosan blend particles [106]. Recently, gelatin/pectin blend was utilized to encapsulate a nanomicellar solution of poor water-soluble drug violacein [76].

Protein-based nanocarriers on the other side, possess an amphiphilic nature that helps them interact with both hydrophobic and hydrophilic drugs. Moreover, the high content of carboxyl and amino groups makes them excellent candidates for surface modification with ligands for targeting. Stability in protein based nanocarriers its normally achieved by crosslinking by formaldehyde, glutaraldehyde, and transglutaminase. Protein particles have been produced from several protein sources. Collagen and silk particles can be used in drug delivery systems, but as we describe later, their mayor interest is around scaffolds [50].

The study of albumin in the construction of drug delivery systems is of special attention, first, because it has hydrophobic pockets with the potential to carry drugs that are not very soluble in water, and second, because some cancerous tumors selectively capture it. Albumin nanoparticles have been used to load both small molecules and other proteins. Abraxane® is a system of these characteristics, albumin nanoparticles coat paclitaxel, an antineoplastic agent. This system was approved by the FDA in 2005 for the treatment of breast cancer and since then, its use has spread to pancreatic cancer and lung cancer treatment [42].

4.2 *Biopolymeric Films*

Biopolymeric films are used in general for the treatment of wounds and burns with or without molecular controlled release systems. For example, BC films were not able to entrap and sustain release small molecules such as antibiotic with MW lower than 1,000 Da. like common antibiotics available in the market. Also, BC because of water retention higher than 99% is not a good support of antibiotics with hydrophobic motif such as levofloxacin (XLogP3 = −0.4, sparingly soluble in water). However, HMP contain not only protonated regions of the polygalacturan polymer but also methylated ones which can serve as hydrophobic pockets to keep molecules with low water solubility. Hybrid BC-HMP developed by in situ technique containing levofloxacin and albumin were developed showing excellent antimicrobial activity against *Staphylococcus aureus*, the main microbial causative agent of skin infection in humans. The BC-HMP film showed an interesting platform for topical delivery of fragile molecules like proteins but also biocide agents [19].

4.3 *Tissue Regeneration*

Biomaterials used in tissue engineering should provide not only the material substrate to tissue to grow but also a combination of physical and chemical clues to proper tissue regeneration. Biodegradable polymers have here a desirable advantage, that is, that once an implant or transdermal patch is in place, no additional intervention is required to remove it.

Poly(lactide-co-glycolic acid) (PLGA), a copolymer of PGA and PLA, has been extensively studied and applied in tissue engineering because since it possess a better biological and mechanical performance compared to PGA o PLA alone [104]. However, Synthetic polymers like PLGA have a mayor limitation, they are not bioactive. Coating, blending, or loading bioactive compounds can help surpass this disadvantage. In bone regeneration, hydroxyapatite has the highest bioactivity. PLGA/hydroxyapatite composites allowed the formation of proper bone tissue in vivo [25]. Moreover, this scaffold can load growth factors like bone morphogenetic protein-2 and mimic extracellular matrix to further improve bone regeneration [107].

BC films are potential tool for tissue engineering because of high tensile strength, similar structure to ECM and importantly made of glucose units trough ß-1,4 glucosidic linkages which cannot be degraded by mammalian cells since they are not capable of producing ß-glucosidases. Recently an application of BC was reported for bone repair using hybrid BC-film. The BC film was modified with BC-COL to obtain a nanocomposite followed by reinforcement with carbonate apatite and containing osteogenic growth peptides. The *ex-situ* synthesis of the BC-COL nanocomposite containing apatite and the osteogenic peptides demonstrate enhanced primary osteoblastic activity in rats [81]. The BC-COL film system showed promissory results for bone regeneration and tissue engineering.

Other composites were recently studied, a hybrid hydrogel of chitosan and human amniotic membranes, a source of collagen, were developed to cartilage tissue regeneration, although in vivo studies are required to prove its efficiency [94].

4.4 Hybrid Polymeric-Lipid Carriers

In order to combine high carrying capacity and drug release profiles of NLC with BC compatibility and strength, [19] developed an implantable device as adjuvant therapy for breast cancer. The device consisted in a BC film loaded with doxorubicin encapsulated in NLC. Once implanted in mice it showed significant reduction of tumor growth, metastasis incidence and local drug toxicities.

On another approach, coating SLN and NLC can modify their surface to generate adhesion on specific tissues. Chitosan coated lipid carriers have been extensively reported as chitosan can increase adhesion to mucosal surfaces and open tight junctions on epithelial cells, his effect has been used to improve lipid carriers retention time in small intestine and thus enhance bioavailability of poorly water soluble drugs [32, 57] on this same rationale chitosan-coated NLC has been used to improve brain delivery via intranasal administration of glial cell derived neurotrophic factor, a protein with possible applications on Parkinson's disease [38]. Hyaluronic acid-coated NLC relied on a similar principle, that is electrostatic interaction, to target cancer cells and deliver paclitaxel [100] and chondroitin sulphate-coated NLC as well to target joints and deliver leflunomide an immunosuppressant drug used in rheumatoid arthritis treatment [103] (Fig. 4c).

On an inverse approach, polymeric nanoparticles (mostly composed of PLGA) be covered with lipidic molecules. In this way, polymeric nanoparticles can achieve larger circulation lifetime and biocompatibility because of biomimetic properties of lipids. These hybrid particles have been tested to carry small molecules like antibiotics and antitumorals but also siRNA and inorganic diagnostic agents. [27, 64].

5 Conclusions

Encapsulation of molecules on biobased structures is a trending strategy in the frame of Green Chemistry approach. The high availability of biopolymers and lipids with different physicochemical and biological properties, plus the recent improvement of analytical tools allow to develop novel carriers for the transport and delivery of different type of molecules, preserving their physical and biological integrity. Also, biopolymers showed stereoselectivity and chiral centers that allowed specific tailoring to specific target.

On the other side, the development of solid lipid carriers such as solid lipid nanoparticles and nanostructured lipid carriers allowed to combine many chemical and biological molecules to make a myriad of hybrid matrices with specific biological activities.

6 Directions of Future Research

The deep advance in the knowledge of structure and function of biomatrices based in the fast progress of analytical and molecular techniques could help with the development of novel multifunctional matrices for personalized therapies.

References

1. Agrawal M, Saraf S, Saraf S et al (2020) Recent strategies and advances in the fabrication of nano lipid carriers and their application towards brain targeting. J Control Release 321:372–415
2. Ahmad M, Manzoor K, Singh S, Ikram S (2017) Chitosan centered bionanocomposites for medical specialty and curative applications: a review. Int J Pharm 529:200–217. https://doi.org/10.1016/j.ijpharm.2017.06.079
3. Akbarzadeh A, Rezaei-Sadabady R, Davaran S et al (2013) Liposome: classification, preparation, and applications. Nanoscale Res Lett 8:102. https://doi.org/10.1186/1556-276X-8-102
4. Akturk O, Kismet K, Yasti AC, Kuru S, Duymus ME, Kaya F, Caydere M, Hucumenoglu S, Keskin D (2016) Collagen/gold nanoparticle nanocomposites: a potential skin wound healing biomaterial. J Biomater Appl 31:283–301. https://doi.org/10.1177/0885328216644536
5. Alcantar NA, Aydil ES, Israelachvili JN (2000) Polyethylene glycol-coated biocompatible surfaces. J Biomed Mater Res 51:343–351. https://doi.org/10.1002/1097-4636(20000905)51:3%3c343::aid-jbm7%3e3.0.co;2-d
6. Amoabediny G, Haghiralsadat F, Naderinezhad S et al (2018) Overview of preparation methods of polymeric and lipid-based (niosome, solid lipid, liposome) nanoparticles: a comprehensive review. Int J Polym Mater Polym Biomater 67:383–400
7. Anitha A, Sowmya S, Kumar PTS, Deepthi S, Chennazhi KP, Ehrlich H, Tsurkan M, Jayakumar R (2014) Chitin and chitosan in selected biomedical applications. Prog Polym Sci 39:1644–1667. https://doi.org/10.1016/j.progpolymsci.2014.02.008
8. Aparicio-Blanco J, Sebastián V, Benoit JP, Torres-Suárez AI (2019) Lipid nanocapsules decorated and loaded with cannabidiol as targeted prolonged release carriers for glioma therapy: in vitro screening of critical parameters. Eur J Pharm Biopharm 134:126–137. https://doi.org/10.1016/j.ejpb.2018.11.020
9. Ashammakhi N, Rokkanen P (1997) Absorbable polyglycolide devices in trauma and bone surgery. Biomaterials 18:3–9. https://doi.org/10.1016/s0142-9612(96)00107-x
10. Bassetti M, Ginocchio F, Mikulska M (2011) New treatment options against Gram negative organisms. Crit Care 15:215. https://doi.org/10.1186/cc9997

11. Bayón B, Rivero Berti I, Gagneten AM, Castro GR (2018) Biopolymers from wastes to high-value products in biomedicine. In: Singhania RR, Agarwal RA, Praveen Kumar R, Sukumaran RK (eds) Waste to wealth. Springer Nature, Singapore, pp1–44. ISBN 978–981–10–7430–1
12. Bielecki S, Kalinowska H, Krystynowicz A, Kubiak K Kołodziejczyk M, de Groeve M (2012) Wound dressings and cosmetic materials from bacterial nanocellulose. In: Gama M, Gatenholm P, Klemm D (eds) Bacterial nanocellulose: a sophisticated multifunctional material. CRC Press, Boca Raton, pp 157–174. ISBN 978-1-4398-6992-5
13. Bueloni B, Sanna D, Garribba E, Castro GR, León IE, Islan GA (2020) Design of nalidixic acid-vanadium complex loaded into chitosan hybrid nanoparticles as smart strategy to inhibit bacterial growth and quorum sensing. Int J Biol Macromol 161:1568–1580. https://doi.org/10.1016/j.ijbiomac.2020.07.304
14. Buie T, McCune J, Cosgriff-Hernandez E (2020) Gelatin matrices for growth factor sequestration. Trends Biotechnol 38:546–557. https://doi.org/10.1016/j.tibtech.2019.12.005
15. Builders PF, Arhewoh MI (2016) Pharmaceutical applications of native starch in conventional drug delivery. Starch 68:64–873. https://doi.org/10.1002/star.201500337
16. Burdick JA, Prestwich GD (2011) Hyaluronic acid hydrogels for biomedical applications. Adv Mater 23:H41–H56. https://doi.org/10.1002/adma.201003963
17. Brzeziński M, Seiffert S (2015) Monodisperse microspheres from supramolecular complexing polylactides. Mat Lett 161: 471–475. https://doi.org/10.1016/j.matlet.2015.09.008
18. Cacicedo ML, Cesca K, Bosio VE, Porto LM, Castro GR (2015) Self-assembly of carrageenin–$CaCO_3$ hybrid microparticles on bacterial cellulose films for doxorubicin sustained. J Appl Biomed 13:239–248. https://doi.org/10.1016/j.jab.2015.03.004
19. Cacicedo ML, Islan GA, Drachemberg MF, Alvarez VA, Bartel L, Bolzan AD, Castro GR (2018) Hybrid bacterial cellulose-pectin films for delivery of bioactive molecules. New J Chem 42: 7457–7467. https://doi.org/10.1039/C7NJ03973E
20. Campo VL, Kawano DF, Braz da Silva Jr D, Carvalho I (2009) Carrageenans: biological properties, chemical modifications, and structural analysis—a review. Carbohydr Polym 77:167–180. https://doi.org/10.1016/j.carbpol.2009.01.020
21. Casalini T, Filippo Rossi F, Castrovinci A, Perale G (2019) A perspective on polylactic acid-based polymers use for nanoparticles synthesis and applications. Front Bioeng Biotechnol 7:259. https://doi.org/10.3389/fbioe.2019.00259
22. Castro GR, Bora E, Panilaitis B, Kaplan DL (2006) Emulsan coated alginate microspheres as a new vehicle for protein delivery. Chapter 2. In: Khemani K, Scholz C (eds) Degradable polymers and materials. American Chemical Society Symposium Series № 939. Oxford University Press, Washington, USA, pp 14–29. https://doi.org/10.1021/bk-2006-0939.ch002. ISBN-10: 084123972X
23. Chen J, Chen L, Xie F, Li X (2019) Drug delivery applications of starch biopolymer derivatives. Springer Nature Singapore Pte Ltd, p 147. ISBN: 978–981–13–3657–7
24. Christen MO, Vercesi F (2020) Polycaprolactone: how a well-known and futuristic polymer has become an innovative collagen-stimulator in esthetics. Clin Cosmet Investig Dermatol 13:31–48. https://doi.org/10.2147/CCID.S229054
25. Cieślik M, Mertas A, Morawska-Chochół A et al (2009) The evaluation of the possibilities of using PLGA co-polymer and its composites with carbon fibers or hydroxyapatite in the bone tissue regeneration process—in vitro and in vivo examinations. Int J Mol Sci 10:3224–3234. https://doi.org/10.3390/ijms10073224
26. Courant T, Bayon E, Reynaud-Dougier HL et al (2017) Tailoring nanostructured lipid carriers for the delivery of protein antigens: physicochemical properties versus immunogenicity studies. Biomaterials 136:29–42. https://doi.org/10.1016/j.biomaterials.2017.05.001
27. Dave V, Tak K, Sohgaura A et al (2019) Lipid-polymer hybrid nanoparticles: synthesis strategies and biomedical applications. J Microbiol Methods 160:130–142
28. Dimond PF (2005) Using nanotechnologies in biotech and medicine. Gen Eng News 25: 21. Website: https://www.genengnews.com/magazine/21/using-nanotechnologies-in-biotech-and-medicine/

29. Divya K, Jisha MS (2018) Chitosan nanoparticles preparation and applications. Environ Chem Lett 16:101–112
30. Doktorovová S, Kovačević AB, Garcia ML, Souto EB (2016) Preclinical safety of solid lipid nanoparticles and nanostructured lipid carriers: current evidence from in vitro and in vivo evaluation. Eur J Pharm Biopharm 108:235–252. https://doi.org/10.1016/j.ejpb.2016.08.001
31. Douglas KL, Piccirillo CA, Tabrizian M (2006) Effects of alginate inclusion on the vector properties of chitosan-based nanoparticles. J Control Release 115:354–361. https://doi.org/10.1016/j.jconrel.2006.08.021
32. Du Q, Chen J, Yan G et al (2019) Comparison of different aliphatic acid grafted *N*-trimethyl chitosan surface-modified nanostructured lipid carriers for improved oral kaempferol delivery. Int J Pharm 568:118506. https://doi.org/10.1016/j.ijpharm.2019.118506
33. Dumont C, Bourgeois S, Fessi H, Jannin V (2018) Lipid-based nanosuspensions for oral delivery of peptides, a critical review. Int J Pharm 541:117–135
34. Duran N, Bosio VE, Frate L, Castro GR (2014) Nanobiotechnology: prospects and perspectives. In: Govil JN (ed) Recent developments in biotechnology, vol 10. Chapter 1024. Studium Press (Tx, U.S.A.), pp 1–48. ISSN: 1–62699–015–8
35. Ehlinger C, Spear N, Doddareddy R, Shankar G, Schantz A (2019) A generic method for the detection of polyethylene glycol specific IgG and IgM antibodies in human serum. J Immunol Methods 474:112669. https://doi.org/10.1016/j.jim.2019.112669
36. Fakruddin Md, Hossain Z, Afroz H (2012) Prospects and applications of nanobiotechnology: a medical perspective. J Nanobiotechnol 10:31. http://www.jnanobiotechnology.com/content/10/1/31
37. Garcês A, Amaral MH, Sousa Lobo JM, Silva AC (2018) Formulations based on solid lipid nanoparticles (SLN) and nanostructured lipid carriers (NLC) for cutaneous use: a review. Eur J Pharm Sci 112:159–167
38. Gartziandia O, Herrán E, Ruiz-Ortega JA et al (2016) Intranasal administration of chitosan-coated nanostructured lipid carriers loaded with GDNF improves behavioral and histological recovery in a partial lesion model of Parkinson's disease. J Biomed Nanotechnol 12:2220–2230. https://doi.org/10.1166/jbn.2016.2313
39. Ghalia MA, Dahman Y (2017) Biodegradable poly(lactic acid)-based scaffolds: synthesis and biomedical applications. J Polym Res 24(5):74. https://doi.org/10.1007/s10965-017-1227-2
40. Göppert TM, Müller RH (2005) Protein adsorption patterns on poloxamer- and poloxamine-stabilized solid lipid nanoparticles (SLN). Eur J Pharm Biopharm 60:361–372. https://doi.org/10.1016/j.ejpb.2005.02.006
41. Gordillo-Galeano A, Mora-Huertas CE (2018) Solid lipid nanoparticles and nanostructured lipid carriers: a review emphasizing on particle structure and drug release. Eur J Pharm Biopharm 133:285–308
42. Gradishar WJ (2006) Albumin-bound paclitaxel: a next-generation taxane. Expert Opin Pharmacother 7:1041–1053. https://doi.org/10.1517/14656566.7.8.1041
43. Guarino V, Gentile G, Sorrentino L, Ambrosio L (2017) Polycaprolactone: synthesis, properties, and applications. In: Mark HF (ed) Encyclopedia of polymer science and technology, pp 1–36. https://doi.org/10.1002/0471440264.pst658
44. Hart ML, Do DP, Ansari RA, Rizvi SAA (2013) Brief overview of various approaches to enhance drug solubility. J Develop Drugs 2:3. https://doi.org/10.4172/2329-6631.1000115
45. Hathout RM, Metwally AA (2019) Gelatin nanoparticles. In: Weissig V, Elbayoumi T (eds) Pharmaceutical nanotechnology. Methods in molecular biology, vol 2000. Humana, New York, NY. https://doi.org/10.1007/978-1-4939-9516-5_6
46. Hodgson J (2001) ADMET—turning chemicals into drugs. Nat Biotechnol 19:722–726. https://doi.org/10.1038/90761
47. Islan GA, Tornello PC, Abraham GA et al (2016) Smart lipid nanoparticles containing levofloxacin and DNase for lung delivery. Design and characterization. Colloids Surfaces B Biointerfaces 143:168–176. https://doi.org/10.1016/j.colsurfb.2016.03.040
48. Islan GA, Saldaña SM, Chevalier MT, Alvarez VA, Castro GR (2018). Nanotechnology and drug delivery—Chapter 7. In: Talevi A, Quiroga P (eds) ADME processes in pharmaceutical sciences. Springer, Switzerland, pp 135–165. ISBN: 978–3–319–99593–9

49. Jackson JK, Letchford K, Wasserman BZ et al (2011) The use of nanocrystalline cellulose for the binding and controlled release of drugs. Int J Nanomed 6:321–330. https://doi.org/10.2147/ijn.s16749
50. Jain A, Singh SK, Arya SK et al (2018) Protein nanoparticles: promising platforms for drug delivery applications. ACS Biomater Sci Eng 4:3939–3961
51. Jaiswal P, Gidwani B, Vyas A (2016) Nanostructured lipid carriers and their current application in targeted drug delivery. Artif Cells, Nanomedicine Biotechnol 44:27–40
52. Kim H, Shin M, Han S, Kwon W, Hahn SK (2019) Hyaluronic acid derivatives for translational medicines. Biomacromolecules 20:2889–2903. https://doi.org/10.1021/acs.biomac.9b00564
53. Kim YM, Lee YS, Kim T et al (2020) Cationic cellulose nanocrystals complexed with polymeric siRNA for efficient anticancer drug delivery. Carbohydr Polym 247:116684. https://doi.org/10.1016/j.carbpol.2020.116684
54. Kularatne SA, Low PS (2010) Targeting of nanoparticles: folate receptor. Methods Mol Biol 624:249–265. https://doi.org/10.1007/978-1-60761-609-2_17
55. Lee SG, Kim CH, Sung SW et al (2018) RIPL peptide-conjugated nanostructured lipid carriers for enhanced intracellular drug delivery to hepsin-expressing cancer cells. Int J Nanomedicine 13:3263–3278. https://doi.org/10.2147/IJN.S166021
56. Li Z, Mei S, Dong Y, She F, Li Y, Li P, Kong L (2020) Functional nanofibrous biomaterials of tailored structures for drug delivery—a critical review. Pharmaceutics 2:522. https://doi.org/10.3390/pharmaceutics12060522
57. Ling Tan JS, Roberts CJ, Billa N (2019) Mucoadhesive chitosan-coated nanostructured lipid carriers for oral delivery of amphotericin B. Pharm Dev Technol 24:504–512. https://doi.org/10.1080/10837450.2018.1515225
58. Liu D, Liu F, Liu Z et al (2011) Tumor specific delivery and therapy by double-targeted nanostructured lipid carriers with anti-VEGFR-2 antibody. Mol Pharm 8:2291–2301. https://doi.org/10.1021/mp200402e
59. Loftsson T, Brewster ME (2010) Pharmaceutical applications of cyclodextrins: basic science and product development. J Pharm Pharmacol 62:1607–1621. https://doi.org/10.1111/j.2042-7158.2010.01030.x
60. Marras-Marquez T, Peña J, Veiga-Ochoa MD (2015) Robust and versatile pectin-based drug delivery systems. Int J Pharm 479:265–276. https://doi.org/10.1016/j.ijpharm.2014.12.045
61. Matougui N, Boge L, Groo AC et al (2016) Lipid-based nanoformulations for peptide delivery. Int J Pharm 502:80–97
62. Miao J, Zhang XG, Hong Y et al (2012) Inhibition on hepatitis B virus e-gene expression of 10–23 DNAzyme delivered by novel chitosan oligosaccharide-stearic acid micelles. Carbohydr Polym 87:1342–1347. https://doi.org/10.1016/j.carbpol.2011.09.022
63. Murata Y, Sasaki N, Miyamoto E, Kawashima S (2000) Use of floating alginate gel beads for stomach-specific drug delivery. Eur J Pharm Biopharm 50:221–226. https://doi.org/10.1016/s0939-6411(00)00110-7
64. Navarro L, Mogosanu DE, Ceaglio N, Luna J, Dubruel P, Rintoul I (2017) Novel poly (diol sebacates) as additives to modify Paclitaxel release from poly (lactic-co-glycolic acid) thin films. J Pharm Sci 106:2106–2114. https://doi.org/10.1016/j.xphs.2017.05.012
65. Nichols JW, Bae YH (2014) EPR: evidence and fallacy. J Control Release 190:451–464
66. Noreen A, Nazli Z-I-H, Akram J, Rasul I, Mansha A, Yaqoob N, Iqbal R, Tabasum S, Zuber M, Zia KM (2017) Pectins functionalized biomaterials; a new viable approach for biomedical applications: a review. Int J Biol Macromol 101:254–272. https://doi.org/10.1016/j.ijbiomac.2017.03.029
67. Odeniyi MA, Omoteso OA, Adepoju AO, Jaiyeoba KT (2018) Starch nanoparticles in drug delivery: a review. Polim Med 48:41–45. https://doi.org/10.17219/pim/99993
68. Oshiro-Junior JA, Sato MR, Boni FI et al (2020) Phthalocyanine-loaded nanostructured lipid carriers functionalized with folic acid for photodynamic therapy. Mater Sci Eng C 108:110462. https://doi.org/10.1016/j.msec.2019.110462
69. Pandey A (2020) Solid lipid nanoparticles: a multidimensional drug delivery system. In: Daima HK, PN N, Ranjan S, et al (eds) Nanoscience in medicine, vol 1. Springer International Publishing, Cham, pp 249–295

70. Parker RN, Wu W, McKay TB, Xu Q, Kaplan DL (2019) Design of silk-elastin-like protein nanoparticle systems with mucoadhesive properties. J Funct Biomater 10(4):49. https://doi.org/10.3390/jfb10040049
71. Parker RN, Cairns D, Wu WA, Jordan KE, Guo C, Huang W, Martín-Moldes Z, Kaplan DL (2020) Smart material hydrogel transfer devices fabricated with stimuli-responsive silk-elastin-like proteins. Adv Healtc Mater 9(11):e2000266. https://doi.org/10.1002/adhm.202000266
72. Picheth G, Pirich CL, Sierakowski MR, Woehl M, Sakakibara CN, de Souza CF, Martina A, da Silva R, Alves de Freitas R (2017) Bacterial cellulose in biomedical applications: a review. Int J Biol Macromol 104:97–106. https://doi.org/10.1016/j.ijbiomac.2017.05.171
73. Pihlajamäki H, Tynninen O, Karjalainen P, Rokkanen P (2008) Enveloping bioabsorbable polyglycolide membrane and immobilization in Achilles tendon repair: a comparative experimental study on rabbits. J Orthop Res 26:264–270. https://doi.org/10.1002/jor.20504
74. Poonia N, Kaur Narang J, Lather V et al (2019) Resveratrol loaded functionalized nanostructured lipid carriers for breast cancer targeting: Systematic development, characterization and pharmacokinetic evaluation. Colloids Surfaces B Biointerfaces 181:756–766. https://doi.org/10.1016/j.colsurfb.2019.06.004
75. Rehm BHA (2009) Alginates: biology and applications. In: Rehm BHA (ed). Springer, Dordrecht, p 269. ISSN 1862–5576
76. Rivero Berti I, Rodenak-Kladniew B, Perez AA et al (2019) Development of biocarrier for violacein controlled release in the treatment of cancer. React Funct Polym 136:122–130. https://doi.org/10.1016/j.reactfunctpolym.2019.01.001
77. Rivero Berti I, Rodenak-Kladniew B, Onaindia C et al (2020) Assessment of in vitro cytotoxicity of imidazole ionic liquids and inclusion in targeted drug carriers containing violacein. RSC Adv 10:29336–29346. https://doi.org/10.1039/d0ra05101b
78. Rodenak-Kladniew B, Islan GA, de Bravo MG et al (2017) Design, characterization and in vitro evaluation of linalool-loaded solid lipid nanoparticles as potent tool in cancer therapy. Colloids Surfaces B Biointerfaces 154:123–132. https://doi.org/10.1016/j.colsurfb.2017.03.021
79. Salvatore L, Gallo N, Natali ML, Campa L, Lunetti P Madaghiele M, Blasi FS, Corallo A, Capobianco L, Sannino A (2020) Marine collagen and its derivatives: versatile and sustainable bio-resources for healthcare. Mat Sci Eng C 113:110963. https://doi.org/10.1016/j.msec.2020.110963
80. Sarkar A, Pérez S (2012) PolySac3DB: an annotated data base of 3 dimensional structures of polysaccharides. BMC Bioinformatics 13:302. https://doi.org/10.1186/1471-2105-13-302
81. Saska S, Teixeira LN, de Castro Raucci LMS et al (2017) Nanocellulose-collagen-apatite composite associated with osteogenic growth peptide for bone regeneration. Int J Biol Macromol 103:467–476. https://doi.org/10.1016/j.ijbiomac.2017.05.086
82. Seabra AB, Bernardes JS, Fávaro WJ et al (2018) Cellulose nanocrystals as carriers in medicine and their toxicities: a review. Carbohydr Polym 181:514–527
83. Selmi TA, Verdonk P, Chambat P, Dubrana F, Potel JF, Barnouin L, Neyret P (2008) Autologous chondrocyte implantation in a novel alginate-agarose hydrogel: outcome at two years. J Bone Joint Surg Br 90:597–604. https://doi.org/10.1302/0301-620X.90B5.20360
84. Schellekens H, Hennink WE, Brinks V (2013) The immunogenicity of polyethylene glycol: facts and fiction. Pharm Res 30:1729–1734. https://doi.org/10.1007/s11095-013-1067-7
85. Shao Z, Shao J, Tan B et al (2015) Targeted lung cancer therapy: Preparation and optimization of transferrin-decorated nanostructured lipid carriers as novel nanomedicine for co-delivery of anticancer drugs and DNA. Int J Nanomedicine 10:1223–1233. https://doi.org/10.2147/IJN.S77837
86. Shirodkar RK, Kumar L, Mutalik S, Lewis S (2019) Solid lipid nanoparticles and nanostructured lipid carriers: emerging lipid based drug delivery systems. Pharm Chem J 53:440–453. https://doi.org/10.1007/s11094-019-02017-9
87. Singhvi MS, Zinjarde SS, Gokhale DV (2019) Polylactic acid: synthesis and biomedical applications. J Appl Microbiol 127:1612–1626. https://doi.org/10.1111/jam.14290

88. Subramaniam B, Siddik ZH, Nagoor NH (2020) Optimization of nanostructured lipid carriers: understanding the types, designs, and parameters in the process of formulations. J Nanoparticle Res 22
89. Sudhakar S, Chandran SV, Selvamurugan N, Nazeer RA (2020) Biodistribution and pharmacokinetics of thiolated chitosan nanoparticles for oral delivery of insulin in vivo. Int J Biol Macromol 150:281–288. https://doi.org/10.1016/j.ijbiomac.2020.02.079
90. Takagi T, Ramachandran C, Bermejo M, Yamashita S, Yu LX et al (2006) A provisional biopharmaceutical classification of the top 200 oral drug products in the United States, Great Britain, Spain, and Japan. Mol Pharm 3:631–643. https://doi.org/10.1021/mp0600182
91. Tang L, Lin F, Li T et al (2018) Design and synthesis of functionalized cellulose nanocrystals-based drug conjugates for colon-targeted drug delivery. Cellulose 25:4525–4536. https://doi.org/10.1007/s10570-018-1904-2
92. Taymouri S, Alem M, Varshosaz J et al (2019) Biotin decorated sunitinib loaded nanostructured lipid carriers for tumor targeted chemotherapy of lung cancer. J Drug Deliv Sci Technol 50:237–247. https://doi.org/10.1016/j.jddst.2019.01.024
93. Tiwari S, Bahadur P (2019) Modified hyaluronic acid-based materials for biomedical applications. Int J Biol Macromol 121:556–571. https://doi.org/10.1016/j.ijbiomac.2018.10.049
94. Toniato TV, Stocco TD, Martins D, Dos S, et al (2020) Hybrid chitosan/amniotic membrane-based hydrogels for articular cartilage tissue engineering application. Int J Polym Mater Polym Biomater 69:961–970. https://doi.org/10.1080/00914037.2019.1636249
95. Ulery BD, Nair LS, Laurencin CT (2011) Biomedical applications of biodegradable polymers. J Polym Sci B Polym Phys 49:832–864. https://doi.org/10.1002/polb.22259
96. Veronese FM, Mero A (2008) The impact of PEGylation on biological therapies. BioDrugs 22:315–329. https://doi.org/10.2165/00063030-200822050-00004
97. Vlasova IM, Saletsky AM (2010) Raman spectroscopy in investigations of secondary structure of human serum albumin at binding of nanomarkers of fluorescein family. Laser Phys 20:1844–1848. https://doi.org/10.1134/S1054660X10170160
98. Weber S, Zimmer A, Pardeike J (2014) Solid Lipid Nanoparticles (SLN) and Nanostructured Lipid Carriers (NLC) for pulmonary application: a review of the state of the art. Eur J Pharm Biopharm 86:7–22
99. Wu H, Liu S, Xiao L, Dong X, Lu Q, Kaplan DL (2016) Injectable and pH-responsive silk nanofiber hydrogels for sustained anticancer drug delivery. ACS Appl Mater Interfaces 8:17118–17126. https://doi.org/10.1021/acsami.6b04424
100. Yang XY, Li YX, Li M, et al (2013) Hyaluronic acid-coated nanostructured lipid carriers for targeting paclitaxel to cancer. Cancer Lett 334:338–345. https://doi.org/10.1016/j.canlet.2012.07.002
101. Yermak IM, Barabanova AO, Aminin DL et al (2012) Effects of structural peculiarities of carrageenans on their immunomodulatory and anticoagulant activities. Carbohydr Polym 87:713–720. https://doi.org/10.1016/j.carbpol.2011.08.053
102. Yucel T, Lovett ML, Giangregorio R, Coonahan E, Kaplan DL (2014) Silk fibroin rods for sustained delivery of breast cancer therapeutics. Biomaterials 35:8613–8620. https://doi.org/10.1016/j.biomaterials.2014.06.030
103. Zewail M, Nafee N, Helmy MW, Boraie N (2019) Coated nanostructured lipid carriers targeting the joints—an effective and safe approach for the oral management of rheumatoid arthritis. Int J Pharm 567:118447. https://doi.org/10.1016/j.ijpharm.2019.118447
104. Zhang F, King MW (2020) Biodegradable polymers as the pivotal player in the design of tissue engineering scaffolds. Adv Healthc Mater 9:1901358
105. Zhang J, Xiao X, Zhu J et al (2018) Lactoferrin- and RGD-comodified, temozolomide and vincristine-coloaded nanostructured lipid carriers for gliomatosis cerebri combination therapy. Int J Nanomedicine 13:3039–3051. https://doi.org/10.2147/IJN.S161163
106. Zhang W, Mahuta KM, Mikulski BA et al (2016) Novel pectin-based carriers for colonic drug delivery. Pharm Dev Technol 21:127–130. https://doi.org/10.3109/10837450.2014.965327

107. Zhao X, Han Y, Li J et al (2017) BMP-2 immobilized PLGA/hydroxyapatite fibrous scaffold via polydopamine stimulates osteoblast growth. Mater Sci Eng C 78:658–666. https://doi.org/10.1016/j.msec.2017.03.186
108. Zhou CZ, Confalonieri F, Medina N, Zivanovic Y, Esnault C, Yang T et al. (2000) Fine organization of Bombyx mori fibroin heavy chain gene. Nucleic Acids Res 28:2413–2419. https://doi.org/10.1093/nar/28.12.2413
109. Zhou X, Huang L, Porter A, Vicente-Gomila JM (2019) Tracing the system transformations and innovation pathways of an emerging technology: Solid lipid nanoparticles. Technol Forecast Soc Change 146:785–794. https://doi.org/10.1016/j.techfore.2018.04.026

Further Reading

110. Poole-Warren L, Martens P, Green R (eds) (2018) Biosynthetic polymers for medical applications. Elsevier, Amsterdam, The Netherlands, p 338
111. Siepmann J, Siegel RA, Rathbone MJ (eds) (2012) Fundamentals and applications of controlled release drug delivery. Springer, NY, USA, p 607
112. Thakur R, Singh R, Laverty G, Donnelly R (eds) (2018) Hydrogels, design, synthesis and application in drug delivery and regenerative medicine. CRC Press, FL, USA, p 353
113. Raza Shah M, Imran M, Ullah S (eds) (2017) Lipid-based nanocarriers for drug delivery and diagnosis. Elsevier, Kidlington, U.K., p 354
114. Castro GR, Kumar A, Nguyen TA, Qi X, Yasin G (eds) (2020) Nanomaterials for biocatalysis. In press. Elsevier, Amsterdam
115. Alonso MJ, Csaba NS (eds) (2012) Nanostructured biomaterials for overcoming biological barriers. Royal Society of Chemistry, Cambridge, UK, p 643

Containers for Food Packaging Application

Bisma Jan, Qurat ul eain Hyder Rizvi, Rafeeya Shams, Aamir Hussain Dar, Ishrat Majid, and Shafat Ahmad Khan

1 Introduction

Since ages food packaging has become important element in the history of mankind's and now packaging technology is used to ensure food to move securely over longer distance in a nutritious and healthy condition at the time of utilization [1]. Packaging is defined as the combination of processing, food science and preservation. Almost 53% sales of the packaging industry's are to the food industry. In present society, the chief aim of food packaging is to prevent the food contamination and external injuries, contain the product to permit for transport, and gives information to consumers about the package contents.

Food packaging is defined as enclosing of *food* in a suitable enclosing system that provides protection, resistance against tampering, basic biological, chemical, or physical requirements. It provides nutrition label facts and new knowledge about *food* being presented for sale. Packaging provides a means to merchandise, distribute protect, preserve and market the foods. Major role of packaging is to make sure that the products reach to the customers in wholesome and safe condition

B. Jan
Department of Food Technology School of Interdisciplinary Sciences, Jamia Hamdard University, New Delhi, India

Q. H. Rizvi
Department of Food Technology, Dr. Khim Singh Gill Akal College of Agriculture, Eternal University, Barusahib, HP, India

R. Shams
Division of Food Science and Technology, Sher-e-Kashmir University of Agricultural Sciences & Technology of Jammu, Jammu, India

A. H. Dar · I. Majid (✉) · S. A. Khan
Department of Food Technology, Islamic University of Science and Technology Awantipora Kashmir, Awantipora, India

J. Parameswaranpillai et al. (eds.), *Micro- and Nano-containers for Smart Applications*, Composites Science and Technology,
https://doi.org/10.1007/978-981-16-8146-2_5

without quality compromising [2]. The most significant function of packaging is the multi-functional protection that it provides in the long path from the manufacturer till the final customer [3]. There is a growing need and demand for changing food packaging due to changes in increased globalization in the food supply, consumer habits, and availability of new technologies, distribution chain developments, new materials, and increasing environmental constraints.

During stone age people used to collect food from their surrounding environment where they found it and eat in the same form devoid of worrying about storing and protecting their food and put leftovers in natural vessels such as hollow tree trunks and ornamental zucchini, hollow stone pieces, shells of sea creatures, trees, leaves and shells. However later on when humans started to settle in communities, containment and storing of food became necessary. The most primitive forms of packaging in ancient times were reed baskets, hollowed-out tree, skins, leaves, limbs, and vessels of earthenware as containers. Big vessels of ceramic and amorphous were utilized from 1500 B.C. to 500 A.D. to transport other products and wine commercially throughout the Mediterranean. Glass-making started in 7000 B.C. as an outcome of pottery, and was developed first in Egypt in 1500 B.C. Metal cans were originally manufactured for snuff, to provide excellent barrier against moisture absorption. Nonetheless metals did not become a familiar packaging substance until other metals, thinner gauges, stronger coatings and alloys were finally industrialized. One of the "new metals' that is allowed to be used in packaging was tin and this method of tin plating was discovered in Bohemia and in the early 1800s canning operation developed by Nicholas Appert to protect food [4]. He used bottles of glass having cork which was secured while heating but the glass bottles seemed to be delicate and were rapidly substituted with metal cans. However, in late 1800s wrapping the product for the consumer was a common practice. Newspaper and cloth sacks were usually used for holding flour or sugar and wrapping meat. Later, cellophane and packaging were introduced into packaging industry. Plastic among packaging materials like metal, glass, and paper is the youngest packaging materials. However, plastics namely styrene vinyl chloride and cellulose nitrate, came into existence in the 1800s however it was not utilized in any packaging untill twentieth century [5]. In the 1950s, some new plastic substances were used in packaging films. Among all plastics polyethylene was the first plastic used for food packaging. "flexible packaging" can be defined as the paper made from papyrus stems in ancient Egypt which is the oldest form. Chinese people used mulberry bark sheets as a flexible packaging material [6]. From mid-twentieth century there have been various main packaging developments that shaped the food preservation and modern packaging world. Some examples of this packaging are microwaveable packaging, MAP, aseptic packaging and processing, and intelligent, smart and active packaging systems. The first food product that was individually packed and sold was biscuits in the 1890s and was developed by the National Biscuit Co.

Today the highly urbane food packaging industry that supplies modern society these days is way better than the primitive packaging activities of previous times. Food processing innovations has sited new demands on packaging and has increased the development of new packages. Such as, the desire to thermally processed low acid foods in rigid plastic cans has lead to the growth of multi-layered plastic substances that maintain their barrier s even after being thermally processed in steam. Generally, the packaging materials are usually grouped into five categories.

- Glass packages (bottle, demijohn, jar, ampoule, vial).
- Metal-based packaging materials (cans, aerosol containers, barrels, lids, rings).
- Paper-based packaging materials examples are wrapping paper, box, cardboard box, cases, tape, etc.).
- Plastic packaging materials (containers, bags, drums, barrels, lids, tapes, etc.) and
- Wood packaging materials (chest, cases, pallet).

1.1 Importance of Packaging

Packaging is an activity closely related to social development and its past extends to the beginning of mankind. Packaging plays a major role in preventing the material from damage caused due to moisture, micro-organisms, toxins, air etc. Packaging mainly prevents leaking or spilling of product. Nutrition, sell-by dates and ingredients are essential for both to the consumer and to grocery stores. Bar codes are used to track sales and inventory. It also has an important function in the marketability and preservation of a product. Co-manufacturer benefit includes speed, cost, innovation and quality. Packaging is an important means for preserving quality of food, lowers wastage of food and preservatives that are used in food. It also serves the essential function in protecting product against damage and gives essential information to marketers and consumers. It is said to be an essential medium where manufacturers only can give information about features of product, ingredients, and nutritional composition. Packaging alternatives made from various substances in different shapes and sizes of forms majority of food and drink products to customers. Companies and consumers take numerous essential issues into consideration while selecting packaging for food like the package capability to retain the food's freshness, quality, correct product identification, pleasant appeal and image, easiness to distribute and store, and environmental effect [7].

Food packaging is not an arbitrary practice by any means. Each and every food item requires packaging suitable to protect its structure. For example, cooked food and fruits require packaging in airtight tins or containers whereas meat is sealed using transparent plastic flexible film. Similarly, jars, pots, and regular boxes are used appropriately for various food items. Food packaging is a very crucial step in food technology and needs to be taken care extensively. There is a proper way in packing the food product and food in such a way that it is prevents contamination of environment and at the same time helps in marketing and distribution. Traditionally design of food packaging has been attained so as to store and transport foods. Packaging of foods intends to detach the food from the external environment by shielding it from external contamination. Packaging hinders food spoilage minimize or decrease damage of product, preserve the valuable components of the product, pro long shelf life, maintain products quality etc. Packaging also provides the information of essential data of food, and assists consumers about purchases [8]. The major purpose of any package is to hold the food and helps in storage distribution and handling throughout the chain from producer to the final customer. Though, there are typically different levels of packaging. Primary function of packaging is to prevent the deterioration of components from various factors like light, heat, oxygen, moisture, enzymes pressure, microbes, odors, dirt, insects and dust. However, the secondary function of food packaging lies within the product marketing and this function of food packaging has a direct effect on the product sales. Interestingly, every traditional packaging aims to reduce the contact between packaging and food, while list preserving the wholesomeness of food [9].

A primary package is that type of package where contained product is in direct contact with the package like glass jars, metal cans or plastic pouches. Ideally primary package must not produce any material which is detrimental to the health of the ultimate consumer. Secondary packaging is the packaging that holds together the individual units of a good. Corrugated box is an example of secondary package. Due to modernization, process of transportation and handling has become more complicated, and this type of packaging is usually secured and palletized by strapping with metal or, generally, by stretch—or shrink-wrapped film to offer another level of packaging called tertiary packaging. These pallet loads are then packed into huge containers of metal like quaternary packaging is used for transportation over longer distances like by sea, land or air. One of the essential functions of container is to prevent the product from any type of wear and tear, spoilage, loss, or damage that maybe caused throughout the distribution chain. Any kinds of physical damage like by vibrational shocks that causes bruises during stacking or transportation in a warehouse could be prevented by proper packaging. Losses caused due to moisture gain or loss, light, and dust, specially UV light, that deteriorates some light-sensitive materials could also be minimize by packaging. It also protects contents in the package against temperature fluctuations during the transportation of frozen and chilled foods.

Beyond the primary roles of food packaging like transportation, protection and preservation, packaging also affects consumer expectations by graphic characteristics (like, labeling, colour) and its structural characteristics (like type of substance, shape) [10]. Some studies have also reported that the visual and haptic characteristics (any form of interaction involving touch) of food packages are capable to affect peoples' sensory like hedonic responses and (taste experience) to the constituents by cross-modal correspondence, i.e., the individuals' capability to make relations amongst different sensory domains, like touch and taste, or taste and vision [11].

2 Different Forms of Packaging Containers Utilized in Food Industries

2.1 Glass Containers

Glass is described as "an amorphous, inorganic creation of fusion that is cooled to a firm and rigid condition devoid of crystallizing" [12]. Even though if glass is regarded as a synthetic substance, originally produced from common elements present in the earth's surface naturally before the world was occupied. Pressed glass was formed when the molten glass was pressed mechanically into an engraved or plain mold by a plunger, facilitating glassmakers to spread the thick, molten glass fastly throughout the mold before it gets solidified. Glass is primarily made up of varying degree of calcium oxide (5–12%), aluminium oxide (0.5–3.0%), sodium oxide (12–17%), silica (59–80%) potassium oxide, barium oxide, magnesium oxide and boric oxide. Due to the occurrence of silica, glass is having high melting point. Containers of glass are often transparent and translucent because they lack free electrons unlike metals. However, colored glasses (amber glass) that selectively pass the light can also be made. Glass, which is usually transparent, has a wide area of use for technological and decorative purposes [13]. A research by international industry analysts said that in the world growth of glass packaging industry is estimated to persist regardless of the emergence of packaging alternatives in North America and Europe [14]. Glass packaging is a kind of packaging commonly used in wide range of foods ranging from heat treated or pressure-packed solid and powdered products to liquid products [15]. Two kinds of glass containers are primarily utilized in food packaging that are bottles and other is jars and pots. Bottles are with narrow necks, jars and pots are with broad openings. Glass closures were once famous for screw action stoppers having prung metal fittings and rubber washers for pressurized bottles, like for vacuumized jars and carbonated beverages, and for heat preserved vegetables and fruits but are not used nowadays. Also, confectioneries were stored in glass jars having ground glass friction fitting stoppers.

Glass is extensively chosen as a packaging material in food industries because of its characteristics such as, recycling, non-reacting, reusing nature. It preserves beverages and foods for longer durations and protects it from contamination. An extensive quantity of foods are packed and stored in glass containers e.g., spices, dairy products, dry mixes, instant coffee, sugar preserves (marmalades and jams), processed baby foods, syrups, spreads, processed fruit, fish, vegetables, meat products, condiments and mustards etc. Glass bottles are also extensively in beverage industries for example for spirits, beers, wines, liqueurs, soft drinks and mineral water. Besides its immense role in food and beverage industries it is widely used in packaging of pharmaceuticals, chemicals, cosmetics etc. The increasing utilization of glass packaging in pharmaceutical and food packaging is crucial factor driving this market. Attributes like recycling, preservation for long-term, reuse are notable factors predictable to drive the glass packaging market for the forecast period. Advanced use of technology and innovative designs are main features of glass packaging. Plastic usage over glass packaging and high cost are the main attributes restraining the global glass packaging market.

2.1.1 Method of Forming Glass Containers

Blowing—in the cavity of a mold the compressed air is used to form the molten glass.

Drawing—in this process through dies molten glass is pulled that forms the soft glass into bottles that can contain drug known as ampoules.

Pressing—mechanical force is used that forces or presses the liquified glass adjacent to the ride of a mold and forms glass.

Casting—centrifugal or gravitational force is employed to commence the development of liquefied glass in the cavity.

Jars, glass jars, containers lids and bottles are prepared of different sizes, shapes and colors as per necessities.

2.1.2 Classification of Glass Containers According to Colour

White flint

This type of glass is also known as clear glass derived from basic composition of all other glasses ie., soda, lime and silica. Its composition comprises of silica (SiO_2) 72%, from high purity sand, soda (Na_2O) 12%, from soda ash, lime (CaO) 12%, from limestone (calcium carbonate), and alumina (Al_2O_3).

Pale green (half white)

This type of glass is formed when faintly fewer pure substances are used. Under this condition iron content (Fe_2O_3) increases and a light green glass is formed. Chromium oxide (Cr_2O_3) is further added to produces somewhat denser blue green colour.

Dark Green

This colour is also attained by the adding of iron oxide and chromium oxide.

Amber

Amber is attained by melting a mixture containing iron oxide with powerfully reduced conditions. Amber glasses are usually employed for light susceptible products and have also UV protection properties.

Blue

By adding cobalt to low iron glass blue glass is produced.

2.1.3 Classification of Glass Containers on the Basis of Shape

Glass packaging is commonly cylindrical in shape used in food packaging. Cylindrical containers may be either bottles with narrow rims or jars with wide mouths. There are also glasses and demijohns, as well as vials and ampoules that are generally used in the field of pharmaceutical industries. In terms of use, the cylindrical shape is close to the spherical shape that provides maximum durability and is easily manufactured.

2.1.4 Classification and Role of Glass in Pharmaceuticals Industries (Nasa Parveen 2014)

According to degree of hydrolytic/chemical resistance to water attack glass containers are categorized into Type I, Type II, Type III and Type IV glasses.

Neutral glass/Borosilicate glass (Type I glass containers)

It contains 10% boric oxide, 80% silica, and little percentage of aluminium and sodium oxide. It is inert chemically due to the occurrence of boric oxide and have elevated hydrolytic resistant. It has very little expansion coefficient and high thermal shock characteristics. It is used for alkalis and strong acids and is also suitable packaging material for most preparations in pharmaceutical industries.

Soda-lime-silica glass/De alkalized soda lime glass/treated soda-lime glass (Type II glass containers)

This one is advanced type of Type III glass container with an eminent hydrolytic resistance resultant from appropriate behavior of the internal surface of a type III glass with sulfur. This type of glass can be easily moulded due to minimal melting point when compared to Type I glass. They are well suited for neutral aqueous and most acidic preparations whether non-parenteral or parenteral.

Regular soda lime glass (Type III glass containers)

This is an unprocessed soda lime glass having average chemical resistance. It possesses 10% calcium oxide, 75% silica, 15% sodium oxide, minute quantity of aluminium oxide, potassium oxideand magnesium oxide. Magnesium oxide minimizes the temperature necessary during moulding while as aluminium oxide impacts chemical durability. They are employed in packaging non-aqueous powders and preparations for parenteral except freeze-dried preparations.

General-purpose soda lime glass/NP glass (Type IV glass containers)

It has minimal hydrolytic resistance. This kind of container is not used for goods which are autoclaved as it enhances reaction rate of erosion. It is utilized to store oral dosage and topical products.

2.1.5 Closures for Glass Containers

Glass jars and bottles need development and closures with consumer-friendly, cost-effective is biggest challenge that packaging industries are facing. Closures are devices that are used to seal or close container like bottles, jar, jug and can be either a cover, cap, plug, lid. For glass packaging containers closures usually used are plastic or metal. However, for wine and spirits cork is still widely. It is very important to choose closure carefully. Eg., very long closure can make leakage because of the force generated upon it from heat produced due to internal gas pressure. The types of closure available are as follows:

Normal seal

These types of seals are utilised for non-pressure/non-vacuum packed products. They consist of composite closures of plastic/aluminium foil for products, like powdered and granular products like coffee dairy products including milk, milk powders and yoghurts.

Vacuum seals

These are metal closures having composite liner to fasten onto the glass rim. It may be capable of being twisted or pressed into place at that time vacuum is produced by using a steam flushing the headspace. They tend to be fairly readily to in-bottle pasteurization and retort sterilization and diameter sizes range from 28 to 82 mm. In case of liquids sizes are normally in the range of 28–40 mm.

Pressure seals

These seals may be of plastic or metal having a complex liner to form the seal, and may also be twisted or pressed into place.

2.1.6 Advantages of Glass

Glass packaging is widely preferred by the consumers due to its abundant positive aspects. Today drink products and most foods are presented to customers in a variety of packaging substitutes prepared from various substances, in different shapes and sizes. The abundance in positive aspects of using glass as food packaging has made it widely preferred by consumers. In one of study to compare the interaction of packaging materials with foodstuffs; it was determined that glass packaging had the highest value in terms of preserving the quality of food and was the only packaging material that could guarantee that there would be no toxic substance transfer from the package to the food. Three food products ie., wine, honey, chocolate cream were packaged in glass and three products yogurt, milk, juice plastic packaged. After blindly tasted by consumers it was found that hedonic response results were eminent for glass packages rather than for plastic packages. Likewise, consumers reported that glass packaged products are more used and preferred than products in plastic packages [16].

Recently extensive efforts have been made to substitute plastic with glass for packaging. Glass is the possible environmental chemically inert in nature. Additionally, it does not react with chemicals and does not corrode. It also does not require additional special coating. From hygiene and health viewpoints, it is said to have an optimal material for packaging food and drink products [17].

Foods in glass containers may remain live comparatively for longer durations than plastics. Fermented product meekiri (yoghurt made from buffalo milk), was kept in glass, clay and plastics. It was observed that the food remained live for longer durations in glass than other two packagings [18]. Glass containers are shaped into various decorative formats, like, silk screening, labelling, printing with ceramic inks, acid etching sleeving, inorganic and organic colour coating.

2.1.7 Disadvantages of Glass

Glass is perceived as healthy and thus preferred by people. According to the statistical research it is has been said that high quality perception is attached to consumers for glass-packaged products. Besides being highly utilized also have its negative sides.

During the production of glass energy resources like electricity and fossil fuels are used in higher quantities that results in 3 times more air pollution when compared to production of plastic. As glass lets the light to pass through it results the discoloration of the food. The fact that it breaks upon effects such as an impact, heat, shock and excessive internal pressure, may lead to problems in production, transportation, storage and sale.

2.2 *Plastics Packaging*

Plastic is defined as the material that is obtained when the bonds between carbon and hydrogen, oxygen, nitrogen and other organic or inorganic elements, called monomers, are broken from their simple structured molecule groups and formed into the long and chained structure called polymers. Plastic materials are shaped into a diversity of valuable products, that are moldable, easy to print and heat sealable [1]. The concept of plastic production started 100 years ago, however, when compared with other materials, plastic is observed to be a relatively younger packaging material and its production has rapidly increased towards the end of the twentieth century. Plastic technology results in multiple production of packages with different colors, types and technical characteristics and almost all foods can be packaged with plastic [19]. Worldwide plastic industries have a growth rate of 4–5% yearly, wheras in India it is 15%. India is likely to become second or third largest plastic consumer from the present 9th rank in the world by the end of this decade. Plastics are used as container compounds, flexible packaging and containers. They are the second most commonly used form of packaging by weight and in provisions of value they stand first. Some examples are listed below:

- Rigid plastic containers, like jars, trays, pots, bottles and tubs.
- Flexible plastic films usually in form of heat-sealable flexible lids, sachets bags, pouches.
- Plastics typically paperboard for beverage packaging cartons.
- Foamed or expanded plastic for applications wherever different sort of rigidity and the capability to resist compression is needed.
- Plastic caps and lids are utilized in such diaphragms, closures on glass and plastic jars to offer tamper evidence and product protection.
- Films used in shrink wrapping and cling.
- Films used as labels for jars, inks and coatings.

2.2.1 Types of Plastic in Food Packaging

Polyethylene (PE)

PE is simplest form of plastic by structure and is prepared by adding of polymerization of ethylene gas in a pressure reactor at elevated temperature. It can be extruded into tough films, with good water vapour and moisture barrier. They have not predominantly high barrier to gases like CO_2 and O_2, oils and fats when co related with other plastics, even though barrier characteristics enhances with density. Its resistance to heat is minimal than other counter plastics utilized in packaging with a melting point of around 120 °C, which rises as the density increase. Polyethylene is poor conductor of electricity and was initially employed as an insulator in the early 1940s.

LDPE

LDPE or low density PE can be effortlessly extruded as a tube and blown to stretch upto three times the original area. This type of plastic is sometimes recycled. It is a tough plastic and tends to be both flexible and durable. Items like plastic grocery bags, sandwich bags, cling-film and squeezable bottles are made from LDPE.

LLDPE

LLDPE or linear low-density PE film has similar density of LDPE. It is superior to LDPE in terms of impact and tensile strength and also in puncture resistance. It is mainly used for pillow packing for liquid foods and liquid milk.

HDPE

HDPE or high-density PE is extruded in the thinnest gauges and is toughest grade. This packing material is utilized for boil-in-the-bag role. To increase its heat sealability co extrusion of HDPE with LDPE is required to attain peelable seals where the polymer layers can be prepared to segregate simply at the interface of the co-extrusion. These plastic products are very safe and do not transmit any chemicals into drinks or foods. These products are also recycled. Items made from this HDPE include motor oil, containers for milk conditioners and shampoos, soap bottles, bleaches and detergents.

Polypropylene (PP)

It is a polymer produced under high heat and pressure utilizing Ziegler–Natta kind catalysts to generate a linear polymer with protruding methyl (CH2) groups. Polypropylene is recycled occasionally. PP is strong enough and can withstand higher temperatures. It is used to make lunch boxes, yogurt pots, margarine containers, prescription bottles and syrup bottles. PP is often used to make plastic bottle caps. The elevated melting point of PP (160 °C) makes appropriate for role wherever thermal resistance is required like in microwave packaging and hot filling. PP is chemically non reactive and resistant many chemicals, both inorganic and organic. It has good barrier properties to water vapour with good has fat and oil resistance.

Polyethylene Terephthalate (PET)

The concept of PET was brought into existence by Pepsi in 1970. The first patent was given to the bottle issued in 1973. PET sometimes absorbs flavors and odors from drinks and foods that are stored in them. These plastic items are recycled. Many common household items are made from PET plastic such as clothing, beverage bottles, rope, medicine, jars and carpet fibre. PET is a medium oxygen barrier but with O_2 and water vapour it becomes high barrier when gets metalized with aluminum. This is also utilized for bag-in-box liquids and vacuumised coffee. Sometimes it is coated with EVA on either side to create extremely efficient seals. Metallised PET is utilized as a susceptor in microwaveable packaging.

Ethylene vinyl acetate (EVA)

EVA is a copolymer of ethylene along with vinyl acetate. It is analogous to PE in several aspects and is utilised when fused with PE in numerous ways. The characteristics of the fusion are based on the quantity of the vinyl acetate constituents. PVdC with EVA is a harsh strong-barrier film that is employed in vacuum packed huge meat cuts and for bag-in-box liners for wine with metallised PET. Advanced EVAs are used as peelable laminates on lidding materials, like OPP, aluminium foil, paper and PET.

Polyamides (PA) polyvinyl chloride (PVC)

PVC is used for all types of tiles and pipes and is sometimes recycled. It is usually found in plumbing pipes. This type of plastic should not come in contact with food items because it is harmful when ingested. Unplasticised PVC (uPVC) has functional characteristics however, is brittle, firm material and alteration are required for it to be used effectively. When plasticizers are added its flexibility is achieved, reduces surface friction with slip agents.

Polyvinylidene chloride (PVdC)

Polyvinylidene chloride (PVdC) PVdC is a copolymer of vinylidene chloride and vinyl chloride. PVdC was introduced initially by Dow Chemical, who named it Saran TM. PVdC has good barrier properties to gases and water vapour, oils and fats and is heat sealable. As a consequence of the excellent odor and gas barrier it is employed to prevent aroma and flavour susceptible foods from ingress of volatile contaminants and loss of flavour. It is utilized in flexible packaging in numerous behaviors.

Polystyrene (PS)

Polystyrene is generally recycled but is hard to perform its recyclability. Items like plastic packing foam, food boxes, plastic cutlery and disposable coffee cups are made from it. It is co-polymerized to enhance its properties. PS has several packaging applications and could be extruded as a monolayer plastic film, co-extruded as an injection moulded, thermoformable plastic sheet and foamed to provide a range of pack types. PS is processed simply by means of foaming to make an inflexible light weight substance that has good thermal insulation properties and impact shield.

Ethylene vinyl alcohol (EVOH)

EVOH is a co-polymer of vinyl alcohol and ethylene. It is interrelated to polyvinyl alcohol (PVOH) that is a water-soluble synthetic polymer having brilliant emulsifying, adhesive and film-forming characteristics. It has good barrier properties against grease, oil, oxygen and organic solvents. Extrusion lamination of paperboard together PE and EVOH is utilised for aseptically packed fruit juices and ultra-high temperature (UHT) milk. EVOH layer offers an O_2 barrier as a substitute for aluminium foil.

2.2.2 Laminates and Co-extrusion

Plastic materials could be formed both as combination of more than one plastic and as a single film. There are various means of combining plastics such as co-extrusion and lamination. Lamination includes fusion of two or more than two plastics to another material like paper or aluminium and that bonding is generally obtained by employing of adhesives. Both films are passed between high pressure rollers when the adhesives are applied to first film. Lamination utilizing laser in place of adhesives has also been employed for thermoplastics [20]. Combining of two or more than two layers of plastics by utilizing adhesives is known as lamination. Numerous kinds of coating processes are accessible to apply plastic coatings by deposition, extrusion, from aqueous or solvent mixtures or by vacuum deposition. Because lamination and co-extrusion combine multiple materials hence recycling is not easy.

2.2.3 Flexible Packaging

Recently high-profile industry survey, 80% of the brand owners agreed that packaging has a major impact on brand value. Packaging has traditionally been a crucial differentiator, a medium to impress and drive home the 'customer first' sentiment. In 2018, the discipline of flexible packaging has emerged as an indispensable resource, particularly in a quality intensive environment such as the beverage and food industry. In the food and beverages industry, where packaging has always presented a risk proposition, flexible food packaging affords a simpler, lighter, durable and eco-friendly alternative to the traditional packaging options. Flexible food packaging technology is exploding with great manufacturing ideas and better results for consumers. Sustainability, convenience goals, cost and appearance are merging like never before. It offers beautiful branding options, improved consumer convenience, less debris hauled to local landfills and better product freshness.

2.2.4 Recent Innovations in Plastic Food Packaging

An integral role is played by food packaging in assuring food safety and maintaining food quality, prolonging of its storage existence, and packaging in plastic has been exceptionally helpful in attaining the functions of food packaging. Moreover, the globe has observed the devastating influence of the utilization of plastic packages on earth in the last decade. Today the drinking water is polluted with microplastics. As per a study by the Ellen MacArthur Foundation, if plastic packaging is not stopped by industries, excess plastic will get enumerated in the sea, by mass more than fish in the year 2050. Amongst the several types of plastic packaging, today one of the most widely used plastic is plastic cling film.

2.2.5 Silk Fibroin Thin Films for Perishables

In edible food sensors and adhesives silk fibroin that is a protein have been widely utilized. It is same as that collagen and comprises of hydrophobic bonds distributed with hydrophilic and acidic regions. The character of silk fibroin gives it exceptional character, like hydrophobicity, conformability and polymorphism. This is used for perishable foods like bananas, apples, strawberries and tomatoes to be well preserved. Researchers at Tufts University formulated a suspension that is water-based which gathers on the exterior of foods that have been immersed into the suspension. This thin film decreases gas diffusivity and dehydration, thus increases its shelf life [21] reported that silk fibroin coatings which are thicker on the food surface led to low carbon dioxide production and oxygen permeability and ensuing in firmer food.

2.2.6 Tree Fiber Films and Crab Shells

Researchers at the Georgia Institute of Technology have established an exclusive way to create an edible film for food packaging by combining cellulose, chitin and biopolymers. The film is prepared by sprinkling various layers of cellulose from trees and chitin from the crabshells to produce a flexible film as compared to PET packaging. In the course of the growth of the packaging film, it is noticed that cellulose nanofibers are negatively charged and chitin nanofibers are positively charged. These characteristics form a good film while employed in discontinuous layers. Contrasting plastic films, crab shells films and tree fiber and are crystalline in nature and make it hard for the diffusion of gas molecules. Tree fiber films and crab shells were found to reduce O_2 penetration by 73%. Though further investigation is required to carry out or to enhance the film's capability to decrease water vapor permeability. This research suggested that these films could provide proper packaging for foods with delicate or rough irregular surfaces.

2.2.7 Gluten Films for Packaging

This film has been studied to exhibit excellent mechanical attributes when the gluten pH is 5 and concentration is 12.5%. In a new research that concentrated on enhancing the mechanical and barrier attributes of gluten films, research was carried out by scientists that shows the impact of oligomeric procyanidins, gum, flaxseed and lauric acid on the mechanical properties, peroxide value and water vapor permeability of the film. The enhanced gluten films i.e., gluten films integrated with these materials make them more uniform, dense and hydrophobic. These films were used in packaging oil, salt, seasonings and vegetables. Plastics that are sourced from renewable feedstocks such as organic polymers, sugar cane and corn that degrade within a year are called as Biobased plastics. European Bioplastics organization defines bioplastics as plastics that depend on recyclable resources or as plastics which are compostable/biodegradable.

Various bioplastics used include:

2.2.8 Corn-Derived Polylactic Acid

Polylactic acid (PLA) which is completely made from corn is among the mostly available used biobased plastics. It is made either by ring opening polymerization of lactide or direct polycondensation of lactic acid. The main manufacturer of PLA is Nature Works. However, many Japanese and European companies also manufacture PLA. It can be employed in the form of cases for electronic devices, filaments for 3D printing, and building materials like wall flooring and coverings. Figure 1 shows the schematic of bioplastics material co-ordinate system.

2.2.9 Thermoplastic Starch (TPS) from Starchy Vegetables

Starch is a simple biodegradable natural resource and extensively available. To generate a plastic starch-based film, plasticizers sorbitol, glycerol is required. These plasticized substances having significance of mechanical and thermal energy) are called thermoplastic starch (TPS) and comprise a substitute for polystyrene (PS) that is a supplementary substance, TPS uses starchy vegetables (like potatoes) as a feedstock. Compared to cellulose that comes in fibers, starch is a branched molecule

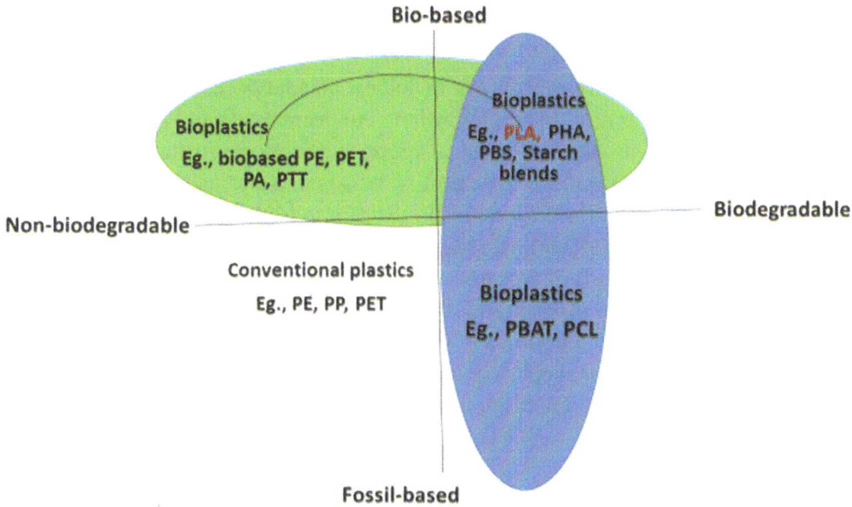

Fig. 1 Bioplastics material co-ordinate system

and comes in crystalline form. In the presence of plasticizer, it destructures the starch molecule and this process is known as gelatinization.

Cellulose

Cellulose is originated by a delignification from cotton linters or wood pulp and is widely spread natural polymer. Being an eco-friendly polysaccharide can be dissolved in a carbon disulphide and sodium hydroxide mixture to attain cellulose xanthate and afterwards recast into an acid solution (sulphuric acid) to create a cellophane film.

2.2.10 Limitations of Bioplastics

Thermal instability, brittleness, difficult heat sealability, reducing melting strength, oxygen and high-water vapor permeability limit the use of PLA films for several food packaging applications [22]. Brittleness (because of melting temperatures and high glass transition, impact resistance, stiffness, thermal and poor instability are also limiting factors for application of PHB/PHAfilms as food packaging [23]. This results in growing exploration on improving the bioplastics utility that is coatings (employing an extra thin layer of an additional substance on top of the biobased films) [24].

2.2.11 Indian Regulations for Plastics

The scientific committee of the Bureau of Indian Standards (BIS) sets the National Standards. The main aim of this organization is to promulgate and prepare National Standards. Assortment and then using of proper packaging substance and the part it exhibits in protection and preservation of non-perishable and perishable goods has turn out to be very significant. In the distribution and transportation of these products, economic factors and safe handling also plays a great function. To yield hygienic and safe food stuff to the customers, the finished products and raw material handling and the use and choosing of a clean and safe packaging substance are of chief significance. For commodities which are perishable in nature like fish, fruits, vegetables and meat, that are kept stored and then shifted under frozen or refrigeration conditions, appropriate safeguarding of cold chain is vital.

2.3 Paper

An entwining association of cellulose fibers derivative of wood are used to make paperboard and papersheet materials by using sulfite and sulfate. These fibers then are bleached or pulped and treated with chemicals like antimicrobial pesticide (slimicides) and intensifying agents to make the paper product. Paperboards and paper are generally employed in milk cartons, corrugated boxes, wrapping papersacks and bags, and folding cartons-. Barrier properties of paper are limited, so to be used as a food container it has to be treated with wax, lacquer or resin. The various kinds of paper utilized in food packaging are listed as below [1].

Kraft papers are formed via sulfate treatment method. This paper is present in various types: unbleached, and bleached white. The ordinary kraft is commonly used for wrapping and bags and is strongest of all papers. It is also utilized for packaging of sugar, flour and dried vegetables and fruits.

Sulfite papers are weaker and lighter as compared to kraft paper, it is glazed to enhance manifestation and to improve its oil resistance and wet strength. It can be used in laminates with foil or plastic and also be coated for advanced print quality. Furthermore, it is employed to make wrappers or tiny bags for packaging confectionary.

Greaseproof paper is paper prepared by a method called as beating where the fibers of cellulose undergo a prolonged hydration period than usual and it results in breaking up of making them gelatinous. These fine thin fibers are compactly packed to make an exterior which only is resistant to oils rather to wet agents. This paper is involved to pack oily foods, candies, snacks and cookies.

2.4 Metals

The abundant identified elements are metals. They are mainly flexible and multipurpose among all types of packaging available. Five essential features of metal make them feasible for packaging, incomparable stiffness (it makes metal packages resistant to probable stress), compactness, high density, malleability (the moulding possibility of metals to any form) and high thermal conductivity (pasteurization sterilization). Moreover, these types of packagings exhibit barrier properties, good formability, decorative potential and physical protection, overall consumer acceptance and recyclability. Various types of metals such as tin plate, aluminum, stainless steel and tin free steel, packaging material which is metal based both semi-rigid and rigid forms like retort pouches, foil wraps and cans are mostly used for food packaging applications. The two metals mostly utilized in packaging are aluminum and steel. Aluminum is light in weight, silvery white in colour metal derivative of bauxite ore, where it is present in combination with oxygen as alumina and is used to make laminated paper, cans, plastic packaging and foil. Addition of manganese and magnesium to aluminum enhances its potency characteristics. Metal containers is the budding packaging material in India. Categories like beverage cans (soft drinks, beer, health drinks etc.) and aerosol packaging (air fresheners, deodorants and insect repellents) are widely packed in metal containers. When metal containers come in contact with a broad range of various beverages and foods a better knowledge of metals as packaging materials is highly needed mainly in significant functions [25]. Packaging materials of metal based present excellent barrier characteristics and thus are used extensively in food packaging applications. They are utilized in several types as closures like for composite cans and glass bottles. However, the main wellbeing and consumer security concerns of metal packaging include bisphenol A migration, iron, aluminium, bulging of- cans, nickel, lead, mercury, cadmium, tin dissolution, corrosion and blackening. These are reactive to food products, so lacquering is done to avoid metal–food migration and dealings of metal component. Metal packaging components have higher recyclability and minimal global warming potential because of its magnetic characteristics that aids in easier separation [26].

2.4.1 Aluminum as a Packaging Material

Aluminum is the 3rd most abundant element found. It contributes about (8.1%) after oxygen and silicon. Process production of aluminium involve the transfiguration of alumina to aluminum hydroxide in a sodium hydroxide solution at temperature of 175 °C. Bauxite is the chief aluminum-containing mineral but various others contain aluminum oxide with oxides of silicon, other metals and iron. Aluminum packaging is the expensive food packaging material to almost all coated steels as its production is not cheap. Eg., 4 kg of mineral is taken by single kg of aluminum needs about seven to tenfold the energy required by the exact mass of iron. Aluminum is well known for its two physical properties that are light in weight. Being light in weight it is mostly used as light weight packaging material in its pure form for soft-drink cans, pet foods and sea foods and its strength is improved by addition of manganese [27]. A prominent role is played by aluminum foil in modern food packaging. In households' aluminum foil is chiefly used as wrappings (like for chocolate), and trays. Aluminum packaging is also used for aseptic cartons, laminated tubes bottle capsules, pouches, wrappings, containers, push-through blisters, trays and lids.

It has excellent mealibility. These properties make it allow to cast in any forms, and mould it into various shapes making aluminum packaging manufacturing very simple. Thermal properties of AL are exceptionally good that makes it useful for food packaging applications. Eg., AL can withstand both low and high temperatures. At sub-zero temperature its aluminum strength improves and does not become fragile even at ultra-freezing temperatures. Furthermore, it has high thermal conductivity and is superior among all other counterpart metals employed in packaging and can withstand high temperature treatments like pasteurization or sterilization treatment. One of the best qualities of aluminum is its easily converted into new products that has recyclability property however, pinholes are present in recycled aluminum and its non-magnetic characteristics create separation glitches.

2.4.2 Metal Drums Caps and Closures

Drum is basically an industrial packaging substance mostly used for transporting food materials and bulk storage in both semi-processed and processed forms. These are mostly used for storing and handling concentrates well corrosive products and powders in food industry. Closures and metal caps provide stability and rigidity and help in portraying a superior product image to the consumers in the market. Moreover, the growing demand and growth in international trade of food and beverage can significantly attributed to effective packaging solutions, such as metal closures and caps. These closures and caps are additionally utilized for sealing of beverage glass bottles i.e., pilfer proof closures (provide protection from leaking or spilling of the liquids inside). In canning industry twist-off, pry off, deep-press and press-twist are the regularly used metal closures. Baby food closures are metal vacuum closures and are designed to ensure easy reseal and easy open. Lids on bottles and glass jar tops are usually composed of 20–70 μ aluminium foils having polyvinyl chloride or polyacrylic coatings. Vacuum closures such as twist-off type, lug caps etc. are utilized for pickles, jams, baby foods and dry powders [28].

2.4.3 Retort Pouches

Aluminum foil packaging is usually used in trays and retort pouches. Aluminum foil laminate pouches are usually made of 3 layers. First layer is an outer layer of polyamide (PA) or polyethylene terephthalate (PET), that confers handling strength then a middle layer of aluminum foil that is light, gas, microorganisms and water vapour impermeable and last an inner layer of polyethylene (PE), polypropylene

(PP) or co-polymers that imparts strength. This layer is readily heat sealed and inert to foods. Main benefits of this laminate of flexible plastic and metal foils to cans are excellent barrier characteristics. Even at lesser densities permits ease of disposal and fast heating could be flattened. Moreover, separation of multiple layers of these retort pouches for reusing and its vulnerability to break during quick cooling following elevated temperature treatments are major problems of these pouches. To overcome the disadvantages recent some advancements are made like recently resealable and transparent retort pouches attained by exchanging aluminium foil layer by oriented polyamide, zippers EVOH and PVDC correspondingly [29].

2.4.4 Corrosion of Metals and Lacquering

Metal Corrosion is an electrochemical interaction between food—metal that is only confined to iron, lead and tin [30]. Corrosion development is based on many factors like composition of food and metals. There are different types of metal corrosions in foods. First one is stress corrosion cracking. This type of corrosion is caused because of hydrogen chloride, sulfide, and sulphur dioxide present in food products and is an excellent example of contained corrosion at stressed zones. Second one is the sulphur staining or sulfide black corrosion is metal containers or product blackening because of reaction between product and metal components like oxygen, phosphorus and sulphur. Third type of metal corrosion is pitting corrosion which is a result of rapid dissolution of iron. Last type of corrosion is filiform corrosion. This type of corrosion is an exterior corrosion that occurs when relative humidity is 60–95% and temperature is 20–35 °C. Despite the increasing use and growing demand of metals (tinplate and iron) by the industry of canning, food products can turn out to be contaminated with iron and tin during the process of corrosion [31]. To stabilize the surface of tinplate and to protect it from corrosion during food- metal contact, the most used substitute is the organics coating applications known as lacquers or enamels. On innerside of food cans lacquering is done to avoid contact among its contents and can which results in product corrosion or lowers performance and overall acceptability of food. Most type of coatings used in food canning industries are conventional lacquers with high organic solvent or solventquantity of about 80% of total weight. However, there is a growing interest of using lacquers with minimal amount of solvent [32]. As per BIS (IS:5818 1988), there are three main categories of can lacquers and decorative finishes. Category A is internal lacquering of food cans, category B as exterior lacquering of cans and category C as attractive decorative finishes.

2.4.5 Safety Issues of Metal Packaging

Migration and interaction are the two major issues of metal packaging that lead to the undesirable effects on health. This means to the movement of constituents of packaging to food stuff or conversely throughout their processing or storage and the most widespread migrants from metal packaging comprises lead, tin, aluminum, bisphenol Zinc chromium, contaminants and coatings from metal. Zn and BPA (and Pb to a lesser extent) exhibits same migration trends. Among all bisphenol is intensively studied. Interaction is the microbiological, physical or chemical behavior at package boundary or food products compatibility with metal containers depending on their chemical composition, coatings on the package, pH, processing treatments, container's material, humidity and storage temperature. Interaction of food and metal results the adverse effects like deterioration, discolouration, pitting, corrosion, perforation and coating loss.

3 Conclusion

Principally, packaging science is materials science so choosing a right packaging is of foremost priority and as a consequence, the precise selection of a package merely depends on the performance of the packaging materials used. Packaging has allowed the world us to have a wide variety of foods round the year that would not be otherwise possible without the protection of the package. Consumers demand for quality food has led to packaging innovation, and innovations in packaging. Plastic packaging have been very useful in a role in maintaining food quality and assuring food safety for the duration of its shelf life, however n the last decade, the world has witnessed the debilitating effects of the use of plastic packages on our planet. And now, portions of the water we drink are contaminated with microplastics. So there is a need to move towards eco-friendly packages that could possibly shape the future of food packaging. The present chapter offered a complete view of the state of the art on biodegradable polymer packages for food application with special emphasis on the application of bioplastics. Metal based packaging materials offer brilliant barrier properties and exhibit lower global warming potential and higher recyclability and are widely in food packaging applications however main health and product safety concerns of metal packaging is migration of bisphenol A, and some metals like lead, cadmium, mercury, aluminium, iron and nickel.

References

1. Marsh K, Bugusu B (2007) Food packaging—roles, materials, and environmental issues. J Food Sci 72(3):R39–R55
2. Raheem D (2013) Application of plastics and paper as food packaging materials—an overview. Emirates J Food Agric 177–188
3. Ahmed A, Ahmed N, Salman A (2005) Critical issues in packaged food business. Br Food J
4. Francis FJ (2000) Wiley encyclopedia of food science, 2nd ed., vol 4. Wiley, New York
5. Risch SJ (2009) Food packaging history and innovations. J Agric Food Chem 57(18):8089–8092
6. Welt B (2005) A brief history of packaging. Institute of Food and Agricultural Sciences, document ABE321, University of Florida, USA

7. Pasqualino J, Meneses M, Castells F (2011) The carbon footprint and energy consumption of beverage packaging selection and disposal. J Food Eng 103(4):357–365
8. Rhim JW, Park HM, Ha CK (2013) Bionanocomposites for food packaging applications. Progress Polym Sci 38(10)(11):1629–1652
9. Khan MI, Sahar A, Ur Rahman U (2016) Nanotechnology in healthier meat processing. In: Novel approaches of nanotechnology in food, pp 313–345. Academic Press
10. Ampuero O, Vila N (2006) Consumer perceptions of product packaging. J Consum Market
11. Tu Y, Yang Z, Ma C (2015) Touching tastes: The haptic perception transfer of liquid food packaging materials. Food Qual Prefer 39:124–130
12. American Society for Testing Materials, ASTM (1965) Standard definition of terms relating to glass products. ASTM Stand C 162–56(13):145–159
13. Borowska (2014) Glass in green buildings. https://prezi.com/hxjrwe9odpkr/glassin-green-buildings/. Accessed 05 January 2016
14. Wells L (2008) Packaging news, 02 September 2008. http://www.packagingnews.co.uk/news/842980/Glass-packaging-sustain-growth/. Accessed on 17 September 2010
15. Yaris A, Ceyhun-Sezgin A (2015) Food packaging: glass and plastic
16. Balzarotti S, Maviglia B, Biassoni F, Ciceri MR (2015) Glass vs. plastic: affective judgments of food packages after visual and haptic exploration. Procedia Manuf 3:2251–2258
17. FSA (2002) Food Standards Agency Report A03029: investigation of the significant factors in elemental migration from glass in contact with food: 2003. http://www.foodbase.org.uk/results.php?freportid=316. Accessed 17 September 2010
18. Jayamanne VS, Adams MR (2004) Survival of probiotic bifidobacteria in buffalo curd and their effect on sensory properties. Int J Food Sci Technol 39(7):719–725
19. BPF (2013) British plastic federation plastic packaging the progressive packaging medium. https://www.linpacpackaging.com/files/Plastics%20Packaging%20Booklet%202013%20web.pdf. Accessed 10 November 2016
20. Kirwan MJ (2003) Paper and paperboard packaging. Food Packag Technol 241
21. Brenckle MA, Kaplan DL, Omenetto FG, Marelli B (2016) silk fibroin as edible coating for perishable food preservation
22. Mensitieri G, Di Maio E, Buonocore GG, Nedi I, Oliviero M, Sansone L, Iannace S (2011) Processing and shelf life issues of selected food packaging materials and structures from renewable resources. Trends Food Sci Technol 22(2–3):72–80
23. Modi SJ (2010) Thesis: assessing the feasibility of poly-(3-hydroxybutyrate-co-3-798 valerate) (PHBV) and poly-(lactic acid) for potential food packaging applications
24. Peelman N, Ragaert P, De Meulenaer B, Adons D, Peeters R, Cardon L, Van Impe F, Devlieghere F (2013) Application of bioplastics for food packaging. Trends Food Sci Technol 32(2):128–141
25. Parkar J, Rakesh M (2014) Leaching of elements from packaging material into canned foods marketed in India. Food Control 40:177–184
26. Deshwal GK, Panjagari NR (2019) Review on metal packaging: materials, forms, food applications, safety and recyclability. J Food Sci Technol 1–16
27. Morris SA (2011) Food and package engineering. John Wiley & Sons
28. Peter KTO, Ulrich N (2007) International life sciences institute report, packaging materials. Printed by ILSI, Europe
29. Chonhenchob V, Tanafranca D, Singh SP (2017) Packaging technologies for pineapple and pineapple products. In: Handbook of pineapple technology: production, postharvest science, processing and nutrition, pp 108–125
30. Montanari A, Zurlini C (2018) Influence of side stripe on the corrosion of unlacquered tinplate cans for food preserves. Packag Technol Sci 31(1):15–25
31. Bastidas JM, Cabañes JM, Catalá R (2000) Corrosion behavior of lacquered tinplate cans in contact with cockles (Cardium edulis) in brine solution. Corrosion 56(4):429–432
32. Furey G (2004) Innovation revives UV coatings for food cans. Canmaker 17:29–30

Containers for Drug Delivery

Sayan Ganguly, Poushali Das, and Shlomo Margel

1 Introduction

Drug delivery is a method or strategy for administering therapeutic cargos inside patient body in an effective way. A Drug's effectiveness can be affected significantly by the method or pathway in which it is delivered [1]. Drug Delivery Systems best elective pathways by virtue of which drug molecules are delivered without affecting the other physiological activities. Drug delivery is a white domain of research adopted by the chemist material scientist and pharmaceutics. There are several strategies by which a drug can be delivered but there are several strategies by which a drug can be delivered in a broad sense it can be categorized into steady, controlled and targeted pathways. As per the recent researches conducted, the therapeutics cargos are administered via skin, buccal, nasal and other routes [2]. Controlled-release technology (CRT) is the most practiced and adopted method followed by the scientists due to its high level of selectivity and efficacy. Depending on the method drug delivery systems (DDS) are sometimes called as 'carrier' of 'vehicle' which actually carry and protect drug molecules until its destined target [3]. Figure 1 is a pictorial representation of the types of carriers which are generally used to impart DDS. Every drug dose to patient body should be so specific that below this concentration/dose drug molecules will not show any impact to patient body and beyond the actually dose, it sometimes become lethal also. Conventional drug delivery is never maintaining these parameters. That's why CRTs are taken for granted since 1950 by several pharmacologists and industries due to their excellent and precise delivery [4]. Statistics says that promotion of a drug, its clinical trials, developments and pharmacokinetics causes in an average of $120 million [1]. In this discussion we want to expose the

S. Ganguly (✉) · P. Das · S. Margel
Department of Chemistry, Institute of Nanotechnology and Advanced Materials, Bar-Ilan University, 529000 Ramat-Gan, Israel

J. Parameswaranpillai et al. (eds.), *Micro- and Nano-containers for Smart Applications*, Composites Science and Technology,
https://doi.org/10.1007/978-981-16-8146-2_6

Fig. 1 Pictorial representation of various drug carriers

present scenario of research of DDS and their method of applications into patient body.

2 Drug Delivery Methods

Drug administration into body follows several routes as shown in Fig. 2. The routes are based on their initiation point of doses. Each pathway has its own merit as well demerits. Herein, we are trying to discuss about the basic features of the various routes by which drug molecules are normally administered into patient body.

2.1 *Buccal Pathway*

Drug administration via buccal mucosa is called buccal drug delivery. It is avoiding fast delivery of drugs but it is little bit challenging to adsorb the drug molecules though membrane [5]. Normally lipophilic drug are easy to adsorb but it is restricted to the hydrophilic drugs and large biopharmaceuticals. The buccal mucosa is also acts as barrier and control release platform for drug molecules. For long term timed

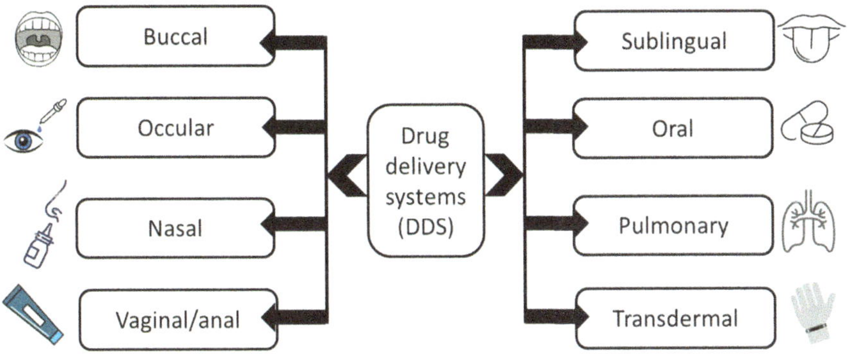

Fig. 2 Several pathways/routes of drug delivery applied for administering drugs into body

release mucosa compatible lipophilicity is the major contributing factor adopted in this method. In this method, drug formulations are formed as tablets, patches, gels and iozenges [6].

2.2 *Ocular Pathway*

The most sophisticated area of our body is eye and that's why ocular delivery of drug is the most difficult task for the material scientists due to its intricate architecture and complexity. Eye drops are the way of administration of ocular drugs which are subjected to target the ocular tissues. But the ocular delivery is sometimes challenging due to limitations by dynamic, precorneal and static ocular barriers. Ocular delivery follows three basic pathways like topical, systemic and local ocular. Local ocular is classified as subconjunctival, retrobulbar, intravitreal and intracameral. Scientists developed some strategies of ocular delivery of drugs through selective influx and efflux transporters.

2.3 *Nasal Pathway*

Nasal drug delivery implies delivery through the nasal cavities. Nasal spray is the most commonly used medication in this case. Nasal spray normally cures infections and allergic rhinitis in the upper respiratory tract. For migraine based treatment, small drug molecules are used in this process [7]. The fast delivery by this process occurs due to highly abundant vascularized architecture of nasal mucosa. Solid state nasal administration is also sometimes used in this case.

2.4 *Vaginal/anal Pathway*

Comparative to oral method, vaginal/anal delivery is the most fats and effective. Rectal medications are normally used for localized actions (laxatives) or whether the inside pathway of drug is compromised. Vaginal administration is better in terms of metabolism where it does not affect the gastrointestinal tract of body. These routes are most of time taken as hormone delivery of women for health issues. Vaginal formulations and carrier design is based on the hydrogels, tablets, suppositories and pessaries [8].

2.5 *Sublingual Pathway*

Drug delivery via sublingual pathway implies administration of drug by placing it under tongue. The drug molecules get transfused into through the tongue's ventral area which is actually situated at the mouth floor. Sublingual administration is fast and prompt action has been made normally because by this process general metabolism pathway is avoided for adsorption of drugs [9]. But the demerits are it restricts some needful activities like talking, eating, and drinking during the medication. Another demerit is this administration is also much selective to patients. For smokers this pathway is not recommended due to vasoconstriction of vessels of the smokers.

2.6 *Oral Pathway*

Oral delivery of drugs is the most common and well-known method among the medical science [10]. The cause of the popularity of the oral delivery is due to their non-invasive nature, less expensive, and high chance of absorption in the gastrointestinal (GI) tract. For oral medical formulation, water solubility is a primary objective. The stability and bioavailability of the drug at the GI tract environment and pH is also mandatory to asses. The main drawback is for this delivery sometimes it is difficult for paediatrics, geriatrics and persons having cognitive impairment.

2.7 *Pulmonary Pathway*

Pulmonary pathway is the delivery of drugs via inhalation through mouth and air channels. This type is normally restricted to the medications for lungs diseases. Pulmonary adsorption is also fast and rapid action takes places due to wide contact of drug molecules to blood stream in the reticular blood vessels. The reticular blood vessels are highly abundant in their alveolar segment of the lungs [11]. The major advantage of this process is it does not comply with metabolism based problems and never varies with the interpatient metabolism activities.

2.8 Transdermal Pathway

Transdermal delivery is the method to send drug molecules through skin without affecting skin. The main thing is that the drug should be enough soluble into the skin because hydrophilic drugs are difficult to penetrate the dermis and epidermis [12]. In transdermal system the drug molecules are adsorbed via dermal microcirculation. The non-invasive type method of delivery is advantageous because it also prohibits the physiological disorders of human beings.

3 Drug Delivery Carriers

3.1 Bioinspired Polymeric Carriers

Bioinspired polymers are physiological activity mimetic materials having excellent biocompatibility and shape morphing character. These are not only applied for healthcare research but also it has a booming applicability into the theranostics. Bioinspired materials consist of synthetic as well as natural polymers like chitin, keratin, collagen, alginate, carrageenan, poly(lactic acid), poly(hydroxyalkanoates) and etc. [13]. These materials are potential drug adsorbent and good carrier but these have disadvantages like poor chemical stability and mechanical strength.

Chitin is a naturally occurring polymer of N-acetylglucosamin and D-glucosamine which is normally abundant in various crustaceans. But chitin is insoluble. That's why a soluble modified variety of chitin has been derived by partial deacetylation of chitin named as chitosan. Both the chitin and chitosan have potential applications in biomedical area especially in drug delivery. *N*-acetylglucosamine which is present in chitosan provides electrical features into the tissues. For this reason chitosan widely used in tissue culture applications. It also shows biocompatibility and biodegradability. Fast adsorption behaviour of drug molecules, bactericidal and nonantigenicity. There are various reported articles which showed how chitosan can be grafted to monomers for preparing chitosan based gels and viscoelastic systems [14]. Gel like viscoelastic polymer assemblies is capable to hold and carry drug molecules followed by sustained release. For conducting applications and pulsatile release device chitosan is also applied widely [15].

Another bioinspired polymer system for drug carrier application is keratin. This polymer adopted the nomenclature form the word 'kera' which means 'horn'. In the year of 1950, keratin came into picture of material scientist research area. Initially several researchers did work over the keratin extraction form natural origin especially form human hairs and animal derives horns, hooves, feathers [16]. These are high thermostable polypeptides either in α-keratin or β-keratin; both are in coiled structured hydrophobic systems crosslinked by disulphide bonds [17]. It is also grafted with other monomers/polymer for drug carrier applications. Polyethylene glycol (PEG) grafted keratin could be a good choice for controlled release of cancer drugs [18]. Keratin based nanoparticles were proposed for anticancer drug carrier where keratin was chemically cross-linked followed by electrostatic attachment with drugs [19]. Keratin derived nanoparticles were pH and glutathione (GSH) responsive in nature. The synthesis of keratin nanoparticles has been depicted in Fig. 3. In another work chicken feather derived keratin was used to prepare micelle for doxorubicin carrier having the maen hydrodynamic diameter of 152 nm [20]. Porous keratin is also another variety of keratin with excellent adsorption quality [21]. Porous keratin shows pH dependent swelling behaviour which controls the drug release behaviour in several media. Human hair derived keratin and minocycline drug loaded biofilm was reported in another work which showed potential dental medication [22].

Native starch is another highly abundant biopolymer which is nontoxic, edible, biodegradable and highly budget friendly [23]. Starch based porous scaffold type materials are very much common as drug entrapment and sustained release vector. Microwave based simple synthesis of porous starch was reported which showed

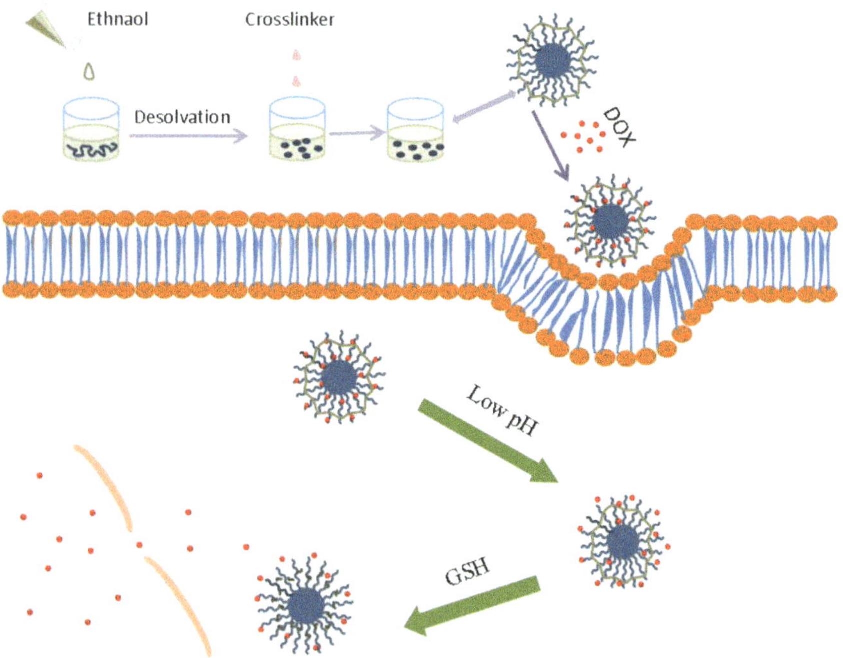

Fig. 3 Preparation of keratin-doxorubicin, internalization of the nanoparticles into cell and their dual responsive character. Reprinted with permission from Ref. [19] © 2017 Elsevier

controlled release of drug and minimizing the burst release effect [24]. Starch was also grafted to acrylic acid (AA) and 2-acrylamido-2-methylpropane sulfonic acid (AMPS) to prepare semi-interpenetrating polymeric network for controlled release of memantine [25]. Starch is a easily soluble in water an forms a gel like materials in high concentration. As polysaccharide, it also consists of large amount of polar groups which is required for functionalization. Magnetic and thermoresponsive starch based nanogel was reported for anticancer drug loading and delivery of therapeutics in a pH dependent fashion [26, 27]. Starch was also been nurtured in several drug carrier formulations like starch-g-poly(AA-co-HEMA) hydrogels [28], methacrylic acid (MA)/starch blend [29], carboxymethyl starch nanoparticles [30], fatty acid grafted starch [31], starch-ε-caprotactone micro-particles [32] and several typical nanocomposites [33, 34].

Another highly abundant polysaccharide is psyllium (*Plantago ovate Forsk*) which is mainly grown in India. It is extracted from ripe seeds of *Plantago psyllium and Plantago indica L.* these are normally used as mucilage. It has a self-gelling behaviour when put in aqueous environment [35]. Psyllium chemically designated as complex heteroxylans of arabinoxylan (based on arabinose and xylose monosaccharides). Arabinoxylans is linear β-D-(1 → 4)-linked xylopyranose (~22.6%) attached to α-L-arabinofuranose (~74.6%) segments [35]. Similar to starch it is also very prone to graft with various vinyl monomers via free radical formation. Acrylic acid and sodium styrene sulfonate (SSNa) were grafted to psyllium for fabrication of semi-IPN hydrogel (Fig. 4) with superabsorbent character [36]. Psyllium and N-vyl pyrrolidone based radiation cured hydrogel which could be used as porous substrate for drug loading [37]. Singh et al. reported how methacrylamide and psyllium were chemically attached for preparing hydrogel and tested for colon specific drug delivery [38]. Sterculiagum and psyllium was blended with acrylic acid and radiation cured for synthesising hydrogel [39]. This hydrogel was loaded with moxifloxacin drug and tested as sustained release carrier of the drug. As per their report, this typical hydrogel showed Fickian diffusion behaviour. The hydrogel was also tested against world which showed sufficient would healing applicability. Psyllium and gelatin based light weight scaffold is fabricated by Poddar et al. which is a porous matrix [40]. Porous matrix is excellent for drug loading and controlled release of drugs. Filler loaded psyllium is also playing as drug delivery vectors [41].

Sodium alginate is an easily water soluble salt form of alginic acid derived from the cell walls of brown algae. Alginates are widely used in food industries for making confectionaries due to its low cost and biodegradability. It consists of large amount polar functionalities tending them to easily crosslink with various chemicals. They have viscous in nature and forms extensive H-bonding in aqueous environment which form spatial arrangement of its macro-chains. Alginate based composite hydrogels are very common class of biomedical material. Generally alginates are suffering from mechanical strength, thus most of times it is covalently modified. Filler loaded composite alginate hydrogels are also common in drug carrier formulations. Methacrylic acid modified alginate was developed for theophylline drug delivery matrix [42]. Alginate microsphere is an potential carrier for drug molecules developed by Ciofani et al. [43]. Alginates microspheres are biodegradable, shape flexible

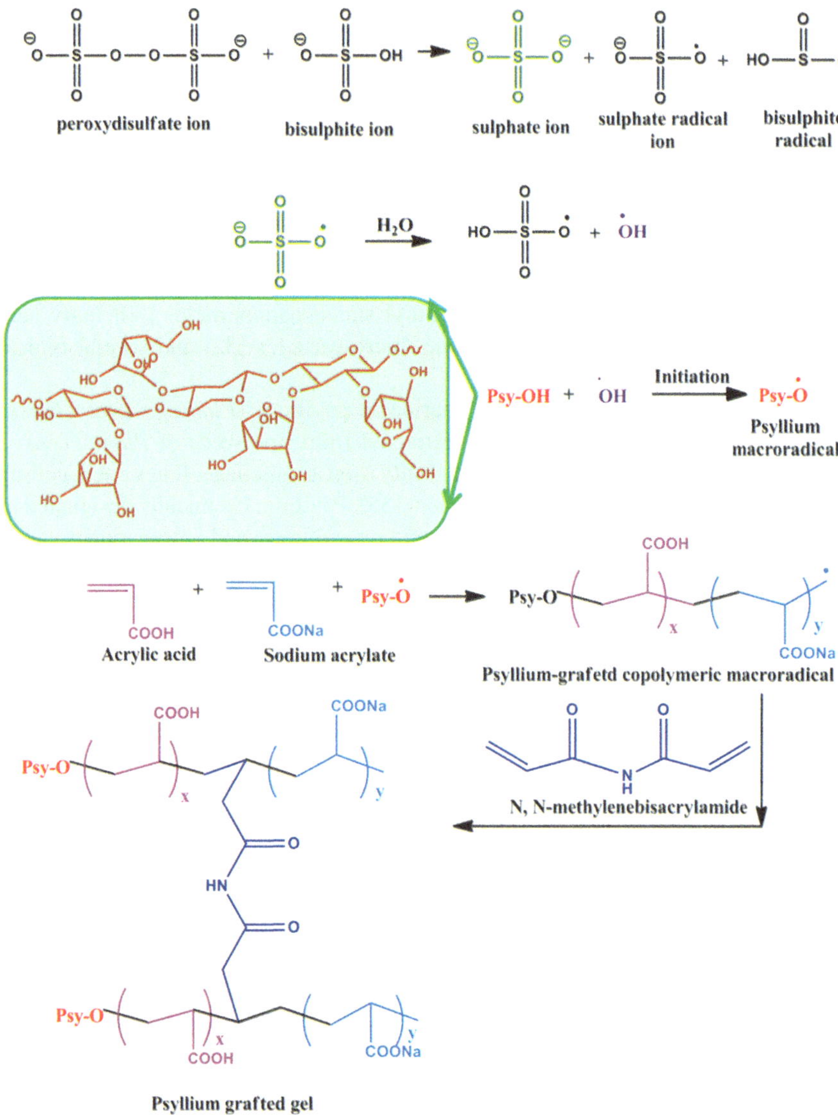

Fig. 4 Schematic representation of the free radical grafting synthesis of psyllium based semi-interpenetrating hydrogel. Reprinted with permission from Ref. [36] © 2018 Elsevier

and good adsorbed of biomolecules. Sometimes alginates were blended with albumin to improve its biocompatibility and used in drug delivery applications [44]. Besides the drug carrier feasibility, alginate based microspheres are also used in growth factor delivery. Kolambkar et al. reported alginate hydrogels for growth factor delivery to regenerate bones by delivering rhBMP-2 [45]. Desai et al. reported alginate nanoparticles for anti-tubercular drug delivery by adopting sustained release method [46].

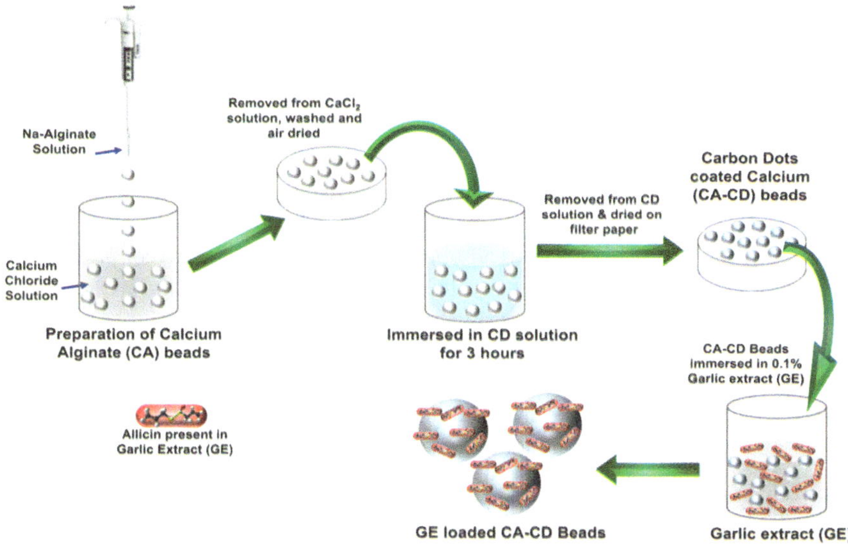

Fig. 5 Schematic of the preparation of carbon dot coated alginate bead and loading of garlic extract into it. Reprinted with the permission from Ref. [50] © 2016 American Chemical Society

Alginate –zein based drug carrier was developed in another work which was used for oral delivery matrix. This composite gel like particle could protect the encapsulated drug molecules from harsh environment [47]. Polyacrylamide grafted alginate was reported by Kulkarni et al. for electro-responsive drug delivery applications [48]. Ionically crosslinked alginate is another crosslinked platform of drug encapsulation and sustained release vehicle [49]. Carbon dot coated florescent hydrogel is also another drug delivery system for stimuli responsive release. Majumdar et al. reported carbon dot coated alginate beads were loaded by garlic extract which contents allicin [50]. They used allicin as a model drug and showed the delivery mechanism (Fig. 5).

Kappa-carrageenan is one of the varieties of high molecular weight sulphated polysaccharides which are extracted from red seaweeds. Kappa carrageenan is widely used in food industries die to its unique physical properties like thickening, gelling and stabilizing activity. Kappa-carrageenan was blended with hydroxyethyl cellulose (HEC) followed by crosslinking with genipin which was used for β-carotene controlled release [51]. Magnetic hydrogel based on kappa carrageenan is also developed for pulsatile release of drug molecules from polymer matrix [52]. Most the cases kappa carrageenan had been used in hydrogel fabrication and carrying drugs for timed release [53, 54].

3.2 Liposome Carriers

Liposomes are a special class of molecular carrier having lipid bilayer type morphology. These are most of the time composed of phospholipids and amphiphilic materials. Liposomes can carry large amount of water. The main potential of liposome carriers is that lipid soluble drugs could be transferred by this method. Phospholipids show a sol–gel transition at a critical temperature by means of melting of the acyl chains. Phospholipids also are susceptible to disrupt its outer layer in presence of acoustic wave/energy. Similar kind of thing was reported by porphyrin based liposomes [55]. Stimuli responsive lysosome carriers are more influential for their better targeting. Anada et al. reported amphipathic bisphosphonate based liposomes for curing metastatic bone disease [56]. They chose this compound which was susceptible to attach hydroxyapatite for triggering bone formation. Sihorkar et al. reported how polysaccharide could be attached to liposomes for better cell adhesion and higher drug carrying performances [57]. Ultrasound sensitive phosphatidylcholine, phosphatidylethanolamine, phosphatidylglycerol, and cholesterol based liposomes were developed by Huang et al. where they used sound waves to prepare liposomes [58]. Graphical illustration gave the sufficient heterogeneity inside the acoustically prepared liposomes having uni-lamellar drug storage as shown in Fig. 6. PEG conjugated liposome was reported for carrying drug molecules and acceleration of blood clearance [59]. Soni et al. developed 5-florouracil conjugated liposomes by cast film method that could cure for brain diseases [60]. Those liposomes were tested to be 17 times better targeted of radioactive suns trances. Liposomes are also suitable for anticancer drug encapsulations. Melanoma treatment was also lead by liposomal delivery of drugs that has been termed as 'protective cage' [61]. Lee et al. synthesized PEG tethered folate targeted liposomes for ovarian carcinoma prevention [62].

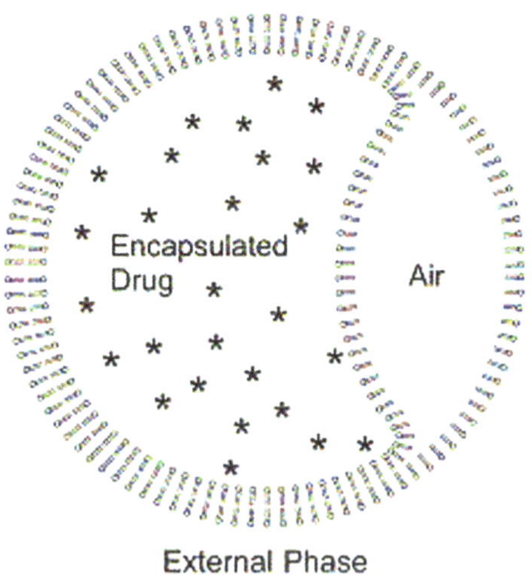

Fig. 6 Structure of acoustically prepared liposome with heterogeneous morphology. Reprinted with the permission from Ref. [58] © 2004 Elsevier

They showed folate conjugated liposomes could entrap doxorubicin, and antisense oligodeoxynucleotides for tumour targeting nano-carrier.

3.3 Nanogels as Drug Carriers

Nanogels are the miniaturised version of highly cross-linked hydrogels. These could be prepared by polymerised by monomer(s) followed by some surface engineering. They are classified as ionic and non-ionic. They are superior in terms of excretory removal from body than other macroscopic systems [63]. Their size ranges from 20–200 nm. Nanogels are also good house of drug molecules in a large amount. They are mostly applicable for genes, proteins, and other biological molecules. As nanogels having high surface charges, it has a tendency to agglomerate to form larger particles. Thus colloidal stability is a must-maintained feature of the nanogels during its fabrication. Most of the physiological conditions provide information those nanogels should have at least ±30 mV of surface change for better adhesion to biological molecules [64]. Nanogels are tailor made with external stimuli responsive behaviours. They can be stimulated in presence of external pH, ionic strength, temperature, ionic strength, magnetic fields, and electric pulses. The environmental factors actually affect their volume or internal architectures that enable to release of the drug/biological molecules from the matrices. Nanogels could be physically cross-linked as well as chemically cross-linked. Physically cross-linked nanogels are called pseudo gels that are dependent on polymer composition, temperature and type of cross-linking agents [65]. Weak forces (van der Waals forcers, H-bonding, electrostatic interactions, and hydrophobic association) are kept them entangled to form 3D assembly. Physical nanogels are created by association of block copolymeric systems and aggregation of them via complication. Supramolecular self-assembly is a common method to prepare micellar nanogels. Ding et al. reported polypeptide nanogels for doxorubicin delivery and easy cell internalization [66]. Chemical nanogels are help tightly from internal covalent linkages that are insoluble and swellable in specific media. Amphiphilic or hydrophilic copolymers are mostly reported fop nanogel preparation. Solvent evaporation strategy was adopted for preparing nanogels from cyclodextrin [67]. Poly(ethyleneglycol) (PEG) based biodegradable nanogel was prepared using lactate spacer. They reported protein loading onto the nanogel matrix and release kinetics [68]. 5 amine-terminated PAMAM dendrimers and alginate were cross-linked to form nanogels (Fig. 7) for doxorubicin drug carrier [69]. The nanogel was reported for fluorescent tracking as well as drug delivery devices. Jayakumar and his co-workers developed chitin-doxorubicin conjugated nanogel for control release of drug molecules [70]. Platinum (Pt)-deposited complex of calcium carbonate and cuccurbit[6]uril-conjugated hyaluronate nanogels were reported by Choi et al. that were suitable for drug delivery [71]. pH–sensitive nanogels were developed from N-[(2,2-dimethyl-1,3-dioxolane)-methyl]acrylamide (DMDOMA) that followed host–guest type interaction to intact their internal 3D network structure in nanoscale domains [72]. Acid degradable

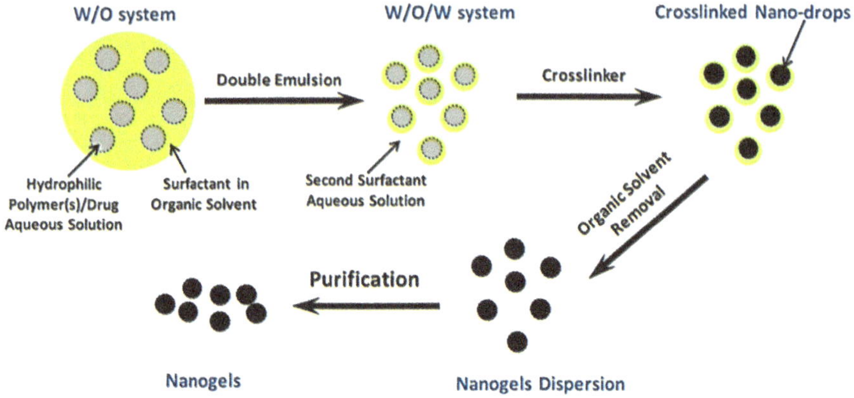

Fig. 7 Schematic illustration of the formation of fluorescent drug-loaded nanogels through a double emulsion method. Reprinted with the permission from Ref. [69] © 2014 American Chemical Society

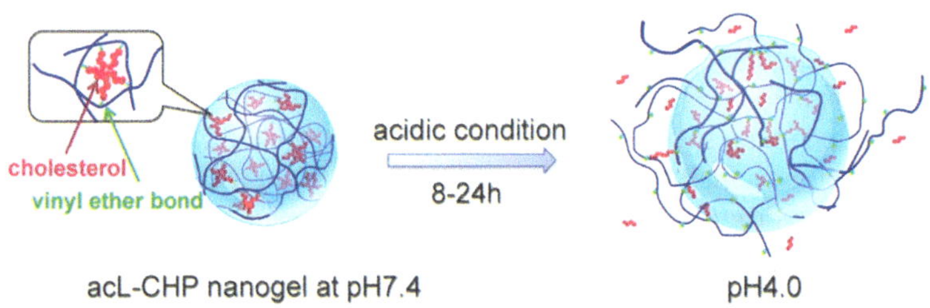

Fig. 8 Graphical illustration of disassembly of nanogels in low pH. Reprinted with the permission from Ref. [73] © 2020 American Chemical Society

nanogel is a new domain of nanogel research where Morimoto et al. developed pH-degradable nanogel (Fig. 8) for tumour targeting and less inflammatory character [73]. At low pH, these nanogels were disassembled to individual chain fragments and reaggregated again with increasing the pH. Poly(sulfobetaine)s are a novel class of zwitter ionic polymers that can be surface functionalized by various polymerisation methods. Poly(sulfobetaine)s based amine-thoil-ene linkage was employed to prepare nanogels having high drug carrying capacity [74]. Zwitterionic nanogels could be used as regulator of protein corona for avoiding immune responses. Carbohydrate containing nanogels were synthesized by post-polymerisation technique where carbohydrate long chains are used as surface functionalization moieties [75]. The implication of carbohydrate polymers are lies on their compatibility with glucose and glycopolymers. These affected their bind capability to drug and other biological molecules.

3.4 *Polyphosphazenes as Drug Carrier*

Polyphosphazines were first synthesized at 1895 by ring opening polymerisation of hexachlorotriphosphazene' [$(NPCl_2)_3$]. The material was colourless, high molecular weight, crosslinked, and rubbery in nature. Later on, Allcock and Kugel reported first time in 1965 about the synthesis of linear polydichlorophosphazene by controlling the reaction parameters. This polymer is hygroscopic and hydrolysed that gives non-toxic by products. This polymer have good thermal, electrolytic, optical and electroactive features. Polyphosphazines are functionalized in various ways. Most common of them are aminated and alkoxy substituted polyphosphazines. Aminated polyphosphazines are biodegradable. Amine functionalization into polyphosphazines is carried out by using amino acid esters and imidazole. Alkoxy substituted polyphosphazines are better stable in moisture containing environment.

The polyphosphazines are used as drug carrier are normally pelletized or films. Pellets and films are fabricated by means of compression moulding or solvent casting method. Polyphosphazines are soluble in wide range of solvents thus it is suitable to choose solvents in a random basis. Micro encapsulation technique was adopted for polyphosphazines by spray drying, solvent evaporation, and emulsification. Naproxane loaded polyphosphazine microsphere was reported by Veronese and his co-workers where polyphosphazines were used as matrix phase prepared by spray drying [76]. The degradation studies of the polyphosphazines imply that the drug diffusion from polyphosphazines matrices could be tuned by modification of the system by inserting polar functionalities into it. The design flexibility is so significant for polyphosphazines that drug release behaviours could be manipulated by means of diffusion, erosion or combination of both. Martinez et al. reported polyphosphazines derived biodegradable drug-conjugated system having narrow size distribution and pH responsive character [77]. The synthesis scheme adopted in this work has been depicted in Fig. 9. Hou et al. devolved dual drug loaded polyphosphazne based 'self-framed' drug-polymer assembly for multi-drug resistance (MDR) tumour curing [78]. This self-framed drug carrier was highly efficient for 80% drug loading. Zhou et al. developed acid responsive degradable polyphosphazine nanoparticles for anticancer drug delivery [79]. They showed that these nanoparticles were stable at pH 7.4 whereas i8t started degradation after lowering of the pH. This is quite significant for cancel cell targeting drug release. Amphiphilic graft polyphosphazenes are also another class of polymers for encapsulation of drugs and specific targeting of them. Methoxy-poly (ethylene glycol) (mPEG) and ethyl-p-aminobenzoate grafted polyphsophazenes were synthesized and applied to cure xenograft tumours [80]. Poly(organophosphazene) hydrogel was reported having hydrophobic and hydrophilic segments that showed sol–gel transition behaviour. This hydrogel was temperature responsive and applied for growth factor delivery [81].

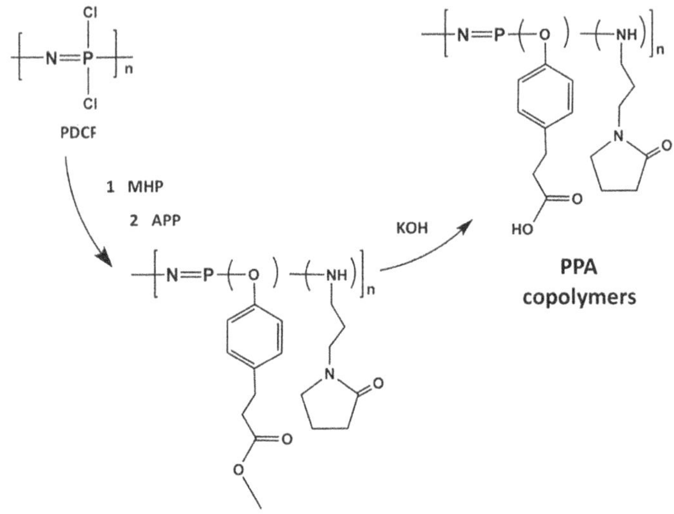

Fig. 9 Schematic representation of Polydichlorophosphazene modification by copolymer formation. Reprinted with the permission from Ref. [77] © 2017 America Chemical Society

3.5 Transfersomes as Drug Carriers

Transfersomes are vesicle type body having purely elastic or viscoelastic nature. These were first developed at 1990s. Elasticity of transfersomes is come after internalization of a terminal promoter in the lipid bilayer assembly. The main constituents of transfersomes are sodium cholate, soya phosphatidyl choline, and a small amount of ethanol. They are utilized in a non-impeded technique thought the skin followed by permeation to stratum. Transfersomes are applied as carriers for drug molecules, peptides and vaccines. As per the patents of Idea AG integral transfersomes could permeate through the skin and passing through the blood stream. There are some structural changes in the stratum corneum that promotes permeation of elastic vesicles via stratum corneum lipid lamellar regions. But after insertion of the elastic vesicles, unruptured vesicles were not found inside the exposed tissues. Edge-activator agents are normally used for bilayer structure where the stability increment is the goal of work.

Transfersomes are behaving like liposomes as described first by Cevc [82]. Conventional liposomes have restrictions to permeate via outer layers of the stratum corneum. This feature sometimes limited the applications of molecular delivery. Transfersomes are used as in vitro or in vivo by skin penetration of by subcutaneous injections respectively. Commercially transfersomes are applied as a successful carrier for drugs, steroids, and NSAIDs. Several research groups reported the efficacy of transfersomes/deformable. Jain and his co-workers reported that the dexamethasone based transfersomes could penetrate through rat skin obeying zero order release kinetics, without any severe phase lag. But for commercial liposome based formulations and ointments have dominant in showing 'lag phase' [83]. This same group also reported elastic transfersomes based release with high flux [84]. Maghraby et al. reported composition dependent phenomena of transfersomes in skin penetration [85]. They reported that transfersomes size is not the sole factor for permeation of drug molecules through skins. They showed how osmotic pressure gradient affected the transport of transfersomes. They also hypothesized that elastic vesicles are superior in penetration than deformable vesicles. Trotta et al. developed dermatitis curable formulations based on transfersomes [86]. They studied penetration of an anti-inflammatory drug, named dipotassium glycyrrhizinate by using deformable vesicles across pig skin. Vaccine delivery via elastic transfersomes is also reported through transcutaneous penetration. Hepatitis B surface antigen (HBsAg)-loaded elastic vesicles were used as medication where formulation was physical mixture of transfersomes and antigen [87]. In vitro cell culture and fluorescence microscopic analysis manifested permeation through skins, biodistribution, and effective antigen delivery to the lymphatics and immunocompetent Langerhan's cells. But, the actually mechanism of permeation through skins are still in research.

3.6 Hydrogels as Drug Carrier

Hydrogels are hydrophilic network of polymers arranged three dimensionally having water absorbing capability of 10–20% to several thousand [88]. Hydrogels can be classified as physical and chemical hydrogels. Physical hydrogels are 3D assembly of hydrophilic polymers or fillers where the components are held together by means of physical interactions [89]. For chemical hydrogel the main structure is formed by covalent bond formation. hydroxymethyl methacrylate (HEMA) hydrogels was first synthesized at 1960 for contact lens application [90]. Because of the porous morphology of the hydrogels, they can store small molecules (drugs, proteins, dyes, growth factors etc.) to a large amount and release in various environment. Hydrogels are stimuli responsive in nature. The externally applied stimuli are pH, temperature, light, electric pulse, ionic strength, solvent, enzymes, and salinity of medium. On basic of the origin of preparation, hydrogels can be synthesized from natural origin of synthetic approaches. Most common natural polymers used for making hydrogels are chitosan, starch (or starch derivatives), guar gum, kappa carrageenan, alginate, and several polysaccharides. Synthetic hydrogels are composed of PEG or polyethylene oxide, polyacrylamide (PAAm), polysaccharides, poly(hydroxyethyl methacrylate), poly(vinyl alcohol) (PVA), etc. [91].

Control delivery of drugs from hydrogels means timed release of drug molecules in specific environment and stimuli. The hydrogels can protect drugs from robust environment and delivery the drugs into targeted domain. Polymer structure is the main factor of hydrogels to be stimuli active. The common feature of temperature-responsive polymers is the presence of hydrophobic groups, such as methyl, ethyl and propyl groups. Some commonly used temperature responsive polymers for drug carrier are poly(*N*-isopropylacrylamide) (PNIPAAm), Poly(*N*,*N*-diethylacrylamide (PDEAAm), and copolymers of them [92]. Block copolymers are also showing temperature dependent sol–gel transition is used in drug carrier. Block copolymers applied for drug delivery is based on poly(ethylene oxide) (PEO) and poly(propylene oxide) (PPO) under the trade names of Pluronics® (or Poloxamers®) and Tetronics®. Most reports on temperature-responsive hydrogels are based on LCST-based systems that show volume fluctuations in ambient condition and at physiological temperature. The volume change is called swelling that admits easy diffusion of drug molecules inside the 3D network in room temperature, while the contracting at a 37 °C implies denser microstructure. This restricts the molecular diffusions in a large extent. Thus, the hydrogel can sustain time dependent drug release in the physiological environment. If temperature would be increased further, the hydrogel network became relaxed resulting release of remaining water and drug from the matrix. Thermodynamically, amenability in the sol–gel transition is critical for each formulation. The transition occurs in a narrower temperature separation and the volume fluctuation is more precise [93]. Dual responsive hydrogels were reported by a group of works from Malaysia where they showed bacterial cellulose and acrylic acid were grafted to prepare hydrogel [94]. Interpenetrating polymeric network (IPN) based hydrogels are another special class of hydrogel where two polymeric chains are intermingled together with crosslinking. Shantha et al. reported IPN based on chitosan, *N*-vinyl pyrrolidone (NVP), and polyethylene glycol diacrylate (PAC) [95]. They tested the hydrogel in gastric fluid environment where it showed ~50% drug entrapment without affecting its behaviour. Polysaccharide grafted hydrogel were reported by Kulkarni et al. that showed pH responsive drug release behaviour [96]. There are several work on pH responsive hydrogels where all were tested good as drug carriers [97, 98]. Electric pulse induced drug delivery is another external stimuli responsive release. Various researchers reported conducting hydrogels having electric pulsatile drug release profile [99, 100]. Conducting hydrogels are based on conducting polymers or conducting fillers. Figure 10 showed the synthesis of in situ prepared graphene based nanocomposite hydrogel [101]. Servant et al. reported multi-walled carbon nanotubes (MWCNTs) and polymethacrylic acid (PMAA) based conductive hydrogel [102]. Conducting hydrogels are tunable as per their requirement. The electrical conductivity was changed by loading of conductive fillers [103]. These hydrogels showed 'on–off' mechanism upon application of external voltage.

Indermun et al. reported semi-IPN hydrogel prepared from polyacrylic acid (PAA) and poly(vinyl alcohol) (PVA) [104]. Here they used poly(ethyleneimine) (PEI) and 1-vinylimidazole (VI) as electro responsible components for external pulse. In another work, biodegradable hydrolysed collagen polypeptide was used to prepare electroconductive hydrogel (ECH) [105]. They modified the collagen by polycaprolactone and used polyaniline (PANI) as conductive filler. The ECH was tested non-cytotoxic and successful for drug delivery. Besides these, there are several stimuli responsive hydrogels for drug delivery applications [106].

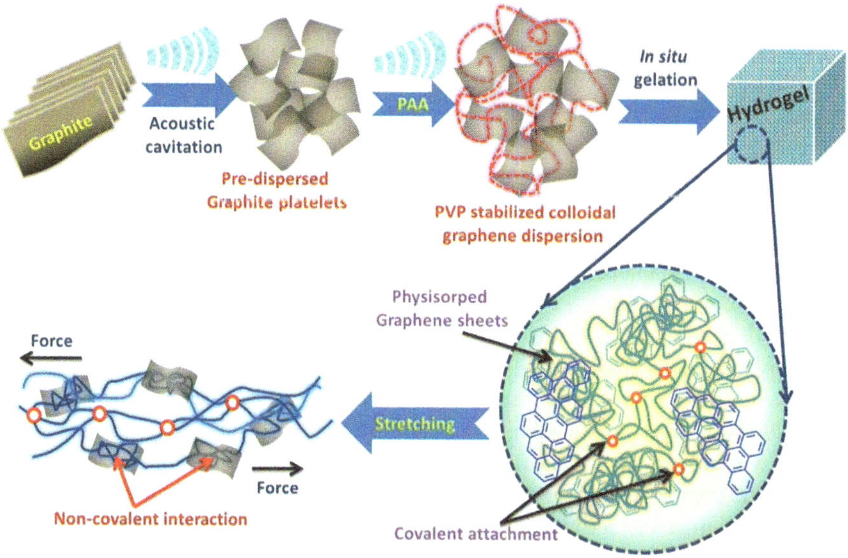

Fig. 10 Graphical illustration of graphene based in situ conductive nanocomposite hydrogel for pulsatile drug release. Reproduced with the permission from Ref. [101] © 2018 Elsevier

3.7 Carbon (Quantum) Dots as Nanocarriers for Drugs

Carbon dots (C-dots) are a special type of recently distinguished nanoscale carbon particles comprised of distinct, quasi-spherical particles having sizes less than 10 nm [107–112]. Like of its famous older cousins; graphene [113, 114], CNTs [115], fullerene, these new member of nanocarbon is developing enormous research endeavours in its own right. C-dots were discovered during purification of SWCNTs fabricated through arc-discharge methods [116]. C-dots can be fabricated inexpensively on a large scale by various approaches like bottom up and top down methods [117]. Compared to traditional semiconductor quantum dots as well as conventional organic dyes, C-dots combine encouraging features namely, resistance to photobleaching, wavelength and size dependent emission behaviour, biocompatibility, ease of bioconjugation [118, 119]. Due to these features they unveiled advantages for application in photocatalysis [120], light emitting diodes [121], sensing [122, 123], antibacterial activity [124], and bioimaging [125, 126] etc.

In recent years increased activities in developing multifunctional nanomaterials are being implemented for simultaneous detection of disease, bioimaging, and drug delivery [127]. For this purpose, different nanoparticles such as iron oxide [128], gold [129], graphene [130, 131], polymer nanoparticles [132] have been explored. But, it should be noted that the fabrication and use of these multifunctional nanoparticles need to be connected with physiological parameters for therapeutic applications. C-dots with excellent biocompatibility, low to no cytotoxicity, inexpensive, non-blinking character and high quantum yield is deliberated as a rising star in nanomedicine [133]. In a report, Singh et al. constructed photoresponsive drug delivery system using C-dots and 7-methoxy quinoline moiety, as phototrigger [134]. The quinoline based phototrigger was chosen due to its mild fluorescence and release performance in both one photon and two photon excitation. The C-dots was prepared from citric acid and urea through domestic microwave oven and their surface were anchored by phototrigger conjugated anticancer drug, 7-(3-bromopropoxy)-2-quinolylmethyl chlorambucil (Qucbl). This drug was released in presence of 632 nm red light followed by photolysis. The HPLC results suggested significant improvement of release data in presence of red light. The authors also illustrated the specific control of photolytic release by observing the chlorambucil release after exposure to light and dark conditions. Zeng and his co-worker reported C-dots from bovine serum albumin by solvothermal synthesis having 6.8 nm in diameter [135]. The C-dots possess a hollow structure of pore size 2 nm which was used as DOX delivery system. The DOX-C-dots conjugate unveiled rapid uptake in cells and pH controlled release of DOX without influencing the drug activity. In another report, Zhao et al. developed pH and redox responsive C-dots for tumor targeted microenvironment activated drug delivery for anticancer activity [136]. This drug delivery system comprised of C-dots as image guided drug nanocarriers, cisplatin(IV) as prodrug, and RGD peptide as active targeting ligand which was PEGylated at physiological pH 7.4 (Fig. 11a, b). This nanocarrier after entering in cancer cells, the cisplatin(IV) was reduced to cytotoxic cisplatin in reductive cytosol of cancer cells to reveal therapeutic effects. Das et al. reported biocompatible C-dots loaded anti-diabetic drug Metformin with drug loading efficiency around 71.6% and examined Metformin release study in different pH [137]. Diltiazem hydrochloride (DTZ) loaded zwitterionic C-dots showed 71.28% drug loading efficiency and controlled delivery of DTZ more prominent in the physiological pH, not in very low or high pH condition [138]. The mechanism of drug attachment to the C-dots surface was pictorially illustrated in Fig. 11c–e. Core–shell MOFs/CDs@OCMC nanoparticles for in vitro diagnosis and cancer therapy was demonstrated by Whittaker and his group [139]. The author illustrated that after 50 h 20% release of DOX was obtained in pH 7.4 whereas complete DOX release was detected in 60 h at pH 3.8. Here, OCMC coating on the C-dots offered pH sensitivity to the composite.

CDs-HA-Hep/DOX nanocarrier with HA and pH dual responsive drug delivery system was reported by Sang and coworkers [140]. Hep and HA were anchored on the surface of C-dots by EDC/NHS method and DOX loading was achieved by electrostatic binding with 67% efficiency. The CDs-HA-Hep/DOX system demonstrated outstanding tumor targeting capability. At tumour microenvironment (pH 5.6, HAase), the CDs-HA-Hep/DOX nanocarrier showed high release efficiency and almost no release under normal physiological condition (pH 7.4, without HAase). This nanocarrier revealed great potential for monitoring of cell activities throughout tumor treatment.

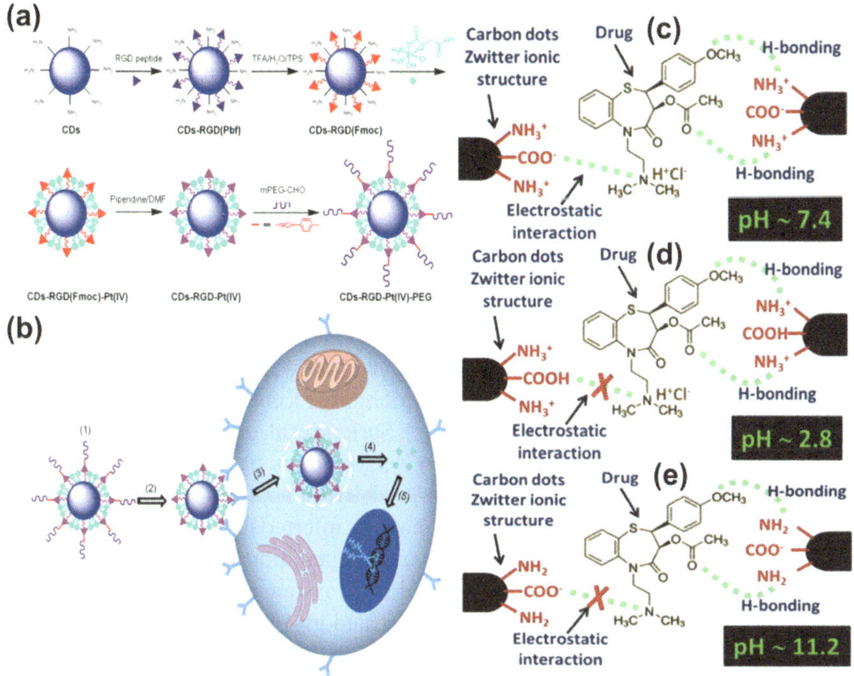

Fig. 11 Schematic illustration for the **a** synthesis of CDs-RGD-Pt(IV)-PEG drug nanocarrier. **b** drug delivery process using CDs-RGD-Pt(IV)-PEG. Reproduced with the permission from Ref. [136] © 2016 America Chemical Society. **c** Interaction of zwitterionic C-dots with DTZ drug molecules in different pH conditions. Reproduced with the permission from Ref. [138] © 2019 Elsevier

3.8 *Glycosylated Carriers for Drug Delivery*

Polysaccharides are the most abundant natural polymers in nature especially carbohydrates. They are building blocks of living cells. Carbohydrates are called glycosyl donors that are prone to attach with hydroxyl moiety or any other molecules. Typically in glycosylation protein molecules are bound to carbohydrates for drug carrier fabrication. Besides carrying drug molecules these materials have another merit of stabilization of drugs form environment. Lectin receptors make interactions to carbohydrate polymers and conjugated to drugs resulting formation of glycosylated carrier or glycoconjugates. Glycosylated molecules have been effectively employed to deliver diverse therapeutic cargos including antiviral drugs like stavudine, azidothymidine, anticancer drugs (methotrexate, doxorubicin), and anti-tubercular drugs (rifampicin, antigens for instance HBsAg, genes etc.) [141, 142]. Lectin receptors are the most widely used biomaterial conjugated with glycosylated polymers. Interaction between lectin and glycosylated polymer had been studied by *Ricinus* communis agglutinin obtained from castor oil based plants and concanavalin

A (Con A) also reported as glycosylated materials [143]. Targeting through polysaccharide acts interactive mechanism with dissimilar sugar molecules [144]. When polysaccharides are coupled to carriers the outcome glycosylated carriers–attached carbohydrate are marked and endocytosed with the help of lectin receptors. Several works have been published that proves that glycosylated carriers are very much susceptible to attach with lectin. Kolomiets et al. have reported glycopeptides dendrimers having C-fucosyl peptide dendrimers with the empirical formula of (CFuc-X(6)X(5)X(4))(4)(LysX(3)X(2)X(1))(2)LysIleHisNH(2) (CFuc ¼ alpha-L-fucosyl acetic acid, X(1–6) ¼ amino acids, Lys ¼ lysine branching) for antibacterial LecB inhibitors [145]. Garg et al. reported special type liposomes based on galactosylated, fucosylated and mannosylated polysaccharides via chemical conjugation method [142]. Frisch et al. developed 'click chemistry' adopted strategy to attach several ligands onto liposome surface [146]. Mannose receptors were applied for anti-HIV activity tracing after glysosylation [147]. Suriano et al. discovered the d-glucose, d-galactose and d-mannose based block glycopolymers by metal-free organocatalyzed ring-opening polymerization for drug delivery application [148]. As of the obtained results from this investigation the authors inferred that these glycopolymers can function as good drug carrier with increased penetration by liver cells. Oda et al. reported conjugated d-galactose and b-cyclodextrin system as a drug delivery system. In this study the authors showed how a conjugated glysosalyed polymer serves the demand of a successful delivery carrier for anticancer drug doxorubicin [149]. Jain et al. prepared the galactosylated functionalized chitosan nanoparticles that showed hepatocyte targeting and sustained drug delivery quality [150].

4 Summary and Future Aspects

The chapter summarizes the mode of drug delivery and the different preparation of drug carriers. In the initial stage of discussion this drug delivery pathways were discussed. In the rest of the part, the fabrication of drug carriers, their merits, demerits, and environment sensitivity were elaborated. From micro to nano domains, fabrication of drug carriers follows some basics rules like choosing the delivery location pH, effect of external stimuli, protection of drug molecules from robust environments, their efficient entrapment, and timed release. Some carriers are quite susceptible to degrade where some are resistant to the environmental systems. Polymers are widely used in macroscopic carriers and nano-vehicles for drug molecules whereas quantum dots are the unique nanomaterial for fast delivery in targeted areas.

References

1. Tiwari G et al (2012) Drug delivery systems: an updated review. Int J Pharm Invest 2(1):2
2. Pal D, Nayak AK (2015) Alginates, blends and microspheres: controlled drug delivery. In: Encyclopedia of biomedical polymers and polymeric biomaterials, 11 Volume Set. CRC Press, pp 89–98
3. Ganguly S, Das NC (2018) Synthesis of Mussel inspired polydopamine coated halloysite nanotubes based Semi-IPN: an approach to fine tuning in drug release and mechanical toughening. In: Macromolecular symposia. Wiley Online Library
4. Jain D, Pathak D, Pathak K (2009) Pharmaceutical product development technologies based on the biopharmaceutical classification system. Die Pharmazie-An Int J Pharmaceut Sci 64(8):483–490
5. Hao J, Heng PW (2003) Buccal delivery systems. Drug Dev Ind Pharm 29(8):821–832
6. Gherman S et al (2016) Enalapril maleate loaded pullulan film for mucoadhesive buccal drug delivery applications. Cellul Chem Technol 50:593–600
7. Lansdowne LE, Drug delivery methods
8. Sahoo CK et al (2013) Intra vaginal drug delivery system: an overview. Am J Adv Drug Deliv 1:43–55
9. Narang N, Sharma J (2011) Sublingual mucosa as a route for systemic drug delivery. Int J Pharm Pharm Sci 3(Suppl 2):18–22
10. Inamuddin A, Mohammad A (2018) Applications of nanocomposite materials in drug delivery. Woodhead Publishing, pp 509–573
11. Labiris N, Dolovich M (2003) Pulmonary drug delivery. Part I: physiological factors affecting therapeutic effectiveness of aerosolized medications. British J Clin Pharmacol 56(6):588–599
12. Alkilani AZ, McCrudden MT, Donnelly RF (2015) Transdermal drug delivery: innovative pharmaceutical developments based on disruption of the barrier properties of the stratum corneum. Pharmaceutics 7(4):438–470
13. Iqbal HM et al (2015) Development of bio-composites with novel characteristics: evaluation of phenol-induced antibacterial, biocompatible and biodegradable behaviours. Carbohyd Polym 131:197–207
14. Bhattarai N et al (2005) PEG-grafted chitosan as an injectable thermosensitive hydrogel for sustained protein release. J Control Release 103(3):609–624
15. Singh N et al (2016) Chitin and carbon nanotube composites as biocompatible scaffolds for neuron growth. Nanoscale 8(15):8288–8299
16. Rouse JG, Van Dyke ME (2010) A review of keratin-based biomaterials for biomedical applications. Materials 3(2):999–1014
17. Brandelli A (2008) Bacterial keratinases: useful enzymes for bioprocessing agroindustrial wastes and beyond. Food Bioprocess Technol 1(2):105–116
18. Li Q et al (2012) Biological stimuli responsive drug carriers based on keratin for triggerable drug delivery. J Mater Chem 22(37):19964–19973
19. Li Y et al (2017) Preparation and characterization of DOX loaded keratin nanoparticles for pH/GSH dual responsive release. Mater Sci Eng C 73:189–197
20. Zhang H, Liu P (2019) Bio-inspired keratin-based core-crosslinked micelles for pH and reduction dual-responsive triggered DOX delivery. Int J Biol Macromol 123:1150–1156
21. Srinivasan B et al (2010) Porous keratin scaffold–promising biomaterial for tissue engineering and drug delivery. J Biomed Mater Res Part B Appl Biomater Offic J Soc Biomater Jpn Soc Biomater Aust Soc Biomater Korean Soc Biomater 92(1):5–12
22. Lee H et al (2015) Human hair keratin-based biofilm for potent application to periodontal tissue regeneration. Macromol Res 23(3):300–308
23. Jafari SM et al (2008) Nano-particle encapsulation of fish oil by spray drying. Food Res Int 41(2):172–183
24. Malafaya P et al (2001) Porous starch-based drug delivery systems processed by a microwave route. J Biomater Sci Polym Ed 12(11):1227–1241

25. Ganguly S et al (2017) Starch functionalized biodegradable semi-IPN as a pH-tunable controlled release platform for memantine. Int J Biol Macromol 95:185–198
26. Massoumi B, Mozaffari Z, Jaymand M (2018) A starch-based stimuli-responsive magnetite nanohydrogel as de novo drug delivery system. Int J Biol Macromol 117:418–426
27. Ganguly S, Margel S (2020) Remotely controlled magneto-regulation of therapeutics from magnetoelastic gel matrices. Biotechnol Adv 107611
28. Sadeghi M (2011) Synthesis of starch-g-poly (acrylic acid-co-2-hydroxy ethyl methacrylate) as a potential pH-sensitive hydrogel-based drug delivery system. Turk J Chem 35(5):723–733
29. Ali AE-H, AlArifi A (2009) Characterization and in vitro evaluation of starch based hydrogels as carriers for colon specific drug delivery systems. Carbohyd Polym 78(4):725–730
30. Saboktakin MR et al (2011) Synthesis and in vitro evaluation of carboxymethyl starch–chitosan nanoparticles as drug delivery system to the colon. Int J Biol Macromol 48(3):381–385
31. Simi C, Abraham TE (2007) Hydrophobic grafted and cross-linked starch nanoparticles for drug delivery. Bioprocess Biosyst Eng 30(3):173–180
32. Balmayor ER et al (2009) Preparation and characterization of starch-poly-ε-caprolactone microparticles incorporating bioactive agents for drug delivery and tissue engineering applications. Acta Biomater 5(4):1035–1045
33. Subramanian SB, Francis AP, Devasena T (2014) Chitosan–starch nanocomposite particles as a drug carrier for the delivery of bis-desmethoxy curcumin analog. Carbohyd Polym 114:170–178
34. Schmitt H et al (2015) Melt-blended halloysite nanotubes/wheat starch nanocomposites as drug delivery system. Polym Eng Sci 55(3):573–580
35. Singh B et al (2007) Synthesis, characterization and swelling responses of pH sensitive psyllium and polyacrylamide based hydrogels for the use in drug delivery (I). Carbohyd Polym 67(2):190–200
36. Ganguly S et al (2018) Design of psyllium-g-poly (acrylic acid-co-sodium acrylate)/cloisite 10A semi-IPN nanocomposite hydrogel and its mechanical, rheological and controlled drug release behaviour. Int J Biol Macromol 111:983–998
37. Singh B, Kumar S (2008) Synthesis and characterization of psyllium-NVP based drug delivery system through radiation crosslinking polymerization. Nucl Instrum Methods Phys Res Sect B 266(15):3417–3430
38. Singh B, Sharma N, Chauhan N (2007) Synthesis, characterization and swelling studies of pH responsive psyllium and methacrylamide based hydrogels for the use in colon specific drug delivery. Carbohyd Polym 69(4):631–643
39. Singh B et al (2019) Polysaccharides Sterculia gum/psyllium based hydrogel dressings for drug delivery applications. Polym Sci Ser A 61(6):865–874
40. Poddar S et al (2019) Fabrication and cytocompatibility evaluation of psyllium husk (Isabgol)/gelatin composite scaffolds. Appl Biochem Biotechnol 188(3):750–768
41. Kaur K, Jindal R, Bandhu M (2020) Monodispersed silica nanoparticles incorporated nanocomposites of gelatin and psyllium for sequestration of noxious pollutants. J Polym Environ 28(1):179–199
42. Ganguly S et al (2018) Polysaccharide and poly (methacrylic acid) based biodegradable elastomeric biocompatible semi-IPN hydrogel for controlled drug delivery. Mater Sci Eng C 92:34–51
43. Ciofani G et al (2007) A drug delivery system based on alginate microspheres: mass-transport test and in vitro validation. Biomed Microdevice 9(3):395–403
44. Kumar Giri T et al (2012) Alginate based hydrogel as a potential biopolymeric carrier for drug delivery and cell delivery systems: present status and applications. Curr Drug Deliv 9(6):539–555
45. Kolambkar YM et al (2011) An alginate-based hybrid system for growth factor delivery in the functional repair of large bone defects. Biomaterials 32(1):65–74
46. Desai J et al (2009) Alginate-based microparticulate oral drug delivery system for rifampicin. Res J Pharm Technol 2(2):301–303

47. Lee S, Kim Y-C, Park J-H (2016) Zein-alginate based oral drug delivery systems: protection and release of therapeutic proteins. Int J Pharm 515(1–2):300–306
48. Kulkarni RV, Setty CM, Sa B (2010) Polyacrylamide-g-alginate-based electrically responsive hydrogel for drug delivery application: synthesis, characterization, and formulation development. J Appl Polym Sci 115(2):1180–1188
49. Li P et al (2017) Encapsulation of autoinducer sensing reporter bacteria in reinforced alginate-based microbeads. ACS Appl Mater Interfaces 9(27):22321–22331
50. Majumdar S et al (2016) Carbon-dot-coated alginate beads as a smart stimuli-responsive drug delivery system. ACS Appl Mater Interfaces 8(50):34179–34184
51. Hezaveh H, Muhamad II (2013) Modification and swelling kinetic study of kappa-carrageenan-based hydrogel for controlled release study. J Taiwan Inst Chem Eng 44(2):182–191
52. Bardajee GR, Hooshyar Z, Rastgo F (2013) Kappa carrageenan-g-poly (acrylic acid)/SPION nanocomposite as a novel stimuli-sensitive drug delivery system. Colloid Polym Sci 291(12):2791–2803
53. Pourjavadi A, Barzegar S, Zeidabadi F (2007) Synthesis and properties of biodegradable hydrogels of κ-carrageenan grafted acrylic acid-co-2-acrylamido-2-methylpropanesulfonic acid as candidates for drug delivery systems. React Funct Polym 67(7):644–654
54. Selvakumaran S, Muhamad II, Abd Razak SI (2016) Evaluation of kappa carrageenan as potential carrier for floating drug delivery system: effect of pore forming agents. Carbohy Polym 135:207–214
55. Wang X et al (2018) Enhanced drug delivery using sonoactivatable liposomes with membrane-embedded porphyrins. J Control Release 286:358–368
56. Anada T et al (2009) Synthesis of calcium phosphate-binding liposome for drug delivery. Bioorg Med Chem Lett 19(15):4148–4150
57. Sihorkar V, Vyas S (2001) Potential of polysaccharide anchored liposomes in drug delivery, targeting and immunization. J Pharm Pharm Sci 4(2):138–158
58. Huang S-L, MacDonald RC (2004) Acoustically active liposomes for drug encapsulation and ultrasound-triggered release. Biochimica et Biophysica Acta (BBA)-Biomembranes 1665(1–2):134–141
59. Mohamed M et al (2020) Liposomes and PEGylated liposomes as drug delivery system. J Adv Biomed Pharmaceut Sci 3(2):80–88
60. Soni V, Kohli DV, Jain SK (2005) Transferrin coupled liposomes as drug delivery carriers for brain targeting of 5-florouracil. J Drug Target 13(4):245–250
61. Tran MA, Watts RJ, Robertson GP (2009) Use of liposomes as drug delivery vehicles for treatment of melanoma. Pigment Cell Melanoma Res 22(4):388–399
62. Lee RJ, Low PS (1997) Folate-targeted liposomes for drug delivery. J Liposome Res 7(4):455–466
63. Oh JK et al (2008) The development of microgels/nanogels for drug delivery applications. Prog Polym Sci 33(4):448–477
64. Sultana F et al (2013) An overview of nanogel drug delivery system. J Appl Pharm Sci 3(8):95–105
65. Shidhaye S et al (2008) Nanogel engineered polymeric micelles for drug delivery. Curr Drug Ther 3(3):209–217
66. Ding J et al (2013) Self-reinforced endocytoses of smart polypeptide nanogels for "on-demand" drug delivery. J Control Release 172(2):444–455
67. Moya-Ortega MD et al (2012) Cross-linked hydroxypropyl-β-cyclodextrin and γ-cyclodextrin nanogels for drug delivery: physicochemical and loading/release properties. Carbohyd Polym 87(3):2344–2351
68. Van Thienen T, Demeester J, De Smedt S (2008) Screening poly (ethyleneglycol) micro-and nanogels for drug delivery purposes. Int J Pharm 351(1–2):174–185
69. Gonçalves M et al (2014) Dendrimer-assisted formation of fluorescent nanogels for drug delivery and intracellular imaging. Biomacromol 15(2):492–499

70. Jayakumar R et al (2012) Doxorubicin-loaded pH-responsive chitin nanogels for drug delivery to cancer cells. Carbohyd Polym 87(3):2352–2356
71. Choi H et al (2020) Degradable nanomotors using platinum deposited complex of calcium carbonate and hyaluronate nanogels for targeted drug delivery. Part Part Syst Charact 37(1):1900418
72. Wei P et al (2020) Straightforward access to glycosylated, acid sensitive nanogels by host-guest interactions with sugar-modified pillar [5] arenes. ACS Macro Lett 9(4):540–545
73. Morimoto N et al (2013) Self-assembled pH-sensitive cholesteryl pullulan nanogel as a protein delivery vehicle. Biomacromol 14(1):56–63
74. Phan QT et al (2020) Synthesis of zwitterionic redox-responsive nanogels by one-pot amine-thiol-ene reaction for anticancer drug release application. React Funct Polym 147:104463
75. Siirilä J et al (2020) Glucose and maltose surface-functionalized thermoresponsive poly (N-vinylcaprolactam) nanogels. Biomacromol 21(2):955–965
76. Veronese F et al (1998) Polyorganophosphazene microspheres for drug release: polymer synthesis, microsphere preparation, in vitro and in vivo naproxen release. J Control Release 52(3):227–237
77. Martinez AP et al (2017) Biodegradable "smart" polyphosphazenes with intrinsic multifunctionality as intracellular protein delivery vehicles. Biomacromolecules 18(6):2000–2011
78. Hou S et al (2020) Polyphosphazene-based drug self-framed delivery system as a universal intelligent platform for combination therapy against multidrug-resistant tumors. ACS Appl Bio Mater 3(4):2284–2294
79. Zhou N et al (2020) Acid responsive and biologically degradable polyphosphazene nanodrugs for efficient drug delivery. ACS Biomater Sci Eng
80. Xu J et al (2014) High loading of hydrophilic/hydrophobic doxorubicin into polyphosphazene polymersome for breast cancer therapy. Nanomed Nanotechnol Biol Med 10(2):349–358
81. Park M-R et al (2010) Sustained delivery of human growth hormone using a polyelectrolyte complex-loaded thermosensitive polyphosphazene hydrogel. J Control Release 147(3):359–367
82. Cevc G, Blume G (1992) Lipid vesicles penetrate into intact skin owing to the transdermal osmotic gradients and hydration force. BBA-Biomembr 1104(1):226–232
83. Jain S et al (2003) Transfersomes—a novel vesicular carrier for enhanced transdermal delivery: development, characterization, and performance evaluation. Drug Dev Ind Pharm 29(9):1013–1026
84. Jain S, Tiwary A, Jain N (2006) Sustained and targeted delivery of an anti-HIV agent using elastic liposomal formulation: mechanism of action. Curr Drug Deliv 3(2):157–166
85. Maghraby GME, Williams AC, Barry BW (1999) Skin delivery of oestradiol from deformable and traditional liposomes: mechanistic studies. J Pharm Pharmacol 51(10):1123–1134
86. Trotta M et al (2002) Elastic liposomes for skin delivery of dipotassium glycyrrhizinate. Int J Pharm 241(2):319–327
87. Mishra D et al (2006) Elastic liposomes mediated transcutaneous immunization against Hepatitis B. Vaccine 24(22):4847–4855
88. Ganguly S et al (2020) Acoustic cavitation assisted destratified clay tactoid reinforced in situ elastomer-mimetic semi-IPN hydrogel for catalytic and bactericidal application. Ultrason Sonochem 60:104797
89. Ganguly S, Das P, Das NC (2020) Characterization tools and techniques of hydrogels. In: Hydrogels based on natural polymers. Elsevier, pp 481–517
90. Wichterle O, Lim D (1960) Hydrophilic gels for biological use. Nature 185(4706):117–118
91. Hamidi M, Azadi A, Rafiei P (2008) Hydrogel nanoparticles in drug delivery. Adv Drug Deliv Rev 60(15):1638–1649
92. Qiu Y, Park K (2001) Environment-sensitive hydrogels for drug delivery. Adv Drug Deliv Rev 53(3):321–339
93. Alvarez-Lorenzo C et al (2005) Temperature-sensitive chitosan-poly (N-isopropylacrylamide) interpenetrated networks with enhanced loading capacity and controlled release properties. J Control Release 102(3):629–641

94. Amin MCIM et al (2012) Synthesis and characterization of thermo-and pH-responsive bacterial cellulose/acrylic acid hydrogels for drug delivery. Carbohyd Polym 88(2):465–473
95. Shantha K, Harding D (2000) Preparation and in-vitro evaluation of poly [N-vinyl-2-pyrrolidone-polyethylene glycol diacrylate]-chitosan interpolymeric pH-responsive hydrogels for oral drug delivery. Int J Pharm 207(1–2):65–70
96. Kulkarni RV et al (2012) pH-responsive interpenetrating network hydrogel beads of poly (acrylamide)-g-carrageenan and sodium alginate for intestinal targeted drug delivery: synthesis, in vitro and in vivo evaluation. J Colloid Interface Sci 367(1):509–517
97. Ganguly S, Das NC (2015) Synthesis of a novel pH responsive phyllosilicate loaded polymeric hydrogel based on poly (acrylic acid-co-N-vinylpyrrolidone) and polyethylene glycol for drug delivery: modelling and kinetics study for the sustained release of an antibiotic drug. RSC Adv 5(24):18312–18327
98. Ganguly S et al (2016) Synthesis of polydopamine-coated halloysite nanotube-based hydrogel for controlled release of a calcium channel blocker. RSC Adv 6(107):105350–105362
99. Murdan S (2003) Electro-responsive drug delivery from hydrogels. J Control Release 92(1–2):1–17
100. Lira LM, de Torresi SIC (2005) Conducting polymer–hydrogel composites for electrochemical release devices: synthesis and characterization of semi-interpenetrating polyaniline–polyacrylamide networks. Electrochem Commun 7(7):717–723
101. Ganguly S et al (2018) Mechanically robust dual responsive water dispersible-graphene based conductive elastomeric hydrogel for tunable pulsatile drug release. Ultrason Sonochem 42:212–227
102. Servant A et al (2013) Design, engineering and structural integrity of electro-responsive carbon nanotube-based hydrogels for pulsatile drug release. J Mater Chem B 1(36):4593–4600
103. Ganguly S et al (2018) Green reduced graphene oxide toughened semi-IPN monolith hydrogel as dual responsive drug release system: rheological, physicomechanical, and electrical evaluations. J Phys Chem B 122(29):7201–7218
104. Indermun S et al (2014) An interfacially plasticized electro-responsive hydrogel for transdermal electro-activated and modulated (TEAM) drug delivery. Int J Pharm 462(1–2):52–65
105. Pourjavadi A, Doroudian M (2015) Synthesis and characterization of semi-conductive nanocomposite based on hydrolyzed collagen and in vitro electrically controlled drug release study. Polymer 76:287–294
106. Satarkar NS, Hilt JZ (2008) Magnetic hydrogel nanocomposites for remote controlled pulsatile drug release. J Control Release 130(3):246–251
107. Bourlinos AB et al (2008) Photoluminescent carbogenic dots. Chem Mater 20(14):4539–4541
108. Das P et al (2019) Graphene based emergent nanolights: a short review on the synthesis, properties and application. Res Chem Intermed 45(7):3823–3853
109. Bourlinos AB et al (2008) Surface functionalized carbogenic quantum dots. Small 4(4):455–458
110. Ganguly S et al (2019) Advancement in science and technology of carbon dot-polymer hybrid composites: a review. Funct Compos Struct 1(2):022001
111. Ganguly S et al (2020) Microwave-synthesized polysaccharide-derived carbon dots as therapeutic cargoes and toughening agents for elastomeric gels. ACS Appl Mater Interfaces 12(46):51940–51951
112. Saravanan A et al (2020) Applications of N-doped carbon dots as antimicrobial agents, antibiotic carriers, and selective fluorescent probes for nitro explosives. ACS Appl Bio Mater 3(11):8023–8031
113. Ganguly S et al (2018) Natural saponin stabilized nano-catalyst as efficient dye-degradation catalyst. Nano Struct Nano Objects 16:86–95
114. Ganguly S et al (2020) Poly (N-vinylpyrrolidone)-stabilized colloidal graphene-reinforced poly (ethylene-co-methyl acrylate) to mitigate electromagnetic radiation pollution. Polym Bull 77(6):2923–2943
115. Mondal S et al (2018) Thermal-air ageing treatment on mechanical, electrical, and electromagnetic interference shielding properties of lightweight carbon nanotube based polymer nanocomposites. Compos A Appl Sci Manuf 107:447–460

116. Xu X et al (2004) Electrophoretic analysis and purification of fluorescent single-walled carbon nanotube fragments. J Am Chem Soc 126(40):12736–12737
117. Zuo P et al (2016) A review on syntheses, properties, characterization and bioanalytical applications of fluorescent carbon dots. Microchim Acta 183(2):519–542
118. Baker SN, Baker GA (2010) Luminescent carbon nanodots: emergent nanolights. Angew Chem Int Ed 49(38):6726–6744
119. Das P et al (2020) Carbon dots for heavy-metal sensing, pH-sensitive cargo delivery, and antibacterial applications. ACS Appl Nano Mater 3(12):11777–11790
120. Cao L et al (2011) Carbon nanoparticles as visible-light photocatalysts for efficient CO_2 conversion and beyond. J Am Chem Soc 133(13):4754–4757
121. Wang F et al (2011) White light-emitting devices based on carbon dots' electroluminescence. Chem Commun 47(12):3502–3504
122. Das P et al (2017) A simplistic approach to green future with eco-friendly luminescent carbon dots and their application to fluorescent nano-sensor 'turn-off' probe for selective sensing of copper ions. Mater Sci Eng C 75:1456–1464
123. Das P et al (2018) Waste chimney oil to nanolights: a low cost chemosensor for tracer metal detection in practical field and its polymer composite for multidimensional activity. J Photochem Photobiol B 180:56–67
124. Das P et al (2019) Surface quaternized nanosensor as a one-arrow-two-hawks approach for fluorescence turn "on–off–on" bifunctional sensing and antibacterial activity. New J Chem 43(16):6205–6219
125. Das P et al (2019) Converting waste Allium sativum peel to nitrogen and sulphur co-doped photoluminescence carbon dots for solar conversion, cell labeling, and photobleaching diligences: a path from discarded waste to value-added products. J Photochem Photobiol B Biol 197:111545
126. Das P et al (2017) Green approach to photoluminescent carbon dots for imaging of gram-negative bacteria Escherichia coli. Nanotechnology 28(19):195501
127. Das P et al (2018) Dual doped biocompatible multicolor luminescent carbon dots for bio labeling, UV-active marker and fluorescent polymer composite. Luminescence 33(6):1136–1145
128. Guo S et al (2009) Monodisperse mesoporous superparamagnetic single-crystal magnetite nanoparticles for drug delivery. Biomaterials 30(10):1881–1889
129. Conde J et al (2013) In vivo tumor targeting via nanoparticle-mediated therapeutic siRNA coupled to inflammatory response in lung cancer mouse models. Biomaterials 34(31):7744–7753
130. Chen M-L et al (2013) Quantum-dot-conjugated graphene as a probe for simultaneous cancer-targeted fluorescent imaging, tracking, and monitoring drug delivery. Bioconjug Chem 24(3):387–397
131. Ganguly S et al (2019) An insight into the physico-mechanical signatures of silylated graphene oxide in poly (ethylene methyl acrylate) copolymeric thermoplastic matrix. Macromol Res 27(3):268–281
132. Yang X et al (2010) Multifunctional stable and pH-responsive polymer vesicles formed by heterofunctional triblock copolymer for targeted anticancer drug delivery and ultrasensitive MR imaging. ACS Nano 4(11):6805–6817
133. Das P et al (2019) Heteroatom doped blue luminescent carbon dots as a nano-probe for targeted cell labeling and anticancer drug delivery vehicle. Mater Chem Phys 237:121860
134. Karthik S et al (2013) Photoresponsive quinoline tethered fluorescent carbon dots for regulated anticancer drug delivery. Chem Commun 49(89):10471–10473
135. Wang Q et al (2013) Hollow luminescent carbon dots for drug delivery. Carbon 59:192–199
136. Feng T et al (2016) Dual-responsive carbon dots for tumor extracellular microenvironment triggered targeting and enhanced anticancer drug delivery. ACS Appl Mater Interfaces 8(29):18732–18740
137. Das P et al (2019) Biocompatible carbon dots derived from κ-carrageenan and phenyl boronic acid for dual modality sensing platform of sugar and its anti-diabetic drug release behavior. Int J Biol Macromol 132:316–329

138. Ganguly S et al (2019) Microwave assisted green synthesis of Zwitterionic photolumenescent N-doped carbon dots: an efficient 'on-off' chemosensor for tracer Cr (+ 6) considering the inner filter effect and nano drug-delivery vector. Colloids Surf A Physicochem Eng Aspects 579:123604
139. Lin C et al (2020) Carbon dots embedded metal organic framework@ chitosan core-shell nanoparticles for vitro dual mode imaging and pH-responsive drug delivery. Micropor Mesopor Mater 293:109775
140. Duan Q et al (2020) Construction and application of targeted drug delivery system based on hyaluronic acid and heparin functionalised carbon dots. Colloids Surf B Biointerfaces 188:110768
141. Budzynska R et al (2007) Antitumor activity of mannan–methotrexate conjugate in vitro and in vivo. Oncol Res Featur Preclin Clin Cancer Therapeut 16(9):415–421
142. Garg M, Dutta T, Jain NK (2007) Reduced hepatic toxicity, enhanced cellular uptake and altered pharmacokinetics of stavudine loaded galactosylated liposomes. Eur J Pharm Biopharm 67(1):76–85
143. Dutta T, Jain NK (2007) Targeting potential and anti-HIV activity of lamivudine loaded mannosylated poly (propyleneimine) dendrimer. Biochimica et Biophysica Acta (BBA)-General Subjects 1770(4):681–686
144. Agarwal A et al (2008) Ligand based dendritic systems for tumor targeting. Int J Pharm 350(1–2):3–13
145. Kolomiets E et al (2009) Glycopeptide dendrimers with high affinity for the fucose-binding lectin LecB from Pseudomonas aeruginosa. ChemMedChem Chem Enab Drug Discov 4(4):562–569
146. Frisch B, Hassane FS, Schuber F (2010) Conjugation of ligands to the surface of preformed liposomes by click chemistry. In: Liposomes. Springer, pp 267–277
147. Garg M, Jain NK (2006) Reduced hematopoietic toxicity, enhanced cellular uptake and altered pharmacokinetics of azidothymidine loaded galactosylated liposomes. J Drug Target 14(1):1–11
148. Suriano F et al (2010) Synthesis of a family of amphiphilic glycopolymers via controlled ring-opening polymerization of functionalized cyclic carbonates and their application in drug delivery. Biomaterials 31(9):2637–2645
149. Oda Y et al (2008) Design, synthesis and evaluation of d-galactose-β-cyclodextrin conjugates as drug-carrying molecules. Bioorg Med Chem 16(19):8830–8840
150. Jain NK, Jain SK (2010) Development and in vitro characterization of galactosylated low molecular weight chitosan nanoparticles bearing doxorubicin. AAPS Pharm Sci Tech 11(2):686–697

Containers for Encapsulation of Fragrances/Aroma/Odour for Textile Applications

Latika Bhatt, Ruchi Kholiya, and Srishti Tewari

1 Introduction

There is a continuous change in the global market and the consumers have not only become aware but their demands have also increased with respect to technologically advanced textiles. Consumers prefer textiles not only for their aesthetics but added functionalities. This has put a pressure on the textile industry to come up with innovative solutions which will strengthen the existing product line. Competition within the textile industry and the rise in consumer demand has opened up various opportunities to innovate and add value to the textile materials. Fragrance/Aroma finishing is one of such emerging area which enhances the value and usability of the textile product by incorporation of aroma into it. The fragrances which are extracted from plant sources i.e. essentials oils not only give off a pleasant smell rather have been utilized as antiseptic, anti-inflammatory, antimicrobial and for the purpose of emotional calming.

Aroma or pure fragrances extracted from plant sources have been used traditionally as supplementary medicine. It has found its application nowadays because people have started believing in sustainable and healthy lifestyles and also prefer

L. Bhatt (✉)
National Institute of Fashion Technology, Mumbai, India
e-mail: latika.bhatt@nift.ac.in

R. Kholiya
Footwear Design and Development Institute, Rohtak, India
e-mail: ruchi.kholiya@fddiindia.com

S. Tewari
Institute of Chemical Technology, Mumbai, India
e-mail: tch19sn.tewari@pg.ictmumbai.edu.in

J. Parameswaranpillai et al. (eds.), *Micro- and Nano-containers for Smart Applications*, Composites Science and Technology,
https://doi.org/10.1007/978-981-16-8146-2_7

products that use natural sources rather chemical. An 'aroma compound' is a chemical compound which gives out a smell or odour. The pre-requisites of a chemical compound to have odour or fragrance are:

(i) the compound should be volatile for it to be transported to the olfactory system.
(ii) that the compound should have sufficient concentration for it to interact with the olfactory receptors present in the nose.

Once received by the olfactory system, the stimuli are transferred to the brain which perceives it as either pleasant or unpleasant odour.

Aroma compounds are complex mixtures which contain various aromatic compounds with different volatile activity. They can be classified into three types: top notes (that last for few minutes), middle notes (that last for 2–5 h and they give the main character to the fragrance) and base notes (these may last more than 8 h). The base notes act as fixatives, which help in increasing the lifetime of the fragrance [30]. Researchers have been trying to impart fragrance/aroma to textiles for many years. Trials have been done to impart aroma as fabric conditioners but the effect did not last long. The textile chemists are often confronted with the challenge of ensuring durability of the aroma in the finished textiles. The volatile nature of fragrance causes it to evaporate away and hence the fragrance finished textiles cannot be put for prolonged usage [5].

Researchers face challenge in developing the fragrance finished textile because most of the aroma compounds being volatile in nature are not very durable. Researchers have been working hard to develop effective mechanism of aroma application, which not only control the release of the aroma but also give better wash durability to the finished textiles. Many techniques have been tried and to develop containers for the aroma in order to ease the finishing process and to make them last longer. The techniques have been discussed in detail in this chapter.

The fragrance finished textile market can bring about a revolution in the textile industry and will attract the consumer because of the well-being and physiological and psychological benefits associated with it. Development of aroma finished textiles products is an innovative approach and user-friendly idea to impart not only fragrance but its healing effects to the user.

1.1 Need for Fragrance/Aroma Application Textiles

The everyday use of textiles, both home and apparel make them smell due to the perspiration, dirt and dust that gets accumulated. Innerwear, sportswear, yoga wear, athleisure wear, shoes and socks are constantly exposed to human sweat which contains a lot of microorganisms resulting in bad odour. Addition of fragrance to such textile will not only add value to them but will increase consumer satisfaction too. The user/wearer will not only feel fresh but also a sense of wellbeing.

The current 'green or sustainable' trend is making both the industry and consumer choose natural products over the chemical and non-sustainable products. The use of essential oils in place of artificial scent is considered to be environmentally responsible. The multiplication of research and the inclination of market for the use of essential oils for fragrance and its application in textiles will find acceptance from the consumers and the industry.

1.2 Limitations and Challenges

Various attempts have been made to develop aroma finished textiles in the last few years. The major issues have always been the volatility which leads to short term effect. Fragrance being volatile in nature and prone to high temperature do not remain of fabrics for longer time and does not last many launderings. During the washing, the combined effect chemical and mechanical action results in reduction of the available fragrance on textiles and the desired effect. There are more factors that can affect the fragrance finish on fabrics like quality water used for washing (hardness/softness, pH), exposure to sunlight, exposure to organic solvents used in washing, ironing, dry cleaning, perspiration etc. The important considerations to obtain the fragrance finished textile are:

(i) To make the fragrance compounds and essential oils without omitting any ingredient in order to ensure its pharmaceutical effects.
(ii) To maintain durability in laundering.
(iii) To retain soft handleof the finished textile.

2 Chemistry of Aromatic/Fragrance Compounds

Aromatic plants have been used for incense, perfumes and cosmetics since long time. These plants have also been successfully used to extract aroma and put to medical and culinary applications. Traditional medicine in the form of essential oils was accepted as an alternative as well as complementary form for health care. The indigenous plants and the extracted oils have found their use in herbal medicine. Essential Oils (EOs) are secondary plant metabolites that can be derived from root, stem, bark, wood, leaves and even flowers, and fruits, etc. A variety of complex organic compounds responsible for fragrance and aroma are present in the oils. These compounds include simple alcohols, mono-terpenes, sesquiterpenes, phenols, ketones, and coumarins [15]. Monoterpenes, sesquiterpenes, and phenols are the chief secondary metabolite compounds found in the plant system [12].

The presence of lower isoprenoids in essential oils is often associated with the repellent properties. Limonene, α-pinene, terpinolene, camphor, citronellol, citronellal, and thymol are few of those constituents which are found in a number of essential oils showing mosquito repellent activity [29].

EOs can be classified into the four main groups (Fig. 1).

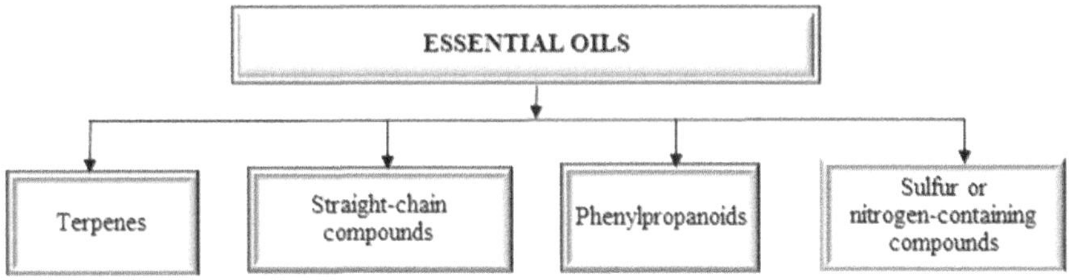

Fig. 1 Types of essential oils

1. Terpenes, related to isoprene
2. Straight-chain compounds not containing any side chain
3. Phenylpropanoids (benzene derivatives)
4. Miscellaneous group having varied structures not included in first three groups (sulfur- or nitrogen-containing compounds)

The essential oils are generally composed of two to three major constituents at higher concentrations (20–70%) while the other components in small amounts. The composition of an essential oil is influenced by number of parameters including the environmental factors such as type of soil, altitude, exposure to sunlight, rains, seasonal variations, storage conditions, methods of extraction, techniques of cultivation and also on the genetic factors. Generally in an essential oil 90–95% of compounds have volatile nature like terpenes, their oxygenated compounds along with aliphatic aldehydes, alcohols and esters [28]. The rest 5–10% compounds are nonvolatile residues containing hydrocarbons, fatty acids, sterols, carotenoids, waxes, and flavonoids. Among the volatile compounds monoterpenoids, sesquiterpenoids, sometimes diterpenoids and phenylpropanoids are most common in the EO [8]. Terpenoids are natural hydrocarbons, based on five-carbon units as their building blocks, these compounds are classified into hemiterpenoids, monoterpenoids, sesquiterpenoids, diterpenoids, tetraterpenoids, triterpenoids and polyterpenoids. The most well-known hemiterpenoid is isoprene (C5H8), which is the building block of all terpenes but not usually found in EO's. Terpenes are the polymers of isoprene (C_5H_8) joined head-to-tail as shown in Figs. 2 and 3 depicts the acyclic and monocyclic monoterpenes.

Monoterpenoid made up of two condensed isoprene units which are abundantly available in oil glands of secretary tissues among higher plants, fungi, insects and also in marine organisms; majority of these compounds exhibit biological activities and fragrance. Sesquiterpenoid is made up of three isoprene units; because of their increased chain length and additional 16 double bond, these compounds are available in cyclic forms, exampled by zingiberene and caryophylleneetc [10]. Diterpenoid is composed of four isoprene units and these molecules are too heavy for evaporation during distillation, so these compounds are rarely found in distilled EOs. Phenylpropanoids are well known for the antimicrobial, anti-inflammatory, sedative and analgesic properties and large group of these compounds also play vital

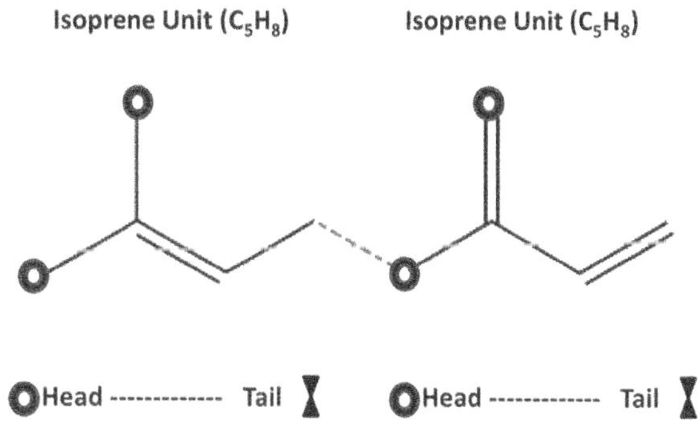

Fig. 2 Link of isoprene molecules

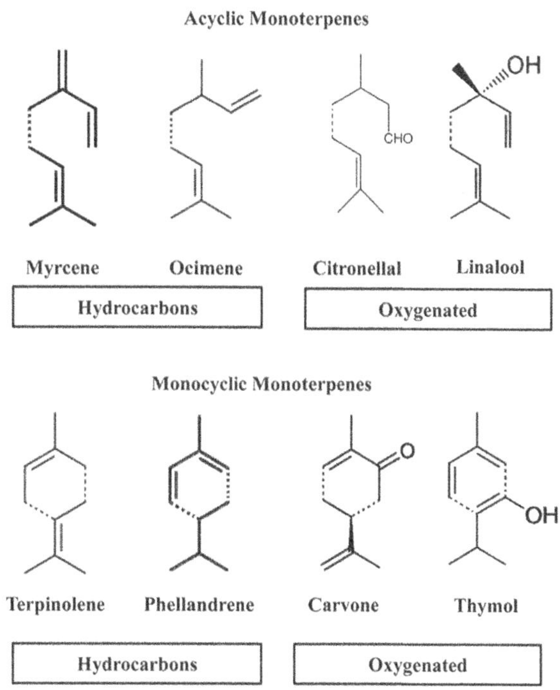

Fig. 3 The chemical structures of acyclic and monocyclic monoterpenes

role in protecting plants from microbial infections and UV radiation, wounding and herbivores [34].

As of date nearly 1,00,000 secondary metabolites have been identified of which 3000 compounds are from EO sources. Among these 3000, only 300 compounds are utilized for food, cosmetic and pharmaceutical industries for commercialization.

3 Methods to Prepare Aroma/Fragrance Containers

3.1 *Microencapsulation*

A method of surrounding one substance with another substance to yield capsules of micrometer order is known as microencapsulation. The morphology of microcapsules is spherical, with core material surrounded by a continuous wall. Figure 4 shows the general structure of microcapsule. The wall may be asymmetrically shaped due to the quantity of core material i.e. droplets which are encapsulated. All the three phases of matter can be microencapsulated. While liquid and gas can be easily encapsulated, solid contents are difficult to handle during the microencapsulation process. The core contents of microcapsule are released under controlled conditions which are suitable to the specific purpose or end use [4].

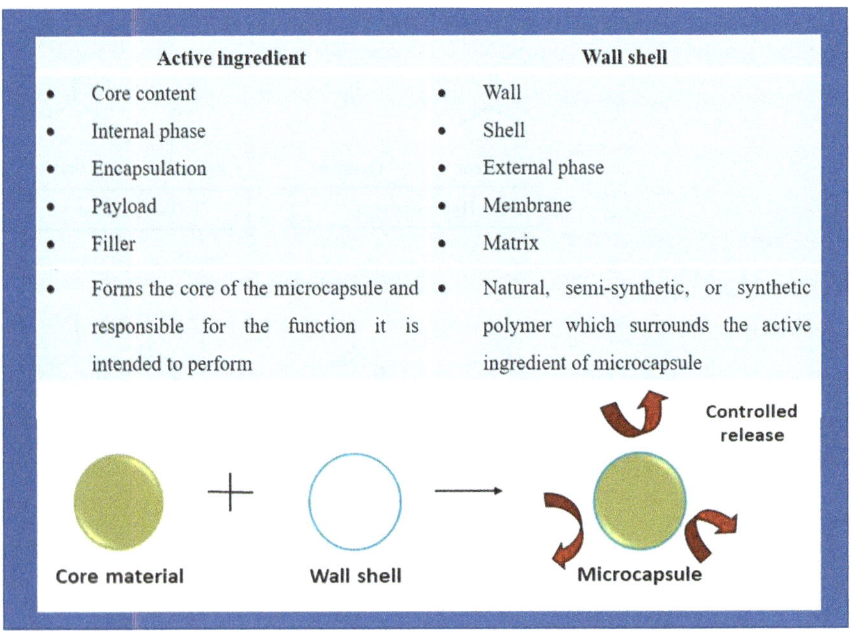

(Cheng *et al.*, 2008).

Fig. 4 General structure of microcapsule [9]

3.1.1 Methods to Prepare Aroma Microcapsules

Abundant studies have reported preparation technologies for the encapsulation of core material. The popular microencapsulation techniques relevant for the encapsulation of aromatic/ essential oils are discussed below.

(a) **Phase separation/coacervation**: This process aims at partial dissolution of a homogenous polymer solution into a polymer rich phase (coacervate) and the poor polymer phase (coacervation medium). Presently, simple and complex processes are the two methods available for coacervation. Amongst both, the mechanism of formation of microcapsule is the same. In simple coacervation a dissolution agent is added for phase separation, whereas complex coacervation involves complexation between two oppositely charged polymers [3]. The three basic steps in complex coacervation are: (i) formation of three immiscible phases, (ii) deposition of the coating (iii) rigidization of the coating.

Coacervation is a unique and promising microencapsulation technology because of very high payloads achievable and controlled release properties based on mechanical stress, temperature and sustained release. Coacervation is typically used to encapsulate oils [21, 22] but can also be adapted for the encapsulation of the nutrients, vitamins, preservatives, enzymes etc. [19]. Figure 5 shows the coacervation process.

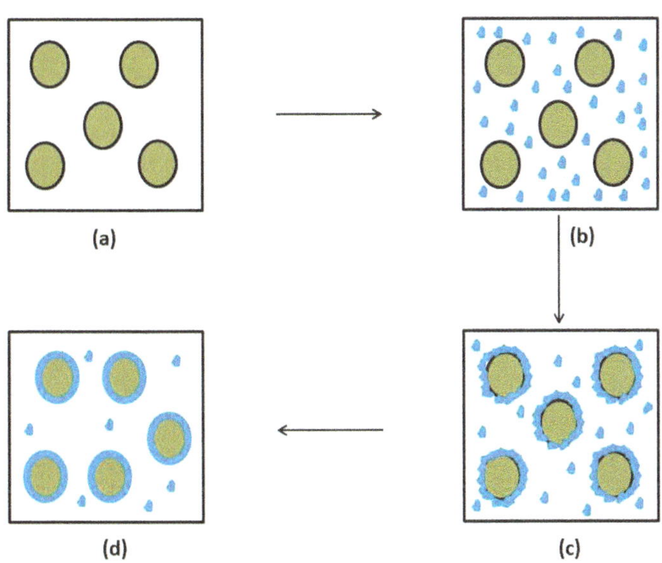

(a) core material dispersion in solution, (b) separation of coacervate from solution, (c) coating of core material by coacervate, (d) coacervate forms continuous shell around core

Fig. 5 Schematic representation of the coacervation process

The concept behind simple or complex coacervation microencapsulation is the phase separation of one or many hydrocolloids from the initial solution and the subsequent deposition of the newly formed coacervate phase around the active ingredient suspended in the same reaction media. The hydrocolloid shell can then be crosslinked, using an appropriate chemical or enzymatic crosslinker. A large number of hydrocolloid system have been evaluated for coacervation microencapsulation but the most studied and well understood system is the gelatin/ gum acacia system [35].

(a) core material dispersion in solution, (b) separation of coacervate from solution, (c) coating of core material by coacervate, (d) coacervate forms continuous shell around core.

(b) **Spray Drying**: It is a low-cost commercial process for microencapsulation which is mostly carried out for the encapsulation of fragrances, oils and flavours. Core particles are dispersed in a polymer solution and then sprayed into a hot chamber. The shell material solidifies onto the core particles with the evaporation of solvent and polynuclear or matrix type microcapsules are obtained. The drawback that has come to light with this process is that the encapsulated particles are aggregated and the use of large amounts of core material can lead to uncoated particles. However, higher loadings of core particles of up to 50–60% have been reported. Water-soluble polymers are purposely used as shell materials because solvent-borne systems produce unpleasant odours and environmental problems. Schematic diagram (Fig. 6) illustrating the process of micro-encapsulation by spray- drying.

(c) **Centrifugal Extrusion**: In this method, liquid drops are encapsulated by a rotating extrusion head having concentric nozzles. A jet of core liquid is covered by a sheath of wall material either in solution or melt form. By the movement of jet through the air, the droplets are broken and then coated with the wall solution. While the droplets are in flight, a molten wall may be hardened or a solvent may be evaporated from the wall solution. most of the droplets are within $\pm 10\%$ of the mean diameter, hence they land in a narrow ring around the spray nozzle. Later, the capsules can be hardened after formation by catching them in a ring-shaped hardening bath if required so. This process is quite efficient for forming particles 400–2,000 μm in diameter. The process involved the formation of dropsby the breakup of a liquid jet, it is only suitable for liquid or slurry [18].

3.2 *Nanoemulsion*

Kinetically stable emulsions containing sub-system of water and oil separated by surfactants. Co-surfactants are sometimes used to maintain the stability of the nanoemulsions. Nanoemulsion is also interchangeably termed as submicron, ultra-fine or miniemulsions [42]. They have appeared as attractive target systems for food,

Fig. 6 Schematic diagram of spray- drying process

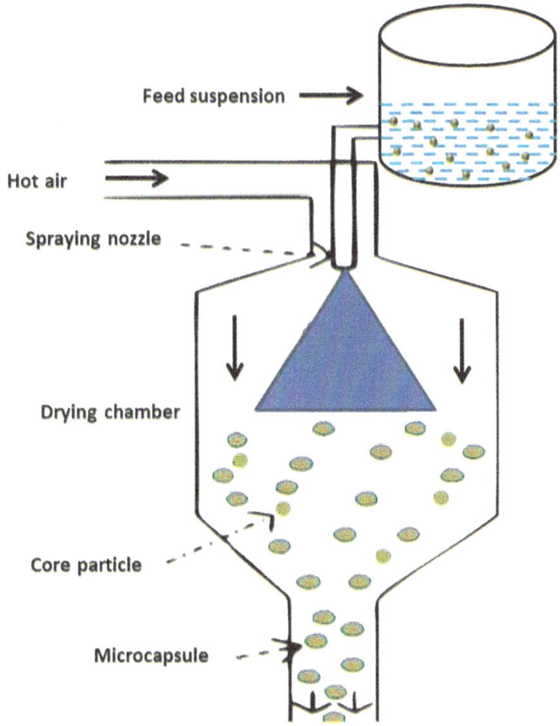

cosmetic and pharmaceutical industry [8]. Figure 7 shows the structure of oil in water (O/W) and water in oil (W/O) nanoemulsion.

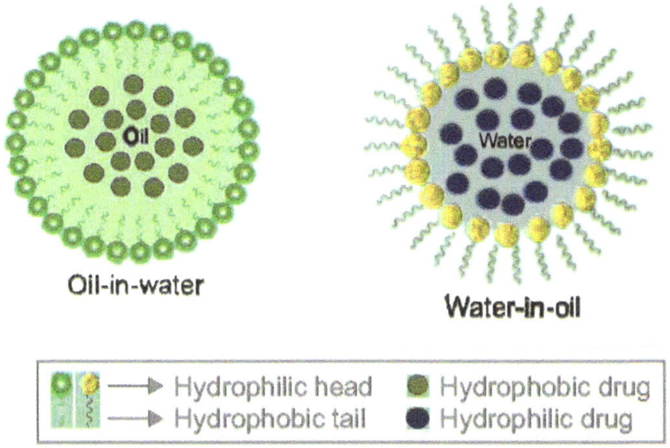

Fig. 7 Structure of oil in water and water in oil nanoemulsion

3.2.1 Nanoemulsion Stability

Some common separation mechanism seen in emulsions are coalescence, flocculation, creaming, and Ostwald ripening. The smaller droplets gradually combine to form larger droplets in coalescence while larger droplets formed in coalescence separate from emulsion suspension in flocculation. The constituents undergo phase separation on the basis of different densities in creaming.

Ostwald ripening is the phenomenon in which the larger droplets grow at the expense of smaller ones. Mc Clements et al., documented that use of extra oils preferably long chain triglycerides or addition of weighting agents can reduce the Ostwald ripening. Being thermodynamically unstable system, nanoemulsions are less prone to decay mechanism like gravitational separation, coalescence and flocculation but decay of nanoemulsion due to Ostwald ripening is more likely to occur [25]. Figure 8 represents the various decay mechanism in a nanoemulsion.

Saulku et al., Encapsulated citronella oil nanoemulsion prepared by high pressure homogenization at varying amounts of surfactant and glycerol. They studied the stability of nanoemulsion for a period of 2 months by monitoring over the change in particle size. After 1 day of storage a phase separation was found in formulations containing glycerol concentration of 100% (w/w) in aqueous phase.

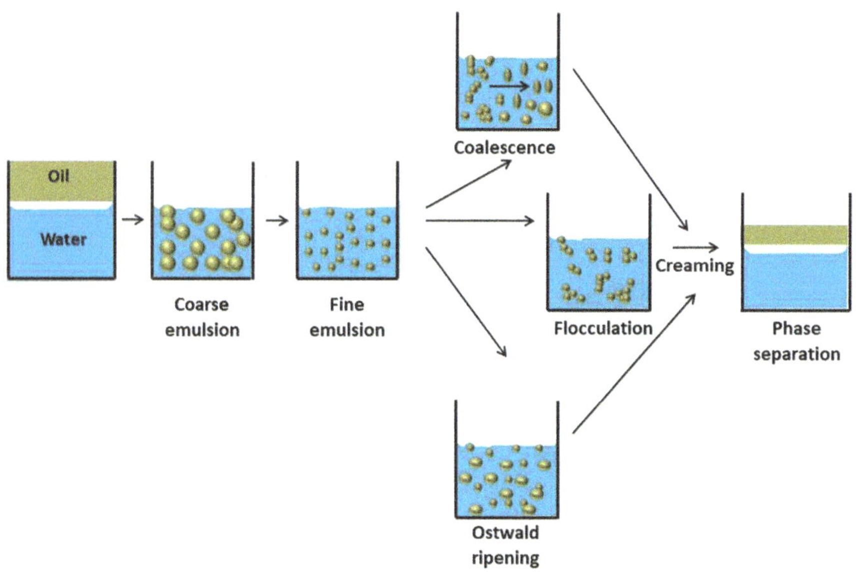

Fig. 8 Different types of decay mechanisms in a nanoemulsion

3.2.2 Methods of Nanoemulsion Formation

Both low energy and high energy methods are implemented in nanoemulsion preparation.

Low energy methods include spontaneous emulsification, membrane emulsification, phase inversion composition (PIC) method, emulsion inversion point (EIP) method, solvent displacement method, and phase inversion temperature (PIT) method. High energy emulsification requires mechanical devices such as ultrasonicator [46], microfluidizer [37], and high pressure homogenizer [14].

Low Energy Emulsification

Low energy methods for formation of nanoemulsion rely on the spontaneous formation of oil droplets due to energy dispersion between oil and aqueous phase. Surfactants play a very important role in nanoemulsion formation and stability and are required in high quantities in low energy emulsification.

Spontaneous emulsification is both cost-effective and energy efficient method on emulsion preparation. It utilizes the chemical energy stored in the system for emulsion preparation [36]. This mechanism happens when two different phases (an organic phase and an aqueous phase) are mixed together spontaneously.

Membrane emulsification comparatively uses less amount of surfactant. Dispersed phase is formed through a membrane into a continuous phase, consequentially reducing the droplet size and less variation in droplet sizes.

PIT (Phase Inversion Temperature) emulsification is a widely used method for the manufacturing of nanoemulsion. This method depends on the changes in solubility of surfactants in the continuous phase with temperature. The basic principle of PIT lies in the ability of the non-ionic surfactants, which can modify their affinities for water and oil both at fixed temperature. Coalescence and instability of emulsions are the main drawbacks of PIT emulsification method [2].

High-Energy Emulsification

High-energy emulsification involves use of vigorous and turbulent forces to break larger droplets into smaller ones [13]. The size of droplets which can be produced depends on the type of device used for high energy approach. The parameters which influence the high energy emulsification include operating conditions (time, energy input, and temperature), composition of sample (oil, emulsifier and their concentrations), and the physicochemical characteristics of the phases (interfacial tension and viscosity). Unstable molecules like proteins and peptides cannot be emulsified using this method. It is more efficient for flavors and fragrances.

3.3 *Nanoparticles/Nanocapsules*

The polymeric nanoparticles (PNPs) are prepared from polymers in size between 10 and 1000 nm where the material is dissolved, entrapped, encapsulated or attached to a nanoparticle matrix. Nanocapsules are systems in which the core material is confined to a cavity surrounded by a unique polymer membrane, while nanospheres are matrix systems in which the core material is physically and uniformly dispersed [27].

The small droplet size and high kinetic stability of nanocapsules improves the solubility, stability and bioactivity of various oil-soluble phytochemicals [38].

Nanocapsulation finds its application in dyeing and printing on textiles, medical and health products, fragrance and aromatherapy, moth and insect proofing, cleaners and detergents, military cloth, photochromic and thermochromic fibers, flame retardant, thermo-regulating fibers and smart textiles [16]. Nanoparticles can be easily absorbed in high amount on natural fibers mainly hair fibers and cotton. NPs have strong resistance to rinsing and therefore they can assure a long lasting fragrance release [6].

3.3.1 Methods of Preparing Nanoparticles/Nanocapsules

A wide variety of methods have been employed to prepare nanoparticles. In Fig. 9, the conventional and non-conventional methods of nanoparticle formation have been listed.

Various approaches have been suggested by various researchers for the preparation of NP's. Solvent Extraction is one of the processes in which aqueous and organic phase is used to prepare emulsion. Surfactants are used to increase the stability of the emulsion. If organic phase is used, it is evaporated at the next step of the process [17].

Another process is spontaneous emulsification which is a better process then the convention solvent evaporation. In this process the oil phase consists of organic solvent and water insoluble solvent. Polymeric nanoparticles are achieved in this process. Emulsification diffusion method which is also known as salting out process is a process which can be used to prepare nano particles without the use of surfactant and chlorine (Moinard-Checot et al. 2006). Here the salting agent plays the vital role and the right choice of salting agent will help in achieving nanoparticles with good durability. Nano precipitation method uses the solvent displacement phenomenon.

All the conventional techniques use either some toxic chemicals or surfactants. An ecofriendly approach to nanoparticles formation is by using the super critical fluid. This approach uses CO_2 in its supercritical form and results in nanoparticles with high purity.

The polymerization process in the emulsion leads to a matrix of nanoparticles called nanospheres. Both conventional emulsion polymerization and non-conventional surfactant emulsion polymerization can be used for nanoparticles

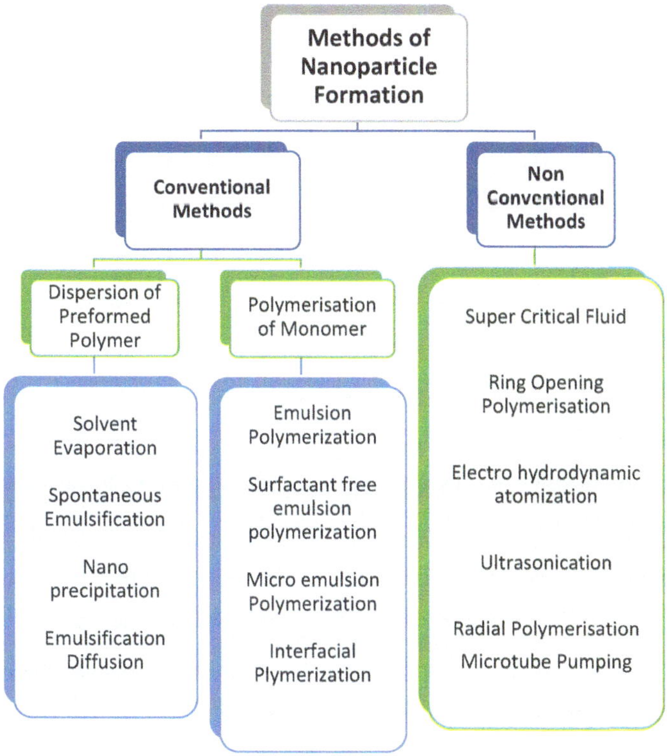

Fig. 9 Method of preparation of nanoparticles

formation. Though, the surfactant free method is costlier and slower. In microemulsion polymerization a high shear equipment like homogenizer or ultrasonicator is used. Micro emulsion polymerization has recently gained a lot of interest as it produces thermodynamically stable nanoemulsions. In interfacial polymerization two reactive monomers are dissolved in two different phases and the reaction takes place in the interface.

3.4 *Release Mechanism of Aroma/Fragrance Containers*

The release mechanism of active ingredients from the shell depends mainly on the wall material and the specific end use. The functional element may be released from the shell either by friction, pressure, change of temperature or diffusion through the polymer wall. Dissolution of the polymer wall and biodegradation are also the factors for releasing the core content of the aroma containers [18]. Figure 10 shows the release mechanism of microcapsules under pressure.

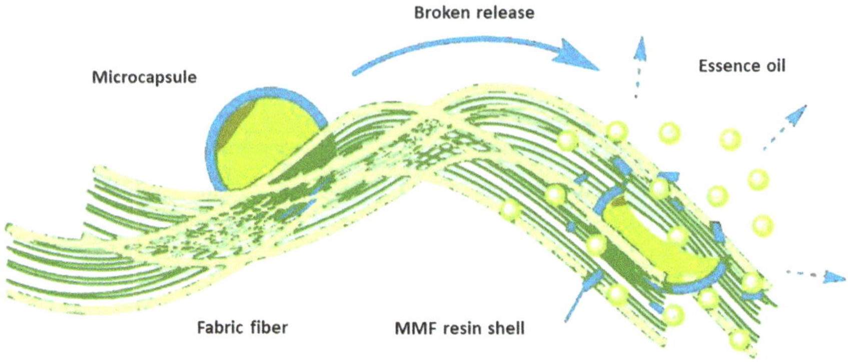

Fig. 10 Release of aroma microcapsules under external force

i. **Mechanical stimuli**: Pressure and friction results in the physical and mechanical destruction of the wall and hence the core content is released in a controlled manner.
ii. **Chemical stimuli**: Dissolution of wall material due to the action of some chemical or solvent is the application of chemical stimuli. Enzymatic action, pH, and temperature are also categorized as chemical stimuli.
iii. **Thermal stimuli**: Action of temperature majorly higher than melting point also leads to the dissolution of the wall polymer hence releasing the core contents.
iv. **Diffusion**: The core contents of the aroma containers get diffused into external liquid such as water. The main factor responsible for diffusion stimuli is vapour pressure of contents inside and the outside of the wall [16].

4 Application Methods of Fragrance Containers on Textiles

Textile materials are found to be one of the best ways to impart the effects of essential oils to the wearer [43]. Therapeutic textiles, antibacterial textiles, medical textiles, insect repellent textiles, mosquito repellent textiles, fragrance finished textiles etc. are various examples where essential oils are incorporated into textiles not only to impart fragrance but other effects as well.

Researchers have tried various methods of application of fragrance/aroma in the textiles. Fragrance either in its pure form, as micro capsules, nanocapsules, nanoparticles, nanoemulsions have been applied on fabrics using different techniques. There are certain important requirements of fixing these fragrances/aromata into the textiles:

- They should not change the visual appeal of the substrate.
- The applied materials should not affect the surface properties of the fabric and should remain comfortable to be worn near skin.
- The applied fragrance/aroma should withstand at least15-30 wash cycles depending upon the end usage.

Below we have compiled various methods through which these aroma containers can be applied on the textiles.

4.1 The Pad Dry Cure Method

The pad dry cure method is the most common method used to finish textiles. In this, the fabric is infused with the finishing solution (fragrance element) by squeezing rollers with low temperature binder. The fabric is then washed to remove unfixed compounds. Pad dry cure is a continuous process in which concentrated dispersion of fragrance/aroma compound is prepared along with binder and softener. First the substrate is dipped in this dispersion and then passed through the pneumatic padding mangle for uniform and adequate pickup. Later the substrate is dried and cured at around 100–120 °C for 1–3 min.

Bhatt and Singh applied lemongrass oil on fabric by two methods: oil microcapsules applied through padding and exhaust method. Lemongrass oil microcapsules were prepared using complex coacervation technique and applied on fabric using pad dry cure technique. The results revealed that fabrics treated with lemongrass oil microcapsules were more effective in fragrance retention as compared to the fabrics treated with exhaust methods. The lemongrass oil microcapsules treated fabric also showed 80% bacterial reduction [5].

4.2 Application in the Form of Polymer Matrices

Polyelectrolyte's either from chemical or natural source can be used to bind aroma compounds to textile surfaces. For polymer matrices both natural and chemical materials can be used. Natural materials like proteins, carbohydrates i.e. alginate, chitosan, gellan gum, collagen, gelatin, fibrin, silk fibroin etc. and synthetic polymers like various polyamides, polyoxides or aliphatic polymers can be used.

4.3 Layer-By-Layer Method

This method involves the synthesis of nano sized organic layers containing functional groups, which can alleviate the surface properties of textile material as desired. The main idea behind layer-by-layer method is subsequent depositing of the oppositely charged polyelectrolyte layers over the substrate (Fig. 11). The electrostatic forces withhold layers with each other. Nanosized layers are deposited on the substrate by subsequent dipping. The layer-by-layer method has gained popularity as it is very easy to useand requires no special equipment. Not only polyelectrolytes but also almost any type of charged nano entities can be used to create the nanolayers. The

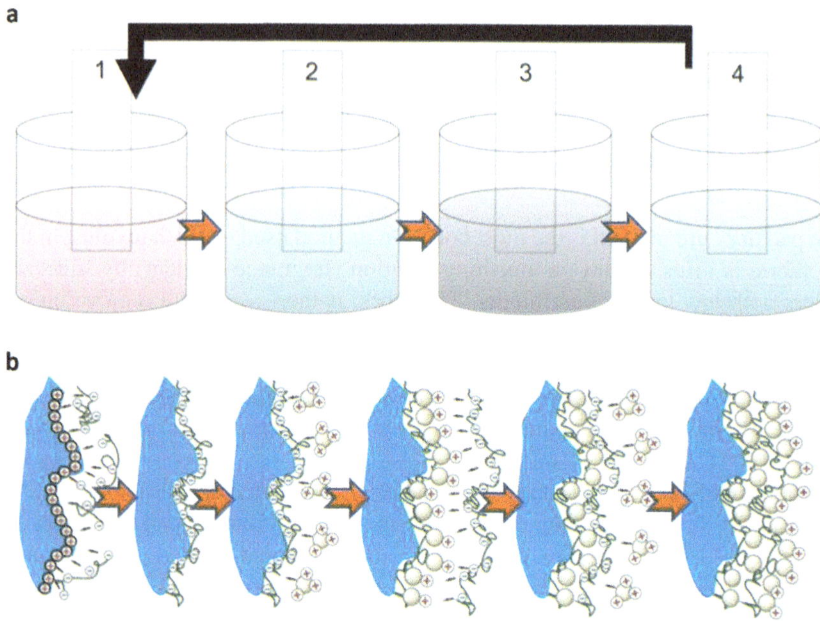

Fig. 11 Schematic representation of polymeric multi layers using layer by layer technique

fixation of the layers to the substrate is depends uponthe properties of the polymer used [11].

4.4 *Using β-Cyclodextrin Inclusion Process*

β-cyclodextrin can react with essential oils to form inclusion compounds. β-Cyclodextrin and aroma compounds can be applied on fabrics with the help of both continuous and batch processes. In the continuous process, aroma compoundand β-cyclodextrin is mixed at a very high-speed in a water alcohol medium followed by pad dry cure process. In the batchprocess, the fabric is first exhausted in a bath which contains β-cyclodextrin, acrylicacid sodium hypophosphite, and potassium persulphate. Acrylic acid polymerholds β-cyclodextrin, and the aroma compounds.

4.5 *Using Pigment Printing Technique*

Recently a single step screen printing technique was developed to apply aroma and pigment together on textiles. Lemongrass microcapsules were applied on silk fabric using the screen-printing technique. This technique requires optimum binder content to achieve the finish. Both aroma finishand pigment print in the single step process were successfully achieved and the results were found to be comparable with aroma finished fabrics using the other techniques [39].

5 Application of Fragrance Finished Textiles

With the growth in textile and fashion industry and the awareness about the smart clothing has led to the novel development of functional and stylish second skin. In the present scenario, the consumers are extremely particular about their demands and their style sense along with the functionality rendered by the clothing. Fragrant textiles are not a new concept; however, the methods and applications have evolved, but aroma finishes have been use in home textiles, garments, and sportswear. Scented flowers and oils have been used in many fabrics to keep the foul smell away.

Generally speaking, varied perfume fabrics create good opportunities for customers to make the cocooning environment they prefer to live in. The application of aromatherapy textile is very diverse, some of the common applications are mentioned below-

(a) **Home textiles**

Home textiles such as sheets, quilt-covers, and curtains are suitable for the attachment of lavender, citrus or cinnamon microcapsules, which are good for hypnogenesis and eliminating fatigue. Patients suffering from high blood pressure feel sedation when they use a pillow made of fabric treated with lavender, basil, and lemon or fennel microcapsules. Similarly, fragrance finish can be imparted to the carpets which can then be utilized for home textiles and also as automotive textiles.

(b) **Sport wear/athleisure and yoga wear**

All types of textiles act as an excellent medium for the transfer of aroma compounds. Sports textiles, athleisure and yoga wear often find fragrance compounds more essential as they keep the wearer energized.

(c) **Kids toys**

Fragrance finished toys can give a nature-like experience to the kids playing with them and help them to relate with the various elements of the environment.

(d) **Apparel**

Especially finished textiles and apparel with aroma have been in market for a good time now. Kanebo, a Japanese company manufactured women's scented tights in the 1960. Hosiery and inner wear have been the most popular range of apparel to be infused with aroma compounds. International companies like Woolmark™ have joined hands with International Fragrances and Flavors Association to extend its R&D initiative with mills and manufacturing units around the globe. Woolmark™ refers to this concept of micro encapsulation of fragrances into fabrics as Sensory Perception Technology™ fabrics. The company is using this novel technology on hosiery fabrics, underwear, intimate clothing as well as outdoor clothing, socks, carpets, and lingerie. LYCRA® Body Care Collection was launched by Invista Company in 2005 which includes use of moisturising and fragrant yarns to manufacture intimate apparel and increase the well-being and comfort of the wearer. The company manufactured micro beads of fragrant compounds and built them into the elastane fibers. As the fabric stretched during the wear, the contents of the micro beads were released. The LYCRA® Body Care Collection was utilised by Olga clothing brand to launch a collection in 2005 itself. Encapsulation methods have also been explored by the Nike. Hermes and Lancme, and other luxurious brands have developed ties, silk scarves and lingerie using encapsulation of fragrances. Apart from these developments there are still some unexplored areas for fragrant textiles like kidswear, women's intimate garments which keep the skin moisturized.

The aroma of rose, citrus, lavender and vanilla were encapsulated into fabric which provided a good option to meet psychological and emotional needs. They also add a touch of physical and sensorial nature to the treated fabrics. The finished fabrics with the aforementioned scents can be used in apparels, daily wears, leisure wear etc. Socks, gloves, t-shirts etc.can also be finished with microencapsulated fragrance and deodorizing finishes for better utility. Jasmine, mint, lemon, rose or lavender may have refreshing effect on the wearer. Kid's apparel like t-shirts with flowery and fruity prints have been manufactured, which give out the smell when the prints are scratched.

(e) **Medical textiles**

Fragrances have been reported to act as mood enhancer, stress reliever and are capable of evoking emotions. Aroma therapy in textiles can help the wearer to invoke alertness and stay calm. This combination might come handy for treatment of patients with Alzheimer's disease by connecting the clothes with specific fragrances and biometric sensors which measure heart rate and stress level. A relaxing aroma can be used to calm a person down or cheering him up. The biometric sensors will help to register scents and release them as and when needed by the individual.

Apart from these aroma/essential oil finished textiles have been used as antimicrobials [7], mosquito repellents [20], treatment of cancer [26], treatment of dermatitis [47].

6 Overview of Research and Development in Fragrance Finished Textiles

In this section we have compiled various researches that have been carried out on the application of aroma on textiles using various techniques. The aroma finish has been done not only to impart scent but other effects associated with the particular aroma compound.

Panchal and Mundkur carried out a survey among young working women of Mumbai to study the awareness of fragranced textiles, preferences and the potential use. In their study they found that majority of the respondent were aware about fragranced textiles. A majority of the respondents expressed a desire to use fragranced textiles on a daily basis on their apparel and in-home textiles. The results indicate that there is potential for a fragranced textiles or fragranced garment in India [32].

Li et al. [24] studied the application of microencapsulated aroma compounds on cotton fabric. Conventional fixing of the aroma microcapsules might degrade the aroma because the curing temperature in this method is higher i.e. 130–170 °C. Aroma compounds being highly volatile will be lost hence depleting the durability of the finish. The research suggested the application of UV resin for fixation of microcapsules. With the utilization of UV resin for fixation, the curing temperature was reduced, hence the time duration also decreased to seconds, and the aroma loss in this process was lesser leading to better durability of the finish. Cotton fabric finished with aroma microcapsule and UV resin gave the wash fastness up to 50 cycles [24].

Limonene and vanillin microcapsules were prepared using chitosan/gum arabic by Sharkawyet al. [40] using complex coacervation method. Tannicacid was used as a hardening agent for the shell material. Efficiency of encapsulation, morphology of microcapsules and release profile were studied for two different emulsifiers viz. polyglycerol polyricinoleate (PGPR), and Span 85. The size of the prepared microcapsule lied in the range of 10.4 and 39.0 μm. The use of Span 85 led to formation of mononuclear morphology and polynuclear morphology was obtained by using PGPR irrespective of the essential oil used a score contents. The microcapsules were grafted on to the cotton fabric by esterification reaction using citric acid and then heat fixed and cured afterwards. The finished fabric was found to have good antibacterial activity [40].

Kumar et al., studied the effects of lavender and aloe-vera finish on cotton fabric home textile (wall hangings) and investigated factors such as fabric performance, durability of the fragrance and laundering properties of the treated fabrics. They found that based on the overall performance aloe-vera fabric is better than cotton fabric for wall hanging (curtain) and in all home textile applications. The fragrance finished fabrics can be used in home textile applications such as wall hangings, table covers, carpets and sofa covers [45].

Sohn et al. [41] developed a durable aroma finishing for PET fabric by using polyvinyl acetate (PVAc) nanoparticles containing lavender oil (LO) as core material. PVAc nanoparticles containing LO in core were prepared by emulsification-diffusion method and their application as an aroma releasing agent for PET fabrics was assessed through the observation of releasing profiles of LO in ethanol for experimental acceleration. PVAc nanoparticles treated on PET fabric showed higher initial releasing amounts. After two days of releasing, PET fabric treated with PVAc nanoparticles showed a slower and more stable releasing profile and reached about 12 ppm of cumulative release after 16 days, which were under two thirds of that with MF microcapsules. They concluded that PVA nanoparticles can be used as a potential agent for durable aroma finishing of PET.

Agarwal and Goel [1] conducted research on microencapsulation. They used citronella oil and gum acacia as core and wall materials respectively and optimized the process for different concentration and temperature. It was found that microcapsules developed in two ratios (1:2 and 2:1) and in both the cases the size varied. Both the larger and the smaller capsule have their own advantages. Their study revealed that microencapsulation helps in controlling the release rate of the fragrance compounds as well as essential oils.

Pan et al. (2017) prepared jasmine oil microcapsule using chitosan for application on jute fabric. Coacervation method was used to prepare the desired microcapsules of dimension 100 micron and pad dry cure method was used for the application of microcapsules on bleached jute fabric. The durability of the finish was very low as the finished fabric could only withstand 5 home launderings [31].

In their study Kumari et al., used geranium oil and gum acacia as core and wall materials for encapsulation and applied them on woven fabric of cotton. The complex coacervation technique was used for the application of these microcapsules. In order to optimize the process, they usedoil, gum and gelatin in the ratio of 1:4:4 at 50 °C temperature and initial and final pH was maintained as 4.5 and 9.0 respectively. They found that microencapsulation technique is very helpful in controlling the release rate of aroma and alsogives a durable fragrance finish to cotton fabric [23].

Tekin et al., in their study developed polyurethane-urea microcapsules containing fabric softener fragrance (TeddysoftTM, by EPS) as active agents by using interfacial polymerization in order to extend the release rate andincrease the persistence of the fragrance. Various parameters such as size, shape and morphology of the microcapsules were studied by using an optical microscope and scanning electron microscope. The optical microscope and SEM pictures confirmed the spherical form of microcapsules. It was also reported that increasing the stirring rate of the emulsion resulted in cracks on microcapsules which may cause quick release of the fragrance from the microcapsules. The agglomeration also increased with increasing stirring rate during emulsion. Optical and SEM results revealed that increasing the stirring rate during the emulsion step results in microcapsules with smaller particle size, but at some time it adversely affects its morphology [44].

Polysulfone microcapsules containing vanillin were prepared by Panisello et al. [33]. The morphology, thermal properties and antibacterial propertied were assessed. The developed microcapsules were applied on cotton fabric by using coating method and results revealed that treated fabric was having poor wash fastness. The microcapsules were stable in the temperature range of 20–100 °C, and the antimicrobial resistance was present at 37 °C for one week [33].

7 Future Prospects Of Aroma Textiles

From the last few decades aroma textiles were being utilized for only skin care and stress management but with recent research advances and growing awareness in the consumers, there is much more to be explored in this particular section. The aroma therapy has the potential to put forth the solutions for hair care and treatment, treatment of medical ailments, and body slimming. The techniques of micro encapsulation of essential oils have been well received by the consumers as well as the manufacturers. The future prospects of micro-encapsulation led to a lot of opportunities which would not only result in an increase of sales but at the same time it will expand its market share. In today's scenario when there is a cut-throat competition in textile market its really crucial for the textile manufacturer to work on the diversification of their products through development and the commercialization of new innovative textiles with a real and functional added value to the customer.

Ease of application, good shelf life and less affective to existing textile properties are few requisites for ideal textile applications of micro capsules. Currently, micro capsules are able to withstand up to 25–30 washing cycles. Conventional heat input processes like ironing and tumble drying and tend to decrease the effect of finish to minimum levels. The micro encapsulation industry needs to address these concerns and look for other possibilities and overcome these barriers to specifically design microcapsules for textile products.

To address the desire for convenience and novelty of the consumers, the textiles need to have fragrance and softness as their inherent property. The consumers usually expect these features to remain theretill the lifetime of the garment, just like other permanent finishes. The routine intervention of adding fabric softeners and other washing aids might divert the attention of the customer.

Microencapsulation might offer an answer to deliver these long-term goals. Textiles which actually interact with the consumer to reduce stress, promote comfort and provide a relaxing feel are possible due to microencapsulation. The yearning for comfortable and healthier lifestyle will act as a catalyst in creating a niche market for textiles that promote well-being of the customers. Since last few decades, the textile industry has started focusing on development of performance fabrics by value addition for sports, outdoor application as well as medical textiles.

Microencapsulation can give a direction to this line of apparel by addition of sensing chemicals to the sports and medical textiles. This will add to the benefit of warning the wearer in case of health hazard. Further in the coming time we can also develop such systems that deliver measured dosage of drugs or other chemicals to combat muscle pain or other injuries. Microencapsulation has a tremendous and very diversified in application in the field of textile.

Value added fabrics not only add to the pool of academic advance but also increase the convenience of our lives. Despite being a novel concept, fragrance finished textile have made their distinct place in our lives. The newest dimensions of application of fragrant fabrics give a much-needed push to the textile industry to explore the vast potential of microencapsulation. The growing demand of comfortable athletic apparels and increased penetration of intimate clothing into daily body wear has opened a new window of opportunities for application of microencapsulation technology to textile products in finishing stage. In order to convert the dream of this kind of fragrant textile into reality a close cooperation is required between laboratory researchers, academicians, industry partners as well as consumers.

References

1. Agarwal M, Goel A (2010) Microencapsulation of Citronella oil for aromatherapic textiles. In: 23rd national convention of textile engineers. New Delhi
2. Anton N, Vandamme TF (2009) The universality of low-energy nano-emulsification. Int J Pharm 377(1–2):142–147. https://doi.org/10.1016/j.ijpharm.2009.05.014
3. Bansode SS, Banarjee SK, Gaikwad DD, Jadhav SL, Thorat RM (2010) Microencapsulation: a review, Int J Phamaceutical Sci Rev Res 1(2):38–43. https://www.globalresearchonline.net/volume1issue2/Article008.pdf. Accessed 14 Sept 2020
4. Benita S (2005) Microencapsulation: methods and industrial applications. 2nd edn. CRC Press. https://www.routledge.com/Microencapsulation-Methods-and-Industrial-Applications-Second-Edition/Benita/p/book/9780824723170. Accessed 14 Sept 2020
5. Bhatt L, Singh, SSJ Rose NM (2016) Durable aroma finish on cotton using microencapsulation technology, J. Cotton Res Dev 30(1):156–160
6. Palmiero UC, Ilare J, Romani C, Moscatelli D, Sponchioni M (2020) Surfactant-free and rinsing-resistant biodegradable nanoparticles with high adsorption on natural fibers for the long-lasting release of fragrances. Colloids Surf B Biointerfaces 190:110926. https://doi.org/10.1016/j.colsurfb.2020.110926
7. Chandrasekaran K, Ramachandran T, Vigneswaran C (2012) Effect of medicinal herb extracts treated garments on selected diseases. Indian J Tradit Knowl 11:493–498
8. Chattopadhyay P, Dhiman S, Borah S, Rabha B, Chaurasia AK, Veer V (2015) Essential oil based polymeric patch development and evaluating its repellent activity against mosquitoes. Acta Trop 147:45–53. https://doi.org/10.1016/j.actatropica.2015.03.027
9. Cheng SY, Yuen CWM, Kan CW, Cheuk KKL (2008) Development of cosmetic textiles using microencapsulation technology. Res J Text Appar 12(4):41–51. https://doi.org/10.1108/rjta-12-04-2008-b005
10. Dambolena JS, Zunino MP, Herrera JM, Romina PP, Areco VA, Zygadlo JA (2016) Terpenes : natural products for controlling insects of importance to human health— a structure-activity relationship study. Psyche A J Entomol. Article ID 4595823. https://doi.org/10.1155/2016/4595823
11. Dawid S (2012) Application of the layer-by-layer method for textiles. In: El Nemr A (ed) Textiles: types uses and production methods. Chapter 16. Nova Publisher, New York
12. DhifiW BS, Jazi S, Bahloul N, Mnif W (2016) Essential oils' chemical characterization and investigation of some biological activities: a critical review. Medicines 3(4):25. https://doi.org/10.3390/medicines3040025
13. El-Din MRN (2011) Study on the stability of water-in-kerosene nano-emulsions and their dynamic surface properties. Colloids Surf A 390(1–3):189–198. https://doi.org/10.1016/j.colsurfa.2011.09.027
14. Forgiarini A, Esquena J, Gonzalez SC (2001) Formation of nano-emulsions by low-energy emulsification methods at constant temperature. Langmuir 17(7):2076–2083. https://doi.org/10.1021/la001362n
15. Gautam N, Mantha AK, Mittal S (2014) Essential oils and their constituents as anticancer agents: a mechanistic view, BioMed Res Int. Article ID 154106 https://doi.org/10.1155/2014/154106

16. Ghayempour S, Montazer M (2016) Micro/nanoencapsulation of essential oils and fragrances: Focus on perfumed, antimicrobial, mosquito-repellent and medical textiles. J Microencapsul 33(6):497–510. https://doi.org/10.1080/02652048.2016.1216187
17. Guterres SS, Alves MP, Pohlmann AR (2007) Polymeric nanoparticles, nanospheres and nanocapsules for cutaneous applications. Drug Target Insights 2:147–157. https://doi.org/10.1177/117739280700200002
18. Holme I (2003) Versatile technology comes of age. Int Dyer 188(8):9–13
19. Junyaprasert VB, Mitrevej A, Sinchaipanid N, Boonme P, Wurster DE (2001) Effect of process variables on the microencapsulation of vitamin A palmitate by gelatin-acacia coacervation. Drug Dev Ind Pharm 27(6):561–566. https://doi.org/10.1081/DDC-100105181
20. Kale RD, Bhatt L (2018) Cymbopogon flexuosus oil nanoemulsion: a natural safe and effective mosquito larvicidal and antibacterial. J Adv Res Dyn Control Syst 10(9):2440–2451
21. Korus J, Tomasik P, Lii CY (2003) Microcapsules from starch granules. J Microencapsul 20(1):47–56. https://doi.org/10.3109/02652040309178048
22. Korus J (2001) Microencapsulation of flavours in starch matrix by coacervation method. Pol J Food Nutr Sci 51:17–23
23. Kumari P, Rose NM, Singh SSJ (2017) Development of fragrant microcapsules for woven cotton fabric. J Appl Nat Sci 9(2):1017–1021. https://doi.org/10.31018/jans.v9i2.1314
24. Li S, Boyter J, Qian L (2005) UV curing for encapsulated aroma finish on cotton. J Text Inst 96(6):407–411
25. Julian McClements D, Henson L, Popplewell LM, Decker EA, Jun Choi S (2012) Inhibition of Ostwald ripening in model beverage emulsions by addition of poorly water soluble triglyceride oils. J Food Sci 77(1):C33–C38. https://doi.org/10.1111/j.1750-3841.2011.02484.x
26. Munin A, Edwards-Lévy F (2011) Encapsulation of natural polyphenolic compounds; a review. Pharmaceutics 3:793–829. https://doi.org/10.3390/pharmaceutics3040793
27. Nagavarma BVN, Yadav H, Ayaz A, Vasudha LS (2012) Different techniques for preparation of polymeric nanoparticles—a review. Asian J Pharm Clin Res 5(SUPPL. 3):16–23
28. Morsy NFS (2017) Chemical structure, quality indices and bioactivity of essential oil constituents. Active ingredients from aromatic and medicinal plants, pp 175–206
29. Nerio LS, Olivero-Verbel J, Stashenko E (2010) Repellent activity of essential oils: a review. Biores Technol 101(1):372–378. https://doi.org/10.1016/j.biortech.2009.07.048
30. Teixeira CSNR (2010) Microencapsulation of perfumes for application in textile industry. Doctoral dissertation, Universidade do Porto (Portugal)
31. Pan NC, Ammayappan L, Khan A, Chakraborty S (2018) Performance of chitosan: jasmine oil microcapsule on jute fabric. Indian J Fibre Text Res 43(3):375–380
32. Panchal J, Mundkur S (2020) Consumer preferences in fragranced textiles. Int J Polym Text Eng 7(1):59–62
33. Panisello C, Peña B, Oriol GG, Constantí M, Gumí T, Garcia-Valls R (2013) Polysulfone/vanillin microcapsules for antibacterial and aromatic finishing of fabrics. Ind Eng Chem Res 52(29):9995–10003
34. Pitasawat B, Champakaew D, Choochote W, Jitpakdi A, Chaithong U, Kanjanapothi D, Rattanachanpichai E, Tippawangkosol P, Riyong D, Tuetun B, Chaiyasit D (2007) Aromatic plant-derived essential oil: an alternative larvicide for mosquito control. Fitoterapia 78(3):205–210
35. Rabišková M, Valášková J (1998) The influence of HLB on the encapsulation of oils by complex coacervation. J Microencapsul 15(6):747–751
36. Saberi AH, Fang Y, McClements DJ (2013) Fabrication of vitamin E-enriched nanoemulsions: factors affecting particle size using spontaneous emulsification. J Colloid Interface Sci 391(1):95–102
37. Salvia-Trujillo L, Rojas-Graü MA, Soliva-Fortuny R, Martin-Belloso O (2013) Effect of processing parameters on physicochemical characteristics of microfluidized lemongrass essential oil-alginate nanoemulsions. Food Hydrocoll 30(1):401–407
38. Sari TP, Mann B, Kumar R, Singh RRB, Sharma R, Bhardwaj M, Athira S (2015) Preparation and characterization of nanoemulsion encapsulating curcumin. Food Hydrocoll 43(2015):540–546

39. Setthayanond J, Lim P (2020) Aroma finishing and pigment printing of Thai silk fabric by 1-step all-in printing process. Mater Sci Forum 1005:123–130
40. Sharkawy A, Fernandes IP, Barreiro MF, Rodrigues AE, Shoeib T (2017) Aroma-loaded microcapsules with antibacterial activity for eco-friendly textile application: synthesis characterization, release, and green grafting. Ind Eng Chem 56(19):5516–5526
41. Sohn SO, Lee SM, Kim YM, Yeum JH, Choi JH, Do Ghim H (2007) Aroma finishing of PET fabrics with PVAc nanoparticles containing lavender oil. Fibers Polymers 8(2):163–167
42. Tadros TF (2013) In: Tadros TF (ed) Emulsion formation, stability, and rheology, emulsion formation and stability, 1st edn. Wiley-VCH Verlag GmbH & Co. KGaA, pp 1–74
43. Tamilvanan S (2009) Formulation of multifunctional oil-in-water nanosized emulsions for active and passive targeting of drugs to otherwise inaccessible internal organs of the human body. Int J Pharm 381(1):62–76
44. Tekin R, Bac N, Erdogmus H (2013) Microencapsulation of fragrance and natural volatile oils for application in cosmetics, and household cleaning products. Macromol Symp 333(1):35–40
45. Kumar DV, Boopathi N, Karthick N, Ramesh P (2012) Aesthetic finishes for home textile materials. Int J Text Sci 2012(3):5–9
46. Yadav S, Gattacceca F, Panicucci R, Amiji MM (2015) Comparative biodistribution and pharmacokinetic analysis of cyclosporine-A in the brain upon intranasal or intravenous administration in an oil-in-water nanoemulsion formulation. Mol Pharm 12(5):1523–1533
47. Yoo KH, Kwon TR, Kim YU, Kim EH, Kim BJ (2020) The effects of fabric containing chamaecyparis obtusa essential oil on atopic dermatitis-like lesions: a functional clothing possibility. Skin Pharmacol Physiol 33(3):82–92

Containers Based on Polymers in Biomedical Devices/Medical Applications

Dania Alphonse Jose, Prabha Prakash, and P. S. Baby Chakrapani

1 Introduction

During the last two decades, significant advancement has been made in investigating biodegradable polymeric materials for biology and medicine. The introduction of biomaterials dates back to history as 'non-biological materials' but currently they are known as 'biomaterials'. According to the National Institute of Health (NIH), 'biomaterial is any substance (other than drug) or a combination of substances, synthetic or natural in origin [1]. Mayan people (600 AD) used nacre teeth from seashells for bone integration; Egyptians used linen sutures for surgical purposes. In the fourth century BC, the idea of artificial heart and organ perfusion came, but its execution came after a long time. Leonardo Da Vinci, in 1508 introduced the concept of contact lenses. In 1891 the first hip replacement surgery was performed using cemented ivory ball implant by a German surgeon Theodor Gluck. The advancement in technology replaced ivory ball implants with chrome-based alloys and stainless steel implants. It provides enhanced mechanical properties to support the skeletal system. In 1809, Maggiolo experimented with dental implants by placing a gold post anchor into the fresh extraction socket. In 1901, John Jacob attempted to remove

D. A. Jose · P. Prakash · P. S. Baby Chakrapani (✉)
Centre for Neuroscience, Cochin University of Science and Technology (CUSAT), Kochi 682022, Kerala, India
e-mail: chakrapani@cusat.ac.in

Department of Biotechnology, Cochin University of Science and Technology (CUSAT), Kochi 682022, Kerala, India

P. S. Baby Chakrapani
Inter University Centre for Nanomaterials and Devices (IUCND), Cochin University of Science and Technology (CUSAT), Kochi 682022, Kerala, India

Centre of Excellence in Advanced Materials, Cochin University of Science and Technology (CUSAT), Kochi 682022, Kerala, India

J. Parameswaranpillai et al. (eds.), *Micro- and Nano-containers for Smart Applications*, Composites Science and Technology,
https://doi.org/10.1007/978-981-16-8146-2_8

toxins from a patient's blood for the first time using a dialysis unit. Dr. Scribner also made a significant contribution in developing a dialysis unit into a functional one. The silicone breast implant, first developed by Thomas Cornin and Frank Gerow in the 1960s. In 1959 in the joined effort of Wilson Greatbatch, an engineer by profession, and WM Chardack, a heart specialist developed a completely purposeful and implantable pacemaker by combining engineering and medicine [2].

Biomaterials can classify into three major classes, namely.

1. Polymers
2. Metals and
3. Ceramics.

In 1920, Herman Staudinger, a German organic chemist, first introduced the concept of polymers. It is 100 years since the discovery and that polymers have revolutionized the quality of people's lives [3]. The term 'polymer' refers to a large molecule built wholly or chiefly by multiple repeating units. Polymers can broadly classify into natural polymers and synthetic/human-made polymers depending on their obtained source. Natural polymers produced by living organisms (plant or animal origin) find application in the pharmaceutical industry because they are economical, readily available, capable of chemical modifications, biodegradable and biocompatible. Certain disadvantages such as microbial contamination (during extraction), batch to batch variation, uncontrolled rate of hydration, slow rate of production, and heavy metal contamination make it challenging to use for biomedical applications [4]. However, synthetic/human-made polymers produced chemically by incorporating monomers with different physical and chemical properties entered the biomedical industry [5].

2 Polymers in Medical Field

Polymers can broadly divide into

1. **Natural polymers**
2. **Synthetic polymers.**

Polymers used for biomedical applications should meet specific requirements to use safely inside the human body. One of the essential characteristics of polymers is their stability when they come in contact with human blood or tissues. Based on the stability, polymers classify into biostable/bioinert polymers and biodegradable polymers. Biostable polymers exhibit stability upon sterilization treatments and hydrolytic and oxidative environments inside the human body [6]. At the same time, biodegradable polymers get degraded into more minor compounds that can be metabolized or excreted from the human body. Generally, natural polymers and some synthetic polymers such as poly (vinyl alcohol), polylactide, polyglycolide, poly (lactic-co-glycolic acid), polycaprolactone, and polycarbonate are biodegradable. However, most synthetic polymers are biostable, including polyethylene, polypropylene, polytetrafluoroethylene, poly (acrylic acid), polyurethanes, polyesters, polyamide, and silicone elastomers. Bioerodibility is another characteristic in which the polymers erode mechanically through biological processes that result in the surrounding tissue's solubilization and absorption. This mechanism is exhibited by hydrophobic polymers where the degradation takes place from the surface in a layer-by-layer fashion. Their primary application is in the drug delivery system. Bioerodible polymers include synthetic polymers such as polyanhydrides, polyorthoesters, poly (lactic acid), and poly (glycolic acid) [7]. Polymers can also be hydrophilic or hydrophobic, depending on the nature of side chains. Hydrophilic or water-soluble polymers can be natural or synthetic. Such polymers dissolve or swell in water, forming gels or emulsions. Synthetic water-soluble polymers include polyethylene glycol, polyvinyl pyrrolidone, polyvinyl alcohol, and polyacrylic acid. In contrast, gelatin is a natural water-soluble polymer [8].

3 Common Properties and Uses of Natural and Synthetic Polymers Used in Fabrication of Containers

3.1 Natural Polymers

3.1.1 Collagen

Collagen is the most abundant protein present in mammals and one of the abundant proteins in the extracellular matrix of cells. The protein constitutes 25–30% of total proteins in the human body [9]. Collagen polypeptide contains specific sites which facilitate cell growth and adhesion [10]. There are 29 types of collagen; type I collagen is found abundantly in human tissues. It is present in bones, tendons, skin, and the vascular system. It comprises 90% of the total collagen found in the human body. Collagen is a triple helix protein with two identical a1 and one a2 chains. Both chains differ slightly in their chemical composition. It is non-antigenic, biocompatible, and can self-aggregate into a hydrogel. These characteristics make collagen a suitable polymer for the fabrication of biodegradable packaging materials for pharmaceutical applications [11].

3.1.2 Gelatin

Gelatin is a water-soluble, biodegradable protein derived by partial hydrolysis of insoluble natural polymer collagen. It is present in melanocytes, keratin, chondroitin, and elastin. Different gelatin-based hydrogels exist based on the source of collagen, polymer concentration, and the hydrolytic treatments it undergoes. Proteolytic enzymes, namely collagenases, gelatinase, and matrix metalloproteases (MMPs) cleaves gelatin into non-toxic byproducts. Gelatin is a thermoresponsive polymer; cooling the solution below 30–35 °C can make it into a gel. The gelation process can be reversed by heating the gel and can be made into hydrogel by photopolymerization at mild conditions. The production of liquid-core microcapsules utilizes the unique thermal properties of gelatin. Gelatin is an inexpensive, biocompatible, cytocompatible, and less immunogenic biopolymer. It has a variety of applications in tissue engineering and drug delivery [12].

3.1.3 Cellulose

Cellulose is a homopolymer of D-glucose. In biomedical science, cellulose has a wide range of applications. It includes the production of microcarriers that supports cell adhesion and proliferation. Cellulose-based microcarriers require appropriate cell surface charges (due to low non-specific protein adsorption) to facilitate this process. Other applications include its use in hemodialysis membranes, wound dressing, scaffolds for tissue engineering, and pharmaceutical production of metabolites [13]. Cellulose is biocompatible, biodegradable, has appropriate water absorption capacity, network structure, capable of chemical modification, has simple preparation, and exhibits good mechanical strength. These features make cellulose a good candidate for cell culture applications.

In contrast, bacterial cellulose has a unique nanofibrous structure. This nanofibrous cellulose finds its application in the field of biomedicine, energy, optoelectronics, and environmental protection. Biomedical application includes its use as a material for wound dressing, vehicles for targeted cell delivery, and living cell-based sensors [14].

3.2 Synthetic Polymers

Synthetic polymers or thermoplastics, also called medical grade plastics, are generally used synonymously and share common characteristics. This section covers some of the essential thermoplastics used for the fabrication of biomedical devices specifically outside the body. The polymer's chemical, physical, and structural properties determine the ideal properties required for a polymer to classify as a biomedical polymer. Polymers' physical properties include critical molecular weight, molar volume, density, degree of polymerization, morphology, crystallinity etc. Thermal properties are melting point and glass transition temperature. Mechanical properties include strength, elongation (ductility), elasticity or tensile strength, toughness etc., Polymers with various specific properties are used for the fabrication of biomedical devices [15, 16]. A polymer selection for containers and sealing material fabrication require some ideal properties and characteristics like:

1. The biomaterial should be stable enough to protect the product from environmental conditions.

2. The packaging material should not react with the pharmaceutical product or does not harm the product.
3. Should not cause a toxic response
4. The plastic material should be easy to fabricate and adaptable to commonly used high-speed packaging equipment's
5. Properties of plastic materials: they should have appropriate chemical, physical and mechanical properties for easy molding and easily sterilizable without changing the material properties.

3.2.1 Polyethylene (PE)

Polyethylene (C_2H_4) is one of the most popular and common thermoplastic materials used as medical-grade polymers; it belongs to the polyolefin family. Its synthesis involves the addition or radical polymerization of ethylene (olefin) monomers. Polyethylene polymer can synthesize with different molecular weights as per the requirement of the healthcare application. Metallocene and Ziegler–Natta is the catalysts used for the polymerization process [17]. This polymer is biologically inert, hydrophobic, and does not degrade in-vivo. It is durable, versatile, and lightweight with a variable crystalline structure, making it the standard choice of medical-grade polymer. This polymer is available in low and high-density Polyethylenes [17, 18].

The common type or classification of polyethylenes based on density and branching includes:

(1) Low-density polyethylene (LDPE)
(2) Linear low-density polyethylene (LLDPE)
(3) High-density polyethylene (HDPE)
(4) Ultra-high-molecular-weight polyethylene (UHMWPE),
(5) Medium-density polyethylene (MDPE),
(6) Very-low-density polyethylene (VLDPE)
(7) High-molecular-weight polyethylene (HMWPE),
(8) Ultra-low-molecular-weight polyethylene (UHMWPE),
(9) Chlorinated polyethylene (CPE).

A high degree of branching leads to a reduction in crystalline regions and this could lower the crystallinity of the polymer. The production of LDPE is by free radical polymerization process at high temperature. The melting point of low-density PE is 105–115 °C; and has good resistance to solvents but lesser resistance to mineral oils, aliphatic, aromatic, and halogenated hydrocarbons. Due to the ease of processing and other unique physical properties like low moisture absorption, strength, flexibility, and sealing ease, LDPE is the preferred synthetic polymer for squeeze bottles. It is unsuitable for steam and autoclave sterilizations due to its low heat deflection temperature. HDPE manufacturing involves slurry polymerization or gas phase polymerization, and its melting point is 120–140 °C.

Compared to other polypropylenes, HDPE has good tensile strength, excellent resistance to most of the solvents like alcohols, chemicals, and dilute acids, and moderate resistance to oils. The common uses are making bottles for solid dosage forms. Few disadvantages include lower stiffness, susceptibility to stress cracking, low UV and heat resistance. The suitable methods of sterilizations for PE are gamma sterilization, ETO, and e-beam sterilization.

3.2.2 Polypropylene (PP)

Polypropylene is a member of the polyolefin family. It is a colorless, odorless, and inert thermoplastic material with good tensile properties. Polypropylenes are cost-effective, highly resistant to chemicals, stress cracking, and can withstand high temperatures. It is one of the most widely used polymers today. Synthesis involves catalyst-mediated polymerization of propene monomer [19]. There are three types of PP that are generally available in markets [18, 19].

(1) polypropylene homopolymer,
(2) polypropylene copolymer
(3) polypropylene impact copolymer.

(1) polypropylene homopolymer

The most commonly used general-purpose grade polypropylene homopolymer contains only a propylene monomer. Inherent properties include stiffness, has high tensile strength, and exists in a semi-crystalline solid form. Its natural state is translucent with excellent contact clarity with liquids, but it has less impact resistance than the other two types. Polypropylene copolymer and impact polymer synthesis are by polymerization of ethene and propane.

(2) Polypropylene Copolymer

These are the most flexible and the optically clearest among the three, and it usually contains up to 6% of ethene. It is low tensile compared to homopolymers but has better room temperature than homopolymers. Both PP homo and copolymer have relatively poor impact resistance at low temperatures.

(3) Polypropylene Impact copolymers

It contains 45–65% ethane and is known as block copolymers. Compared with the other two PP polymers, Impact polymers have intermediate stiffness, tensile strength, and cloudiness. If more is the ethylene monomer, higher is the impact resistance but lower stiffness and tensile strength.

Additional additives like clarifiers, flame retardants, glass fibers, minerals, conductive fillers, lubricants, pigments etc., significantly improve the physical and mechanical properties. The blending/special formulations with additives help increase polypropylene life; for example, modified propylene has increased radiation resistance. The medical application of polypropylene includes disposable syringes, medical vials, diagnostic devices, Petri dishes, intravenous bottles, specimen bottles, food trays, pans, tablet containers, etc. Commonly recommended sterilization techniques are gamma radiation and autoclave.

3.2.3 Polystyrene (PS)

Polystyrene is the simplest plastic made by polymerization of monomer named styrene. It is versatile; hard-solid plastic with limited flexibility; pure solid polystyrene is colorless. Polystyrene can produce in various colurs by combining with colorants and additives [20]. The three common polystyrene types are:

1. General purpose PS (GPPS),
2. High impact (HIPS)
3. Syndiotactic (SPS).

Polystyrene is economical to use and has a wide range of biomedical applications. The application includes making medical devices, test tubes, Petri dishes, trays for conducting tissue culture tests, etc., and making diagnostic test equipment, medical cups, vaginal dilator speculum etc. Steam and autoclave sterilization is not advisable for styrene, but it is resistant to sterilization by UV light.

3.2.4 Polyester

There was a paradigm shift from biostable polymers to biodegradable polymers and in specific Poly-a-esters in recent years. It is a widely used and useful material in both the medical and non-medical fields. The unique properties of polyester, like its synthetic versatility, immense diversity, and durability, made it the top among the synthetic polymers. Polyesters can be developed from a variety of monomers and synthesized by polymerization techniques. It is predominantly produced by random, ring-opening, and block copolymerization techniques [8, 21]. The most commonly used polyesters are (1) polylactic acid (PLA), (2) poly(lactic-co-glycolic acid) (PLGA), (3) poly(ε-caprolactone) (PCL), (4) poly-3-hydroxybutyrate (or poly-β-hydroxybutyric acid, PHB), (5) poly(3-hydroxybutyrate-co-3-hydroxy valerate) (PHBV), (6) poly(propylene carbonate) (PPC), (7) poly(butylene succinate) (PBS) and (8) poly(propylene fumarate) (PPF). In general, biocompatible and biodegradable poly-a-esters are used in biomedicine. Plastic containing aromatic groups is mostly biocompatible polymers. It includes polycarbonates (PC), poly(ethylene terephthalate) (PET, Dacron). Polyesters containing linear aliphatic groups are the most common biodegradable polymers, poly(glycolic acid) (PGA), poly (L lactic acid)(PLLA), and poly (D-lactic acid) (PDLA).

3.2.5 Polycarbonates (PC)

Poly-carbonate is a polyester produced by bis-phenol and carbonic acid and a widely used polymer for containers in biomedical applications. It is robust, rigid, and transparent by nature and can be processed to produce different grades of polycarbonate. The specialty grades of polycarbonates are highly durable, can be transparent or colorant, tough at low temperatures, glass-like clarity, highly resistant to heat and UV applications. It has a wide range of applications, including semi-permeable membrane cartridges for renal dialysis, cardiac surgery, blood oxygenators, blood reservoirs, blood filters, intravenous connectors, and surgical instruments [22, 23]. Poly-carbonate can be cleaned using different sterilization methods like ethylene oxide (EtO), irradiation, and steam autoclaving. One of the significant advantages of PC is that it can withstand several types of sterilization; it can easily re-use, and still, the properties of PC remain unchanged.

3.2.6 Poly Lactic Acid (PLA)

PLA is biodegradable and compostable derived by fermentation of natural sources like corn starch or sugarcanes by microorganisms. This polymer is one of the widely used polymers, and hence it blends extensively by incorporating different organic and inorganic components for various biomedical applications. PLA, obtained from LA, can be made by chemical synthesis or by fermentation process [24]. Three steps generally produce PLA.

(1) By microbial fermentation,
(2) Purification of LA followed by its cyclic dimer (lactide) preparation
(3) By polycondensation of LA or by ring-opening polymerization (ROP) of lactides.

PLA and its copolymers are biodegradable and biocompatible. The main applications include tissue-engineering scaffolds, delivery system materials, covering membranes, medical implants, and suturing material in dermatology and cosmetics.

3.2.7 Poly Glycolic Acid (PGA)

PGA is one of the common synthetic biodegradable polymers is used for medical applications. It is biodegradable and biocompatible, thus can be effectively used for many biomedical applications. Environmental friendly properties like biodegradability make PGA unique among the semi-synthetic polymers. It is also one of the simplest linear aliphatic plastic and is highly amorphous. Its manufacturing processes include (1) polycondensation of glycolic acid, (2) ring-opening polymerization of glycolide, (3) solid-state polycondensation of halogen acetates, and (4) Bronsted acid-catalyzed polymerization of carbon monoxide and formaldehyde [25]. Applications include stabilization of fractures, bone grafting, reattachment of ligaments,

PGA polymers reduce risks during pre-implant osteoporosis and also exhibit a slow release of bioactive agents is used in the field of orthopedics as drug carriers. The recommended PGA sterilization methods are gamma radiation, ethylene oxide (EtO), plasma etching/electron beam irradiation techniques. Dry heating and autoclaving are not recommendable for PGA devices as it may alter its physical and mechanical properties.

3.2.8 Polyethylene Terephthalate (PET)

Polyethylene terephthalate is a thermoplastic polymer. Its synthesis involves trans-esterification of ethylene glycol with either terephthalic acid or dimethyl terephthalate. PET is colorless, highly flexible, and semi-crystalline in its natural state. Based on its processing or blending, PET can be hard, stiff, dimensionally stable, and recyclable; hence PET is a promising material for biomedical applications. PET is commercially available in the market with the name Dacron. PET is relatively impermeable to liquids and most gases, and this property can be improved by either increasing its crystallinity or blending it with other monomers. Hydrophobic aromatic groups make it biostable. A high degree of crystalline nature restricts it from hydrolytic breakdown, so PET does not absorb water easily once it is molded and set; resistant to dissolution by ordinary solvents. The main biomedical application of PET includes sutures, tubing for angioplasty balloons, and bottles for pharmaceutical preparations. It has poor resistance to steam and dry heat; hence the recommended sterilization techniques for PET are ETO and gamma radiation.

3.2.9 Polyvinyl Chloride

Polyvinyl chloride is one of the top leading thermoplastics used in medical devices and packaging applications. Its unique range of properties makes it highly demanding in the health care industry. It is glossy, transparent, flexible, durable, has good resistance to chemicals, scratch-resistant, biocompatible, and recyclable. PVC replaced rubber in the medical field, primarily where the need for single-use of pre-sterilized medical components increased. PVC synthesis involves the polymerization of vinyl chloride monomer. Chlorination of ethylene and pyrolysis of ethylene 1,2-dichloroethane produce vinyl chloride monomer. For the bio-based PVC, bio-ethylene is synthesized from bioethanol. The rigid PVC can be broadly classified into three, Type I, Type II, and CPVC. Type I is the most widely used PVC, it has excellent corrosion resistance, but lower impact resistance, and type II has lower corrosion and high impact resistance. Chlorinated polyvinyl chloride or CPVC has high-temperature resistance, and manufacturing involves the chlorination of PVC resin. This plastic, which is unblended with other additives, is considered unplasticized because they are less flexible than modified formulations. PVC mixed with the additives like stabilizers, plasticizers, and lubricants typically fabricate biomedical devices with desired properties. The additives can be classified based on the way they modify the properties of PVC [26, 27].

The First group additives include substances that help modify PVC's physical properties: plasticizers, lubricants, mold release agents, macromolecular modifiers, reinforcing fillers, reinforcing agents, and coupling agents. By adding heat stabilizers, antioxidants, UV protecting agents, Gamma ray's resistance's etc gives the polymer protection against aging and degradation. The most common plasticizers are phthalate plasticizers, phosphate plasticizers, Trimellitate plasticizers, Low-temperature plasticizers, citrate plasticizers, and miscellaneous. The plasticizers mainly help enhance the properties of PVC to get the desirable characteristics for biomedical applications like low toxicity, transparency, flexibility, strength, elongation, stability at high and low temperature, Etc. Example of heat stabilizers is barium-zinc additives and calcium-zinc formulations. Medical grade PVC did not use barium-zinc additives in its synthesis. Tinuvin®P and 2-(2Hbenzotriazol-2-yl)-p-cresol are examples of UV protecting agents in PVC.

PVC is the most commonly used thermoplastics for the medical packaging of biomedical devices and drugs. The primary role of this packaging process is to protect the medical or pharmaceutical product till it reaches the end consumer. A few PVC-based containers are dialyzing bags, IV bottles, urine bags, catheters, drip chambers, diaphragms. Also, infusions and transfusions systems, bags for blood and blood products, blister packs, solution containers, insulin pens, drip chambers, pull rings, oxygen facemasks, and gloves. PVC-based biomedical devices are sterilized using steam, radiation or ethylene oxide, or low-temperature plasma. According to the market statistics, around 25% of all plastic medical products are PVC-based. The main reasons for the medical industry to love this polymer are its low cost, safety, chemical stability, and other unique properties, as mentioned before. For over six decades, the experience and knowledge from healthcare and international environmental authorities show that PVC is safe.

Polyacrylate

Polyacrylates are synthetic, bioinert, biodegradable and their physicochemical properties depend on the type of chemical substitutions in the polymer. Based on the acrylate group substitutions, it produces polymers with different properties. It makes polymethyl acrylate more of a rubber-like texture. In contrast, the polymethyl methacrylate combination gives a hard-transparent glass-like material. Poly 2-hydroxyethyl methacrylate forms a hydrogel. The applications include manufacturing of breathing apparatus accessories, chest drainage units, urological accessories, check valves, Y-sites, IV adaptors, filter housings, IV pump housings, blood handling components, catheter accessories, and medical cassettes. The suggested sterilization methods are gamma radiation, ETO, and e-beam radiation, but wet ETO and steam sterilization methods are not recommended [17].

Polyethersulfone

Polyethersulfone comes under the class of high-temperature thermoplastics; it is amorphous and transparent. This family of thermoplastics contain diphenyl sulfone as its standard structural unit. The production involves polysulfonation /polyester synthesis. The other members of this family are polyarylsulfone, standard PSU and polyphenyl sulfone [28]. The basic properties include its remarkable capacity to withstand high temperatures (a glass transition temperature of 225 °C), excellent mechanical, electrical and flame resistance. It also has good chemical and hydrolytic resistance. The applications include manufacturing of dialysis filtration cartridges, reusable syringe injectors, sterilizer trays, prosthesis packaging dental tools, in making of an outer shell of implantable catheter ports in oncological screening, heart valve sizers, pharmaceutical filter housings, for making the transparent portion of anesthesia masks, surgical instrument cases and trays. The recommended sterilization techniques include general sterilization techniques like cold disinfectants, sterilants, and germicides, gamma and electron beam radiation, autoclaving or steam sterilization. PES can withstand more than 1,000 steam sterilization cycles, so repeated sterilizations did not cause any significant change in clarity or performance.

1. Applications of Polymeric Containers in Biomedical Field

A wide variety of polymers are used in various fields of medicine. Each polymer used in the biomedical application will have unique properties and specific formulations. The primary properties like tensile strength, elasticity, stiffness, biocompatibility, and toxicity have to be considered while opting for a polymer for container manufacturing. Because medical packaging/containers often require sterilization before use. A wide array of containers/medical packaging materials are available in the market designed to uphold the highest medical standards for performing its purpose with complete efficacy (Figs. 1 and 2). Table 1 discuss polymer-based containers which we use commonly for healthcare applications.

4 Conclusion

Numerous polymers are currently in use in various fields of biomedical application. Due to their unique properties, polymers are extensively used for various applications inside and outside the body. Polymers are macromolecules and which can blend with other materials with low and high molecular weight to obtain different properties for any application. So synthetically-blend polymers/medical grade polymers have considerable potential in medical application, and they replaced conventional polymers to an extent in the healthcare industry. Thermo plastics or medical-grade plastics are used specifically for the packaging of drugs and devices. They also act as the primary tool for the controlled delivery of drugs from the formulations. This chapter provides basic concepts about major polymers widely used in clinical settings for the fabrication of containers. The chapter begins with a general introduction about

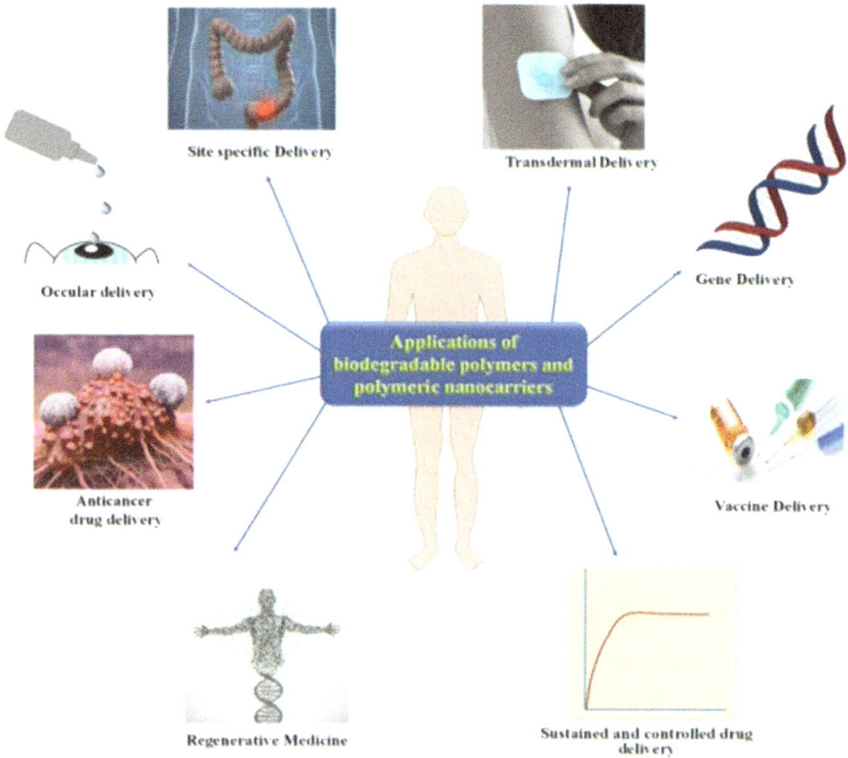

Fig. 1 Applications of biodegradable polymers in biomedical field [29] (Copyright permission taken, License Number: 5154650346139)

a brief history of entry of polymers into the field of medicine, basic concepts and classification of polymers, and natural and synthetic polymers used in the fabrication of containers, its synthesis and properties, and various applications.

5 Future Directions

Currently, thermoplastic applications in the biomedical field are found to be significantly more effective, safe, less toxic, and relatively simple. But it is necessary to spend time and money on R&D in natural and synthetic polymers to overcome the existing limitations to meet the vast requirements of the future healthcare industry. The development of new unique polymeric materials with emerging technologies like 3D printing can profoundly impact the current era. The research should generate cost-effective and versatile polymeric materials and utilize them for mainstream application to improve various healthcare applications, like drug delivery.

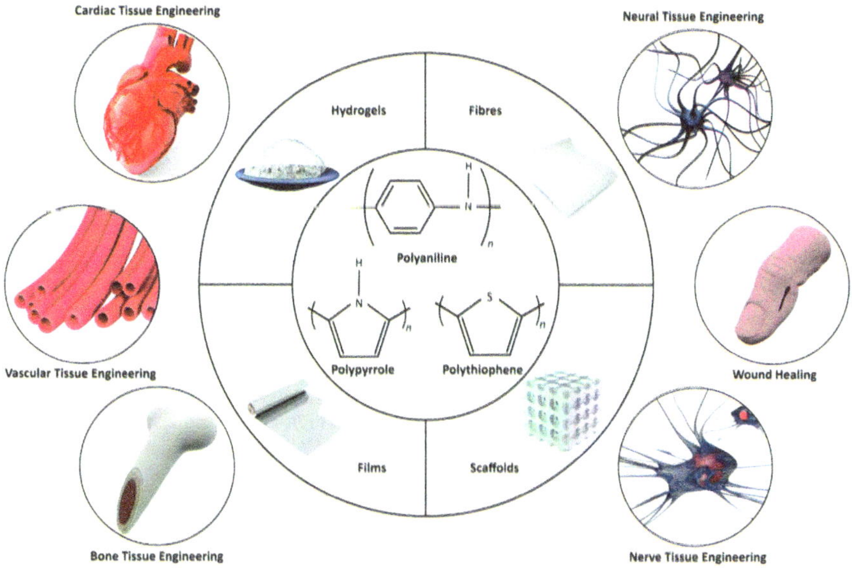

Fig. 2 Bioactive synthetic polymers for biomedical applications [30]

Table 1 Applications of polymeric containers in biomedical field

Materials	Biomedical applications
I. Natural polymers	
a. Collagen	• Drug delivery • Cosmetic surgery [31] • Promogran (Systagenix) is an FDA approved material for diabetic and ulcer wounds [32] • Integra—Dermal Regeneration template is an FDA approved skin substitute [33] • DuraGen—FDA approved material for spinal dural repair and regeneration [34] • Sulzer-spine—FDA approved sealant for cardiovascular and spinal surgeries • CoStasis—a surgical haemostat and Floseal—high viscous hemostatic agent • Sumycin, Collatamp—gentamicin delivery vehicles [35] • Used as an inactive component in solutions, gels, films, inserts, particles, shields, sponges and injections • Used for producing liquid, semi-solid and solid drug forms • Lignocaine used as anesthetic to reduce pain [36] • Used to stabilize liposomes for drug delivery • Used as protective membranes and inserts in ophthalmology to treat corneal infections [37] • Stents coated with collagen is used to treat restenosis • Collagen shields are used as a protection for contact lenses, treat local infections, injuries, glaucoma and immunosuppression • Collagen sponges with drugs are used to promote wound healing
a. Gelatin	• Wound dressings [38] • Used for making hard and soft capsules, tableting (used as a binder), tablet coating, granulation, encapsulation, and microencapsulation [39]
b. Cellulose	• Cellulose and its derivatives (Cellulosics) are used in oral solid dosage forms • Used for topical and parenteral administration
II. Synthetic polymers	
a. Polyethylene terephthalate (PET)	• Used in catheters, sutures, and also for the preparation of bottles in the pharmaceutical industry [40]
b. High Density Poly Ethylene (HDPE)	• Used to make drug containers, blister package, and medical device housing [41]

(continued)

Table 1 (continued)

Materials	Biomedical applications
c. Low Density Poly Ethylene (LDPE)	• Used to produce single-dose ampoules, intravenous fluid bottles, and tubes • Used in the manufacture of lids, caps, closures, and also in diagnostic packaging [42]
d. Lincar Low Density Poly Ethylene (LLDPE)	• Used to produce container tubes in the pharmaceutical industry [43]
e. Polyvinylidene chloride (PVDC)	• Used in pharmaceutical blister packaging [44]
f. Polypropylene (PP)	• Used for making bottles, jars, blister, ampoules, vials, strips, and strips in the pharmaceutical industry [45]
g. Polyvinyl chloride (PVC)	• Used to make blood bags and transfusion sets, oxygen mask, catheters, endotracheal tubings, IV infusion sets, nasal cannulas, ostomy bags, surgical examination gloves, blister packaging, containers for urine continence [46]
h. Polycarbonate (PC)	• Used to make connectors for intravenous fluid lines such as stopcocks, y-injection sites, cannulae, check valves, filter housings, male and female Luer fittings [47]
i. Polystyrene (PS)	• Used in making sterile packaging of disposable medical items
j. Polyethylene (PE)	• Used in making ampoules, vials, and cartridges in the pharmaceutical industry
k. Polyethersulphone (PES)	• Used in pharmaceutical industries to make cartridges, instrument holders, reusable syringe injectors, prosthesis packaging, sterilizer trays, and dental tools [48]
l. Polychlorotrifluoroethylene (PCTFE)/PVC laminates	• Used as blister packaging for drugs in the pharmaceutical industry [49]
m. Styrene	• Used in pharmaceutical industries to make vials [50]
n. Poly(ethylene–vinyl acetate) (PEVA)	• Used to make tubes for pharmaceutical preparations [51]
o. Orientated polyamide (OPA)	• Used mainly for making blister packaging for drugs [52]

References

1. Wiliams DF (1987) Definitions in biomaterials : proceedings of a consensus conference of the European Society for Biomaterials, Chester, England, 3–5 Mar 1986
2. Bhat S, Kumar A (2012) Biomaterials in regenerative medicine. J Postgr Med Educ Res 46:81–89
3. Hartmann L, Staffilani M, Unterlass MM (2020) Polymers for the future. Macromol Chem Phys 221:2000077

4. Advantages of polymer components over metallic materials for machined parts. https://aipprecision.com/the-advantage-of-plastics-over-metals-for-custom-parts/
5. Maitz MF (2015) Applications of synthetic polymers in clinical medicine. Biosurf Biotribol 1:161–176
6. Simionescu BC, Ivanov D (2014) Natural and synthetic polymers for designing composite materials. In: Handbook of bioceramics and biocomposites. 1–54
7. Uhrich KE, Abdelhamid D (2016) 3—biodegradable and bioerodible polymers for medical applications. In: Biosynthetic polymers for medical applications. 63–83
8. Ulery BD, Nair LS, Laurencin CT (2011) Biomedical applications of biodegradable polymers. J Polym Sci B Polym Phys 49:832–864
9. Lee CH, Singla A, Lee Y (2001) Biomedical applications of collagen. Int J Pharm 221:1–22
10. Batorsky A, Liao J, Lund AW, Plopper GE, Stegemann JP (2005) Encapsulation of adult human mesenchymal stem cells within collagen-agarose microenvironments. Biotechnol Bioeng 92:492–500
11. Langmaier F, Mokrejs P, Kolomaznik K, Mladek M (2008) Biodegradable packing materials from hydrolysates of collagen waste proteins. Waste Manage 28:549–556
12. Mariod A, Fadul H (2013) Review: Gelatin, source, extraction and industrial applications. Acta Sci Pol Technol Aliment 12(2):135–147
13. Li B et al (2015) Past, present, and future of microcarrier-based tissue engineering. J Orthop Transl 3:51–57
14. Ramezani Kalmer R, Mohammadi M, Karimi A, Najafpour G, Haghighatnia Y (2019) Fabrication and evaluation of carboxymethylated diethylaminoethyl cellulose microcarriers as support for cellular applications. Carbohydr Polym 226:115284
15. Ozdil D, Aydin HM (2014) Polymers for medical and tissue engineering applications. J Chem Technol Biotechnol 89(12):1793–1810
16. Qiang W, Deng N-N, Guo J, Deng J (2018) Synthetic polymers for biomedical applications. Int J Biomater 2018. Article ID 7158621
17. Maitz M (2015) Applications of synthetic polymers in clinical medicine. Biosurf Biotribol 1:161–176
18. Modjarrad K, Ebnesajjad S (2013) Handbook of polymer applications in medicine and medical devices
19. Plastic materials. Free online database for plastic industry. https://omnexus.specialchem.com
20. Van Lierde S (2004) Latest medical applications of polypropylene. Med Dev Technol 15:33–34
21. Encyclopedia of materials: science and technology, 1st edn. https://www.elsevier.com/books/encyclopedia-of-materials-science-and-technology/mahajan/978-0-08-043152-9
22. Holland SJ, Tighe BJ, Gould PL (1986) Polymers for biodegradable medical devices. 1. The potential of polyesters as controlled macromolecular release systems. J Control Release 4:155–180
23. Brannigan RP, Dove AP (2017) Synthesis, properties and biomedical applications of hydrolytically degradable materials based on aliphatic polyesters and polycarbonates. Biomater Sci 5:9–21
24. Pawar RP, Tekale SU, Shisodia SU, Totre JT, Domb AJ (2013) Biomedical applications of poly(Lactic Acid). Recent Patents Regen Med (Discontinued) 4:40–5
25. Göktürk E, Erdal H (2017) Biomedical applications of polyglycolic acid (PGA). Sakarya Üniversitesi Fen Bilimleri Enstitüsü Dergisi 21(6):1237–1244
26. Cameron RE, Kamvari-Moghaddam A (2009) Synthetic bioresorbable polymers—durability and reliability of medical polymers
27. Polyvinyl Chloride (PVC) plastic: uses, properties, benefits & toxicity. https://omnexus.specialchem.com/selection-guide/polyvinyl-chloride-pvc-plastic
28. Hong KZ (1996) Poly(vinyl chloride) in medical device and packaging applications. J Vinyl Add Tech 2:193–197
29. Prajapati SK, Jain A, Jain A, Jain S (2019) Biodegradable polymers and constructs: a novel approach in drug delivery. Eur Polym J 0014–3057

30. Bolívar-Monsalve EJ, Alvarez MM, Hosseini S, Espinosa-Hernandez MA, Fernando C, Ceballos-González, Sanchez-Dominguez M, Shin SR, Cecen B, Hassan S, Di Maiog E, Trujillo-de Santiago (2021) Engineering bioactive synthetic polymers for biomedical applications: a review with emphasis on tissue engineering and controlled release. Mater Adv 14(2):4435–4884
31. Liu B et al (2005) The use of type I and type III injectable human collagen for dermal fill: 10 years of clinical experience in China. Semin Plast Surg 19(3):241–250
32. Vin F, Teot L, Meaume S (2002) The healing properties of Promogran in venous leg ulcers. J Wound Care 11:335–341
33. Pomahac B, Svensjö T, Yao F, Brown H, Eriksson E (1998) Tissue engineering of skin. Crit Rev Oral Biol Med Off Publ Am Assoc Oral Biol 9:333–344
34. Narotam PK, José S, Nathoo N, Taylon C, Vora Y (2004) Collagen matrix (DuraGen) in dural repair: analysis of a new modified technique. Spine 29:2861–2867. Discussion 2868–2869
35. Gruessner U et al (2001) Improvement of perineal wound healing by local administration of gentamicin-impregnated collagen fleeces after abdominoperineal excision of rectal cancer. Am J Surg 182:502–509
36. Balakrishnan K, Ebenezer V, Dakir A, Kumar S, Prakash D (2015) Bupivacaine versus lignocaine as the choice of locall anesthetic agent for impacted third molar surgery a review. J Pharm Bioallied Sci 7:S230–S233
37. Kaul S, Kothiyal GP (2012) An insight into ocular insert. Int J Pharm Sci Res 1905–1912
38. Gaspar-Pintiliescu A, Stanciuc A-M, Craciunescu O (2019) Natural composite dressings based on collagen, gelatin and plant bioactive compounds for wound healing: a review. 854–865
39. The importance of gelatin in pharmaceutical and medical applications. https://www.rousselot.com/biomedical/media/blog/testimonials/the-importance-of-gelatin-in-pharmaceutical-and-medical-applications
40. Siegert J (2002) Applications of PET in pharmaceutical R & D. Eur J Clin Pharmacol 58
41. Balakrishnan H, Husin MR, Wahit MU, Kadir MRA (2013) Maleated high density polyethylene compatibilized high density polyethylene/hydroxyapatite composites for biomedical applications: properties and characterization. Polym-Plast Technol Eng 52(8):774–782
42. LDPE grades for blow fill seal applications in pharmaceutical packaging. http://www.plastemart.com/plastic-technical-articles/ldpe-grades-for-blow-fill-seal-applications-in-pharmaceutical-packaging/1451
43. Jin-hua T, Guo-qin L, Huang C, Lin-jian S (2012) Mechanical properties and thermal behaviour of LLDPE/MWNTs nanocomposites. Mat Res 15:1050–1056
44. Paisley K (2004) PVDC—new developments, new opportunities. Division of Solvay advanced polymers
45. Sanjay PD, Vasantrao PP, Pandit DP, Dattaprasad DS, Shivaji MS (2016) Polymers used in pharmaceuticals: a brief review. 2:6
46. Van Dooren AA (1991) PVC as pharmaceutical packaging material. Pharm Weekbl Sci Ed 13:109–118
47. Rogalsky S, Moshynets O, Lyudmila L, Tarasyuk O (2014) Antimicrobial polycarbonates for biomedical applications. EPMA J 5
48. Pillai SA, Chobisa D, Urimi D, Ravindra N (2016) Filters and filtration: a review of mechanisms that impact cost, product quality and patient safety. J Pharm Sci 8:8
49. Pasbrig E (2007) Covering foil for blister packs
50. Miller RR, Newhook R, Poole A (1994) Styrene production, use, and human exposure. Crit Rev Toxicol 24:S1–S10
51. Schneider C, Langer R, Loveday D, Hair D (2017) Applications of ethylene vinyl acetate copolymers (EVA) in drug delivery systems. J Control Release 262:284–295
52. Winnacker M (2017) Polyamides and their functionalization: recent concepts for their applications as biomaterials. Biomater Sci 5(7)

Containers for Self-healing/Self-repairing Polymers

Negin Farshchi

1 Encapsulation Methods

The revolutionary publication of White et al. [1], was a great motivation for the next investigations about the design and fabrication of containers encapsulating self-healing agents/materials. The aim of these studies is the restoration of materials' mechanical properties, electrical conductivity, or the inhibition of corrosion.

Initial work inspired by other disciplines [2] resulted in the development of self-healing polymers consisting of hollow fibers containing the reactive agent and embedded reactive microcapsule in their matrix. The basic concept of these systems to achieve self-healing polymers is the spillage of the reactive agent from hollow fibers (or microcapsules) into the polymer matrix and then its crosslinking acting as an internal glue. Based on this concept, microvascular networks presented interconnected hollow fibers filled with healing agents, which introduce the ability to refill the hollow channels with healing agents upon weakening, enabling multiple damage cycle repairs in the same area [3].

One of the most used methods is capsules, which are core-shell structures with a liquid self-healing agent core. This is due to the simple preparative procedure and their high potential for improvement. On the other hand, using commercial materials as a matrix to embed the capsules make it possible to use the current industrial processes. Another advantage of this system is the great potential for further modifications and developments.

Although there is no need for an external energy source in these systems (acting as internal glue), they show various shortcomings. In addition, the fact that introducing liquid capsule/hollow fibers affects the mechanical properties of polymer

N. Farshchi (✉)
Faculty Mechanical Engineering, Institute of Material Science, Technische Universität Dresden, 01062 Dresden, Germany
e-mail: negin.farshchi@mailbox.tu-dresden.de

J. Parameswaranpillai et al. (eds.), *Micro- and Nano-containers for Smart Applications*, Composites Science and Technology,
https://doi.org/10.1007/978-981-16-8146-2_9

matrix by weakening it is a big concern, even though many applications have been reported. However, the concept alone stimulated innovative approaches to achieve self-healing (self-repairing) properties, such as directly embedding reactive particles into polymer networks without capsulation or fibrous networks. Further details concerning the advantages and the limitations of these systems have been extensively outlined elsewhere [4–6].

The focus of this chapter is on the recent developments in encapsulation techniques used for self-healing composites. Throughout this chapter different methods regarding encapsulation of reactive agents in form of microcapsules, nanotubes, and hollow fibers have been considered in a non-comprehensive manner.

2 Microcapsules

There are several methods to prepare microcapsules, which include self-assembly, phase separation, polymerization, and templating synthesis. The durability and stability of capsules are dependent on the parameters of these encapsulation processes. Choosing the right method to a high extend depends on the final application as each offers different advantages along with the specific applications adapted and to the range of capsule characteristics [7–11].

2.1 Suspension Polymerization

So far, most of the reported works considering self-healing materials through microcapsules, utilized suspension polymerization as the synthesis method of microcapsules. Various types of polymerization could occur in micro-/milli-droplets created by shearing a liquid phase in another immiscible liquid phase containing stabilizer. Depending on the polarity of the healing agent, one can choose the right solution to surround the liquid droplets.

Important parameters for choosing the right material for capsulation such as good performance of seal, flexibility, penetration resistance, high strength, and impermeability also affect the method of encapsulation. As an example Poly (urea-formaldehyde) (PUF) one of the most used materials as microcapsule wall is generally utilized through encapsulation of healing agent by in situ polymerizations of urea and formaldehyde. As a solvent-free method, it also has several advantages such as simple procedure, low costs, and ease of industrialization when compared to the other encapsulation techniques [12].

In this method, the encapsulation process follows as the urea and formaldehyde form colloidal cross-linked particles by reacting in the water phase (Fig. 1), which is contained in the emulsified healing agent droplets. In this manner, the urea-formaldehyde (UF) shell wall is covering the healing agent [13]. This can lead to the

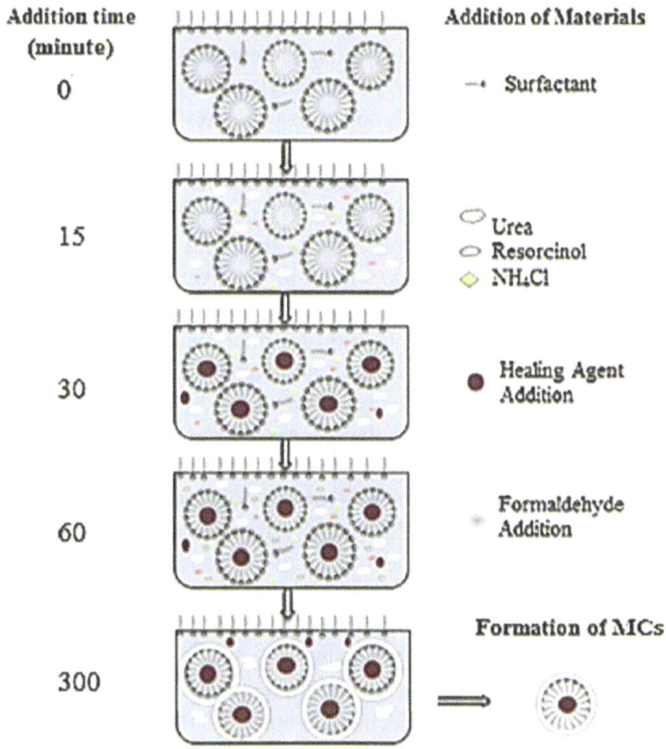

Fig. 1 PUF encapsulation process. Reprinted with permission from Ref. [24]. Copyright (2018) Elsevier BV

controllable size and shell thickness of the microcapsule, which is another advantage of this method.

Although, it is common to use dicyclopentadiene (DCPD) as a healing agent. However, the disadvantage of this system is the essential dispersion of the catalyst in the matrix [14]. While drying oils through the oxidation of their unsaturated bonds (auto-oxidation) are great alternatives for catalyst involving systems [15]. This application in anti-corrosion coatings is well reported in the literature [11, 16–23].

Çömlekçi et al. investigated the surfactant amount (PVA) and stirring rate on the mechanical properties such as capsule size, thermal stability, and morphology of the new healing agent LO-Alkyd and Linseed oil to have a comparison. Through the response surface methodology, they showed that the relation between the size of the microcapsules and the parameters concerned is inverse. Both capsules in freeflowing powder form showed enough good thermal stability to be utilized in coating applications. In addition, the increasing PVA amount and stirring rate result in a smooth surface of capsules. In addition, the type of healing agent, surfactant amount, and stirring rate have an impact on the thickness of the capsule wall (0.5–2.5 μm) [24].

Considering the encapsulation of isocyanates, it has been restricted mostly to the solid-state or blocked form [25, 26]. Although, there are studies also considering the encapsulation of liquid isocyanates, hexamethylene diisocyanate (HDI), and IPDI in the polyurethane shell walls by the oil-in-water (O/W) emulsion method [27, 28]. Recently, Di Credico et al., established an innovative method for the microencapsulation of liquid IPDI by double-layered polyurethane/poly(urea-formaldehyde) (PU/PUF) microcapsules in a single batch process via polymerization in an O/W emulsion. In this process, PUF outer layer is hard and preserves the liquid IPDI. While the soft inner PU layer introduces flexibility and tightness to the microcapsules [29].

Another widely investigated application for microencapsulation is the encapsulation of protein trapping agent (vaccine adjuvant) in PLGA microspheres by a water/oil/water emulsion–solvent evaporation method [30]. Desai et al. proposed an interesting method by co-encapsulating protein-sorbing vaccine adjuvants and polymer plasticizers. Resulted in optimal active self-healing microencapsulation of vaccine antigens in PLGA. In order to avoid the agglomeration of particles, PVA was utilized as an emulsifier (Fig. 2). In this system, the encapsulation is controlled by the inner adjuvant/trehalose phase to outer oil phase ratio during the formation of the initial emulsion [31].

Considering the composition of shells, nanocapsules used in self-healing materials can be divided into three different categories as polymer, inorganic and hybrid nanocapsules. The polymeric nanocapsules are widely used in self-healing materials but considering the thermal and mechanical stability, they are usually less steady compared to inorganic ones. In the case of polymer nanocapsules, there is always the risk of collapsing of shell or formation of large aggregates above their glass transition temperature. Introducing an inorganic layer in the form of a thin coat on the polymeric capsules or direct encapsulation of a healing agent in the inorganic nanocontainer can solve this problem.

One of the major applications of nanocapsules is the encapsulation of the corrosion inhibitors in self-healing coatings. One of the shortcomings of using this method is its negative influence on the barrier properties of the coating layers. However, using a suitable method this could be resolved, as Choi et al. had utilized multi-stage

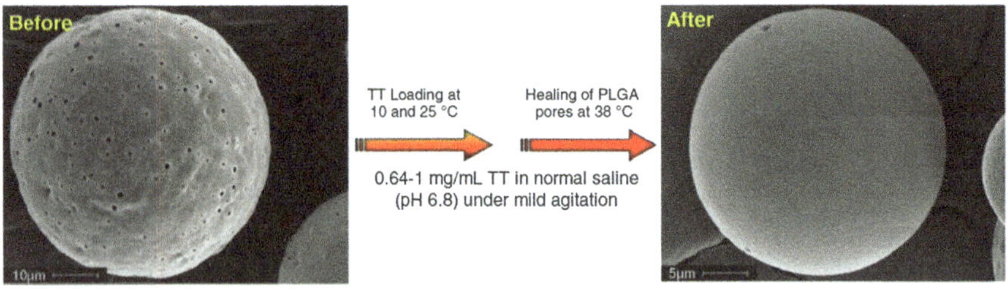

Fig. 2. Active self-healing encapsulation of tetanus toxoid (TT) in PLGA microspheres. The surface morphology of 3.2 wt% $Al(OH)_3$–PLGA–3.5 wt% trehalose–5 wt% diethyl phthalate. Reprinted with permission from Ref. [31]. Copyright (2013) Elsevier

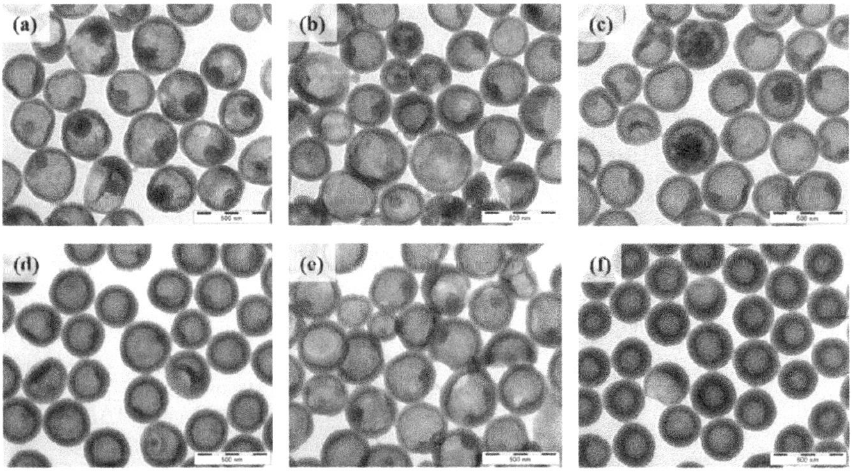

Fig. 3 Transmission electron micrograph of the nanocapsules sensitized through multi-stage emulsion polymerization. Reprinted with permission from Ref. [32]. Copyright (2013) Elsevier BV

emulsion polymerization to synthesize polymeric nanocapsules containing six types of amine corrosion inhibitors (Fig. 3) [32].

2.2 *Miniemulsion*

Concerning the key parameters for successful self-healing through encapsulation good dispersion of capsules is an important one. As it is well known from reinforced material, a convenient method is the functionality of additive. When considering capsules with the functional surface, Miniemulsion polymerization is a method of choice [33]. Miniemulsion as a tool to produce functional nanoparticles has been widely investigated, which is suitable for encapsulation of both hydrophobic [10, 34–40] and hydrophilic [41–44] materials.

Utilizing an Aqueous continuous phase containing a bit of hydrophilic functional monomer for copolymerization of a hydrophobic monomer is the most direct method to obtain the polymer nanoparticles with functional groups at their surfaces dispersed in water. These functional groups can be various groups such as carboxylic acid [45], sulfonate [46], or phosphate groups [47].

Moreover, this encapsulation method can be used for both liquid [10, 34, 36, 40–42, 44] and solid [37, 38, 45–48] substances. Although, only the liquid ones seem suitable for self-healing application as the need for mobility for a self-healing agent to diffuse to the damaged area. Utilizing miniemulsion encapsulation to develop core-shell nanoparticles was a novel idea performed by Fickert et al. [49], which has yielded hollow nanocapsules instead of a solid ones. They reported the synthesis of

Table 1 Characterization of the capsules and the healing materials

Capsules material	Capsules size	Self-healing material	Refs.
PUF	51 μm 97 μm	LO	[24]
	56 μm 261 μm	LO-Alkyd	
Glass	3.35 mm	PU	[50]
PVA	0.6–1 μm	Reactive magnesium oxide (MgO)	[51]
	1–2 μm	Silica fume	
	2–4 μm	Bentonite	
Polystyrene	300–600 nm	Diethanolamine (DEA) Ethanolamine (ETA) propylamine (PPA) Triethanolamine (TEA) dipropylamine (DPA) 5-Amino-1 pentanol (5AP)	[32]
PLGA	20–63 μm	$Al(OH)_3$	[31]
	63–90 μm	$CaHPO_4$	
PU	40–400 μm	IPDI	[27]
PU	5–350 μm	HDI	[28]
(PU/PUF)	50–500 μm	IPDI	[29]
Styrene	290 nm	DCPD	[39]
Silica	260 and 230 nm	ROMP	[49]

silica nanocapsules, containing amino- and thiol-functionalized alkoxysilanes. Base on the number of monomers used in the miniemulsions, it is possible to tune the capsule thicknesses. While the amount of surfactant used to emulsify the dispersed phase is controlling the size of the nanocapsules.

Table 1 listed the containers and healing agent materials as well as their size used in recent studies containing micro/nanocontainers for self-healing.

3 Hollow (Glass) Containers

Nature has been always a great source of inspiration for engineers and scientists, as the healing process of mammals such as bleeding, bruising, and healing of the bone fracture inspired scientists to try to mimic it. The very first attempt to induce this kind of self-healing in engineering was performed through composites containing hollow glass capillaries/tubing filled with self-healing agents [52].

Glass capsules have also been used in combination with the different healing agents, which can have variant chemical nature [53]. In successful healing, two important processes should take place. First, the healing agent should remain in the fluid phase (to spread efficiently in the crack by capillarity) and the second is the reaction progression of the agent (to ensure a compact filling of the damaged area).

To accelerate the procedure, one attempt is to use two closely jointed capsules, one inholding the healing agent and the other an accelerator [54, 55]. However, there are some limitations for the success of this method, as both capsules should break simultaneously once a crack reaches them, and then the proper mixing of both fluids should take place to trigger the reaction efficiently. In the case of self-healing concrete using polyurethane agent in glass capsules monitoring the process is a complex task, as many factors (namely the mixing of the fluids, formation of the foam, and filled area by the healing agent) remain unclear. In addition, as the mechanical approach, the strength of this repaired area is unknown [50].

One of the major shortcomings of this method is that the handling of such systems is troublesome. On the other hand, the microscopic hollow fibers or spherical capsules usually have diameters over 100 μm, which may unfavorably affect the superior specific properties of polymer composites. In the case of glass capsules the general disadvantages of capsule-based systems such as high effective cost of the encapsulation processes, the poor distribution of capsules, and the lack of interaction between the capsules and matrix, and the low availability of healing agents for multiple healing cycles [56] are also should be considered. Although, using hollow fibers and, microvascular networks were recommended to improve their problems, but researchers demonstrated that such systems also undesirably influence the mechanical properties of the material as a consequence of presenting local stress concentrations and disorganizing the material structure to hold the fibers. However, decreasing the diameter of the fibers will moderate the mentioned adverse effects [57].

4 Electrospinning

Although encapsulation is an imperative aspect in the synthesis of self-healing materials, it should always be considered that the encapsulation method must provide stable storage of the healing agent in a way that prevents a negative impact on the initial material properties as mentioned earlier. In contrast to microencapsulation methods that are related to chemical reactions, electrospinning has the potential for accelerating new chemistries by being basically physical, being dependent on electric fields and solvent evaporation, instead of chemical. Also, the relatively small size of fibers (ranging from nm to μm) is in favor of the integration of healing components within a composite. The schematic design of the coaxial electrospinning is represented in Fig. 4.

Park et al. [58] in order to encapsulate a dimethylsiloxane healing agent in a self-healing coating, the healing agent was electrospun into poly(vinyl pyrrolidone) fibers resulting in beads of healing agent on a polymer string. In this method, the preparation of multiple solution feed systems makes it possible to simultaneously electrospun two or more polymer solutions from coaxial capillaries. In the first attempt, various inert oils were encapsulated within electrospun fibers [59], also it is possible to utilize this method to encapsulate a variation of materials such as bioactive agents, and other polymers [60] such as encapsulation of epoxy pre-polymer in PVA shell [61].

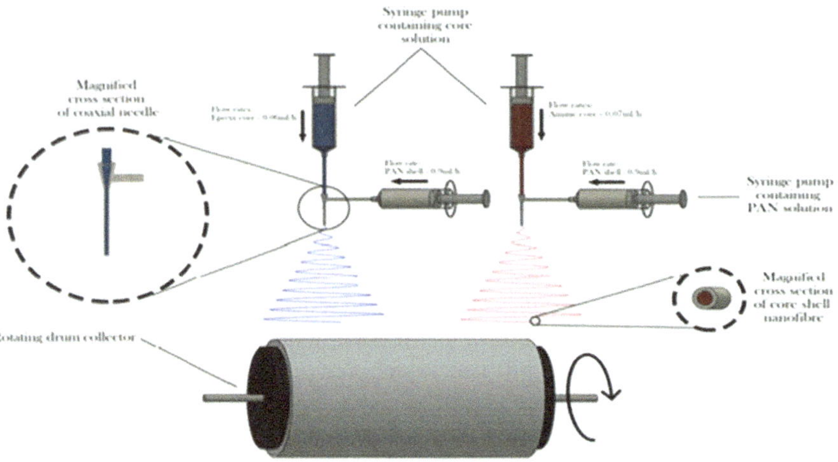

Fig. 4 Schematic illustration of the coaxial electrospinning. Reprinted with permission from Ref. [62] Copyright (2016) RSC Advances

Another approach would be using the inter interfacial toughening technique based on entangling small fibers at interfaces (interlayers) [63] for self-healing application through co-electrospinning, emulsion electrospinning. It has several advantages such as economical manufacture, low fiber concentration, and ease of merging into the customary composites processing. Since many failures in materials occur at the first stage in the nano-scale and then propagate to the micro one, nanovessels such as nanofibers are excellent candidates for the healing of cracks or delamination at their exact early stages of failure [64].

It has been used along with other methods for encapsulation of healing agents such as dicyclopentadiene (DCPD) and isophorone diisocyanate in polymer fibers like [polyacrylonitrile (PAN)] and amorphous turbostratic carbon nanotubes (CNTs). In addition, the self-sustained diffusion technique was utilized to introduce the liquid monomers (for self-healing) into CNTs. The relevant toughening mechanisms and deliverability of the healing agent through this method are validated [65]. There are several reports on the successful fabrication of self-healing polymers through including core-shell nanofibers containing siloxane-based healing agents in PAN nanofibers fabricated through the emulsion and coaxial electrospinning approaches [66–69].

In a similar attempt, an epoxy self-healing system based on encapsulated epoxy resin and amine-based curing agent in polyacrylonitrile (PAN) shell using coaxial electrospinning technique was introduced. With the capacity of 24 and 37 wt% of the epoxy resin and amine curing agent of the PAN nanofibers [62]. This method can be successfully applied to achieve self-healing coatings. Through a two-step electrospinning method, it is possible to encapsulate two parts dimethylpolysiloxane in nanosized fiber channels of polyvinylpyrrolidone (PVP) for self-healing of acrylate polyurethane coating [58] as the similar attempt for encapsulation of two healing

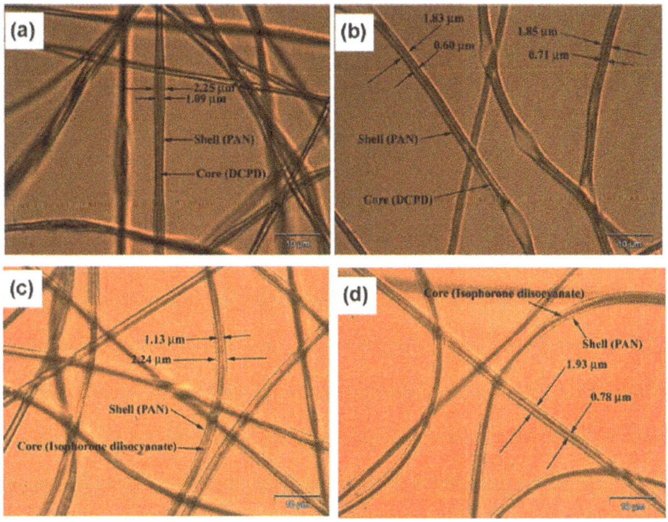

Fig. 5 Optical images of core-shell fibers electrospun from emulsion are shown. Panels (**a**) and (**b**) show core-shell fibers obtained from the emulsion of DCPD in PAN and DMF, while panels (**c**) and (**d**) show core-shell fibers obtained from the emulsion of PAN and isophorone diisocyanate in DMF. The scale bar in all images is 10 mm. Reprinted with permission from Ref. [65] Copyright (2012) the Royal Society of Chemistry

agents, dicyclopentadiene (DCPD) and isophorone diisocyanate (IPDI) in PAN nanofibers (Fig. 5) [65].

Using hybrid composites for self-healing carbon/epoxy composite by inserting core-shell nanofibers within the carbon fabric layers is achieved. A Styrene acrylonitrile (SAN) copolymer is utilized as the shell material, while the healing agents were two constituents: a low viscosity epoxy resin and its amine-based curing agent. The healing can happen up to three cycles at room temperature in the absence of any external driving force (Fig. 6) [70].

Halloysite nanotubes (HNT) are interesting candidates as nano-container for self-healing applications. Their size can be different around 1–15 μm in length and 10–150 nm in inner diameter, which depends on the deposit [71]. Being economical and ease of the availability in natural deposits are also interesting for mass production applications. It has been successfully used for encapsulation of solvent for self-healing of the epoxy system [72]. The fiber material, size, and the type of the healing agent used in recent investigations of self-healing strategies are summarized in Table 2.

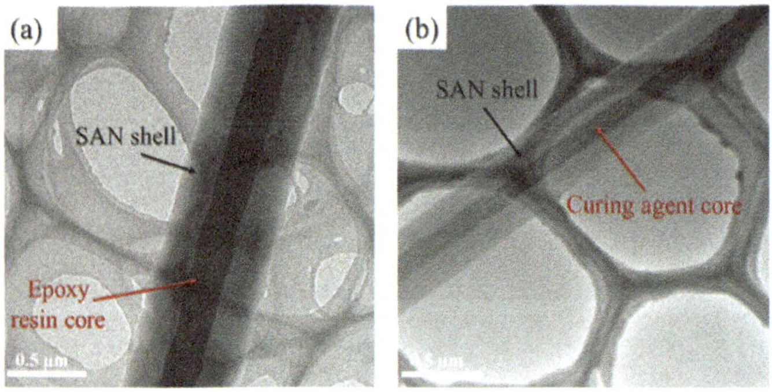

Fig. 6 TEM images of the core-shell nanofibers (**a**) epoxy core, (**b**) amine-based curing agent core. Reprinted with permission from Ref. [70] Copyright (2017) Elsevier BV

Table 2 Characterization of fibers/tubes used for encapsulation of self-healing materials

Fiber/tubes material	Fiber/tubes diameter	Self-healing materials	Refs.
Poly(vinyl pyrrolidone)	2–10 μm	Dimethylsiloxane	[58]
EVA	30 nm	Epoxy	[61]
CNT	50–100 nm	DCPD	[65]
	50–100 nm	Isophorone diisocyanate	
PAN	1.5–3 μm	DCPD	
	1.5–3 μm	Isophorone diisocyanate	
SAN	750 nm	Epoxy	[70]
	670 nm	Amine-based curing agent	
PAN	483 nm	Epoxy	[76]
	406 nm	Amine-based curing agent	

5 Spray Coating

Using spray coating as an encapsulation method is also investigated, as Alghamri et al. had developed coated pellets as a self-healing system in cement-based materials. They used three different mineral powders as healing agents (reactive magnesium oxide (MgO), silica fume, and bentonite), which are encapsulated using polyvinyl alcohol (PVA) based film. The thickness of the coating ranged between 10 and 50 mm. This approach is promising as the coating material has acceptable integrity and stability in water and high pH solutions [51].

6 Cell Encapsulation

"Cell encapsulation" is the term used widely in studies concerning hydrogels with the capability of cell carriers. The most investigated material for this means is natural carbohydrates for emerging injectable hydrogels for cell delivery, which should show good biocompatibility, biodegradability, and non-toxicity [59, 60, 73–75] Lately, Lee et al. establish that integration of type-II collagen or chondroitin sulfate into chitosan injectable hydrogels promotes better proliferation and cartilage matrix construction through encapsulated chondrocytes and mesenchymal stem cells (MSCs) [77].

Despite the good biocompatibility, and promotion of cell persistence and proliferation of these natural carbohydrates the development of a real system using injectable hydrogels for cell encapsulation to be utilized in medical application is so far in its initial phase. Due to potential shortcomings such as probable toxicity of the used agents, undesirable side products from gel degradation, the complex synthesis process, and is uneconomical. Recently, Lü et al. have developed an injectable hydrogel derived from natural polymers, namely chondroitin sulfate and chitosan. The main disadvantages of these systems that impede their applications are the complicated synthetic procedures (photo-crosslinking [78], enzymatic crosslinking [79], grafting reaction [76, 80], and click chemistry [81]), and the probable cytotoxicity of the used agents. However, using an inexpensive method, they formulate hydrogels using n-succinyl-chitosan and chondroitin that chondroitin sulfate was oxidized with sodium periodate by cleaving the vicinal glycols to obtain multiple aldehyde groups enabling chondroitin sulfate to function as a macromolecular cross-linker for n-succinyl-chitosan sulfate without using any crosslinking agents. In addition, it is self-healing under physiological conditions without any external stimulus. However, it is not relevant to the concept of this chapter and interested readers can follow it in literature [82, 83].

7 Conclusions and Outlook

Self-healing materials have been growing during the last years and significant advances have been made. These materials will bring many advances and novel technology also in the future. Combining the optical, magnetic, and electric properties along with self-healing abilities in engineered materials sounds promising for future technological advances.

It will also be required to develop further suitable analytical tools that can evaluate the molecular processes responsible for the relationship of self-healing behavior and the containers used for the delivery of the healing agent.

In principle, the methods for the production of self-healing materials through containers can be divided into purely chemical approaches through polymerization and synthetic steps and on the other hand, the mechanical strategies that focus on mechanical properties before and after repair. The key for understanding the molecular interactions and relating the structure-property attributes to the complex process of self-healing via micro/nanocontainers, would be a combination of both strategies with theoretical considerations.

The main difference between biological self-healing and manmade self-healing is the absence of the metabolism in the latter one. This can be defined in the self-healing materials containing containers as a method to refill or recycle the healing agent autonomously. This would be a great challenge for encapsulating systems through spherical capsules but theoretically, it may be possible in vascular systems to be obtained through reversible reactions. Also in the case of recovery/recycling of shell materials, using reverse reactions chemistry in the synthesis of self-healing micro/nanocapsules will be potential.

Considering the application of self-healing materials containing encapsulated self-healing agent, although there is already several advanced but the combination of different containers containing various healing agents in one system can lead to advanced applications for it as energy storage ability, smart packages, antifouling surfaces, etc.

References

1. White SR et al (2001) Autonomic healing of polymer composites. Nature 409(6822):794–797
2. Dry C (1996) Procedures developed for self-repair of polymer matrix composite materials. Compos Struct 35(3):263–269
3. Trask R, Williams G, Bond I (2007) Bioinspired self-healing of advanced composite structures using hollow glass fibres. J R Soc Interface 4(13):363–371
4. Caruso MM et al (2009) Mechanically-induced chemical changes in polymeric materials. Chem Rev 109(11):5755–5798
5. Keller MW (2013) Encapsulation-based self-healing polymers and composites. Healable polymer systems. The Royal Society of Chemistry Cambridge, UK, pp 16–61
6. Trask R, Williams H, Bond I (2007) Self-healing polymer composites: mimicking nature to enhance performance. Bioinspiration Biomim 2(1):P1
7. Yow HN, Routh AF (2006) Formation of liquid core–polymer shell microcapsules. Soft Matter 2(11):940–949
8. Cui J et al (2010) Monodisperse polymer capsules: tailoring size, shell thickness, and hydrophobic cargo loading via emulsion templating. Adv Func Mater 20(10):1625–1631
9. Esser-Kahn AP et al (2011) Triggered release from polymer capsules. Macromolecules 44(14):5539–5553
10. Zhao Y et al (2012) Encapsulation of self-healing agents in polymer nanocapsules. Small 8(19):2954–2958
11. Nesterova T et al (2012) Microcapsule-based self-healing anticorrosive coatings: capsule size, coating formulation, and exposure testing. Prog Org Coat 75(4):309–318
12. Qin R et al (2012) Preparation and characterization of a novel poly (urea–formaldehyde) microcapsules with similar reflectance spectrum to leaves in the UV–Vis–NIR region of 300–2500 nm. Mater Chem Phys 136(2–3):737–743
13. de la Paz Miguel M et al (2016) Effect of the preparation method on the structure of linseed oil-filled poly (urea-formaldehyde) microcapsules. Prog Org Coat 97:194–202
14. Mangun C et al (2010) Self-healing of a high temperature cured epoxy using poly (dimethyl-siloxane) chemistry. Polymer 51(18):4063–4068

15. Suryanarayana C, Rao KC, Kumar D (2008) Preparation and characterization of microcapsules containing linseed oil and its use in self-healing coatings. Prog Org Coat 63(1):72–78
16. Samadzadeh M et al (2011) Tung oil: an autonomous repairing agent for self-healing epoxy coatings. Prog Org Coat 70(4):383–387
17. Juita J et al (2012) Low temperature oxidation of linseed oil: a review. Fire Sci Rev 1(1):3–3
18. Boura SH et al (2012) Self-healing ability and adhesion strength of capsule embedded coatings—micro and nano sized capsules containing linseed oil. Prog Org Coat 75(4):292–300
19. Behzadnasab M et al (2014) Preparation and characterization of linseed oil-filled urea–formaldehyde microcapsules and their effect on mechanical properties of an epoxy-based coating. Colloids Surf A 457:16–26
20. Siva T, Sathiyanarayanan S (2015) Self healing coatings containing dual active agent loaded urea formaldehyde (UF) microcapsules. Prog Org Coat 82:57–67
21. Hasanzadeh M, Shahidi M, Kazemipour M (2015) Application of EIS and EN techniques to investigate the self-healing ability of coatings based on microcapsules filled with linseed oil and CeO_2 nanoparticles. Prog Org Coat 80:106–119
22. Szabó T, Telegdi J, Nyikos L (2015) Linseed oil-filled microcapsules containing drier and corrosion inhibitor–their effects on self-healing capability of paints. Prog Org Coat 84:136–142
23. Lang S, Zhou Q (2017) Synthesis and characterization of poly (urea-formaldehyde) microcapsules containing linseed oil for self-healing coating development. Prog Org Coat 105:99–110
24. Çömlekçi GK, Ulutan S (2018) Encapsulation of linseed oil and linseed oil based alkyd resin by urea formaldehyde shell for self-healing systems. Prog Org Coat 121:190–200
25. Querat E, Tighzert L, Pascault J-P (1996) Microencapsulation of isocyanates. Characterization and storage stability of microcapsules in a polyester α, ω-ol. JCT J Coat Technol 68(854)
26. Cheong IW, Kim JH (2004) Synthesis of core–shell polyurethane–urea nanoparticles containing 4,4′-methylenedi-p-phenyl diisocyanate and isophorone diisocyanate by self-assembled neutralization emulsification. Chem Commun 21:2484–2485
27. Yang J et al (2008) Microencapsulation of isocyanates for self-healing polymers. Macromolecules 41(24):9650–9655
28. Huang M, Yang J (2011) Facile microencapsulation of HDI for self-healing anticorrosion coatings. J Mater Chem 21(30):11123–11130
29. Di Credico B, Levi M, Turri S (2013) An efficient method for the output of new self-repairing materials through a reactive isocyanate encapsulation. Eur Polymer J 49(9):2467–2476
30. Reinhold SE et al (2012) Self-healing microencapsulation of biomacromolecules without organic solvents. Angew Chem 124(43):10958–10961
31. Desai K-GH, Schwendeman SP (2013) Active self-healing encapsulation of vaccine antigens in PLGA microspheres. J Control Release 165(1):62–74
32. Choi H, Kim KY, Park JM (2013) Encapsulation of aliphatic amines into nanoparticles for self-healing corrosion protection of steel sheets. Prog Org Coat 76(10):1316–1324
33. Crespy D, Landfester K (2010) Miniemulsion polymerization as a versatile tool for the synthesis of functionalized polymers. Beilstein J Org Chem 6(1):1132–1148
34. Fickert J et al (2012) Design and characterization of functionalized silica nanocontainers for self-healing materials. J Mater Chem 22(5):2286–2291
35. Bathfield M, Graillat C, Hamaide T (2005) Encapsulation of high biocompatible hydrophobe contents in nonionic nanoparticles by miniemulsion polymerization of vinyl acetate or styrene: influence of the hydrophobe component on the polymerization. Macromol Chem Phys 206(22):2284–2291
36. Crespy D, Musyanovych A, Landfester K (2006) Synthesis of polymer particles and nanocapsules stabilized with PEO/PPO containing polymerizable surfactants in miniemulsion. Colloid Polym Sci 284(7):780–787
37. Darwish M et al (2011) Bi-layered polymer–magnetite core/shell particles: synthesis and characterization. J Mater Sci 46(7):2123–2134
38. Erdem B et al (2000) Encapsulation of inorganic particles via miniemulsion polymerization. In: Macromolecular Symposia. Wiley Online Library

39. Li H et al (2008) Incorporation of polyoxometalates into polystyrene latex by supramolecular encapsulation and miniemulsion polymerization. Macromol Rapid Commun 29(5):431–436
40. Staff RH et al (2011) Phase behavior of binary mixtures of block copolymers and a non-solvent in miniemulsion droplets as single and double nanoconfinement. Soft Matter 7(21):10219–10226
41. Crespy D, Landfester K (2007) Preparation of nylon 6 nanoparticles and nanocapsules by two novel miniemulsion/solvent displacement hybrid techniques. Macromol Chem Phys 208(5):457–466
42. Crespy D et al (2007) Polymeric nanoreactors for hydrophilic reagents synthesized by interfacial polycondensation on miniemulsion droplets. Macromolecules 40(9):3122–3135
43. Wang Y et al (2011) Facile one-pot preparation of novel shell cross-linked nanocapsules: inverse miniemulsion RAFT polymerization as an alternative approach. Soft Matter 7(11):5348–5352
44. Wu D et al (2006) Aqueous-core capsules via interfacial free radical alternating copolymerization. Macromolecules 39(17):5848–5853
45. Charoenmark L et al (2012) Preparation of superparamagnetic polystyrene-based nanoparticles functionalized by acrylic acid. Macromol Res 20(6):590–596
46. Herrmann C et al (2012) Re-dispersible anisotropic and structured nanoparticles: formation and their subsequent shape change. Macromol Chem Phys 213(8):829–838
47. Ramos J, Forcada J (2011) Surfactant-free miniemulsion polymerization as a simple synthetic route to a successful encapsulation of magnetite nanoparticles. Langmuir 27(11):7222–7230
48. Manzke A et al (2007) Etching masks based on miniemulsions: a novel route towards ordered arrays of surface nanostructures. Adv Mater 19(10):1337–1341
49. Fickert J et al (2012) Efficient encapsulation of self-healing agents in polymer nanocontainers functionalized by orthogonal reactions. Macromolecules 45(16):6324–6332
50. Gilabert F et al (2017) Integral procedure to assess crack filling and mechanical contribution of polymer-based healing agent in encapsulation-based self-healing concrete. Cement Concr Compos 77:68–80
51. Alghamri R et al (2018) Preparation and polymeric encapsulation of powder mineral pellets for self-healing cement based materials. Constr Build Mater 186:247–262
52. Bleay S et al (2001) A smart repair system for polymer matrix composites. Compos A Appl Sci Manuf 32(12):1767–1776
53. van der Zwaag S (2007) Self-healing materials: an alternative approach to 20 centuries of material science. Springer, Delft
54. Van Tittelboom K et al (2011) Self-healing efficiency of cementitious materials containing tubular capsules filled with healing agent. Cement Concr Compos 33(4):497–505
55. Tsangouri E et al (2013) Detecting the activation of a self-healing mechanism in concrete by acoustic emission and digital image correlation. Sci World J 2013
56. An S et al (2015) Highly flexible transparent self-healing composite based on electrospun core–shell nanofibers produced by coaxial electrospinning for anti-corrosion and electrical insulation. Nanoscale 7(42):17778–17785
57. Kousourakis A, Mouritz A (2010) The effect of self-healing hollow fibres on the mechanical properties of polymer composites. Smart Mater Struct 19(8):085021
58. Park JH, Braun PV (2010) Coaxial electrospinning of self-healing coatings. Adv Mater 22(4):496–499
59. Roche ET et al (2014) Comparison of biomaterial delivery vehicles for improving acute retention of stem cells in the infarcted heart. Biomaterials 35(25):6850–6858
60. Hoemann C et al (2005) Tissue engineering of cartilage using an injectable and adhesive chitosan-based cell-delivery vehicle. Osteoarthr Cartil 13(4):318–329
61. Mitchell TJ, Keller MW (2013) Coaxial electrospun encapsulation of epoxy for use in self-healing materials. Polym Int 62(6):860–866
62. Neisiany RE et al (2016) Encapsulation of epoxy and amine curing agent in PAN nanofibers by coaxial electrospinning for self-healing purposes. RSC Adv 6(74):70056–70063
63. Dzenis Y, Reneker D (2001) United States Patent 6265333: delamination resistant composites prepared by small diameter fiber reinforcement at ply interfaces. Interfaces

64. Crespy D, Zhao Y (2013) Preparation of nanocapsules and core–shell nanofibers for extrinsic self-healing materials. Self-healing polymers: from principles to applications, pp 247-271
65. Sinha-Ray S et al (2012) Encapsulation of self-healing materials by coelectrospinning, emulsion electrospinning, solution blowing and intercalation. J Mater Chem 22(18):9138–9146
66. Lee MW et al (2015) Self-healing nanofiber-reinforced polymer composites. 1. Tensile testing and recovery of mechanical properties. ACS Appl Mater Interfaces 7(35):19546-19554
67. Lee MW et al (2015) Self-healing nanofiber-reinforced polymer composites. 2. Delamination/debonding and adhesive and cohesive properties. ACS Appl Mater Interfaces 7(35):19555-19561
68. Lee MW et al (2014) Hybrid self-healing matrix using core–shell nanofibers and capsuleless microdroplets. ACS Appl Mater Interfaces 6(13):10461–10468
69. Lee MW et al (2014) Self-healing transparent core–shell nanofiber coatings for anti-corrosive protection. J Mater Chem A 2(19):7045–7053
70. Neisiany RE et al (2017) Towards the development of self-healing carbon/epoxy composites with improved potential provided by efficient encapsulation of healing agents in core-shell nanofibers. Polym Testing 62:79–87
71. Zheludkevich M (2009) Self-healing anticorrosion coatings. Self Heal Mater
72. Melo JDD et al (2014) Encapsulation of solvent into halloysite nanotubes to promote self-healing ability in polymers. Adv Compos Mater 23(5–6):507–519
73. Frith JE et al (2013) An injectable hydrogel incorporating mesenchymal precursor cells and pentosan polysulphate for intervertebral disc regeneration. Biomaterials 34(37):9430–9440
74. Kim YM et al (2014) Adipose-derived stem cell-containing hyaluronic acid/alginate hydrogel improves vocal fold wound healing. Laryngoscope 124(3):E64–E72
75. Lü S et al (2011) Thermoresponsive injectable hydrogel for three-dimensional cell culture: chondroitin sulfate bioconjugated with poly (N-isopropylacrylamide) synthesized by RAFT polymerization. Soft Matter 7(22):10763–10772
76. Cao Y et al (2007) Poly (N-isopropylacrylamide)–chitosan as thermosensitive in situ gel-forming system for ocular drug delivery. J Control Release 120(3):186–194
77. Choi B et al (2014) Cartilaginous extracellular matrix-modified chitosan hydrogels for cartilage tissue engineering. ACS Appl Mater Interfaces 6(22):20110–20121
78. Li Q et al (2004) Photocrosslinkable polysaccharides based on chondroitin sulfate. J Biomed Mater Res Part A An Off J Soc Biomater Jpn Soc Biomater Aust Soc Biomater Korean Soc Biomater 68(1):28–33
79. Jin R et al (2009) Injectable chitosan-based hydrogels for cartilage tissue engineering. Biomaterials 30(13):2544–2551
80. Wiltsey C et al (2015) Thermogelling bioadhesive scaffolds for intervertebral disk tissue engineering: preliminary in vitro comparison of aldehyde-based versus alginate microparticle-mediated adhesion. Acta Biomater 16:71–80
81. Fan M et al (2015) Cytocompatible in situ forming chitosan/hyaluronan hydrogels via a metal-free click chemistry for soft tissue engineering. Acta Biomater 20:60–68
82. Lü S et al (2015) Injectable and self-healing carbohydrate-based hydrogel for cell encapsulation. ACS Appl Mater Interfaces 7(23):13029–13037
83. Chen Y et al (2018) Bioinspired self-healing hydrogel based on benzoxaborole-catechol dynamic covalent chemistry for 3D cell encapsulation. ACS Macro Lett 7(8):904–908

Containers with Anti-Corrosion Agents for Metal Protection Paints

Sahar Amiri

1 Introdution

Corrosion is a harmful and extremely costly metal corrosion problem which is caused by chemical reactions between it and the atmosphere around it. Type of metals and the environmental conditions, determine the form and rate of corrosion. Corrosion processes grow rapidly after the protective barrier has been broken and a reaction affects the structure of coated surface and the surrounding environment so that anti-corrosion activity is more realistic and feasible than total removal. Anti-corrosion and self-healing coatings is a promising route in protecting metals, but coating performance is dependent to various parameters. Despite of this, the use of corrosion inhibitors will protect the surface and prevent their assault on metallic materials. Several elements need to be recognized when choosing an inhibitor, such as cost and quantity, availability and safety for the ecosystem and its organisms. Direct mixing of coatings with inhibitors of corrosion can result in the aggregation of corrosion inhibitors and ineffective protection, so an appropriated carrier used to cover and protected the corrosion inhibitor for successful delivery of corrosion inhibitors where corrosion damage has occurred that develops a smart metal protective anti-corrosion coating. Various containers or reservoirs can be used for encapsulation of corrosion inhibitors as active corrosion protection systems such as halloysite nanotubes, titania containers, mesoporous silica, cyclodextrins, polyurethane and polyelectrolyte shell.

The self-healing capacity may be an especially crucial property for coatings planned to ensure the fabric of a development against erosion and to duplicate their unique abilities, and astuteness as a cleverly and low-cost defensive materials. The coating was stored on gentle steel substrate by sol–gel handle, in which thin layers

S. Amiri (✉)
Department of Polymer Engineering, Science and Research Branch, Islamic Azad University, Tehran, Iran
e-mail: S.amiri@srbiau.ac.ir

J. Parameswaranpillai et al. (eds.), *Micro- and Nano-containers for Smart Applications*, Composites Science and Technology,
https://doi.org/10.1007/978-981-16-8146-2_10

containing metal oxides which are coated the metal surface and utilizing a few sorts of oxides and the consolidation of a natural compound progress the execution of the movies. Applying sol–gel coatings permits the homogenous distribution of inhibitors and healing agents, making a curiously and flexible way to get a coating. In this chapter, defensive coating by means of sol–gel prepare and containing inhibitor erosion Nano- or miniaturized scale holders, natural or inorganic inhibitor erosion, recuperating nano- and microcapsules.

2 Basic of Corrosion

Erosion is caused by breaking down of basic properties in a fabric due to chemical responses with its environment which suggests a misfortune of electrons of metals responding with moisture and O_2 molecules. Erosion of coated substrate caused by jump to higher energy state or reaction with O_2 and water, and driven to erosion. Erosion causing side surrenders and deterioration of its mechanical properties within the exceptionally long term which may be a common prepare [1, 2].

In the presence of moisture, water and O_2, chemical and electrochemical reactions occurred which is a combination of cathodic and anodic reactions [3–5]. Protection of special metal or alloys substarte is a good candidate to improve life time of substrate and economical challenge to have a better resistance which led to saved 25% of product cost [5, 6].

One of the foremost pertinent strategies for assurance of metals is applying polymeric coating containing corrosion inhibitor or healing agents which protect the surface against erosion and act as auto-protected and self-healing substrate. Parameters such as chemical structure and composition, nuclear structure, electronic properties and environmental parameters are critical for protection of the surface from corrosion [6–8].

Corrosion is classified in eight main forms (Fig. 1):

(a) Uniform, or common assault may be a result of coordinate chemical assaults of metals surface and O_2 is the driving force for this kind of erosion of steels and other metals and combinations. Increment metal thickness, utilizing paints or coatings on the metal surface or utilizing erosion inhibitors can be utilized for dodge this handle.

(b) Galvanic erosion is another sort of erosion which take put between two different metals with coordinate contact since as a result of electrical interaction between a metal or nonmetallic conductor in a corrosive electrolyte. This kind of erosion may be a degree of resistance to erosion and oxidation in sodden discuss and choice of metals/alloys as near together as conceivable within the galvanic arrangement, protected different metals wherever down to earth and apply coatings with caution may anticipate this phenomenon.

(c) Pitting Erosion is one of the foremost harming forms of corrosion and may be a localized corrosion of a metal surface which creats of little gaps within the metal

Fig. 1 **a** Uniform corrosion, **b** Galvanic corrosion, **c** pitting corrosion, **d** Crevice corrosion, **e** Intergranular corrosion, **f** Graphitic corrosion, **g** Stress-corrosion and **h** Erosion corrosion

or cavities. Control of pH around 7, Cl concentration, halogen concentration, O_2 content and temperature can be done for anticipate event of this erosion [9–11].

(d) Crevice erosion may be a localized assault on a metal surface via formation of hole in the interface of two joining substrate where a stagnant arrangement is display. This sort of erosion occurred at interphase between two metals or a metal and non-metallic substrate.

(e) Intergranular erosion or selective leaching caused by ductile stretch and splitting may occure along grain boundaries area, while the bulk of metal or substrate isn't attacked.

(f) Graphitic erosion is the weakening of gray cast press (Fig. 1f) [12–14].

(g) Stress-corrosion splitting is related to hydrogen and comes about from the linkage activity of a touchy fabric, and didn't taking after particular component for stretch erosion breaking within the writing. The stretch level, working temperature and the electrochemical potential are basic parameters in this sort [12–14].

(h) Erosion is as a rule found at tall stream rates around tube and metal surround. Protection of metals and combinations may be a promising modern strategies for increment the life of metals. A later think about assessed the entire yearly fetched of erosion in China at over 310 billion US dollars, speaking to 3.34% of the country's GDP [15]. In case the same rate is connected to worldwide economy, the full fetched of erosion around the world sums to ~2.5 trillion US dollars per year. Natural defensive coatings are the foremost broadly utilized strategy to anticipate erosion, composed of polymeric coating which have special and unique properties due to their chemical compositions [16–18]

Polymeric coatings shaped chemical holding and mechanical interlocking with useful bunches of metal surface and act as an obstruction against the water and other natural absconds. Composition of the polymer, the ability of formation of covalent or non-covalent crosslinking straightforwardly influenced on physiochemical properties and execution of the binder [19–21].

Defensive natural coating containing inhibitor or other particles with defensive, functional groups, composition, measure, structure, amount and dissemination of nanoparticles play imperative parts within the coating execution. Polymer coatings appear multi-responsive properties such as UV-protective, water-repellent, anti-microbial coating, anti-corrosion and anti-scratch coatings [22–27].

The bond breakage from natural components counting UV, warm, oxygen, dampness and particles may too actuate more plainly visible corruption of coatings' common defensive. Without viable repair exhausted the coating structure, harms make channel for the quick admissions of destructive stimulus and inevitably caused coating failures [23–27]. Self-healing properties allude to heal the coating harms with negligible or no outside mediation and the recuperating impacts ordinarily act by means of reestablishing the physical coating boundaries by deformity fixing or closing, or by hindering the erosion responses at the coating absconds. Erosion inhibitors are good candidate to stacking within the structure of defensive coating to get self-healing coatings [26, 27].

2.1 Anti-Corrosion Methods

The foremost profitable and effective strategies for anticipates erosion prepare is applying a protective polymer nanocomposite coating onto a desired substrate which restricted damaging effects of surface with unique properties such as resistance to environmental conditions, chemical solutions, resistance to physical of dampness, good adhesion and cohesion [28, 29].

Coatings are separated into 3 fundamental bunches included natural, inorganic and half breed coatings. The foremost broadly utilized anti-corrosion and self-healing coatings is natural coatings which are tough defensive coating connected to a substrate for enriching or particular specialized properties. The most component component mindful for the creation of well-adhering following film (layer) are natural compounds such as polymers, oligomers, monomers, or blends therefor.

The sorts of natural coatings are groundworks, cement cements, topcoats with tall resistance to outside variables (varnish, finish, paints). Natural coatings act as a defensive boundary against erosion, scratch and oxidation. Natural coatings are strong coatings connected to a substrate for their enriching or particular specialized properties, depend basically on their chemical idleness and impermeability. Natural coatings are apply with showers, rollers, plunges, or by electrostatic implies and cures or dries by vanishing or misfortune of dissolvable, polymerization and oxidation [25–28].

Inorganic coating is containing metallic and ceramics particles and gives higher security erosion and are able of being made by a compound act, either with or without the help of power. Crossover coating is containing organic–inorganic gum that combined for common coating properties upgrade [19, 30]. Connected inorganic coating secure the coated surface by means of shaping a film of either metallic oxide or another compound. Arrangement of holding between an inorganic coating and metal surface, will secure the surface of your extend and offer assistance the application of any consequent layers. There are four primary sorts of inorganic coatings that are broadly utilized incorporate acrylic gum, ceramic, intumescent (a fire retardant coating) and zinc-rich metal powder is mixed with an inorganic silicate tint binder. Inorganic coatings have different benefits such as impenetrable to oxygen and water, tall electrical assurance, extraordinary temperature erosion continuance, protects well against coastal components and has solid bright radiation assurance. Half breed coating is containing organic–inorganic tar that combined for common coating properties improvement [19, 30] having a nanophasic morphology, where an natural stage is entirely interconnected with an inorganic one. The properties of these half breeds are entirety of the person commitments of both natural and inorganic stages, and related to morphology of the interfacing and nanophases. When natural and inorganic components are blended together, distinctive sorts of intelligent are shaped such as van der Waals, hydrogen, or ionic bonds, solid covalent and iono-covalent bonds.

3 Self-Healing Coatings

Metals and alloys are used widely in various engineering structures and applications which need to effectively protect against corrosion by coatings which is a good method to obtain a better life-time with anti-corrosion activity based on a dense barrier against corrosive species. Self-healing and anti-corrosion coatings are cost effective and are able to increase the corrosion resistance and durability of metallic structures due to real healing of the defect [31, 32].

Self-healing coatings can ensure metal surface as two techniques, to begin with step is reestablishing the mechanical or obstruction properties and moment step is assurance effectively against erosion of the fundamental substrate by diminishing the erosion action within the imperfection. Savvy coating with self-healing capacity offer a modern course to get longer-lasting items and components which are divided into three categories: capsule based, vascular, and intrinsic [24, 25]. These strategies contrast by the component which is utilized to confine the recuperating specialist until activated by harm and dependent to the harm volume that can be mended, the recoverability, and the recovery efficiency for each route is a vital parameter for self-healing property [26–28].

3.1 Self-Healing Based on Capsule

Encapsulation of active healing agents or corrosion inhibitors in good core materials is an effective method to improve long term protection of metals and alloys. By start micro-crack propagation, physical damaging of surface starts and although release of healing agent or inhibitor corrosion is critical to protection of surface. Capsulated agents or inhibitor corrosion (core) are incorporated into the appropriated coating matrix (shell) and when the surface expose to physical damage or moisture, micro-crack production starts, capsules rupture and healing agent or corrosion inhibitors release, propagates through the coating and reaction initiated in the fracture plane and involved into damage sites in the coating and healing the crack or damage (Fig. 2) [33, 34].

Encapsulation of reactive materials (healing agents or inhibitor corrosions) are obtained via various methods which are based on controlled release of healing agent from capsule until healing damage sites [33]. For perfect response of self-healing coating to corrosion and physical damaged, encapsulation should be done via strong bond with the healing agent to remain during usage and rupture when the polymer is damaged for considerable time to provide long shelf-life the capsules.

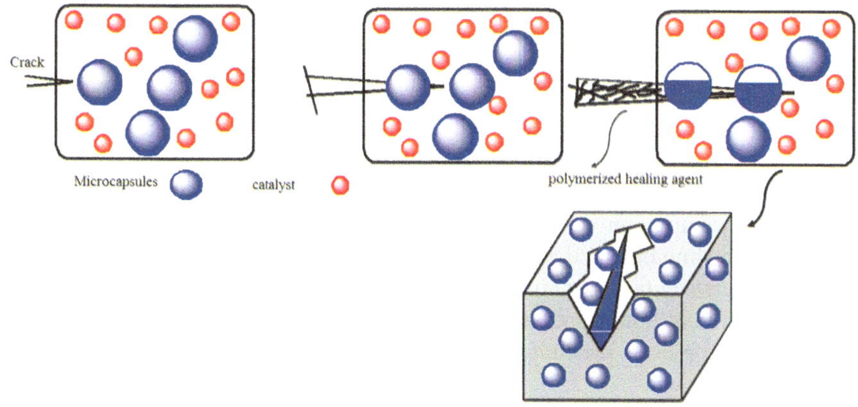

Fig. 2 Capsule based healing process

3.2 *Vascular Self-Healing Materials*

Vascular self-healing coating is based on a network of hollow tubes or fibers, containing a pressurized healing agent, is located in the concrete matrix and break down with damage or stress to release healing agent during damage and produced crack, flow into the damage, and heal it [35]. Vascular self-healing is based on releasing healing agent in a series of parallel long tubes until damage occurred which the network may be recovered via formation of new interaction between healing agent and polymer chain so local healing occurred (Fig. 3) [33, 36].

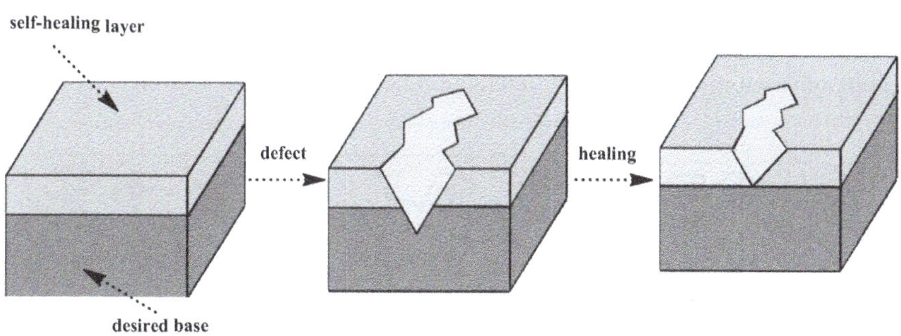

Fig. 3 Schematic mechanism of self-healing coating

3.3 Intrinsic Self-Healing Systems

Intrinsic self-healing strategies are based on reversible bonding's such as non-covalent bonding and hydrogen bonding. Driving force for this process is external forces like as pressure, light or heat that enables crack healing under certain stimulation (Fig. 4) which led to release of healing agent [37–39]. For fully or partially healing manner and reproduce ability of coating, intrinsic healing mechanism are used which healing obtained via formation of new and strong chemical or physical interaction between polymer chain and healing agent [40–42]. Due to formation of reversible covalent bonds and supramolecular interactions, the intrinsic healing is useful in thermoplastic, thermoset polymers and elastomers [43, 44].

4 Self-Healing Coatings on Metal Surface

The protective coatings are promising route for construction anti-corrosion surface which involved metal oxides, organic or inorganic or hybrid polymer layers applied via sol–gel method. Sol–gel method is a safe and easy method for synthesis and applying smart coating applications. Obtained coatings from sol–gel are flexible and formed strong chemical bonds with good adhesion to the metallic surfaces. Presence of organo-functional groups in the structure of hybrid nanocomposite led to formation of strong covalent bonding between obtained coating and interface and by increasing protective coatings thickness, barrier properties increased. External mechanical impacts, formation of pinholes and cracks may rapture of barrier and corrosion starts, so for obtaining an effective long-term protection, corrosion inhibitors or healing agents employing in coating formulation via a formation of reversible bonds. By physical damage or a crack formation, encapsulated active agents released and formed new interaction between polymer coating matrix and

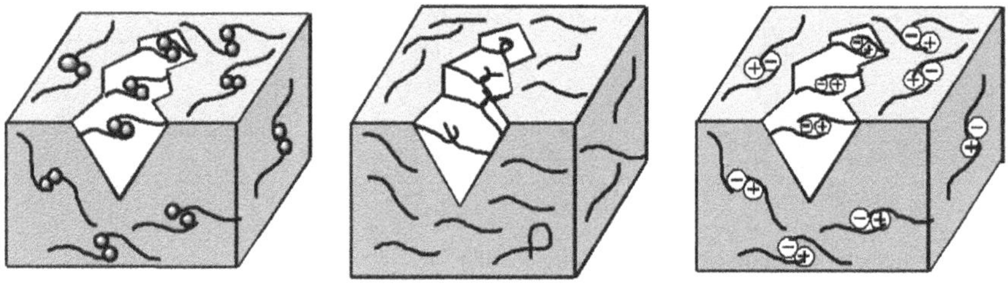

(a) (b) (c)

Fig. 4 Intrinsic self-healing materials (**a**) Reversible bonding schemes make use of the reversible nature of certain chemical reactions (**b**) Chain entanglement approaches utilize the mobility at crack faces to entangle chains that span the crack surfaces (**c**) Noncovalent self-healing systems rely on reversible hydrogen bonding or ionic clustering which manifests as reversible cross-links in polymers substrate and led to long-term effective protection of coated metallic structures [20–26].

4.1 An Introduction of Sol–Gel

Sol–gel method is a chemical technique based on reaction between inorganic and organic transition metal oxides. The obtained hybrid nanocomposite is pure and homogeneous with tailored properties because of formation of covalent bond between organic polymers and inorganic components. This method is based on hydrolysis and condensation reactions of metal alkoxides in relative low temperature, so the control on molecular weight and purity is considerable. Due to reaction condition and low reaction temperature and pressure, organic functional groups incorporated in the inorganic materials, led to good flexibility and adhesion [20, 33].

For synthesis of anti-corrosion coating via sol–gel, organic inhibitor introduced into inorganic sol–gel networks, caused crack-free hybrid coating which slow down corrosion. Sol–gel derived coating can be hydrophobic or hydrophilic, which hydrophobic coating prevents diffusion of polar solvents towards the metal/coating interface, act as anti-corrosion coating and improvement of general passive protective properties. Structure of organic inhibitor and organic/inorganic components ratio affect coating performance [20–26].

4.2 Addition of Corrosion Inhibitor or Healing Agents to Sol–Gel Coating

By direct addition of healing agent and corrosion inhibitor into hybrid nanocomposite coating, control release of active agent within the coating matrix is so difficult and also corrosion inhibitor may form interactions with active sites from coating which may change barrier activity of obtained coating. By encapsulation of healing agent, possible interactions of healing agent and the matrix decreased and release of the inhibitor is controlled for obtain a long-term protection of metal surface from environmental parameters such as moisture and oxygen [22–26]. Desired functional groups, corrosion inhibitor or healing agents can be incorporated in the organic–inorganic hybrid coating to achieve smart coating with specific properties. These compounds are encapsulated for long-term effectiveness which dependent to low solubility and ratio of active agents at damaged site [24–27].

Most of the inhibitors have functional groups of nitrogen such as amines, amides, quaternary, ammonium salt and specially imidazole's and other derivatives of these compounds. Anti-corrosion efficiency of inhibitors directly affected by chemical and molecular structure, interaction formation between metal surface to formation a protective and anti-corrosion layer. Cr(VI) based corrosion inhibitors are the best inhibitors with good efficiency but their usage is restricted because of toxic effects and environmental hazards. The green corrosion inhibitors with excellent corrosion protection in the applications are developed without danger degradation products with minimum carbon dioxide production [45].

Quaternary compounds showed good inhibitor activity which is due to formation of electrostatic interaction between positive ions and metals as negative ions. Inhibitor corrosions released at the appropriate time and strongly decreases the corrosion rate, in other hand when absorbing rate of inhibitor is slow and low, efficiency of inhibitor decreased [46].

In the presence of corrosion inhibitors, polarization (aniodic and catiodic) increased so corrosion process decreased. In anodic inhibition mechanism (such as nitrite, chromate and nitrate, phosphate and molybdate), adsorption of the inhibitor on the anodic areas led to lower current density need for achieve to critical passive potential. Cathodic inhibitor by decreasing H_2 or O_2 gas, led to alkaline environment for the cathodes and corrosion potential is shifted to more negative values [33]. The most common route to create long-term protective anti-corrosion layer is entrapment of corrosion inhibitors, healing agent or other active agent into hybrid coatings which is based on formation of strong cross-linking between the corrosion inhibitor within the coating material and metal surface [46]. Controlled release of corrosion inhibitor from capsules obtained via formation of reversible interaction between the inhibitor with the coating applied on substrate. Several organic compounds, such as mercapto benzothiazole, mercapto benzimidazole mercapto benzimidazole sulfonate, and thiosalicylic acid, have been selected to evaluate the effectiveness of these two approaches [19, 29, 30, 47, 48].

5 Anti-Corrosion Coatings Containing Organic Micro-/Nanocontainers

Addition of active agents (healing agent or corrosion inhibitors) into coating matrix, led to uniform active protection of the substrate with active self-healing aspects. Micro/nanocontainers have hollow, porous, or core/shell shape in different size from nanometer to micrometer.

Encapsulation is a chemical process which active ingredients covered by a polymeric shell and core content protected from surrounding environment and other corrosive parameters and allows controlled release at specific times or a triggering event. Encapsulation of corrosion inhibitors or healing agent act as nano-reservoir and improved their applicability on self-healing anticorrosion coatings caused protection of metal/coating interface with controlled released of active substances of inhibitor for long term protection [30, 36]. Crack formation or mechanical impact, damaged shell layer and active inhibitor corrosion released with a controlled rate in a target time using custom stimuli, viz., pH, temperature, pressure, and chemical, physical or mechanical changes.

5.1 *Self-Healing Coatings Containing Inorganic Micro-/Nanocontainers of Inhibitors*

Containers including inorganic inhibitors have been used as additive to hybrid nanocomposites due to increase mechanical properties of the coatings. Inorganic corrosion inhibitors such as oxide particles or ion-exchange containers are incorporated in coatings as corrosion resistance layer.

5.1.1 Oxide Nano-/Microparticles

Nanostructured materials and nanomaterials are good corrosion inhibitors with high protective efficiency with low cost and easy production so can act as excellent carriers for corrosion inhibitors. Release of corrosion inhibitors in nanomaterials is controlled with a specific rate for having long-term protection. Metal oxide nanoparticles increase the hindrance efficiency of the coating and provide self-healing coating with resistance to photo degradation. The most frequently used nanoparticles are Al_2O_3 [49], ZrO_2 [50], TiO_2 [51], Y_2O_3 [52], NiO [53], ZnO [54], Zn [55], and SnO_2 [56] which could be incorporated into various coating systems and improve barrier properties of coating against corrosion.

5.1.2 Self-Healing Systems Containing CeO2 Nanocontainers

CeO_2 nanocontainers loaded with dodecylamine (DDA) and n-methyl thiourea (NMTU) and used as effective capsules for different corrosion inhibitors. Bode and phase angle plots of the EIS spectra for the blank sample and coatings containing epoxy/CeO_2 are shown in Fig. 5 which anti-corrosion activity of CeO_2 indicated by the high impedance modulus at the low frequency |Z0.1 Hz| value [57, 58]. CeO_2 is a good corrosion inhibitor for mild steel in different acid mediums (HCl & H_2SO_4) and improved barrier and corrosion protection properties of the Si-based hybrid coatings [59]. Results for protective hybrid coating containing CeO_2 showed that Ceric oxide nanoparticles act as good candidate for synthesis of anti-corrosion coatings with improved physical properties in the silane coatings.

Research's in the case of cerium based coatings indicated that by modifying the silane solution with cerium nitrate drastically improved barrier properties and anti-corrosion activity of obtained hybrid nanocomposites. Cerium nitrate provided active sites for new interactions with coating and the metal surface so coating act as self-healing coating and the corroded areas leech (Fig. 6). Salt spray testing after 144 h exposure indicated that at the initial stages, scratched area led to devotional dissolution of zinc, but after specific time zinc coatings degraded completely and formed white rust [58, 60].

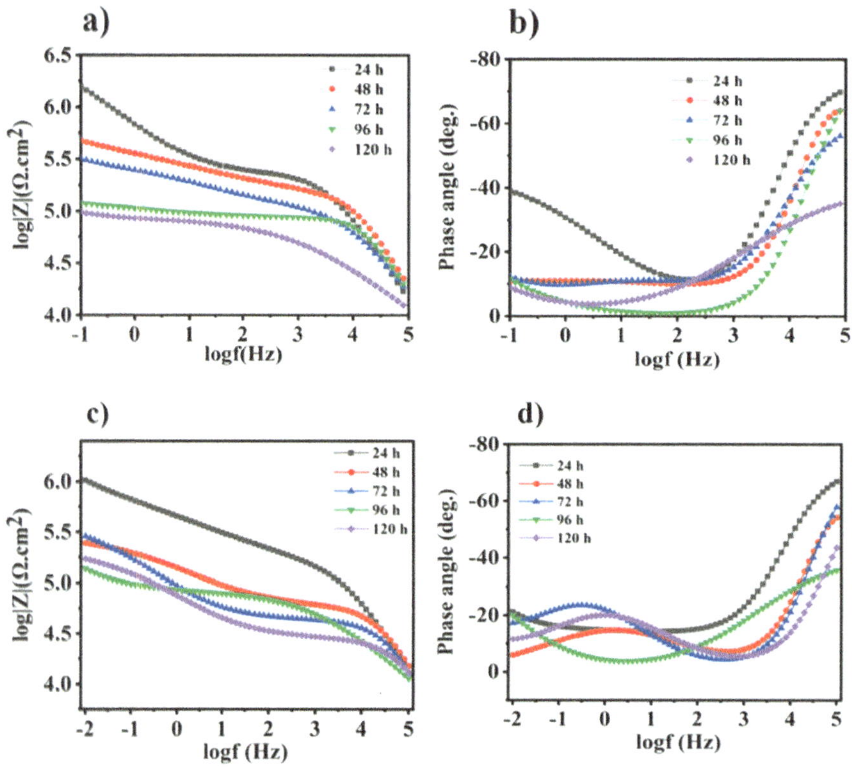

Fig. 5 **a**, **c** Bode and **b**, **d** the corresponding phase angle plots for the scratched blank and epoxy/CeO_2 coatings, respectively, after different immersion times in 3.5 wt.% NaCl at room temperature

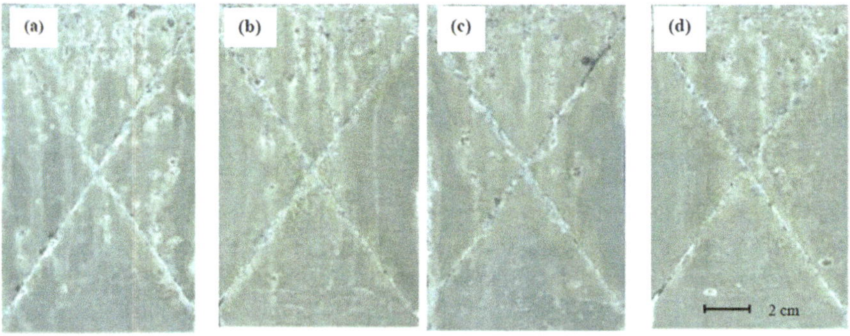

Fig. 6 Photographs of the HDG steel samples coated with blank silane (**a**), $Ce(NO_3)3.6H_2O$ (**b**), CeO_2 nanoparticles (**c**), and CeO_2 + $Ce(NO_3)3.6H_2O$ (**d**), after 144 h of salt spray exposure

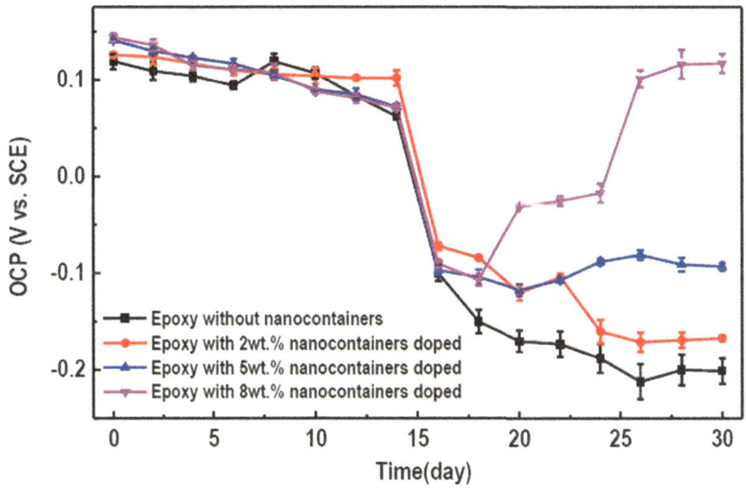

Fig. 7 Time dependence of OCP of the steel electrode coated with epoxy coatings without and with various contents of BTA-encapsulated nanocontainers in 0.1 M NaCl solution

5.1.3 Coatings Containing Mesoporous Silica Nanocontainers

Smart self-healing and anti-corrosion coatings containing corrosion inhibitors with controlled release rate showed long term-protection coating. Benzotriazole (BTA) is a corrosion inhibitor which formed an adsorptive protective film on the steel surface and can be encapsulated and used as self-healing smart coatings [61]. Encapsulation of BTA within low mesoporous silica spheres was reported by Chen and Fu [62] which controlled release of BTA from silica spheres, loaded in epoxy coating and release based on the change of local pH due to corrosion of metals. Figure 7 showed the open-circuit potential (OCP) of the steel coated with epoxy coatings without and with various contents of BTA-encapsulated. OCP for epoxy coating showed a decrease until reach a relatively steady value of -0.2 V (SCE) after 26 days of immersion. In the case of epoxy coating involved 2 wt.% BTA nanocontainers, Self-released of BTA from the nanocontainers with a controlled manner showed more positive steady-state OCP which indicated protective efficiency of obtained coating [61, 62].

5.1.4 Halloysite Nanocontainers

Halloysite is an environmentally friendly, naturally occurring clay mineral available in hydrated form which is similar to kaolinite except that halloysite has an additional water monolayer present between the interlayered spaces. Halloysite nanotubes are good candidate for encapsulation of active agents such as corrosion inhibitors. However, these active agents can get adsorbed on the external and internal walls of the tubes, may be intercalation, and the substances loaded into the halloysite lumen, which get crystallized into the pores [63, 64]. HNTs showed specific applications due to their structure and specific shape, so can be used as nanoreactors and entrapped various active agent to deliver at controlled rate and manner. For design self-healing coating with appropriated physical and mechanical properties, HNTs were used for encapsulation of corrosion inhibitor such as benzotriazole and release this agent with a controlled manner to achieve anti-corrosion coating in a controlled way.

Self-healing ability of coating containing HNTs containers is due to smart and controlled release of inhibitor from container. HNTs due to their structure and having an internal lumen and presence of alumina oxygen octahedrons entrapped inhibitor corrosion in their empty spaces present in the multilayered walls of the tube. Encapsulation of 2-mercaptobenzothiazole as corrosion inhibitorin HNTs and incorporation in into ZrO_2-SiO_2 hybrid coating indicated controlled release of inhibitor by pH changes when a crack starts [64]. Figure 8 showed EIS spectrum of ZrO_2-SiO_2 and ZrO_2-SiO_2 containing HNTs nano-containers after 2 weeks of accelerated immersion test in 3% NaCl solution. In the presence of HNTs nano-containers, higher barrier properties and higher low-frequency impedance indicated better corrosion protection. This manner is due to slow and controlled release of 2-mercaptobenzothiazole from HNTs nano-containers and achieving self-healing properties which block the initial corrosion processes and curing microdefects in the coating [64].

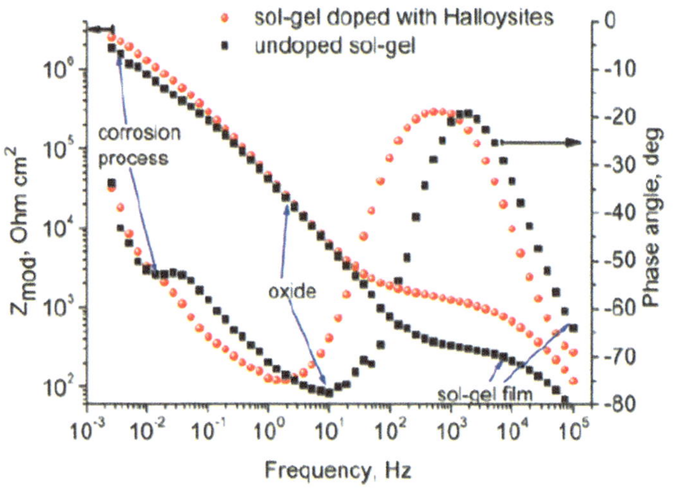

Fig. 8 Impedance spectra of undoped and halloysite doped solgel coatings after a 2-week immersion test in 3% NaCl

5.1.5 Sol–Gel Coating with LbL-Assembled Nanocontainers

Layer-by-layer (LbL) deposition is a versatile route to achieve intrinsic self-healing anticorrosion coatings via pairs of polyelectrolytes. The shell of polyelectrolyte containers is sensitive to environmental parameters such pH, temperature, physical and chemical conditions which controlled release inhibitor corrosion. LbL-coated nanocontainers can drastically improve the release rate of corrosion inhibitor under a certain "trigger" action [65]. LbL-sol–gel anticorrosion coatings are challenge for an effective self-healing coating to prohibit matrix deformation with a uniform distribution of inhibitor capsules in the coating surface. LbL-assembled benzotriazole-containers loaded in epoxy-functionalized ZrO_2/SiO_2 coating and applied with poly(ethylene imine)/poly(styrene sulfonate) (PEI/PSS) polyelectrolyte layers and EIS spectra after 48 h of immersion in NaCl solution (Fig. 9). Results showed effective barrier against corrosive species for undoped coating and film loaded with nano-reservoirs [65].

In another work, self-healing anticorrosion coatings with rapid release of inhibitor from multilayer structure containing the cerium layer, graphene oxide, and branched poly(ethylene mine) (PEI)/ poly(acrylic acid) (PAA) multilayers on the magnesium alloy substrate which incorporation of graphene oxide into PEI/PAA coating led to good self-healing activity of coating (Fig. 10) which showed with the electrochemical impedance spectroscopy (EIS) [66].

PEI/PAA based coating incorporated with Ce(IV) and Ce(IV)/PEI/GO showed healing ability and anti-corrosion activity and healing time is 48 h and 10 min, respectively. This healing rate is 12 h for a scratched PEI/PAA system prepared on aluminum. Decreasing system pH from 9.5 to 4.5, increasing self-healing ability of coating due to decreasing electrostatic interactions and crosslinks [66].

Halloysite nanocontainers are good candidate for loading different types of inhibitors, for example self-healing ability of benzotriazole-halloysite nanocontainers coatings were investigated by SVET [66]. In another study, halloysite nanocontainers loaded with 2-mercaptobenzothiazole inhibitors improved anticorrosion activity of HNTs based hybrid coatings [67]. Encapsulation of dodecylamine in halloysite as smart self-healing coatings responsive to pH showed controlled rate of inhibitor release and healing the damaged sites [68]. To investigate self-healing and anti-corrosion ability of coatings, salt spray test was done which results showed that by applying 10 wt.% of dodecylamine-loaded HNTs, protection against corrosion improved and controlled release of dodecylamine from containers, led to long-term life and prevents crack propagation (Fig. 11) [66].

5.1.6 Sol–Gel Coating with Organic Nanocontainers

By encapsulation of organic inhibitor corrosion in special molecules as containers of capsules, anti-corrosion ability of polymeric coating may be changed and bulk material which coated with anticorrosion polymeric coatings showed intrinsic self-healing ability. Encapsulation of inhibitors protect them from environmental and

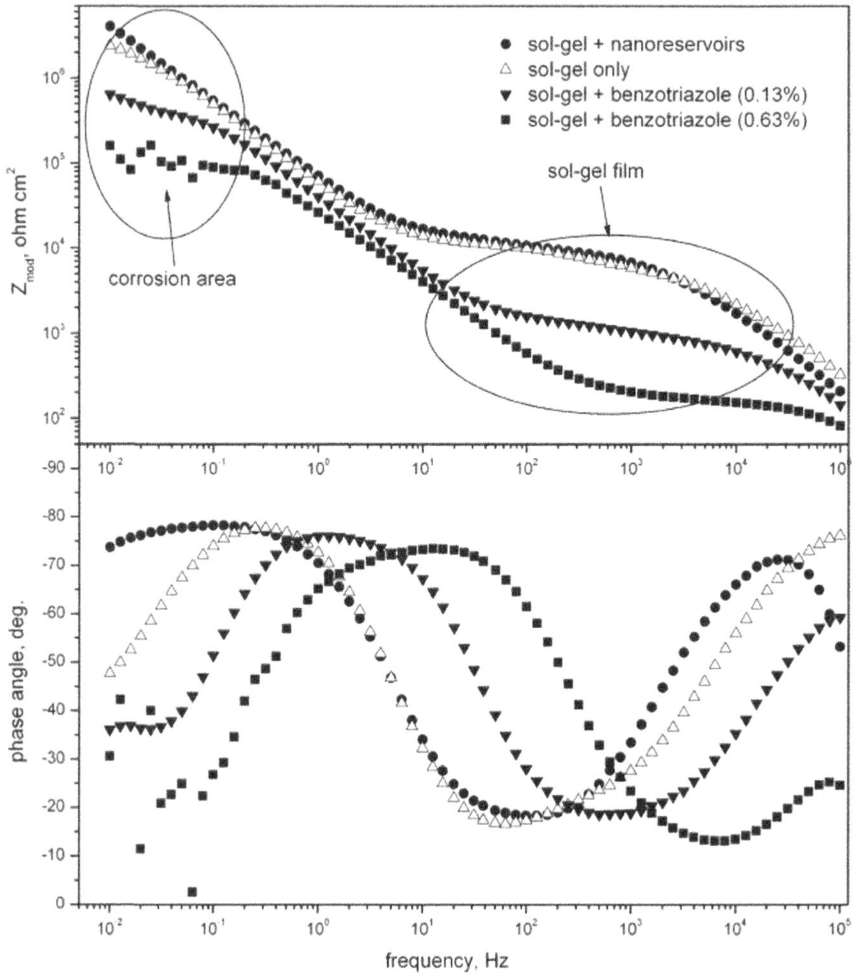

Fig. 9 Bode plots of the AA2024 aluminum alloy coated with different sol–gel films after 48 h of immersion in 0.005 M NaCl solution. Zmod is an impedance module

chemical corrosive which led to high quality self-healing ability at damaged sites with slow and controlled release of active agent, so toughness of material recovered. By increasing containers size, higher amount of active agent may be entrapped in the capsule, so higher anti-corrosion ability may be achieved. The reasonable parameter in encapsulation is proportional fit between capsule cavity size and inhibitor size [68, 69].

Corrosion inhibitors based on thiazole, imidazole, thiazole, imidazolesulfonate and their derivatives, thiosalicylic were encapsulated in Cyclodextrins(CDs) as smart capsules and incorporated into hybrid coatings. Results showed that CDs are good candidate for formation of self-healing coating with controlled release rate and

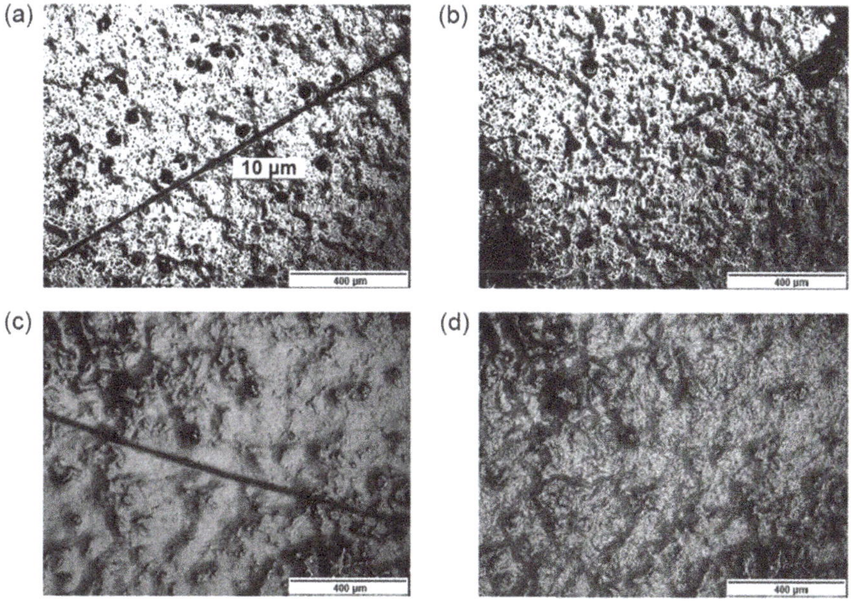

Fig. 10 Self-healing tests of the polymer coated samples with and without graphene oxide in the presence of water: **a** the cut on the surface of Ce(IV) /(PEI/PAA)10, **b** the healed surface of Ce(IV)/(PEI/PAA)10 in 10 min in the presence of water, **c** the cut on the surface of Ce(IV)/PEI/GO/(PEI/PAA)10, and **d** the healed surface of Ce(IV)/PEI/GO/(PEI/PAA)10 in 10 min in the presence of water

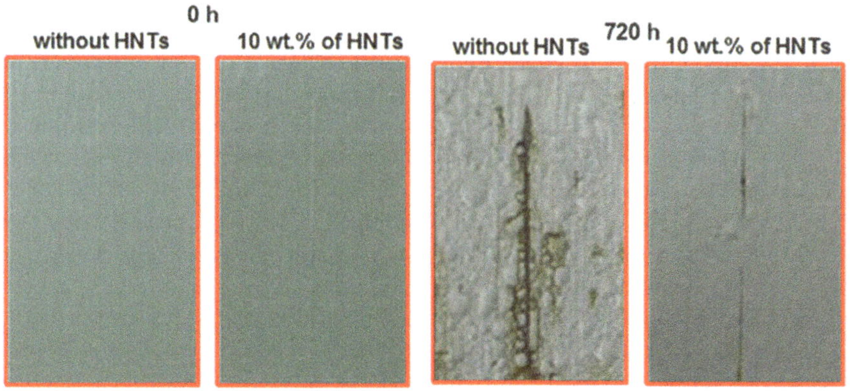

Fig. 11 Aspect of coated samples with two layers of alkyd primer containing 0 and 10 wt.% of halloysite nanotubes before and after 720 h of exposure in the salt spray chamber

controlled corrosion rate and act as smart capsules to deliver active agent to target sites and stopped corrosion activity [46].

The other containers with good adhesion and incorporation in coating structure, is polyurethane microcapsules (PU-MC) which are loaded with organic or inorganic corrosion inhibitor. PU-MC are synthesized via reaction between diol and cyanate functional groups. Maia and coworkers reported PU-MC via oil-in-water microemulsion polymerization using diethylenetriamine (DETA) and 2,4-toluene diisocyanate (TDI) and encapsulated 2-mercaptobenzothiazole as inhibitor corrosion and incorporated into SiOx/ZrOx-epoxy hybrid coating and applied onto AA2024 alloy substrates [70].

When inhibitor directly add to hybrid matrix, decrease barrier properties of the coatings which is related to un-wanted reactions between inhibitor and surrounded medium. When inhibitor encapsulated and add to obtain coating, unwanted reactions limited and inhibitor delivered to target sites and acct as healing agent. The most important parameter is homogenous distribution of capsules in the hybrid coating to provide a good compatibility [70].

5.1.7 Benzotriazole Loaded Nanocontainers

Benzotriazole (BTA) is colorless and polar act as a heterocyclic inhibitor which is due to presence of three nitrogen functional groups. BTA restricts undesirable reactions of substrate surfaces such as transition metals, copper and its alloys against corrosion in the presence of neutral and alkaline solutions for a longer period which is due to formation a protective barrier layer in aqueous solutions [71].

Novel smart coatings containing TiO_2 nanotubes encapsulated with dodecylamine (DOC) was reported by Shakoora. Encapsulated DOC in TiO_2 containers decrease corrosion rate and also improved mechanical properties of metal surface which is related to formation of strong interactions. Figure 12 showed EIS spectrum of TiO_2/DOC containers self-healing ability indicated that by increasing TiO_2 content, anticorrosive performance increased due to higher amount of DOC at target sites. Release rate and DOC amount could be determined with UV–vis spectroscopy with change in pH as trigging parameter and showed critical rule in self-healing and anti-corrosion ability of smart coatings [71].

Also results showed that in acidic and alkaline pH may changes release rate of DOC from TiO_2 nanotubes and the amount of DOC released in the media directly effect on the self-healing ability of coatings [71].

5.1.8 2-Mercaptobenzothiazole(2-MBT) Loaded Nanocontainers

Derivatives of thiazole and triazole groups can be used as active corrosion inhibitors for various metals and reduced anodic reaction of metals surface so caused anti-corrosion activity and improved life time of substrate which is commercially important [72]. Novel active protective coatings based on silica nanocapsules (SiNC)

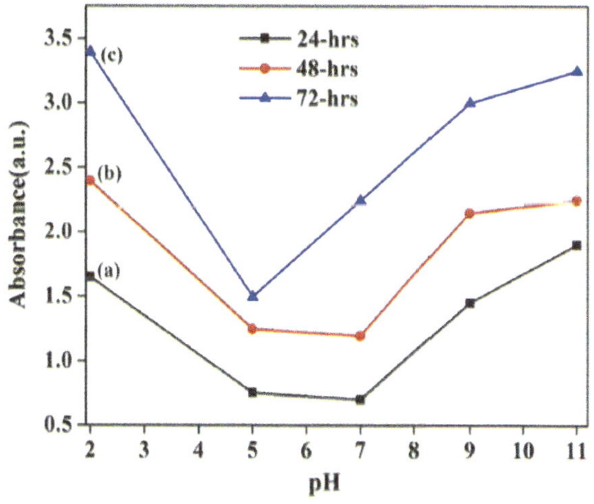

Fig. 12 UV–vis spectra of the dodecylamine encapsulated nanocontainers after (**a**) 24 h (**b**) 48 h and (**c**) 72 h of immersion in 0.1 M NaCl solutions with various pH values

containing 2-mercaptobenzothiazole (MBT) was used as a long-term anticorrosion coating via encapsulation of MBT via continuous oil-in-water emulsion polymerization. Release of MBT will be controlled with pH and corrosive Cl concentration. At acidic pH or high concentrations of NaCl, higher number of capsues may be damaged and so higher amount of MBT may be released where a prolonged release is required so higher anti-corrosion activity obtained [72].

Reversible self-healing smart coating are good candidates for encapsulation of inorganic and organic inhibitor corrosion. This property led to healing cracks via inverse reaction such as recombination of the broken molecules or reversible bonds. Cyclodextrins (CDs) are special copounds with specific structure which may formed reversible interactions with various compounds such as hydrophobic and hydrophilic compounds. Driving force in formation of inclusion complexes between CDs and guest molecules is reversible non-covalant bonding, such as formation of hydrogen bonding, Vander Waals interaction or charge transfer interaction [33].

To achieve sufficient self-healing via reversible hydrogen bonds which formed new interactions between surface and polymeric coating chains and increase active sites for reaction between chains and heal the damaged sites. MBT was encapsulated within CDs by Amiri et al [33].

The supramolecular nanocontainers are useful in formation of self-healing coating which prepared by homogenous distribution of the active capsules into the coating. Amiri et al reported encapsulation of MBT in CDs cavity via formation of reversible interaction between MBT and CDs which led to a hybrid nanocomposite coating with high self-healing ability and long-term protection of Al based substrate [33]. Formation of hydrogen bonding help to healing process and inhibitor entrapped in CDs cavity release with a controlled rate to obtain a slow and continuous inhibitor release, which led to long time protection against corrosion. Reversible hydrogen bonding improved interaction between damaged chains on the metal surfaces and led to self-healing property. Encapsulation of MBT corrosion inhibitor in β-CD cavity

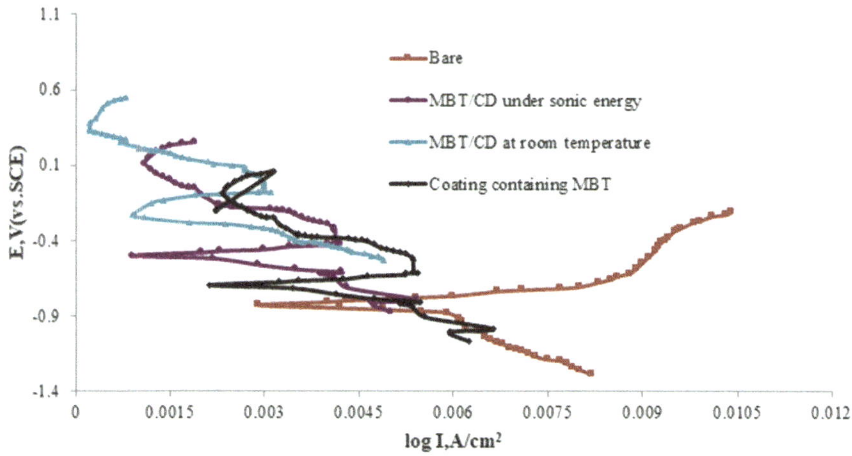

Fig. 13 Potentiodynamic scans for coating containing encapsulated and free inhibitor at different conditions in dilute 5% NaCl solution

in different conditions were characterized which may act as smart anti-corrosion reservoirs for release repairing agent and chemical initiator, to obtain smart protective system [22–28].

EIS results of MBT nano-containers loaded in Silane based coating were showed in Fig. 13. which coatings were located under extreme environmental conditions (exposed 5% NaCl solution) and results indicated that entrapped MBT from CDs containers led to high protection of Al against corrosive conditions and improved resistance of metals compare to bare sample without inhibitor containers [73, 74, 75, 76]. At room temperature, capsules have larger size so have entrapped higher amount of inhibitor in the cavity, so higher self-healing ability is seen. In the other hand, higher amount of MBT led to formation higher number of reversible bonding which controlled release of MBT in an appropriate rate which led to good anti-corrosion activity. In contrast, when capsules are prepared under sonic energy, small capsuled obtained and containers involved lower content of inhibitor and presents a lower protection [76].

5.1.9 Mercaptobenzimidazole Loaded Nanocontainers

2-Mercaptobenzimidazole(2-MBI) consists of thiol functional groups which is good candidates for anticorrosion properties and if the release rate controlled, self-healing ability may be obtained.

2-MBI are containing of thioamide group (-N–C = S), with nitrogen or sulfur atoms which may undergo reactions with other compounds with these substituent and led to higher protection of stainless steel in aqueous solutions of NaCl, mild steel and zinc in phosphoric acid [77–80].

2-MBI has a wide range applications such as medicinal chemistry, non-biological application such as plant growth regulators [77] and as corrosion inhibitor for mild steel in sulfuric acid solution, stainless steel in aqueous solutions of NaCl, mild steel and zinc in phosphoric acid [79, 80].

Encapsulated 2-MBI inhibitor within graphene-structured was synthesized and corrosion protection of coating containing these capsules coated on mild steel was studied with EIS techniques. For this reason, samples were immersed in the NaCl solution and results indicated an anodic protection of the MBI/CHS extract (Fig. 14) and shift OCP to more positive values. OCP values for coated samples reach to -630 mVSCE after 250 min immersion, while for uncoated samples after 750 min OCP value reached to -710 mVSCE which indicated self-healing ability of obtained coatind containg 2-MBI containers [81].

Amiri et al reported encapsulation of MBI in α-CD cavity based on hydrogen bonding which are reversible bonding sensitive to pH of medium and become active in corrosive electrolytes. Due to higher amount of hydrogen bonding between CDs cavity and inhibitors, active sites for formation of new interaction increased and inhibitors may react with damage sites after controlled release and prolong inhibiting effect of corrosion. Higher reversible interaction between inhibitor and coating controlled release rate and restricted agglomeration of inhibitor release, so inhabitation is takes longer time. Entrapment of inhibitors and active agents in good core material led to smart applications of polymeric coatings ensures the long-term delivery of corrosion inhibitor and thus the healing of a scratched coating on materials. Trigging agents such as pH, UV-light, curing, temperature and redox reaction are critical for release of active agent to healing the surface [21–25]. Reversible interaction formation and breakage is a driving force for creating active sites for formation new bonding between damaged end chains and coating containing inhibitor or healing agent, formed new bridge and recovered as few minutes at ambient temperature. Figures 15 and 16 showed anti-corrosive behavior of Al coated with coating containing inhibitor corrosion (MBT and MBI) encapsulated in CDs cavity and presence of containers

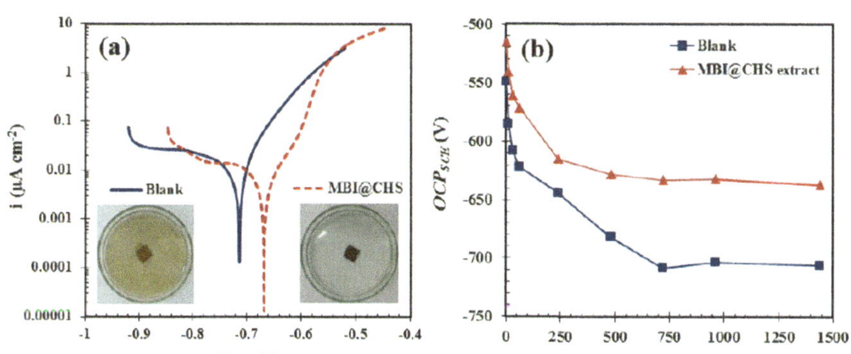

Fig. 14 Polarization curves of the mild steel samples after 24 h immersion (**a**) and OCP values (**b**) in the blank and MBI/CHS extract solutions

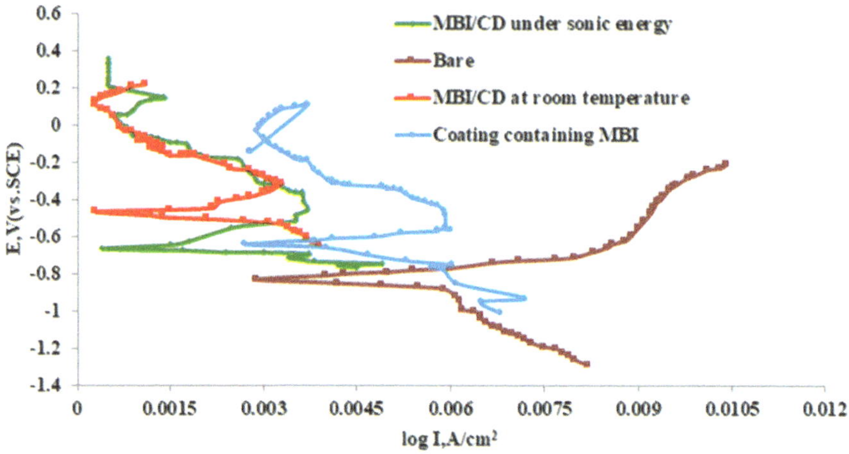

Fig. 15 Potentiodynamic scans for coating containing encapsulated and free inhibitor at different conditions in dilute 5% NaCl solution

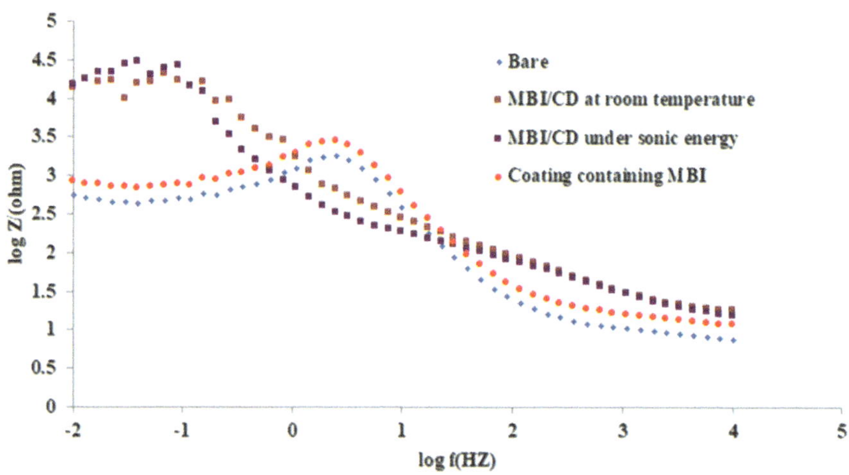

Fig. 16 The electrochemical impedance spectra for scribed coatings with different inhibitor nanocontainers in 5% NaCl solution

decrease the corrosion current density for coating containing inhibitor containers compare to bare sample without inhibitor containers. Results indicated that by encapsulation of inhibitor corrosion, efficiency of inhibitor improved and led to efficient protection of the alloy surfaces and shift the current density values to small values in the polarization curves. Containers which prepared at room temperature are more effective than other which prepared under sonic energy which is related to containers size and entrapped inhibitor [22, 24]. Tafel polarization showed that coatings containing corrosion inhibitors containers showed passivation regions and controlled release of inhibitor from containers retained inhibition activities [22, 24].

5.1.10 Hydroxyquinoline Loaded Nanocontainers

One of the other organic compounds which can be used as corrosion inhibitor is 8-Hydroxquinoline(8-HQ) which is composed of a phenol ring fused with pyridine ring. Hydroxyquinoline derivatives are different is in the placement of the hydroxy group on various carbons throughout the ring system. 2-hydroxyquinoline, 4-hydroxyquinoline, and 8-hydroxyquinoline are common isomers that simply have the hydroxy group connected at different points on the ring (Fig. 17) [82–85].

Inhibition effect of 8-Hydroxyquinoline on the corrosion of aluminium alloy in 3.5% NaCl solution was studied by potentiodynamic polarization and electrochemical impedance spectroscopy (EIS) methods [82–85]. EIS results of epoxy based coatings containing TiO_2/8-HQ nanocontainers which applied on aluminium alloy (AA) 2024-T3 indicated a good protection of coating (Fig. 18). TiO_2 improved physical properties of coating and when loaded with 8-HQ as inhibitor, act as a smart

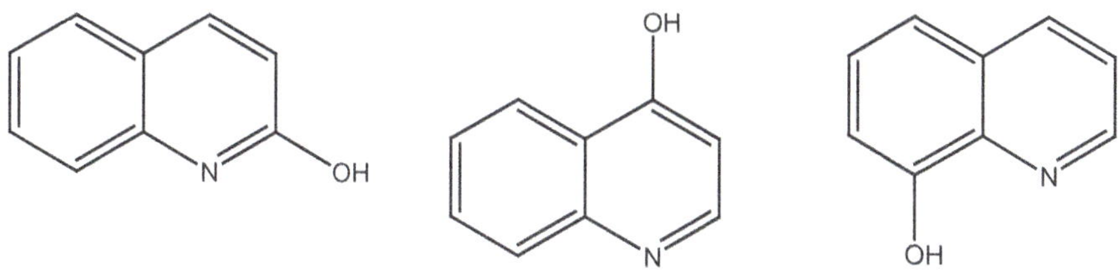

2-Hydroxyquinoline 4-Hydroxyquinoline 8-Hydroxyquinoline

Fig. 17 Structures of various derivative of hydroxyquinoline

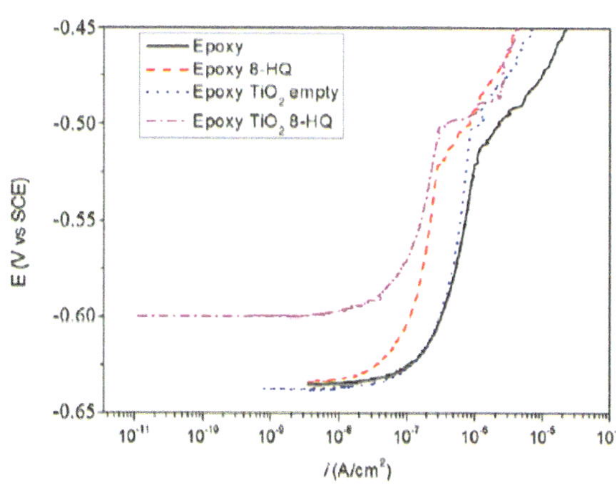

Fig. 18 Anodic polarization curves for all the coatings after exposure in 0.05 M NaCl solution for 72 h

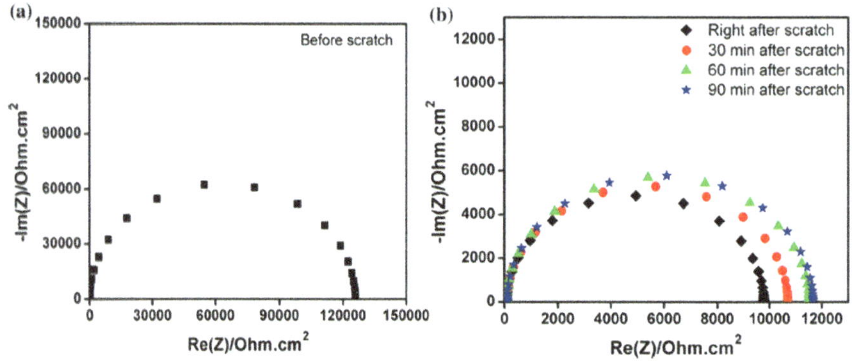

Fig. 19 Electrochemical impedance study, Nyquist plot of **a** before scratch, **b** after scratch of 8-HQ-encapsulated GO/PS container modified epoxy coating

container and decreased the anodic current density so shifting the corrosion potential towards more positive values which indicated positive protection of coating containing 8-HQ [86].

In another work, ceria nanocontainers were loaded with 8-hydroxyquinoline (8-HQ) and structure of obtained nanocontainers was analyzed via various analysis. Incorporation of 8-hydroxyquinoline nanocontainers to coating led to formation of smart coating which sufficiently release inhibitor as a response to environmental triggers such as pH or crack. A wide field of implementations from surface modification to corrosion etc. can be explored in the direction of smart coatings [87].

In another study, 8-HQ was encapsulated into graphene oxide/polystyrene (GO/PS) as container and the corrosion protection was estimated after 30 days of exposure to NaCl solution. Due to slow and continuous release of 8-HQ, after 30 days no corrosion occurs and the self-healing ability of obtained coating concluded (Fig. 19). When the epoxy coating scratched and a defect introduced to metal surface, Rp value does not change significantly because of release of inhibitor to the medium and healing the damage part [88].

Molybdate-Based Corrosion Inhibitors Nano-Containers

Molybdate salts are good candidate as environmentally friendly and nontoxic corrosion inhibitors for anti-corrosive coating. Sodium molybdate, Na_2MoO_4, is a non-toxic, environment-friendly and anodic corrosion inhibitor which can beorm loaded into coating surface and protect surface against corrosion. The efficiency of corrosion inhabitation is dependent on the concentration, temperature, pH, and the oxidizing agent. Nanocontainers containing molybdate-based inhibitor corrosion act as smart containers to protect inhibitor from undesirable reactions and to controlled release to target sites and release them with a controlled rate from capsules and led to a long term protection [89–91]. Mesoporous silica used as carrier for encapsulation of corrosion inhibitors such as molybdate which release active agent in acidic or alkaline media. EIS results of epoxy coating containing mesoporous silica/sodium molybdate after 8 weeks of exposure to NaCl solution showed that mesoporous silica controlled release of inhibitor from container in the response to corrosion damage. This manner attributed to the formation of protective compound induced by release of molybdate from mesoporous silica in the attacked zone [92].

6 Concluding Remarks

One of the most important and commercial routs to protect metal surface from corrosion and scratch is applying protective coating on the surface which containing inhibitor containers. Smart coating could be synthesized via direct incorporation of corrosion inhibitor into polymeric coating or paints. Various structures may be used as core material for encapsulated of inhibitor or active agent or healing agent to obtain a self-healing coating. Release of inhibitor in a controlled rate is promising for new routes to achieve smart healing agents to protect metal and alloys surface from corrosive parameters. In this chapter, some chemical structure which can be used as containers investigated to obtain a self-healing coating for recover the damage in the metal surface.

References

1. Li Y, Kalia RK, Nakano A, Vashishta P (2015) Reactive molecular dynamics study of oxidation of aggregated aluminum nanoparticles. In: MRS Proceedings, Cambridge Univ Press, 1758:mrsf14–1758-vv1703–1703
2. Davis JR (1999) Corrosion of aluminum and aluminum alloys. Asm International
3. Vasudevan AK, Doherty RD (2012) Aluminum alloys-contemporary research and applications: contemporary research and applications. Elsevier
4. Doberdò I, Löffler N, Laszczynski N, Cericola D, Penazzi N, Bodoardo S, Kim GT, Passerini S (2014) Enabling aqueous binders for lithium battery cathodes–Carbon coating of aluminum current collector. J Power Sources 248:1000–1006
5. Nguyen T, Foley R (1980) The chemical nature of aluminum corrosion III. The dissolution mechanism of aluminum oxide and aluminum powder in various electrolytes. J Electrochem Soc 127:2563–2566
6. Perez N (2004) Electrochemistry and corrosion science. Springer Science & Business Media
7. Pardo A, Merino M, Coy A, Arrabal R, Viejo F, Matykina E (2008) Corrosion behaviour of magnesium/aluminium alloys in 3.5 wt.% NaCl. Corros Sci 50:823–834
8. Paul MC, John PB (2011) Cathodic protection of steel in concrete and masonry, Second Edition
9. Feliu S, Barajas R, Bastidas JM, Morcillo M (1989) Mechanism of cathodic protection of zinc-rich paints by electrochemical impedance spectroscopy. J Coatings Technol 61:71–76
10. Jones DA (1992) Principles and Prevention of Corrosion, 2nd edn. Macmillan Publishing Company, New York, p 398
11. Richards JW (1896) Aluminium, its history, occurrence, properties, metallurgy and application, including its alloys, Philadelphie

12. Arman S, Ramezanzadeh B, Farghadani S, Mehdipour M, Rajabi A (2013) Application of the electrochemical noise to investigate the corrosion resistance of an epoxy zinc-rich coating loaded with lamellar aluminum and micaceous iron oxide particles. Corros Sci 77:118–127
13. Ali A, Megahed H, Elsayed M, El-Etre A (2014) Inhibition of acid corrosion of aluminum using Salvadore persica. J Basic Environ Sci 1:136–147
14. Umoren S, Eduok U, Solomon M (2014) Effect of polyvinylpyrrolidone–polyethylene glycol blends on the corrosion inhibition of aluminium in HCl solution. Pigm Resin Technol 43:299–313
15. Committee on Corrosion Loss in Japan. Report on corrosion loss in Japan. Boshoku-Gijutsu (Corros. Eng) 26:401–512 (1977)
16. Committee on Corrosion Loss in Japan. Survey of corrosion cost in Japan. Zairyoto-Kankyo (Corros. Eng) 50:490–512 (2001)
17. Hoar TP (1976) Corrosion of metals: its cost and control. Proc R Soc 348:1–18
18. Bennett LH (1978) Economic effects of metallic corrosion in the United States: a report to the congress
19. Feng W, Patel SH, Young MY, Zunino JL, Xanthos M (2007) Smart polymeric coatings recent advances. Adv Polym Technol 26:1–13
20. Amiri S, Rahimi A (2016) Sol-gel Technology and various hybrid nanocomposite coatings applications. Iran Polym J 25:559–577
21. Amiri S, Rahimi A (2014) Preparation of supramolecular corrosion inhibitor nanocontainers for self-protective hybrid nanocomposite coatings. J Polym Res 21:556. https://doi.org/10.1007/s10965-014-0566-5
22. Amiri S, Rahimi A (2014) (2014) Self-healing hybrid nanocomposite coatings containing encapsulated organic corrosion inhibitors nanocontainers. J Polym Res 22:624. https://doi.org/10.1007/s10965-014-0624-z
23. Amiri S, Rahimi A (2015) Synthesis and characterization of supramolecular corrosion inhibitor nanocontainers for anticorrosion hybrid nanocomposite coatings. J Polym Res 22:66. https://doi.org/10.1007/s10965-015-0699-1
24. Amiri S, Rahimi A (2015) Anti-corrosion hybrid nanocomposite coatings with encapsulated organic corrosion inhibitors. J Coat Technol Res 12:587–593
25. Amiri S, Rahimi A (2016) Self-healing anticorrosion coating containing 2-mercaptobenzothiazole and 2-mercaptobenzimidazole nanocapsules. J Polym Res 23:83. https://doi.org/10.1007/s11998-014-9652-1
26. Ershad Langroudi A, Rahimi A (2003) DGEBA TEOS ormosil coatings for metal surfaces. In: Proceedings of the 6th Iranian Seminar on Polymer Science and Technology (ISPST 2003) Tehran, Iran, 331, 12–15 May
27. Zandi-zand R, Ershad-langroudi A, Rahimi A (2005) Silica based organic–inorganic hybrid nanocomposite coatings for corrosion protection. Prog Org Coat 53:286–291
28. Rau SR, Vengadaesvaran B, Ramesh K, Arof AK (2012) Studies on the adhesion and corrosion performance of an acrylic-epoxy hybrid coating. J Adhes 88(4–6):282–293
29. Wicks J, Zeno W, Jones FN, Pappas SP, Wicks DA (2007) Organic coatings: science and technology. Wiley
30. Murphy EB, Wudl F (2010) The world of smart healable. Mater Prog Polym Sci 35:223–251
31. Zhang HL, Lorente S, Bejan A (2007) Vascularization with trees that alternate with upside-down trees. J Appl Phys 101:094904
32. Matejka L, Dukh O, kolarik J (2000) Reinforcement of crosslinked rubbery epoxies by in-situ formed silica. Polymer 41:1449–1459
33. Amiri S, Amiri S (2017) Cyclodextrins: properties and industrial applications, 1 edn. Wiley
34. Davis DA, Hamilton A, Yang J, Cremar LD, Van Gough D, Potisek SL, Ong MT, Braun PV, Martinez TJ, White SR, Moore JS, Sottos NR (2009) Force-induced activation of covalent bonds in mechanoresponsive polymeric materials. Nature 459:68–72
35. Hinton BRW, Wilson L (1989) The corrosion inhibition of zinc with cerous chloride. Corros Sci 29:967–985

36. Verma N, Singh WR, Tiwari SK, Singh RN (1990) Influence of minor additions of La, Ce and Nd on the corrosion behavior of aluminum bronze in sulphuric acid solution Br Corros J 25:131–142
37. Zhao J, Frankel G, McCreery R (1998) Corrosion protection of untreated AA-2024-T3 in chloride solution by a chromate conversion coating monitored with raman spectroscopy. J Electrochem Soc 145:2258–2264
38. Xia L, McCreery RL (1998) Chemistry of a chromate conversion coating on aluminum alloy AA2024-T3 probed by vibrational spectroscopy. J Electrochem Soc 145:3083–3089
39. Hawkins K, Isaacs HS, Heald SM, Tranquada J, Thompson GE, Wood GC (1987) An investigation of chromate inhibitors on aluminium using fluorescence detection of X-ray absorption. Corros Sci 27:391–399
40. Cotton JB, Scholes IR (1967) Benzotriazole and related compounds as corrosion inhibitors for copper. Br Corros. J. 2:1–5
41. Gerengi H, Darowicki K, Bereket G, Slepski P (2009) Evaluation of corrosion inhibition of brass-118 in artificial seawater by benzotriazole using dynamic EIS. Corrosion Sci 51:2573–2579
42. Morito N, Suetaka W (1973) In situ ultra-violet high-sensitivity reflection studies of the corrosion inhibition of Cu by Benzotriazole. J Japan Inst Metals 37(1973):216–221
43. Wilson GO, Caruso MM, Reimer NT, White SR, Sottos NR, Moore JS (2008) Evaluation of ruthenium catalysts for ring-opening metathesis polymerization-based self-healing applications. Chem Mater 20:3288–3297
44. Wilson GO, Porter KA, Weissman H, White SR, Sottos NR, Moore JS (2009) Stability of second generation Grubbs' alkylidenes to primary amines: formation of novel ruthenium-amine complexes. Adv Synth Catal 351:1817–1825
45. Bardez E, Devol I, Larrey B, Valeur B (1997) Excited-state processes in 8-hydroxyquinoline: photoinduced tautomerization and solvation effects. J Phys Chem B 101:7786–7793
46. Amiri, S., Amiri, S.(2020) Inhibitors-loaded nanocontainers for self-healing coatings, Corrosion Protection at the Nanoscale, Micro and Nano Technologies, 379–402.
47. Blaiszik BJ, Sottos NR, White SR (2008) Nanocapsules for self-healing materials. Compos Sci Technol 68:978–986
48. Tittelboom KV, Belie ND (2013) Self-Healing in cementitious materials—a review. Materials 6:2182–2217
49. Dhoke SK, Khanna AS (2012) Electrochemical impedance spectroscopy (EIS) study of nano-alumina modified alkyd based waterborne coatings. Prog Org Coat 74:92–99
50. Lee KM, Shin KR, Namgung S, Yoo B, Shin DH (2011) Electrochemical response of ZrO2-incorporated oxide layer on AZ91 Mg alloy processed by plasma electrolytic oxidation. Surf Coat 205:3779–3784
51. Atta AM, Al-Lohedan HA, El-saeed AM, Al-Shafey HI, Wahby MH (2017) Epoxy embedded with TiO2 nanogel composites as promising self-healing organic coatings of steel. Prog Org Coat 105:291–302
52. Popoola API, Aigbodion VS, Fayomi OSI (2016) Anti-corrosion coating of mild steel using ternary Zn-ZnO-Y2O3 electro-depositon. Surf Coat Technol 306:448–454
53. Deepa K, Venkatesha T (2018) Synthesis of NiO-ZrO2 mixed metal oxide nanoparticles and their application in Zncomposite coating on mild steel. Anal Bioanal Electrochem 10:890–900
54. Torknezhad Y, Khosravi M, Assefi M (2018) Corrosion protection performance of nanoparticle incorporated epoxy paint assessed by linear polarization and electrochemical impedance spectroscopy. Mater Corros 69:72–480
55. Hovestad A, Janssen LJJ (1995) Electrochemical codeposition of inert particles in a metallic matrix. J Appl Electrochem 25:519–527
56. Fayomi OSI, Popoola API (2015) Anti-corrosion and tribo-mechanical properties of co-deposited ZneSnO2 composite coating. Acta Metall Sin (Engl Lett) 28:521–530
57. Sehrish H, Eman F, Muddasir N, Adnan K, Rana AS, Ramazan K, Aboubakr A (2020) Cerium dioxide nanoparticles as smart carriers for self-healing coatings. Nanomaterials 10:791. https://doi.org/10.3390/nano10040791

58. Low CTJ, Wills RGA, Walsh FC (2006) Electrodeposition of composite coatings containing nanoparticles in a metal deposit. Surf Coat Technol 201:371–383
59. Haddadi SA, Mahdavian M, Karimi E (2015) Evaluation of the corrosion protection properties of an epoxy coating containing sol-gel surface modified nano-zirconia on mild steel. RSC Adv 5:28769–28777
60. Zandi Zand R, Verbeken K, Adriaens A (2013) Evaluation of the corrosion inhibition performance of silane coatings filled with cerium salt-activated nanoparticles on hot-dip galvanized steel substrates. Int J Electrochem Sci 8:4924–4940
61. Lu YC, Ives MB (1995) Chemical treatment with cerium to improve the crevice corrosion resistance of austenitic stainless steels. Corros Sci 37:145–155
62. Cheng YL, Li JF, Zhang JQ, Cao CN, Cao FH, Zhang Z (2003) Electrochemical features during pitting corrosion of ly12 aluminum alloy in different neutral solution. Acta Metallurgica Sinica (English Letter) 16:319–326
63. Abdullayev E, Lvov YM (2011) Halloysite clay nanotubes for controlled release of protective agents. J Nanosci Nanotechnol 11(11):10007–10026
64. Dmitry GS, Lamaka SV, Yasakau KA, Zheludkevich ML, Ferreira MGS, Molhwald H (2008) Active anticorrosion coatings with halloysite nanocontainers. J Phys Chem C 112(4):958–964
65. Fan F, Chunyu Z, Xu W, Jerzy S (2015) Layer-by-Layer assembly of a self-healing anticorrosion coating on magnesium alloys. Appl Mater Interf 7:27271–27278
66. Skorb EV, Andreeva DV (2013) Layer-by-Layer approaches for formation of smart self-healing materials. Polym Chem 4:4834–4845
67. Andreeva DV, Fix D, Möhwald H, Hack T, Shchukin D (2008) Self-healing anticorrosion coatings based on pH-sensitive polyelectrolyte/ inhibitor sandwichlike nanostructures. Adv Mater 20:2789–2794. https://doi.org/10.1002/adma.200800705
68. Yuan P, Southon PD, Liu Z, Kepert CJ (2012) Organosilane function-alization of halloysite nanotubes for enhanced loading and controlled release. Nanotechnology 23:1–5. https://doi.org/10.1088/0957-4484/23/37/375705
69. Khramov AN, Voevodin NN, Balbyshev VN, Donley MS (2004) Hybrid organo-ceramic corrosion protection coatings with encapsulated organic corrosion inhibitors. Thin Solid Films 447–448:549–557
70. Maia F, Yasakau KA, Carneiro J, Kallip S, Tedim J, Henriques T, Cabral A, Venâncio J, Zheludkevich ML, Ferreira MGS (2016) Corrosion protection of AA2024 by sol-gel coatings modified with MBT-loaded polyurea microcapsules. Chem Eng J 283:1108–1117
71. Fareeha U, Nazal N, Shakoora RA, Ramazan K (2019) Effect of concentration of DOC loaded TiO2 nanotubes on the corrosion behavior of smart coatings. Ceram Int 45:10492–10500
72. Frederico M, Joa˜o, T, Aleksey DL, Andrei NS, Mikhail LZ, Mario GSF (2012) Silica nanocontainers for active corrosion protection. Nanoscale 4:1287–1289
73. Khramov AN, Voevodin NN, Balbyshev VN, Mantz RA (2005) Sol-gel-derived corrosion-protective coatings with controllable release of incorporated organic corrosion inhibitors. Thin Solid Films 483:191–196
74. Feliu S, Barajas R, Bastidas JM, Morcillo M, Feliu S (1993) Electrochemical impedance analysis and interpretation. American Technical Publishers Ltd., Philadelphia
75. Holness RJ, Williams G, Worsley DA, McMurray HN (2005) J Electrochem Soc 152(2):B73–B81
76. Amiri S, Rahimi A (2016) Anticorrosion behavior of cyclodextrins/inhibitor nanocapsule-based self-healing coatings. J Coat Technol Res 13:1095–1102
77. Korotkikh N, Aslanov A, Raenko G (1995) Hetrocyclization of 2-Allylthiobenzimidazole under the action of bromine. Russ J Chem 31:721
78. Narkhede H, More V, Dalal D, Mahulikar P (2008) Solid supported synthesis of 2-Mercaptobenzthiazole derivatives using microwaves. J Sci Ind Res 97:374–376
79. Wang L (2001) Evaluation of 2-Mercaptobenzimidazole as corrosion inhibitor for mild steel in phosphoric acid. Corros Sci 43:2281–2289
80. Wang L, Pu J, Luo H (2003) Investigation into the evolution of corrosion product layer (CPL) of 1018 C-steel exposed to multiphase environment using FIB and EIS techniques. Corros Sci 45:59–80

81. Haddadi SA, Ramazani SAA, Mahdavian M, Taheri P, Mol JMC (2018) Fabrication and characterization of graphene-based carbon hollow spheres for encapsulation of organic corrosion inhibitors. Chem Eng J. https://doi.org/10.1016/j.cej.2018.06.063
82. Ji G, Dwivedi P, Sundaram S, Prakash R (2013) Inhibitive effect of Chlorophytum borivilianum root extract on mild steel corrosion in HCl and H_2SO_4 Solutions. Ind Eng Chem Res 52:10673–10681
83. Rajendran S, Vaibhavi S, Anthony N, Trivedi DC (2003) Transport of inhibitors and corrosion inhibition efficiency. Corros Sci Sect 59:529–534
84. Sagüés AA, Pech-Canul MA, Shahid Al-Mansur AKM (2003) Corrosion macrocell behavior of reinforcing steel in partially submerged concrete column. Corros Sci 45:7–32
85. Solmaz R (2014) Investigation of corrosion inhibition mechanism and stability of Vitamin B1 on mild steel in 0.5 M HCl solution. Corros Sci 81:75–84
86. Balaskasa AC, Kartsonakisa IA, Tzivelekaa LA, Kordasa, GC (2012) Improvement of anti-corrosive properties of epoxy-coated AA 2024–T3 with TiO2 nanocontainers loaded with 8-hydroxyquinoline. Prog Org Coat 74:418–426
87. Ioannis K, Daniilidis I, Kordas G (2008) Encapsulation of the corrosion inhibitor 8-hydroxyquinoline into ceria nanocontainers. Sol-Gel Sci Technol 48:24–31
88. Sneha D, Mamata P, Raja VS, Smrutiranjan P (2016) Encapsulating 8-hydroxyquinoline in graphene oxidestabilized polystyrene containers and its anticorrosion performance. J mater Sci 51:10262–10277
89. Ilevbare GO, Burstein GT (2003) The inhibition of pitting corrosion of stainless steels by chromate and molybdate ions. Corros Sci 45:1545–1569
90. Mu GN, Li XH, Qu Q, Zhou J (2006) Molybdate and tungstate as corrosion inhibitors for cold rolling steel in hydrochloric acid solution. Corros Sci 48:445–459
91. Qu Q, Li L, Jiang SA, Bai W, Ding ZT (2009) Effect of sodium molybdate on the corrosion behavior of cold rolled steel in peracetic acid solution. J Appl Electrochem 39:569–576
92. Keyvania M, Yeganehb HR (2017) Application of mesoporous silica nanocontainers as an intelligent host of molybdate corrosion inhibitor embedded in the epoxy coated steel. Progre Nat Sci: Mater Int 27(2):261–267
93. Obot IB, Umoren SA, Johnson AS, Sunlight-mediated synthesis of silver nanoparticles using honey and its promising anticorrosion potentials for mild steel in acidic environments. J Mater Environ Sci 4(6):1013–1018

Containers with Lubricating Agents for Friction and Wear

Qurat Ul Ain, H. S. Ashrith, Manjesh Kumar Singh, and T. P. Jeevan

1 Introduction

A separate realm of modern technology is colonized by elements with encapsulated species. The creation and processing of capsules originated in 1930s and later on elevated in 1970s. Encapsulation technology has entered its significant progress during past two decades Regardless of progress, this technique has been adopted by a limited field of spheres. Encapsulation methods are designed to shield ingredients (such as active agents in food industry and lubricant particles in lubrication process) from damaging and detrimental contacts while improving their functionality and bioavailability [1]. The arrival of the controlled release technology has a pronounced bearing in the field of agriculture, food industries, pharmaceuticals, lubricant supplements. There is a huge gap in the proposed characteristics of encapsulation technology and their possible utilization in various fields [2]. Encapsulation covers a wide spectrum as it developed into the base for carbonless copy technology in printing industry and is currently utilized in the food handling and processing methods [3], pharmacological technology [4], agro-chemical [5] and cosmetic production [6]. Likewise, the encapsulation takes up vital part in the design and production of clothing. This method has divulged a wide variety of long-term qualities to fabric manufacturing industries, antibiotic and antimicrobial resistance drugs, pesticides, fire intensity reducing agents, lubricating agents and temperature-controlled phase-change resources [7–9].

Q. U. Ain
Department of Mechanical Engineering, National Institute of Technology, Srinagar, India

H. S. Ashrith · T. P. Jeevan (✉)
Department of Mechanical Engineering, Malnad College of Engineering, Hassan, India

M. K. Singh
Department of Mechanical Engineering, Indian Institute of Technology, Kanpur, India

J. Parameswaranpillai et al. (eds.), *Micro- and Nano-containers for Smart Applications*, Composites Science and Technology,
https://doi.org/10.1007/978-981-16-8146-2_11

Around late 1930s, the growth of microencapsulation originated with producing capsules comprising colors or dyes, a cleanser for paper and ribbons made of carbon. In 1950s, a small amount of gelatin with dyes are integrated into paper for replication thereby replacing carbon paper [10]. Later this technology is explored in different areas of research. The encapsulated elements are used in agricultural science, drugs, foodstuffs, cosmetics and perfumes, fabrics, paper, dyes, coats and glues, printing and several other fields are included [11–13]. The benefit of controlled release in medicine delivery is utilized; however microencapsulated vitamins are secured before their release. The food constituents and flavors are encapsulated to prevent them against degradation and shield from bad taste of constituents. Nowadays, textile industries customize microencapsulation techniques to improve the properties of produced fabric. The significant application lies in the integration of phase change materials in encapsulation. This method has been proved to be beneficial in the surface treatments. Lubricant-filled microcapsules are prepared by a mini-emulsion process. Recently shielding coatings comprising of different self-repairing nano-capsules have been extensively investigated [14–16].

The process of encapsulation is retaining three different states of matter (particles) forming an envelope made of solid or liquid, or any alternative immiscible material. The two main forms of encapsulation are as core enveloped in wall or molecules in matrix. The core, internal phase or fill is an element inside the capsule, whereas the wall surrounding the core is the shell material. In the core of shell, content is in the continuous phase walled by membrane resembled cover which is referred as shell, while particle in matrix form has the micro or nano elements, a dispersed segment, and are homogenously dispersed into a consistent matrix medium. The external wall in micro-capsule is known as shell, coating, envelope, membrane, or external phase. Encapsulation in solid phase is often reflected as a transformation from liquid form to powdered particulate form. These constituent parts are vaporized for added transportation or processing. The material within core is transferred through wall via diffusion or dissolution and by volatilization or eliminating the wall by other means [17].

Encapsulation method improves the application properties of lubricating oil when the pressure is applied and the collapsed microcapsules containing lubricating oil are used in different areas for lubrication. The microcapsule shells are highly resistant to abrasion with high temperature resistance, and should break or collapse when exposed to high pressure during force applied processes. The ideal form of capsule is sphere of small sizes. The microcapsules consisting liquid lubricants are more feasible in injection force, apparent density and shrinkage, and are related to commonly used lubricants in tensile strength and transverse rupture strength. Microcapsules enclosing lubricants proved to be appropriate and beneficial in several studies, one such example is during electrochemical deposition of nickel from $NiSO_4$ solution on steel substrates. The strength of encapsulation technique lies in formation of even and steady coating of shell material which encapsulates the internal phase by physical or chemical actions. The processing and different characteristics of encapsulation are influenced by the physical, chemical and mechanical

Table 1 Classification of encapsulation techniques

Method	Categorized methods
Physical methods	Air suspension, envelop-combination, extrusion, supercritical solution, porous centrifugal, electrostatic binding, solvent extraction/evaporation, rotary separation
Chemical methods	Interfacial polymerization, in situ polymerization, piercing Solidifying
Physicomechanical methods	Spray drying, spray cooling, spray coating, fluidized bed coating, pan coating, spinning disc and co-extrusion
Physicochemical methods	Simple co-acervation, complex co-acervation/phase separation, encapsulation by polyelectrolyte multilayers, phase inversion, hot melt, drying bath, powder bed grinding, melting-dispersion condensation, capsule-core exchange

features of the encapsulant used. Depending on various mechanisms and environmental conditions microencapsulation techniques can be categorized into different methods, namely physical methods, chemical methods, physicomechanical methods and physicochemical methods shown in Table 1 [18].

The parameters which represent the encapsulation technology, independent of the processing technique used are mentioned below:

Encapsulation efficiency (EE): It is the ratio of mass m_f of fill to the total mass m_{dc} of the encapsulation material.

Encapsulation yield (EY): It is the ratio of mass m_p of produced capsules to the total mass of core m_{dc} and shell m_{ds} material.

Loading capacity (LC): It is the ratio of mass m_{ec} of encapsulated core material to the mass m_p of produced capsules.

Payload: Payload is an estimation of the amount of core material enclosed in a wall material (shell) or specified matrix.

2 Tribology and Its Importance

The working of different types of machineries (artificial and natural mechanisms) involve interaction of surfaces. The discipline of analyzing friction, wear and lubrication during sliding/rolling interactions in natural and artificial systems is known as tribology. This is a highly interdisciplinary area that requires expertise from various fields such as mechanical, production, material science, physics, chemistry and chemical engineering, mathematics to name a few. Tribology involves understanding the core mechanisms of relative motions between interacting surfaces which requires expertise from various research fields. The progress in tribology has strengthened and benefited the world of science and engineering. This field encompasses almost

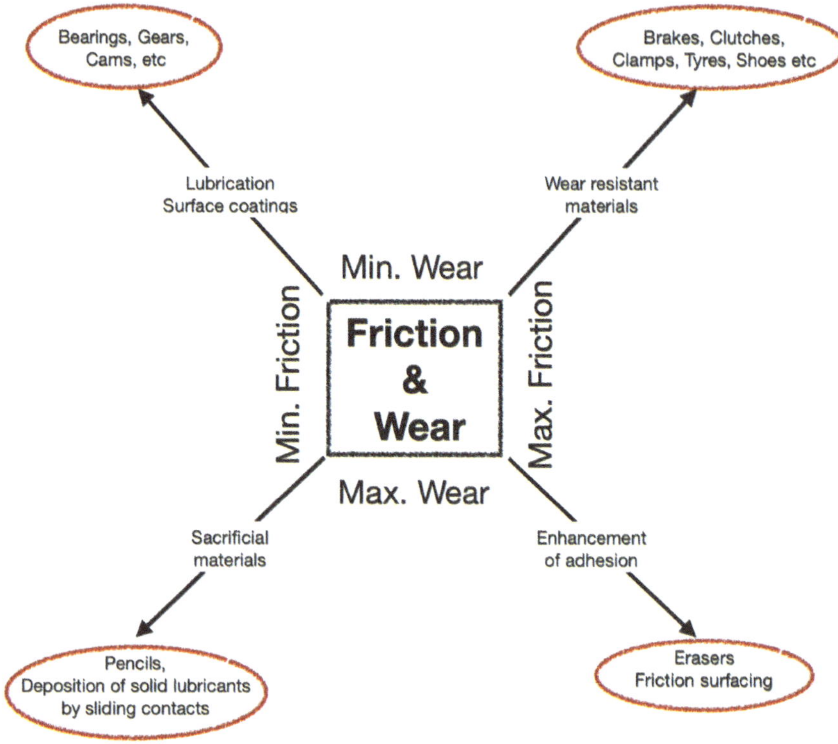

Fig. 1 Role of friction and wear in different applications

every feature and characteristic of science and engineering, where in the surfaces interact with/against other surfaces, such as from automotive and electricity production to biomedical fields, and edible technology to miniature cosmetic products [19]. Figure 1 shows the role of friction and wear in various disciplines.

The focus of this field is not just to investigate and analyze friction, rather friction has a crucial role in determining different governing mechanisms of tribo/mechanical-elements. There might be cases where less friction is needed but in an appropriate manner. The proper functioning and working of human joints (hip, knee and elbow), door junctions, link connections etc. require a low value of coefficient of friction (COF). The low COF may not always be desirable, for example in brakes and clutches, suitable and appropriate frictional force is indispensable which helps in dispel of kinetic energy and transmission of torque; likewise, high value of frictional force is essential during motion of vehicle wheels on the surface and walking on the ground [20].

As the surfaces interact and relatively move against one another, chances like losses in the form of removal of material may result in interacting surfaces, mostly deformation or volume/mass damage to the material. Usually the wear rate is low, however rate could be extremely high as well. Wear damages the interacting surface material and is destructive in most of the situations, thereby leading to uneven surfaces, undesirable degrees of freedom and results in decrease in accuracy. These phenomena may originate vibrations, mechanical stresses and in fact higher wear rate leads to fatigue failure. The low wear rate may increase and reach the point where its rate is sufficient for complete damage to the machine/system. The studies report that sometimes the main reason for the breakdown of engineering components lies in the core of high wear rates. In fact, higher wear rates are desirable in shaving, grinding and polishing which are required as pre-requisites in some machining processes to remove excess material using an appropriate method. Different processes like running-in and press-fitted sealing favor and expect the removal of material in initial phase of operation for smooth working of the machinery [21].

Therefore, friction and wear properties must be optimized and must be in control for specific applications as these define the system properties. These are controlled by applying lubricants (friction and wear controllers) which are well-defined and interrelated to friction, wear and lubrication processes. The presence of different constituents (moisture content and oxygen in atmosphere) or contaminants (like dust particles) in the nearby environment may aid in lubrication process and are vitally important in the interacting elemental surfaces [22].

The research studies in the field of tribology are constantly developing due to the growth of scientific methods and newer technologies. This field has entered the discipline of science in real and applied form which is interrelated to the desires and needs of nations through manufacturing and production technologies. During the initial phase of eighteenth century, it was Amontons and Coulomb who suggested the mathematical models in the form of classical equations for friction. This type of research was based on the previously performed experiments and investigations. During the late nineteenth century, the theoretic basis of lubrication mechanism was established by Reynolds. He discovered the mechanism of basic hydrodynamic lubrication and further derived an equation (popularly known as Reynolds equation) based on lubrication of bearings using viscous medium of fluids. As a result, this led to a novel research established on continuum mechanics. With further developments in manufacturing science and technology during the twentieth century, the progress of research in this very particular field extended. With the course of time, Hardy projected the principle of boundary lubrication and revealed different mechanisms of physico-chemical adsorption layer formation on the interacting surfaces. This particular theory upheld the research in lubrication along with additives. Further, the origin of friction (sliding) from energy (probably the bond energy) transformation perspective was described by Tomlinson using different molecular attraction and repulsion forces. Following this theory, Bowdon and Tabor came up with well-known adhesion junction theory of friction explained on the basis of plowing process. All these developments not only extended the spectrum of tribology field but also stimulated this field to possess a multitude dimension of various fields and build an interdisciplinary area of study [23].

The most vital part in understanding the potential of tribology is with respect to economy. The research and investigations in this field advanced from micro to nano level, from extent (quantity of research) to excellence and from unidirectional to multidirectional disciplines. The current research trend has reached to the point where the study of characterization and different properties of tribology has stretched to the investigation and control of these properties. Even nowadays the specific properties are kept under control by aiming different goals. In the earlier times, primary objective of the research in this area only concentrated on product and machine elemental maintenance. Presently, the focus is to generate pioneering designs for production and manufacturing of machines and other tribo-systems. The progress in tribology is determined by various research and development which are of vital importance and keep track of objectives in growth of tribology. The difficult mechanisms have been analyzed and resolved, thus some of the complex systems are now recognized easily due to growth in computable analysis. The computational analysis precisely enlightens and explains the complex mechanisms. Thus, the mathematical techniques and models are utilized in simulating lubrication processes. They pick up the studies using dynamic aspects which affect the functioning and principle design of present machines. The extensively utilized equipment's suitable for analysis of surface, physical, mechanical and tribological properties consisting of morphology, topography, structural, elemental and compositional characteristics include optical/electron microscopy, micro/nano-analyzing mechanisms [24].

In present times, the progress in the field of material engineering and technologies led to expansion in discovery of novel materials and surface engineering techniques. The analysis of tribological characterization has expanded to non-metallic materials, composites and nanocomposites of ceramics and polymers. The in-situ techniques used in characterization of nanostructured materials have rapidly enhanced the research methodologies. The surface engineering methods are utilized to alter the features and properties of material surfaces and are considered as one of the most evolving filed of tribology. The methods include coating, texturing of surfaces and encapsulating micro/nano-particles. The field of nanotechnology develops novel subjects interconnected to science, one of the prominent fields is nano-tribology. This field came into existence due to the interlink of underlying mechanisms of tribology with mechanics of micro/nano-structures and the molecular dynamics performance of material surfaces. Nano-mechanics and nano-tribology offer an innovative way of research from micro- to atomic-level to comprehend the phenomena of friction, wear and lubrication process. This developed a connection in micro-level and nano-level characteristics, hence depicting the molecular structural configurations and dynamics at atomic scale. The arrival and rise of nano-tribology represents that the elementary phenomena in the study of tribology are moving towards a novel phase of technology with unique properties. At present the tailoring of tribology with modern science and technology signifies an exceptional phase with objectives aiming on sustainable, ecofriendly, and cost-effective performance of machines. With these targets, the dynamic part of research in tribology is not only to control friction and wear in systems but to improve the service lifespan of machineries [19].

Different type of tribometers used study the tribological properties of encapsulated materials are presented in Table 2.

Table 2 List of tribometers used to study the tribological behavior of encapsulated materials

Type of tribometer used	References
Reciprocating (TE77) tribometer	[25, 26]
Four-ball wear tester	[27]
Pin-on-disc configuration	[28–30]
Ball-on-disk tribometer	[31–34]
Block-on-ring apparatus (M-200)	[35–37]
Fretting test method	[38]
Disc on-pin friction-testing machine (CBZ-1 tribometer)	[39, 40]
Ball-on-disk microtribometer (CSM)	[41]
Crossed-cylinders type tribometer	[42]
Universal tribometer (UMT-3, Bruker Corporation)	[43]

3 Friction and Wear Behavior of Polymer and Their Composites Using Encapsulation Technology

The varied features of polymers place them in a significant class of engineering materials, possessing properties low weight, ease in manufacturing and production, high specific strength, highly anti-corrosive, design flexibility, self-lubrication properties, low COF and wear resistant properties [44–47]. The mechanical characteristics of polymer materials are adversely affected by the environmental conditions and the fillers used in lubrication. The two primary methods exist which enhance different aspects of characteristics in polymer composites. The first method includes fillers as solid lubricants like PTFE, MoS_2 and graphite in polymer matrix [48–51]. The lubricant filling particles possess self-lubricating properties which would enhance the tribological and mechanical properties such as friction coefficient, stiffness, strength, corrosive resistance, and fatigue which results in outstanding wear resistance [52]. The second method uses liquid lubrication where the properties like wear and friction reduction are attributed to drop in contact temperature and formation of interfacial films [53–58]. Thus, liquid lubrication can advance tribological features of polymer-based composites. However, compatibility of liquid lubricants usually restricts the application in sensitive or contamination free operative condition. In order to purge above mentioned problem, liquid lubricants can be encapsulated to produce organic or inorganic shell microcapsules, and these would be reinforced in polymer matrices which gives ultra-low friction coefficient and high wear resistance since the lubricant after rupture of capsules come into action [31, 59].

Encapsulation technologies used in self-lubricating composites have attracted a lot of attention and are widely exploited in current research trends [35, 60–62]. As examined in various investigations, the lubricant required for self-lubrication should be encapsulated prior to reinforcement into the self-lubricating polymer composites or coating. As the self-lubricating particles are ruptured by motion or friction, the surrounding microcapsules are also damaged and the lubricant is released, formation of transfer film is responsible for lubricating the interacting interfaces. Thus, self-lubricating microcapsules should be thermally stable for a wide range of temperature, better encapsulation efficiency and improved tribological properties. Several lubricating agents like lubricant oils [29, 31], wax [34, 63–65] di(iso)cyanate, [42, 66] paraffin and ionic liquids have been utilized and investigated in recent studies [33, 67, 68]. These lubricants improves friction and wear properties and the cooling action from the lubricating oil takes away the heat generated due to friction. Thermal and oxidation stability of oils is essential as these properties can reduce service life and storage time of lubricants. The oil might turn to vapors or catch fire from heat produced due to friction. So, low volatile and noninflammable fluids are desired for lubricants [69]. The ionic liquids when compared to commonly used lubricants possess properties such as high thermal stabilities, low volatility and noninflammability; categorized as high-potential lubricants [70–73]. Nevertheless, as the enveloped internal fill substantial is in the form of ionic liquid (IL), its efficiency would be less since the highly viscous fluid plus the nature being bipolar restricts the dispersion in a continuous phase during microencapsulation [74–76]. Encapsulation technique enhances the wear resistance of polymer based composites and shields the matrix from defects like distortion, twist or failures under stress due to the additional competence of shell material produced from the composite matrix to prevent some part of deformation [37].

Bulk polymeric materials and coatings are generally used in different tribological applications. The bulk polymers provide better durability as they have adequate interacting surface accessible for wear, however the misalignment may increase during interaction. Surface treatment like coating of polymers has low durability as they do not provide much interacting surface available during contact, but they prevent major contact misalignments due to the low overall thickness of the coatings and also provide the mechanism with full rigidity of the substrate materials. The polymer based research studies in context with tribological characterization are mostly related to bulk materials. The self-lubricating polymer coatings with increased rate of performance are in huge demand as the machineries required must be efficient, robust and cost effective. The self-lubricating coatings are utilized during severe environments like combined starved lubrication and dry sliding condition, operations under vacuum at very high or low range of temperatures and extreme pressure conditions. These conditions present significant challenge of coating development for engineers and designers. The idea and theory of a self-lubricating or solid–liquid coating mechanism [77, 78] where a liquid part of the lubricant collapses due to the interaction of surfaces which results due to wear of solid material. With this method the exterior interference and maintenance is not required, as it will possibly decrease the COF while aiding durability and strength to the material. One such method is encapsulation, which forms a rational track for the idea behind liquid added solid surface treatment or coating [42].

A detailed discussion on some of the research findings on tribological behavior of polymers and their composites (thermosetting and thermoplastic) and other materials with encapsulation technology is presented below. The researchers have analyzed

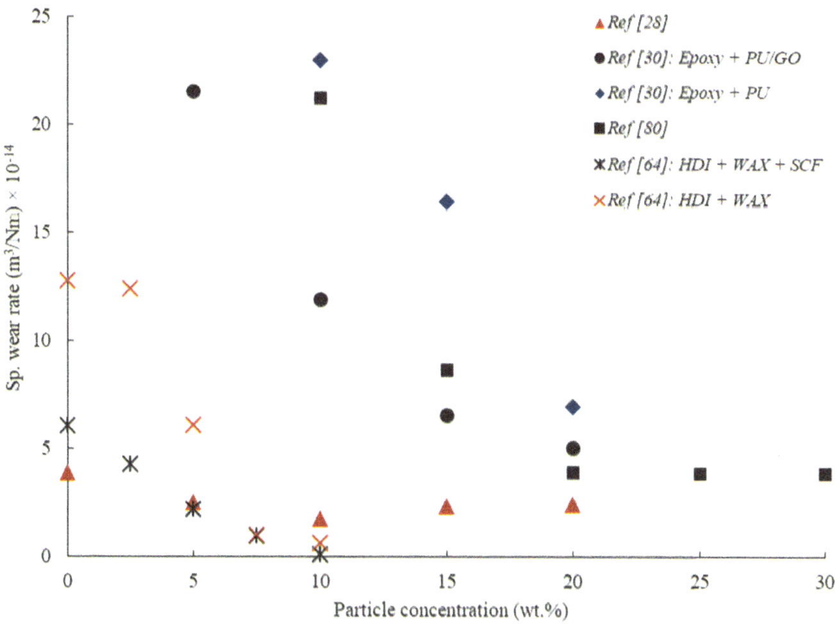

Fig. 2 Specific wear rate w.r.t microcapsule concentration extracted from literature

the tribological performance of the encapsulated materials with respect to various parameters like specific wear rate, wear rate, wear width, wear depth and coefficient of friction. Variation of specific wear rate (Fig. 2), wear rate (Fig. 3), wear width (Fig. 4), wear depth (Fig. 5) and coefficient of friction (Fig. 6) with respect to microcapsule concentration are extracted from literature and graphs are plotted.

3.1 Polymers and Their Composites

3.1.1 Thermoset Polymers

Li et al. [30] have fabricated and studied the tribological properties of epoxy composites containing polyurethane/graphene oxide (PU/GO) microcapsules. The coefficient of friction was found unstable for the pure epoxy sample and the epoxy [BMIm]BF4 microcapsules, which remained stable, was mainly due to [BMIm]BF4 lubrication released during testing from microcapsules and improved the tribological properties of epoxy composite effectively. GO's exceptional two-dimensional laminar structure, exceptionally low friction coefficient, super hardness and excellent thermal-conductivity is due to the excellent friction resistance of epoxy composites with PU/GO microcapsules. The interfacial relation between microcapsules and

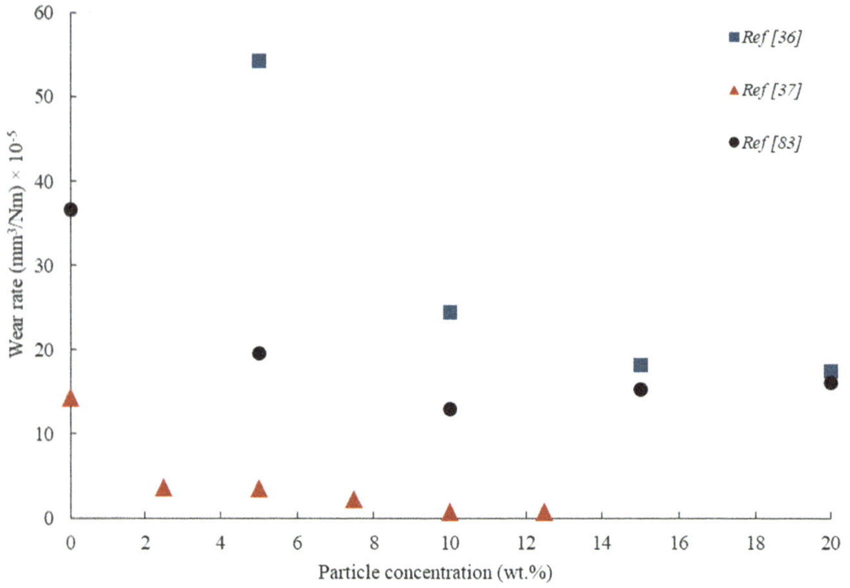

Fig. 3 Wear rate w.r.t microcapsule concentration extracted from literature

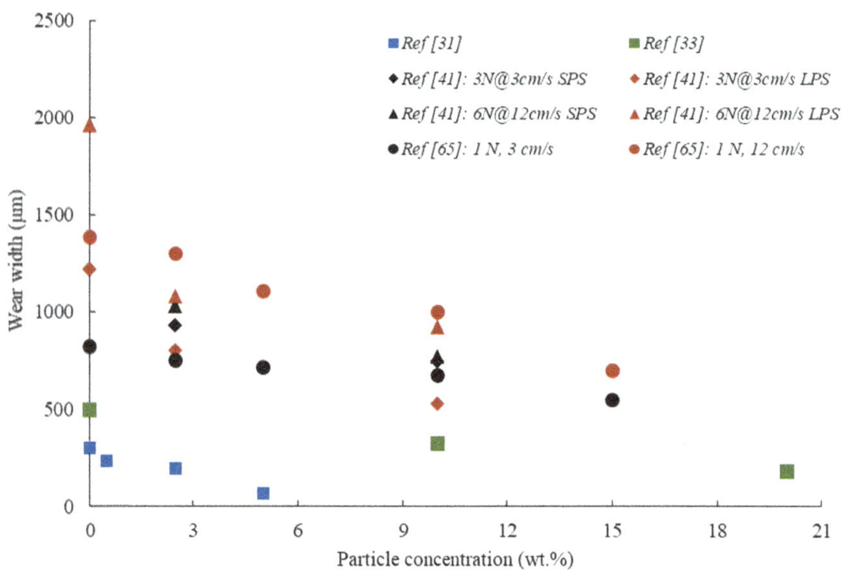

Fig. 4 Wear width w.r.t microcapsule concentration extracted from literature

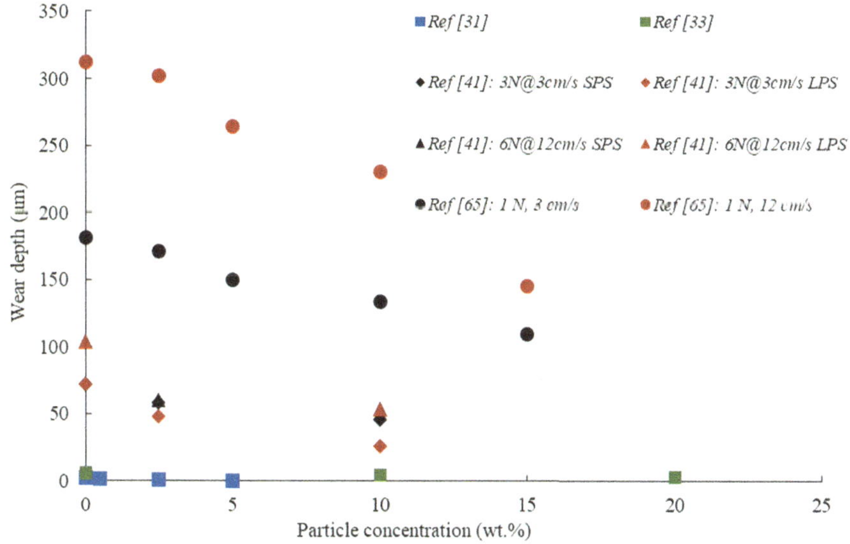

Fig. 5 Wear depth w.r.t microcapsule concentration extracted from literature

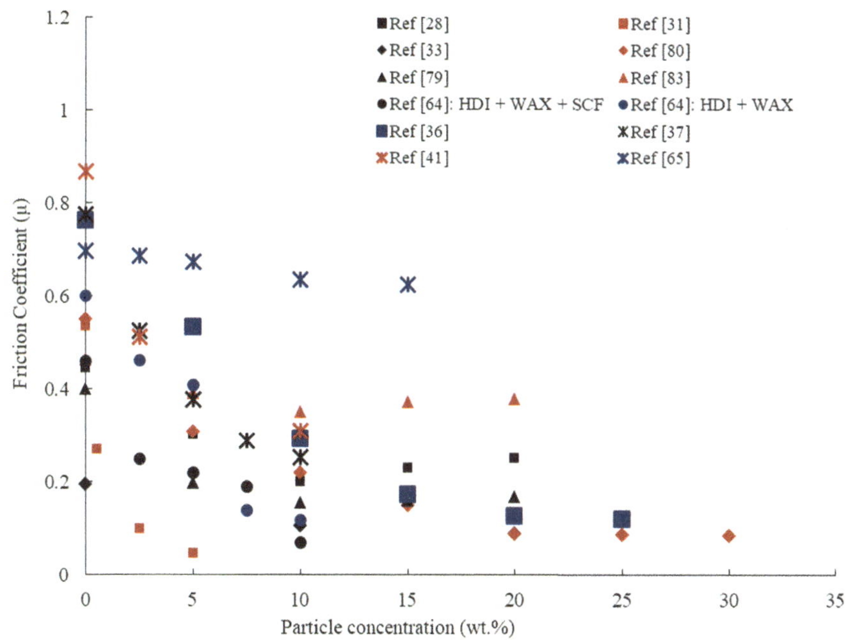

Fig. 6 Friction coefficient w.r.t microcapsule concentration extracted from literature

epoxy-resin could be improved by a large number of functional groups such as the hydroxyl group and carboxy group within GO.

Guo et al. [35] obtained an enhanced tribological property by incorporating small amounts of oil-loaded microcapsules, grafted nano-SiO_2 and short carbon fibers into epoxy. The dual roles of both solid lubricant and reinforcement played by short carbon fibres and the grafted nano-SiO_2 and release of trace quantity of lubricant to rubbing surface results in effective reduction in wear and friction coefficient. Subsequently, the limitations with respect to mechanical property weakening due to increased oil-loaded microcapsules content can be overawed by addition of the grafted nano-SiO_2 and short carbon fibers.

The incorporation of microcapsules containing lubricant oil with Poly(melamine-co-formaldehyde) as outer layer into epoxy composites results in materials with very-low friction and wear behavior. It is observed during sliding wear tests, the asperities of the counter face damaged the microcapsules, freeing the oil to area of contact. The lubrication effect of the oil released and the entrapment of the wear debris in the craters left by the ruptured microcapsules leads to decrease in frictional coefficient and the specific wear rate [59].

Imani et al. [61] investigated the tribological properties of the silica nanoparticles and/or microcapsules containing lubricant wax (WMCs) of epoxy and its composites. The combination of a suitable silica nanoparticles and WMCs showed that the friction coefficient of epoxy materials was enhanced at the same time as the real wear rate had been increased. The WMCs developed a thin and continuous transfer of films on contact surfaces, resulting in reduced wear and friction, with a 3-orders magnitude reduction in coefficients of friction and a tenfold reduction in rubber coefficient in silica/WMC/epoxy composites, compared with those of neat epoxy. Imani et al. investigated the tribological features of epoxy composites with graphene nanoplatelets (GNP) and WMCs. The results showed that specific wear resistance and COF of nano-composites were improved proportionally by the addition of GNP. Moreover, the incorporation of WMC to GNP/epoxy nanocomposites leads to dramatically decrease in the COF and the specific wear resistance. Throughout wear testing, the WMCs were damaged by the severity of the counterface, releasing the wax to the contact area caused the formation of thin transfer film. Therefore, it caused a significant reduction in specific wear rate and COF [43].

Wax-filled microcapsules were applied to the unitary epoxy composite hardness lower than hexamethylene diisocyanate (HDI) filled microcapsules as wax-filled microcapsule's with larger dimensions are present. Tribological study revealed that the unitary wax composites possess lesser friction coefficient and wear rate than those of microcapsules filled with HDI and observed that the released wax lubricant exhibit greater effect on overall working efficiency (mostly tribological properties) than that with released HDI core liquid. The results showed that the unitary wax composite composites had less wear. The decline in the frictional force and wear of epoxy (binary and ternary) composites is clearly observed as the content of microcapsules filled with wax is enhanced, this observation was further verified by the added quantity of wax lubricant generated. However, by addition of SCFs via strong grading and free rolling effect, the frictional force and wear rates of the binary and ternary epoxy composites was lower than earlier [64].

Li et al. [79] prepared double-component lubricant oil-containing microcapsules with SiO_2 nanoparticles. The COF along with specific wear resistance in microcapsules/EP composites are 60.8 and 93.3% lower, respectively, than in pure epoxy. SiO_2 nanoparticles played a positive role in improving self-lubricating composites' tribological properties. This research offered a new way of improving the frictional characteristics of self-lubricating composites based on microcapsules. Furthermore, there was no strong impact on the tribological impact of the compatibility of silica and lubricant oil.

Multi-layered shell type microcapsules filled with [BMIm]PF6 were synthesized by Li et al. [80]. A very low COF and high wear resistance of self-lubricating composites was attained through addition of microcapsules into the composite. The COF of composites was reduced by 83.6% and specific wear rate improved around 218 times as compared to pure epoxy resin. The ultralow friction performance is credited due to boundary lubricating film achieved by the release of [BMIm]PF6 lubricant due to friction at the interacting surfaces. The [BMIm]PF6 bipolar characteristic improved the lubricating film and could effectively avoid interaction between the matrix and the counterface. The uneven PUF external shell of the multilayer microcapsules led to value-added interfacial properties concerning the microcapsules and the resin (matrix), followed by an improvement in friction efficiency.

The tribological characteristics of several composites with microcapsulated epoxy resin and amine-containing etched glass hubbies were systematically studied by Khun et al. [81]. The findings showed that the higher content of epoxy composites as a healing agent seemed to decrease their COF and wear rate through the self-lubricating composites with released core fluids. As the wear rate of epoxy composites increased due to higher normal load, but the COF declined due to the increased wear of the epoxy composites. The simultaneous effect of increased normal load resulted in increased wear rate and surface roughness.

Mu et al. [36], synthesized self-lubricating microcapsules by encapsulating methyl silicone oil as core material with melamine–formaldehyde resin. These microcapsules as antifriction additive are employed to expand the tribological properties of binary and ternary epoxy (EP) composites, which were made-up by integrating microcapsules and potassium hexatitanate whiskers (PTWs) into the EP matrix. The self-lubricating effect of the methyl silicone oil resulted in diminished wear rate and friction coefficient of microcapsules/EP binary composites with the increase of microcapsule content. For microcapsules/EP/PTWs ternary composites, the synergistic effect of PTWs and microcapsules led to the reduction in friction and antiwear effect of the ternary composites.

The mean friction coefficient of the epoxy composites consequently decreased due to presence of high quantity of microcapsules which clearly depicts the COF values for the epoxy composites are lesser than that for epoxy without microcapsules. The cracking of microcapsules due to wearing out of the composite during testing and interaction of surfaces causes the successive release of wax lubricant which lubricates the interacting medium in contact which eventually restricts the direct surface interactions [31].

Khun et al. [41] evaluated the self-lubricating and wear resilient epoxy composites united with microencapsulated wax. The coefficients of friction of epoxy composites in high load are somewhat below than the composites verified under low normal load, as the use of epoxy composites linked to an elevated normal load releases more wax lubricants during sliding to effectively self-lubricate composites. Using epoxy composites, the irregular surfaces of these composites discharge the wax lubricant which lubricates the touching contact medium. Both epoxy composite wear widths and depths are less than those of those measured under lower loads at higher loads.

A wear resistant epoxy composites is formed by incorporating Hexamethylene diisocyanate (HDI) loaded microcapsules in epoxy matrix. The friction coefficient value for epoxy composites decreased to some extent as the content of microcapsules was enhanced, the HDI fluid released due to surface interactions which further lubricated the rubbing surfaces so as to prevent a direct contact between the surface, Moreover, new polyurea layers on the rubber surfaces were produced by the rapid reaction of the HDI liquid with moisture to minimize friction of epoxy composites by preventing direct contact. The wear of epoxy composites decreased significantly as the self-lubricating properties come into play and the improved self-healing efficiency in order to produce new polyurea layers for the whole process and to compensate for the potential poor wear resistance of decreased mechanical properties [65].

Yang et al. [82] added polymer microcapsules to matrix material as compositional modifications, to improve the tribological properties of the bearing as it was worn against a cast copper bushing. These microcapsules appeared to reduce the phase splitting between the low-melting-point lubricant and the matrix. Urea formaldehyde (UF) microcapsules filled with two lubricants included N,N′-ethylene Bis-stearamide and Erucic Amide as core materials and then blended with high-density polyethylene (HDPE) to form the composites. The results revealed that the tribological properties of the composites were improved as the quantity of microcapsules in the composite was increased. The composites containing the microcapsules produced lesser sliding friction which exhibited an even surface morphology, which due to the lubrication provided by Erucic Amide in the microcapsules. As shell of microcapsules was worn away during the test, the encapsulated Erucic Amide was released into the friction pairs, which established a dense lubrication film.

3.1.2 Thermoplastic Polymers

Li et al. [28] observed reduced frictional coefficient with Polysulfone (PSF) and silica hybrid double-wall microcapsules filled with lubricant. The friction coefficient of about 54.9% and wear rate of about 54.3% lesser than pure polypropylene (PP), which is attained with 10 wt%. of microcapsule. The incorporated microcapsules ruptured and released the lubricant forming a transfer thin film thereby reducing the value of frictional force by avoiding through interactions between the steel counterpart and pure polypropylene. Thus mechanical properties along with tribological properties tends to decrease with increased (>10 wt%) microcapsule content.

The synergistic effects of short carbon fibres, treated by liquid-phase oxidation and by chemical grafting, have been combined by Mu et al. [37] and simultaneously adopted it as reinforcing additives to extend the tribological properties of the polyurethane. The coefficient of friction and wear of polyurethane composites without the use of SCFs significantly decreased due to releases of methyl siliconic oils, with increased concentration of the self-lubricating microcapsule and with enhanced mechanical strength and thermal stability of the polyurethane composites and SCFs.

Li et al. [83], synthesized polysulfone tung oil microcapsules. The microcapsules containing tung oil were integrated into an epoxy matrix for the multifunction coating. In comparison with pleasant epoxy, the frictional coefficient and wear rate of the self-lubricating cover decreased significantly. The creation of a transfer film from the release of tung oil and the trap in cavities left by the ruptured microcapsules constituted the key mechanism of anti-friction.

Yang et al. systematically studied tribological behavior of the polyurethane composite comprising ionic liquid, 1-butyl-3-methylimidazolium hexafluorophosphate (BMIMPF6) coated with silica gel shell microcapsules. A lower coefficient of friction and wear, than that of without microcapsule is exhibited by microcapsules embedded with polyurethane composite coating. This is confirmed with reference to the ILs coming out of fragmented microcapsules lubricates the surfaces in direct contact between steel ball and coating, resulting in lesser coefficient of friction and a lesser wear width and depth [33].

Mitchell et al. studied the friction and wear behavior of poly (methyl methacrylate) (PMMA) comprising of with and without encapsulated organic friction modifier. Addition of particles to a dodecane model oil, either solid PMMA particles, PMMA particles with a methanol core, or friction modifier filled PMMA particles, decreased the coefficient of friction lower than that of pure dodecane, thus indicating a importance of increasing the friction modifier concentration in the oil by the process of encapsulation. The addition of 1.5 wt% particles tends to reduce the coefficient of friction and volumetric wear less than dodecane saturated with friction modifier [25].

Microcapsules containing shell friction modifier (both organic friction modifier and molybdenum friction modifier) core and poly (methyl methacrylate) (PMMA) were prepared by Xu [27] to decrease the friction and prolong the service time of engine lubricant. Intense reduction in friction coefficient is observed in lubricant combined with microcapsules against virgin friction modifier. Better compatibility of molybdenum friction modifier and organic friction modifier is seen during the encapsulation process.

An improved tribological performance is observed by integration of oil-loaded PSF microcapsules in epoxy composites. The 25 wt%, concentration of microcapsules reduced the frictional coefficient by 2.3 times and specific wear rate by 18.3 times against the neat epoxy. This was credited to the lubrication effect of the oil let-off from microcapsules and the entrapment of the wear debris in the craters left by the ruptured microcapsules [29].

Yang et al. [39] investigated the tribological properties of urea formaldehyde (UF) microcapsules loaded with MoS_2 in high-density polyethylene (HDPE) matrix composites. Experimental results revealed that, encapsulated MoS_2 performance was better than the composite without encapsulation. The shield benefited by surrounding walls resulted in low value of sliding friction and an even surface morphology with MoS_2 microcapsules. The UF walls of microcapsules defer the oxidation of MoS_2 crystals by isolating the steam formed by water and friction. The encapsulated MoS_2 tends to release between the friction pairs, when walls of microcapsules were fractured, thus reducing the wear due to adhesion and coefficient of friction.

Yang et al. [40] incorporated palmityl palmitate microcapsules coated with a shell of urea–formaldehyde resin with ultra-high molecular weight polyethylene (UHMWPE) composites. The COF of composites tends to decrease with addition in microcapsules content, as microcapsules were directly dispersed to the contact point of friction pairs, thus efficiently preventing the direct contact between surfaces.

Bandeira et al. [42], reported a PTFE-based self-lubricating coating containing polysulphone (PSF) microcapsules with ionic liquid (IL) lubricant. Incorporation of IL microcapsules led to 12% decrease in COF in comparison to baseline formulation and 70% decrease in wear rate under high loading and low sliding speed conditions. The studies predict presence of IL decrese the abrasiveness at contact points during interactions, assisting the theory of boundary lubrication film formation.

3.2 *Other Composites*

The tribological behavior of graphene encapsulated silicon carbide nanoparticles incorporated into aluminum matrix composite were investigated by Zhang et al. [32]. The coefficient of friction of the composite is seen to be 35.9% lesser than the reference sample which has silicon carbide nanoparticles incorporated. The better tribological features may be due to the uniform distribution of graphene nano-layers present in the bulk composite, thus accumulating on the wear scar and procedures graphene-rich micro thin tribofilms during the sliding wear.

Khun et al. [34] systematically investigated the tribological performance of WMCs incorporated with silicone composite coatings deposited on aluminum alloy substrates. The increased microcapsule content in the silicone composite coating greatly decreased the friction coefficient of the coated sample because the rubbing surface was lubricated by the wax lubricant produced by broken microcapsules due to wear. It has also been discovered that silicone composite coated samples, embedded with a higher concentration of microcapsules, were more likely to wear abrasively because the microcapsules protruded from the coating surfaces are easier to crack.

A new approach of encapsulating spherical shaped copper particles/elements by GO shells to improve the tribological properties of Cu alloys has been developed by Zhai et al. [38]. It is observed that the torque resistance (frictional torque) of the composite could be streamed or kept constant during fretting against the steel, with a load-insensitive behavior in comparison to the conventional copper materials. The improved tribological characterization is attributed to combined influences which incorporate strong metal/GO interfacial coupling, uniformly dispersed GO, and the GO-assisted tribo-film made on peaks and valleys of wear paths.

4 Conclusions

This chapter gives brief summary about the encapsulation technology, its tribological importance and research findings of authors across the globe related to tribological behavior of various polymer composites and metal matrix composites using encapsulation technology. Thus, it is noticed from the all the above discussions that, encapsulation technology has engrossed much consideration and has been extensively examined and studied in recent years as it can modify and increase the interfacial features between micro/nano-capsules and the matrix, resulting in better and enhanced tribological properties of the composites.

References

1. Voloshin Y, Belaya I, Krämer R (2016) The encapsulation phenomenon: synthesis, reactivity and applications of caged Ions and molecules. Springer
2. Lakkis JM (2007) Encapsulation and controlled release technologies in food systems. Wiley Online Library
3. Gaonkar AG, Vasisht N, Khare AR, Sobel R (2014) Microencapsulation in the food industry: a practical implementation guide. Elsevier
4. Benita S (2005) Microencapsulation: methods and industrial applications. CRC Press
5. Knowles A (2012) Chemistry and technology of agrochemical formulations. Springer Science & Business Media
6. Rosen M (2005) Delivery system handbook for personal care and cosmetic products: technology, applications and formulations. William Andrew
7. Nelson G (1991) Microencapsulates in textile coloration and finishing. Rev Prog Color Relat Top 21(1):72–85
8. Christensen FN, Bertelsen P (1997) Qualitative description of the Wurster-based fluid-bed coating process. Drug Dev Ind Pharm 23(5):451–463
9. Šiler-Marinković S, Bezbradica D, Škundrić P (2006) Microencapsulation in the textile industry. Chem Ind Chem Eng Q 12(1):58–62
10. Green BK, Lowell S (1956) Manifold record material. Google Patents
11. Xiao Z, Liu W, Zhu G, Zhou R, Niu Y (2014) A review of the preparation and application of flavour and essential oils microcapsules based on complex coacervation technology. J Sci Food Agric 94(8):1482–1494
12. Alagusundaram M, Chetty MS, Umashankari K, Badarinath AV, Lavanya C, Ramkanth S (2009) Microspheres as a novel drug delivery system—a review. Int J Chem Tech Res 1(3):526–534
13. Andersson C, Järnström L, Fogden A, Mira I, Voit W, Zywicki S, Bartkowiak A (2009) Preparation and incorporation of microcapsules in functional coatings for self-healing of packaging board. Packag Technol Sci Int J 22(5):275–291
14. Ghosh SK (2006) Functional coatings and microencapsulation: a general perspective. Funct Coat 1–28

15. Wang W, Xu L, Li X, Yang Y, An E (2014) Self-healing properties of protective coatings containing isophorone diisocyanate microcapsules on carbon steel surfaces. Corros Sci 80:528–535
16. Grigoriev D, Akcakayiran D, Schenderlein M, Shchukin D (2014) Protective organic coatings with anticorrosive and other feedback-active features: micro- and nanocontainers-based approach. Corrosion 70(5):446–463
17. Szalóki-Dorkó L (2017) Encapsulations nanotechnology in the agri-food industry, vol 2. Acta Aliment 46(1):123–126
18. Fu F, Hu L (2017) Temperature sensitive colour-changed composites. In: Advanced high strength natural fibre composites in construction. Elsevier, pp 405–423
19. Mang T, Bobzin K, Bartels T (2011) Industrial tribology: tribosystems, friction, wear and surface engineering, lubrication. Wiley
20. Bhushan B (2011) Nanotribology and nanomechanics I: measurement techniques and nanomechanics, vol 1. Springer Science & Business Media
21. Yamaguchi Y (1990) Tribology of plastic materials: their characteristics and applications to sliding components, vol 16. Elsevier
22. Hutchings I, Shipway P (2017) Tribology: friction and wear of engineering materials. Butterworth-Heinemann
23. Czichos H, Klaffke D, Santner E, Woydt M (1995) Advances in tribology: the materials point of view. Wear 190(2):155–161
24. Wen S, Huang P (2012) Principles of tribology. Wiley Online Library
25. Mitchell K, Neville A, Walker GM, Sutton MR, Cayre OJ (2018) Synthesis and tribological testing of poly (methyl methacrylate) particles containing encapsulated organic friction modifier. Tribol Int 124:124–133
26. Williamson T, Buren MFV, Massucco AA, Lindstrom RS (1993) Cast material with encapsulated lubricant. Google Patents
27. Xu S (2020) Preparation and tribological behaviour of poly (methyl methacrylate) (PMMA) microcapsules containing friction modifiers. J Microencapsul 37(4):314–323
28. Li H, Ma Y, Li Z, Ji J, Zhu Y, Wang H (2017) High temperature resistant polysulfone/silica double-wall microcapsules and their application in self-lubricating polypropylene. RSC Adv 7(79):50328–50335
29. Li H, Wang Q, Li M, Cui Y, Zhu Y, Wang B, Wang H (2016) Preparation of high thermal stability polysulfone microcapsules containing lubricant oil and its tribological properties of epoxy composites. J Microencapsul 33(3):286–291
30. Li H, Chen S, Li Z, Feng Y, Zhang M (2020) Preparation of PU/GO hybrid wall microcapsules and their self-lubricating properties for epoxy composites. Colloids Surf A Physicochem Eng Asp 124729
31. Khun NW, Zhang H, Yang J, Liu E (2013) Mechanical and tribological properties of epoxy matrix composites modified with microencapsulated mixture of wax lubricant and multi-walled carbon nanotubes. Friction 1(4):341–349
32. Zhang J, Yang S, Chen Z, Wu H, Zhao J, Jiang Z (2019) Graphene encapsulated SiC nanoparticles as tribology-favoured nanofillers in aluminium composite. Compos B Eng 162:445–453
33. Yang M, Zhu X, Ren G, Men X, Guo F, Li P, Zhang Z (2015) Tribological behaviors of polyurethane composite coatings filled with ionic liquid core/silica gel shell microcapsules. Tribol Lett 58(1):9
34. Khun NW, Zhang H, Yang J, Liu E (2012) Tribological performance of silicone composite coatings filled with wax-containing microcapsules. Wear 296(1–2):575–582
35. Guo QB, Lau KT, Rong MZ, Zhang MQ (2010) Optimization of tribological and mechanical properties of epoxy through hybrid filling. Wear 269(1–2):13–20
36. Mu B, Li X, Yang B, Cui J, Wang X, Guo J, Bao X, Chen L (2020) Microstructure and tribological properties of epoxy composites modified by methyl silicone oil-loaded microcapsules and/or reinforced with potassium hexatitanate whiskers. High Perform Polym 32(3):334–343

37. Mu B, Li X, Yang B, Cui J, Wang X, Guo J, Bao X, Chen L (2017) Tribological behaviors of polyurethane composites containing self-lubricating microcapsules and reinforced by short carbon fibers. J Appl Polym Sci 134(43):45331
38. Zhai W, Lu W, Chen Y, Liu X, Zhou L, Lin D (2019) Gas-atomized copper-based particles encapsulated in graphene oxide for high wear-resistant composites. Compos B Eng 157:131–139
39. Yang Z, Guo Z, Yuan C (2019) Effects of MoS2 microencapsulation on the tribological properties of a composite material in a water-lubricated condition. Wear 432:102919
40. Yang Z, Guo Z, Yang Z, Wang C, Yuan C Study on tribological properties of a novel composite by filling microcapsules into UHMWPE matrix for water lubrication. Tribol Int 153:106629
41. Khun N, Zhang H, Yue C, Yang J (2014) Self-lubricating and wear resistant epoxy composites incorporated with microencapsulated wax. J Appl Mech 81(7)
42. Bandeira P, Monteiro J, Baptista AM, Magalhães FD (2015) Tribological performance of PTFE-based coating modified with microencapsulated [HMIM][NTf 2] ionic liquid. Tribol Lett 59(1):13
43. Imani A, Zhang H, Zhao J, Yang J, Owais M, Zhang Z (2017) Significant improvements in wear and friction of epoxy composites filled with graphene nanoplatelets and wax-containing microcapsules. In: ICCM international conferences on composite materials
44. Liu P, Huang T, Lu R, Li T (2012) Tribological properties of modified carbon fabric/polytetrafluoroethylene composites. Wear 289:17–25
45. Sharma M, Bijwe J, Mitschang P (2011) Abrasive wear studies on composites of PEEK and PES with modified surface of carbon fabric. Tribol Int 44(2):81–91
46. Ye XJ, Song YX, Zhu Y, Yang GC, Rong MZ, Zhang MQ (2014) Self-healing epoxy with ultrafast and heat-resistant healing system processable at elevated temperature. Compos Sci Technol 104:40–46
47. Ren G, Zhang Z, Zhu X, Men X, Liu W (2014) Influence of lubricant filling on the dry sliding wear behaviors of hybrid PTFE/Nomex fabric composite. J Mater Sci 49(10):3716–3724
48. Ren G, Zhang Z, Zhu X, Yang M, Men X, Jiang W, Liu W (2015) WS 2-filled hybrid PTFE/Nomex fabric composites with improved antiwear property. J Mater Sci 50(3):1065–1070
49. Liu B, Pei X, Wang Q, Sun X, Wang T (2011) Effects of proton and electron irradiation on the structural and tribological properties of MoS2/polyimide. Appl Surf Sci 258(3):1097–1102
50. Shivamurthy B, Bhat KU, Anandhan S (2013) Mechanical and sliding wear properties of multi-layered laminates from glass fabric/graphite/epoxy composites. Mater Des 44:136–143
51. Dubey MK, Bijwe J, Ramakumar S (2013) PTFE based nano-lubricants. Wear 306(1–2):80–88
52. Suresha B, Ramesh B, Subbaya K, Kumar BR, Chandramohan G (2010) Influence of graphite filler on two-body abrasive wear behaviour of carbon fabric reinforced epoxy composites. Mater Des 31(4):1833–1841
53. Song H-J, Zhang Z-Z, Men X-H (2008) The tribological behaviors of the polyurethane coating filled with nano-SiO2 under different lubrication conditions. Compos A Appl Sci Manuf 39(2):188–194
54. Ren G, Zhang Z, Zhu X, Men X, Jiang W, Liu W (2014) Tribological behaviors of hybrid PTFE/nomex fabric/phenolic composite under dry and water-bathed sliding conditions. Tribol Trans 57(6):1116–1121
55. Martin J-M, Grossiord C, Le Mogne T, Igarashi J (2000) Transfer films and friction under boundary lubrication. Wear 245(1–2):107–115
56. Saurín N, Sanes J, Bermúdez M (2014) Effect of graphene and ionic liquid additives on the tribological performance of epoxy resin. Tribol Lett 56(1):133–142
57. Sanes J, Carrión F, Bermúdez M (2010) Effect of the addition of room temperature ionic liquid and ZnO nanoparticles on the wear and scratch resistance of epoxy resin. Wear 268(11–12):1295–1302
58. Mąka H, Spychaj T, Kowalczyk K (2014) Imidazolium and deep eutectic ionic liquids as epoxy resin crosslinkers and graphite nanoplatelets dispersants. J Appl Polym Sci 131(12)

59. Guo QB, Lau KT, Zheng BF, Rong MZ, Zhang MQ (2009) Imparting ultra-low friction and wear rate to epoxy by the incorporation of microencapsulated lubricant. Macromol Mater Eng 294(1):20–24
60. Zhang G, Xie G, Si L, Wen S, Guo D (2017) Ultralow friction self-lubricating nanocomposites with mesoporous metal-organic frameworks as smart nanocontainers for lubricants. ACS Appl Mater Interfaces 9(43):38146–38152
61. Imani A, Zhang H, Owais M, Zhao J, Chu P, Yang J, Zhang Z (2018) Wear and friction of epoxy based nanocomposites with silica nanoparticles and wax-containing microcapsules. Compos A Appl Sci Manuf 107:607–615
62. Aruna S, Arunima S, Latha S, William Grips V (2016) Preparation of oil-encapsulated microcapsules and tribological property of Ni composite coating. Mater Manuf Processes 31(1):107–111
63. Win Khun N, Zhang H, Tang X-z, Yoon Yue C, Yang J (2014) Short carbon fiber-reinforced epoxy tribomaterials self-lubricated by wax containing microcapsules. J Appl Mech 81(12)
64. Khun N, Zhang H, Sun D, Yang J (2016) Tribological behaviors of binary and ternary epoxy composites functionalized with different microcapsules and reinforced by short carbon fibers. Wear 350:89–98
65. Khun NW, Sun D, Huang MX, Yang JL, Yue CY (2014) Wear resistant epoxy composites with diisocyanate-based self-healing functionality. Wear 313(1–2):19–28
66. Sun D, Chong YB, Chen K, Yang J (2018) Chemically and thermally stable isocyanate microcapsules having good self-healing and self-lubricating performances. Chem Eng J 346:289–297
67. Li H, Wang Q, Wang H, Cui Y, Zhu Y, Wang B (2016) Fabrication of thermally stable polysulfone microcapsules containing [EMIm][NTf2] ionic liquid for enhancement of in situ self-lubrication effect of epoxy. Macromol Mater Eng 301(12):1473–1481
68. Minami I (2009) Ionic liquids in tribology. Molecules 14(6):2286–2305
69. Ye C, Liu W, Chen Y, Yu L (2001) Room-temperature ionic liquids: a novel versatile lubricant. Chem Commun 21:2244–2245
70. Liu W, Ye C, Gong Q, Wang H, Wang P (2002) Tribological performance of room-temperature ionic liquids as lubricant. Tribol Lett 13(2):81–85
71. Bermúdez M-D, Jiménez A-E, Sanes J, Carrión F-J (2009) Ionic liquids as advanced lubricant fluids. Molecules 14(8):2888–2908
72. Zhou F, Liang Y, Liu W (2009) Ionic liquid lubricants: designed chemistry for engineering applications. Chem Soc Rev 38(9):2590–2599
73. Chen DX, OuYang XK, Wang YG, Yang LY, Yu D (2012) Polysulfone microcapsules containing ionic liquid. Advanced Materials Research. Trans Tech Publ, pp 273–276
74. Gao H, Xing J, Xiong X, Li Y, Li W, Liu Q, Wu Y, Liu H (2008) Immobilization of ionic liquid [BMIM][PF6] by spraying suspension dispersion method. Ind Eng Chem Res 47(13):4414–4417
75. Weiss E, Dutta B, Kirschning A, Abu-Reziq R (2014) BMIm-PF6@ SiO2 microcapsules: particulated ionic liquid as a new material for the heterogenization of catalysts. Chem Mater 26(16):4781–4787
76. Yang W, Lu Y, Xiang Z, Luo G (2007) Monodispersed microcapsules enclosing ionic liquid of 1-butyl-3-methylimidazolium hexafluorophosphate. React Funct Polym 67(1):81–86
77. Armada S, Schmid R, Equey S, Fagoaga I, Espallargas N (2013) Liquid-solid self-lubricated coatings. J Therm Spray Technol 22(1):10–17
78. Espallargas N, Vitoux L, Armada S (2013) The wear and lubrication performance of liquid–solid self-lubricated coatings. Surf Coat Technol 235:342–353
79. Li H, Shi N, Ji J, Wang H (2018) Preparation of microcapsules containing double-component lubricant and self-lubricating performance of polymer composites. Mater Res Express 5(5):055302
80. Li H, Ma Y, Cui Y, Li Z, Wang H (2019) Ultralow tribological properties of polymer composites containing [BMIm] PF6-loaded multilayer wall microcapsule. Macromol Mater Eng 304(4):1800791

81. Win Khun N, Zhang H, Yang J (2015) Wear resistance of polymers with encapsulated epoxy-amine self-healing chemistry. J Appl Mech 82(5)
82. Yang Z, Guo Z, Yuan C (2019) Tribological behavior of polymer composites functionalized with various microcapsule core materials. Wear 426:853–861
83. Li H, Cui Y, Wang H, Zhu Y, Wang B (2017) Preparation and application of polysulfone microcapsules containing tung oil in self-healing and self-lubricating epoxy coating. Colloids Surf A 518:181–187

Hydrogen Encapsulation and Storage as an Alternative Energy Source

Masoud Darvish Ganji and Atyeh Rahmanzadeh

1 Introduction

In general, the sustainable development of many countries is mainly founded on energy pathways. As stated in the report of Brundtland or WCED[1] commission in 1987, sustainable development is described as creating a logical balance in preserving the amount of resources utilized by humans for current demands while considering the needs of the next generations [1–3]. Accordingly, it is expected that energy-based development provides the five principles, namely, integration of protection and advancement; meeting the elemental biological needs of people; social justice; autonomy and cultural diversity, and maintenance in ecological integrity [4, 5]. However, serious dependence on nonrenewable energy resources has created pervasive problems in the course of global advancement. For instance, oil-led development of great powers[2] has two major global effects: first, continuous extraction and production of oil and its derivatives have caused a severe concerning about the restriction of nonrenewable resources, in a way that it is estimated by the end of the present century ~46 TW of power will be utilized all over the world as a regular demand of humans. However, people of many developing countries despite having abundant sources of

[1] World Commission on Environment and Development Definition.

[2] Great power is a country that is known as possessing the expertise and capability to exert its supreme influence on a worldwide scale.

This chapter reviews on containers for hydrogen encapsulation and storage as an alternative energy source.

M. D. Ganji (✉)
Department of Chemistry, Qaemshahr Branch, Islamic Azad University, Qaemshahr, Iran

A. Rahmanzadeh
Department of Nanochemistry, Faculty of Pharmaceutical Chemistry, Tehran Medical Sciences, Islamic Azad University, Tehran, Iran

J. Parameswaranpillai et al. (eds.), *Micro- and Nano-containers for Smart Applications*, Composites Science and Technology,
https://doi.org/10.1007/978-981-16-8146-2_12

non-renewable energy such as oil, gas, etc., are deprived of their concessions owing to the poor management, lack of modern extraction/production equipment, lack of sufficient relevant and up-to-date knowledge, lack of human capital and so forth. Hence, this destitution has caused the economic and political lords to use these sources to their advantage, whilst the main owners, namely, the people of the Middle East, Africa, etc., are in poverty and deprived of energy [6–9]. The irreparable detrimental effects of nonrenewable sources on the eco-environment as well as the health of human and living species is the second influence of global utilization of these sources of energy. Furthermore, the exploration, production processes, and transportation of petroleum and its derivatives are usually accompanied by serious oil spillage in various environments such as thalassic regions which leads to devastating effects on environments and heavy economic losses to the equipment of mechanical operations [10–12].

In spite of the fact that the commencement of utilizing renewable energy date back to 200 BC in Europe by inventing the brutal form of energy means waterwheel, which is considered as the infrastructure of the renewable sources of energy, this cutting edge science has continued until today and very recently it has cast a brilliant shadow on the critical challenges of nonrenewable resources [13, 14]. Renewable energy is defined as a sort of energy that is piled up from renewable resources and is naturally replenished within the human life course, including waves, sunlight, geothermal heat, wind, and so on [15–17]. The most underlying renewable sources of energy are summarized in Fig. 1. Indisputably, the enormous potential of renewable energy sources is capable of exponentially increasing the requisition of the planet's energy, hence this category of energy possesses a pivoting role in the current and future of energy issues throughout the world. Moreover, in addition to the wide spectrum of applications, the development of the technology of renewable sources has provided a magnificent opportunity for diminishing the emission of greenhouse

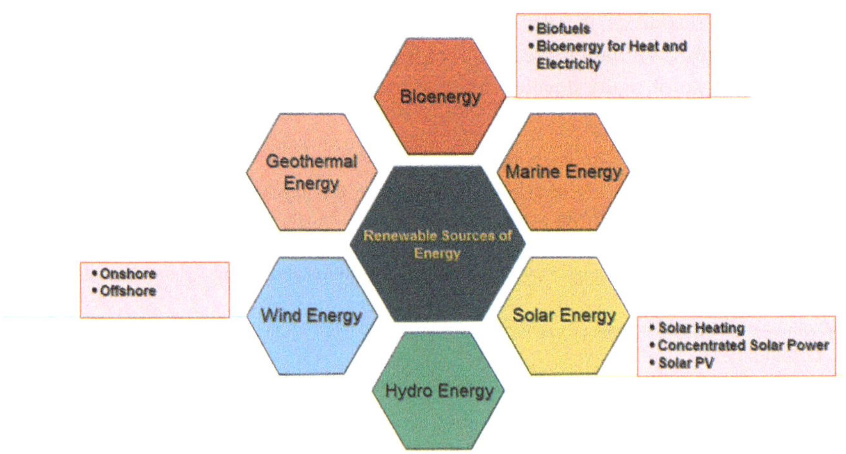

Fig. 1 Main renewable sources of energy

gases, consequently minimizing the extreme climatic and weather impacts, and certifying profitable, trustworthy, and seasonable energy for the people all over the world [18–20].

Since in current global community, sustainable advancement of integrated renewable systems of energy is urgent for the future of mankind and according to the scientific consensus, the present climate system has no more potential for adsorption of CO_2, hence the future generation of energy must be founded on fundamental infinite resources which render three main benefits [21, 22]:

1. The consumption of hydrocarbon-based reservoirs will be drastically reduced
2. Almost zero emissions of various GHG[3] and CO_2 is estimated
3. The motto "clean and cheap energy for all the inhabitants of the planet" could be realized.

In order to produce, store and harvest a great quantity of energy according to increasing demands, the intermittent renewable resources, including hydrogen must be utilized as a versatile reservoir of energy. Hydrogen has the capability of producing infinite grid-scale power by representing a possibility for transforming buffer the grid and also transport the harvested power to target demands. The cutting-edge technology of hydrogen vehicles is well expandable to an extensive spectrum of vehicles and applications with approximately zero emissions [23]. Hydrogen can be widely utilized as the carriers of energy, for instance, fuel cells conjunction for electrifying the transportation functions when the recharge period, range, and power necessities are higher than the technologies of grid-fed batteries. Furthermore, although hydrogen can play an effective role in the power supply of massive industries, such as aviation and automotive, computer, etc., it can also meet the heating necessities and stationary domestic powers supplies [24–26]. Hydrogen is the most well-known ubiquitous element in the whole universe, in a way that the huge amount of visible mass of our universe, as well as, the Sun is comprised of hydrogen (nearly 75%). According to studies, only an insignificant amount, $5 \times$ of 10^{-5} mol fraction of pure hydrogen presents in the atmosphere of our planet. This paucity originates from this axiom that molecular H_2 has an adequate velocity (1.77 km/s) at ambient temperature and a considerable fraction of distribution attributed to the molecular velocity is more considerable than the velocity of Earth's escape which is around 11 km/s. Therefore, the fate of any molecular hydrogen in the planet's ambient is leaking into space after a prolonged time and the natural incidence of molecular H_2 is considerably diminished due to the superb reactivity of hydrogen which binds up in the atomic form in methane, water, numerous organic compounds and so on. Accordingly, howbeit H_2 is known for its intrinsic property, flammability, but it is important to note that it cannot exist in adequate affluence above or within the Earth to be described as a fuel and it cannot be readily burned or accumulated to supply power [27–30]. The most important properties of hydrogen are summarized in Fig. 2. The integrated of all these favorable properties has made the hydrogen as one of the most strategic substances. Hydrogen has no color, taste, and odor, and similar to methane has no

[3] Green House Gas.

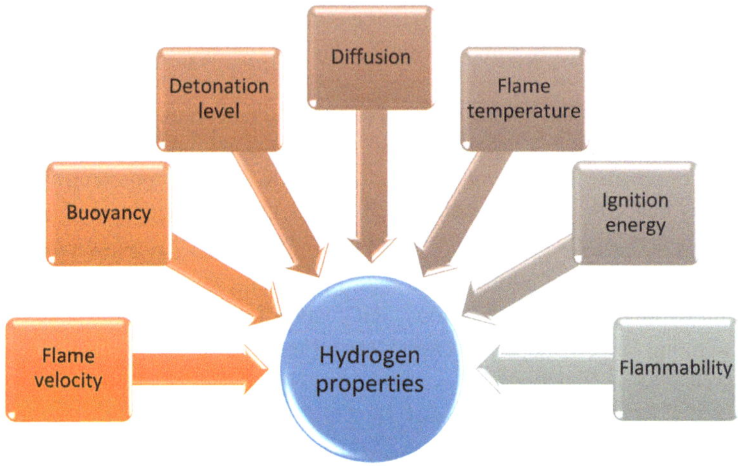

Fig. 2 The most important properties of hydrogen

toxicity. The ignited flame from hydrogen-air is cooler than gasoline and hotter than methane-air at the stoichiometric circumstances. In this situation, the temperature of the air is around 2207 °C in comparison with 2307 °C for gasoline and 1917 °C for methane [31, 32]. Hydrogen has an acceptable flame velocity compared to other flames. The flame velocities of hydrogen, gasoline vapor and methane are 1.85, 0.42, and 0.38 m/s, respectively. Hydrogen flammability is directly associated with the level of concentration and is obviously higher than other fuels such as methane [33]. Hydrogen is able to burn within a minor-visibility flame. The flammability restricts the hydrogen mixture with various oxidizers including oxygen, air, etc., which possess strong dependency on temperature, available diluents, the energy of ignition, pressure, and also the characteristics of equipment like size, facility, and configuration. The flammability restriction of hydrogen at ambient circumstances is 5–75% [34, 35]. Furthermore, the flammability limit of gasoline in air and gasoline in air are 4.5–15%, and 1.5–7.5%, respectively. Hydrogen could be ignited whenever its concentration reaches the range of flammability through utilizing a small quantity of energy by virtue of the low energy of ignition, which is about 0.02 mJ for hydrogen and 0.24 and 0.28 mJ for gasoline and methane, respectively [36–38]. Further, hydrogen has explosive property over a vast range of concentrations in a confined condition. Although alike to other prevalent fuels, it is almost impossible to be exploded if not in a limited space [39]. Regarding the densities of methane, propane and gasoline vapor, which are about 1.33, 4.22, and 5.83 kg/m^3 (at standard circumstances), hydrogen could rise faster and has a significant buoyancy [40]. In addition, compared to other gas-phase fuels, diffusion of hydrogen through the air is much faster and this rapid rate of dispersion indicates the safety of hydrogen as a desirable fuel [41, 42]. Hydrogen is mostly established as a versatile energy carrier or in a better word, unparalleled energy storage material and in order to use the H_2

molecules for specific functions, it must be extracted from diverse chemical materials and compounds which demand so much energy and efforts [43, 44].

To date, hydrogen storage technology has been developed based on two main techniques:

1. Physical methods
2. Chemical methods.

The Next sections will be earmarked to a general overview of the developed techniques for hydrogen storage, and then, the pivoting role of various nanostructured materials in this field will be thoroughly discussed.

2 Hydrogen Storage Techniques

2.1 Physical Methods

Three principal sub-techniques for physical methods of hydrogen storage have been advanced:

- Compressed Gaseous Hydrogen (CGH2), ambient temperature, 35–70 MPa
- Liquid Hydrogen (LH2), −253 °C, 0.1–1 MPa
- Cryo-adsorption on the materials with high surface area, −193 °C, 0.5 MPa
- Hybrid materials consisted of ionic liquid (IL) and Metal–organic framework (MOF).

In physical methods, no potent chemical bonding could be formed (ionic or covalent interactions) amongst the hydrogen and host substances. According to the most recent investigations, physical of methods of hydrogen storage particularly CGH2 and LH2 are classified as the most developed techniques, in such a way these two physical-based methods are extensively used in preparation of prototypes for the automotive fuel-cell. It is worth to state that 70 MPa of CGH2 is the most-modern technology of hydrogen storage [45–49].

2.1.1 CGH2

Considering the hydrogen density at ambient circumstances is beyond any technical importance as an effective carrier of energy. The value of lower heating or value of net calorific is around 120 MJ per kg of molecular hydrogen. In order to obtain vehicles supporting a capacity around of 500 km, it is essential to preserve nearly 5–6 kg of H_2 onboard the automotive. Since the existed space is restricted, operating at a 35–70 MPa pressure level is inevitable. A lower level of pressure is useless owing to the inadequate content of energy. Further, the deviations from the behavior of ideal gas are enormous at the pressures higher than 70 MPa. Hence, the augmented needs for pressure containers could not be fulfilled through a limited rise in the content of energy [50, 51] In addition to the chemical content of energy for hydrogen in comparison, the content of mechanical energy for the system of CGH2 has drawn many attentions since this energy is attributed to the amount of energy which is necessary for compressing from ambient circumstances to ~35 or ~70 MPa, by considering the value of lower heating. The complete refilling of an empty system of CGH2 is feasible within less than three minutes. So, the technology of CGH2 is founded as the brilliant benchmark of all other alternative and conventional systems of hydrogen storage. This strategy is the cutting-edge technology for fuel-cell automotive for almost all car industries [52–54].

2.1.2 LH2

Until the recent decade, many vehicle manufacturers evaluated LH2 as an achievable on-board option of storage. The great mass density of hydrogen at 0.1 MPa and 2538 °C compared to systems of CGH2 with 70 MPa of pressure. Further, the flow of heat from ambient into the vessel of the tank results in hydrogen evaporation. As a conclusion, the interior pressure of the vessels augments, and after a specified uttermost pressure is gained in order of about 1 MPa, the venting of hydrogen has to be carried out. Dormancy is the time period from putting an automotive into parking or idle mode and try to vent the hydrogen. The period of dormancy takes several days to be accomplished in an ideal case [55, 56]. Following the gaining at the boil-off gas point, the molecules of hydrogen will be lost to the ambient. The amount of hydrogen losses is substantial and must be repressed as much as is viable. Woefully, due to the very low enthalpy of phase change which is estimated around 0.45 MJ per kg of molecular hydrogen among the gaseous and liquid state, even the concepts of developed engineering, have not been capable of addressing the issues of diminishing the input of heat into the system adequately [57–59]. The process of heat input into the considered system takes places through three main approaches:

1. Convention
2. Thermal radiation
3. Thermal conduction by cables, pipes and mountings.

Among the aforementioned approaches, thermal conduction and radiation have the most dominant effect. A very low ratio of surface-to-volume (S/V) as a desirable criterial is required, the overall low values of heat flows could only be obtained by using the tank in a cylindrical structure. Although the utilizable density of volumetric storage is a bit greater in comparison with CGH2 systems, most of the automotive industries have apperceived that there are no potent benefits in tanks of liquid hydrogen that could gratify the aforementioned drawbacks. Besides, the required energy for liquefying hydrogen use almost 30% of the stored chemical energy. Hence, approximately all of the largest automotive companies all over the world, aim to concentrate on 70 MPa as an efficient target in hydrogen storage strategies [60, 61] (Fig. 3).

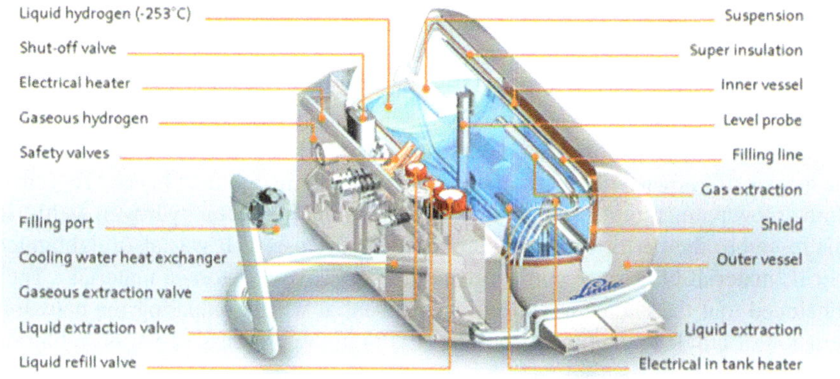

Fig. 3 The vessel of liquid hydrogen (Reproduced by permission of Journal of Power Sources)

2.1.3 Cryo-adsorption

In general, adsorption is considered as borderline condition between chemical and physical storage. Most of the conventional adsorbents interact weakly with hydrogen and as hydrogen adsorbed molecularly and subsequently not dissociated, it is typically classified as the technology of physical storage. At ambient pressure and room temperature, the target values of the storage capacity of the prevalent adsorbents are not sufficient and convincing. Thus, the efficient solutions of adsorptive storage could be pressure and cryogenic systems [62, 63].

2.1.4 Hybrid Materials

Ionic Liquids (ILs)

The peculiar characteristics of ionic liquids (ILs) play a key role in the development of many aspects of science owing to their environmentally friendly behaviors. ILs illustrate a plethora of distinguished physical and chemical characteristics, including negligible vapor pressure, low melting point, extensive electrochemical window, thermal stability, and considerable solvation. Moreover, through modification of anion and cations, ILs could be adjusted to obtain an essential feature, namely miscibility with high polarity solvents for instance ethanol and water, or low polarity organic solvents such as toluene, hexanes, and supercritical CO_2. A broad range of fundamental applications for example biotechnology and synthesis have been reported for ILs as favorable "greener" solvents. The most important applications of ILs are as follow, special solvents for homogeneous and heterogeneous for catalysis, selective solvents for the elimination of heavy metal pollutant, and dispersive agents for stabilization of nanostructured materials and electrolytes in different electrochemical procedure and instruments. But, very recently many researchers have scrutinized in hydrogen storage ability of these materials. Doroodian et al. investigated methylguanidinium borohydride as an ionic liquid (m.p. $-$ 5 °C, density ς (20 °C) $= 0.95\,\text{gmL}^{-1}$) and found out that this material has desirable potential for hydrogen storage capacity [64]. Sahler et al. studied about Ionic liquid base H_2 storage for H_2 generation from NH_3BH_3 derivatives as one of the most well-known hydrogen storage agents [65]. Ionic liquids enhanced H_2 generation at low temperature, with good reaction rate and high total hydrogen yields. In this research, the performance of carbenes was discussed. It was also highlighted that IL material blends are competitive with pervasive H_2 storage materials. They concluded that the choice of the cation is essential when comparing the dehydrogenation of ethylene diamine bisborane (EDB) in the presence of various chloride ILs.

Metal–Organic Framework (MOFs)

The preferable crystalline and porous structures of metal–organic frameworks (MOFs) are known as an outstanding template for hydrogen storage. Several MOF-derived nanostructured materials have marvelous mechanical and chemical stabilities, well-established binding sites for hydrogen, vast specific surface areas, diverse functionalities, and adjustable porous structures [66]. MOFs have almost simple to synthesize with cost-effective raw materials and as well as uncomplicated methods. Several approaches to enhancing the hydrogen binding energy to a MOF have been presented [67]. Small-scaled pore size enhances the binding of hydrogen and permits a single molecule of H_2 to interact with myriad aromatic rings. Chen et al. observed that the open metal sites have a desirable influence on the capacity of hydrogen sorption related to the MOF-505, additional assessment was demanded to confirm it [68]. Another possible definition for the high capacity of hydrogen uptake of MOF-505 is a facile correlation among specific surface area (As) and adsorption. Ganji et al. investigated the capacity of hydrogen storage related to the cluster of Li-decorated titanium metallocarbohedryne (Ti met-car) utilizing DFT calculations [69]. It is divulged that Li atom tends to be adsorbed above the site of trigonal hollow among C and Ti atoms containing average energy of binding of around -1.75 eV. They revealed that the milieu generated via 12 Li atoms coated on the cluster could store up to 36 molecules of hydrogen leading to the gravimetric density of around 10.55 wt%. The binding energy of about -0.37 eV/H_2 for this type of system permits H_2 recycling at the room circumstances. Additionally, the adsorption performance of 36 H_2 molecules about the 12Li-decorated Ti met-car system has been studied via the ab initio MD simulation at ambient temperature. Their outcome indicated that the molecules of hydrogen eloped from the nanocage, which shows that the considered system simplifies the adsorption of hydrogen at room circumstances for various applications.

2.2 Chemical Methods

Chemical methods of hydrogen storage are based on various methods and materials which are exhibited in Fig. 4.

Comparative studies of various chemical and physical methods of hydrogen storage indicate that until now, the performance of none of the developed technologies has gained a level of satisfactory performances. Currently, the prototypes of vehicles equipped with fuel-cell power are based-on storage reservoirs with high-pressure. However, the required weight and space have restricted the advancement of the well-studied prototypes. Comprehensive investigations on the mobile applications of a PEM (proton exchange membrane) fuel cells based on diverse methods of hydrogen storage have been carried. These types of systems consume the waste released heat of the fuel cells to heat up or decompose the systems of storage simultaneously by using storage materials [70, 71]. For almost all systems that the release and storage are not virtually thermoneutral, the management of heat is a pivotal issue. When short times of refueling is required, the capacity of the heat exchanger must be around hundreds of kW, in the situation that the hydrogenation enthalpy is about

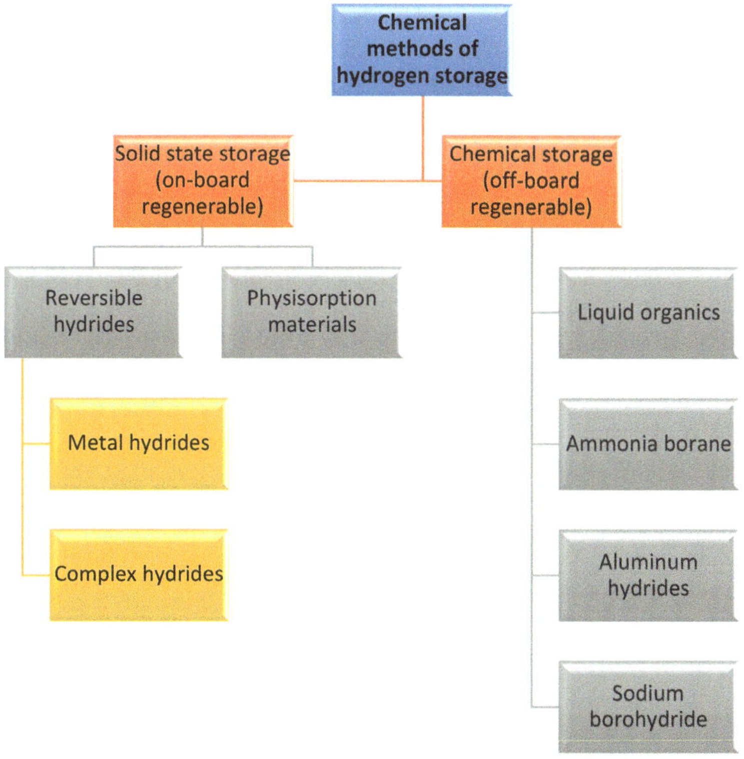

Fig. 4 Chemical methods for hydrogen storage

40 kJ per mol for molecular hydrogen. This proceeding creates crucial engineering difficulties and restricts the utilization of various options of technology. To store in large volumes, hydrogen in the liquid phase could be the most brilliant option and under this circumstance, the losses of evaporation are not considered as a drastic challenge [72–74]. For the storage applications on the smaller-scale, for instance in cars, storage by taking advantage of pressure has been found to be the most promising solution, although it could alter with possible development in other fields of technology. To attain the target performance for application in transportation reforming solutions have no sufficient capability, however, they could play an important role in the hydrogen supply of leisure sectors, APU[4]s, etc. Hydrogen storage by using an absorbent material on the ultrahigh-surface area has gained a level that comes almost close to the expected limit from physical methods, so much more advancement cannot be obtained within this field [75, 76].

It is hard to predict the developments of hydrogen storage by using hydrides materials and chemically, the greater capacities of storage could be attained. Currently, the thermodynamic and kinetic behavior of these strategies has not been divulged thoroughly. Solid-state materials have promising potential for outperforming the storage based-on physical methods, including high-pressure and cryo-storage technologies. Although, on the scale of short-time, these recent technologies appear to own the greatest large-scale practical applicability. Despite the fact that numerous experimental investigations have been allocated into the efficient storage of hydrogen by using different materials and methods, theoretical studies have created excellent developments over the last decades and could advance into a marvelous approach to conduct the researches for novel materials and novel options of hydrogen storage. It is essential to point out that the whole components of an energy system have to be appraised, and not one special fragment of the chain of technology individually. This technique implies that certain disadvantages related to the technologies of physical storage for vehicle applications no more appear to be undesirable. As an example, for liquid-phase hydrogen, the consumption of energy for liquefying was mentioned as one of the major issues. Further, it should be noted that although an infrastructure chain has to be provided via hydrogen and it is founded on liquid-phase hydrogen, there exist no excessive steps for liquefying to fuel a car, and there exists no excessive penalty of energy. Moreover, if the infrastructure of a system has created based on liquid hydrogen, a great quantity of efforts must be dedicated to technologies in order to regain the utilized energy in the liquefying step. This is not probably viable in a vehicle, but consideration of filling the fuel station makes this possible and the balance in the entire system could then modify for greater efficiency. The aforementioned straightforward consideration illuminates that an efficacious prediction of the leading energy system in future is though and directly depend on development in various sectors of technology [77–80].

[4] Accelerated Processing Unit.

3 The Role of Nanostructured Materials in Hydrogen Storage Applications

3.1 *Introduction*

Over the last few decades, substantial efforts have been dedicated to the development of hydrogen storage systems. Dating back to 1997, when new versatile nanostructured material, namely, carbon nanotube (CNT) was emerged to overcome the hydrogen storage challenges and advance this field of science, a great number of experimental and theoretical researches have been assigned to scrutinize the luminous role of CNTs in hydrogen storage systems, and subsequently, an extensive range of nanostructures was devised and synthesized for this target. In this section, different types of nanostructures and their behavior in hydrogen energy systems will be discussed. The interaction nature among host nanomaterials and hydrogen molecules, as divulged in theoretical simulation and modeling, lead us to apperceive the mechanisms of hydrogen storage, fundamentally [81–84].

3.2 *Nanostructures Materials for Hydrogen Storage Systems*

3.2.1 Carbon Nanotube (CNT)

From the time that Iijima outlined the synthesis of CNTs in 1991, this nanostructure has been considered as an excellent alternative for the storage of hydrogen and 6 years later, Dillon et al. announced and published the initial experimental evidence for the capability of hydrogen storage application of CNTs. The initial research by this research group was the prelude to conducting more and more experimental researches in this field and remarkable developments were made. In the primary investigation, it was exhibited that CNTs are able to store substantial amounts of hydrogen, event at the temperature of the room [85, 86]. Chen et al. represented that alkali-doped CNTs have great ability of hydrogen adsorption. They studied lithium- and potassium-doped CNTs and realized adsorption of hydrogen of about 14–20 wt% between ambient temperature and 400 °C [87]. It was explained by Yang that the notable capacity of hydrogen adsorption is mostly related to the weight and moisture attained via reactions the species of alkali within the alkali-metal-doped nanotubes at last and thus the allotment of the storage of pure hydrogen was restricted [88]. Recent experimental investigations, including the research of Kajiura et al., have illuminated that the performance of hydrogen storage of multi-walled nanotubes (MWNTs), single-walled nanotubes (SWNTs), and nanofibers (CNFs) at room temperature could reach ~8 MPa, is not able to surpass 0.43 wt% [89]. Similarly, Ritschel et al. surveyed the capacity of hydrogen storage in various carbon materials such as MWNTs, SWNTs, as well as CNFs. The pure SWNTs displayed capacity of reversible storage of around

0.63 wt% at the temperature of room and pressure of 45 bar and the achieved efficiency was greater than CNFs and MWNTs [90]. It should be notice that multiple of outcomes had been appeared to be controversial and they were not reproduced or confirmed via other scientists, which may attribute to the type of experiments require highly complicated circumstances and assessment processes and any side effects of the procedures and can simply influence on the total efficiency. In this respect, the great quantity of hydrogen desorption primary found in various experimental investigations may be ascribed to the side effects. Regarding the abovementioned cases, the DOE[5] commercialization targets have commenced a serious competition to discover the ultimate materials for hydrogen storage, which has been concluded quickly. This is promising to state that despite the controversial outcomes and rough situation of experiments, novel literature has reported tremendous efficiency of the hydrogen storage systems based on CNTs (purified and doped) [91, 92]. According to achieved data, CNTs are able to reserve hydrogen but merely in cryogenic circumstances which makes them improper for application in mobile devices. This because the interaction among CNTs and H_2 is slight (~1 kcal/mol). Theoretical and experimental researches have indicated that the energies of interaction among robust walls of CNTs and adsorbed molecules are not sufficient for reserving huge amount of hydrogen at low pressures and ambient temperature. Optimal energies of interaction must be between chemisorption (e.g. metal Hydrates) and physisorption (e.g. CNTs). Elemental thermodynamic assessment illustrates that the optimum energy of interaction for considerable reversible storage in the ambient conditions is about 7 kcal/mol. For this situation, adequate hydrogen could be reserved in t moderate pressures and at room temperature. Accordingly, the procedure of adsorption has a negligible amount of barrier of energy which enables overcoming with insignificant heating. Two major approaches for enhancing the energy of interactions are suggested based on the doping/incorporating heteroatoms and light metallic atoms and modifying the charges of host substances. Therefore, the energy of binding related to the hydrogen molecules will be enhanced owing to the charged-induced interactions. Furthermore, plenty of theoretical researches have been carried out to investigate the hydrogen storage capacity of carbon nanotubes in various conditions, for instance pure, doped, in presence of other substances, and so forth [93, 94]. Ganji et al. surveyed the capacity of hydrogen storage of S, P, and Si-doped CNTs and C_{60} fullerenes by using Density functional theory computations [95]. The energies of binding were computed and the hydrogen binding behavior of various dopants was thoroughly compared. The outcomes revealed that all of the dopant atoms can create bindings with high stability within the studied nanotube and nanocage. It was divulged that up to 8 molecular hydrogens could be adsorbed on the S-doped nanotube and create stable complexes, whilst up to 6 molecular hydrogens could be attached to the Si-doped fullerene (C_{60}). Besides, it was concluded that molecular hydrogens tend to be adsorbed onto the outer surface of the doped nanostructures compared to the inward sidewall. Anikina et al. investigated Li-doped CNTs and their performance as a desirable milieu for the storage of hydrogen by utilizing computational

[5] Department of Energy (US).

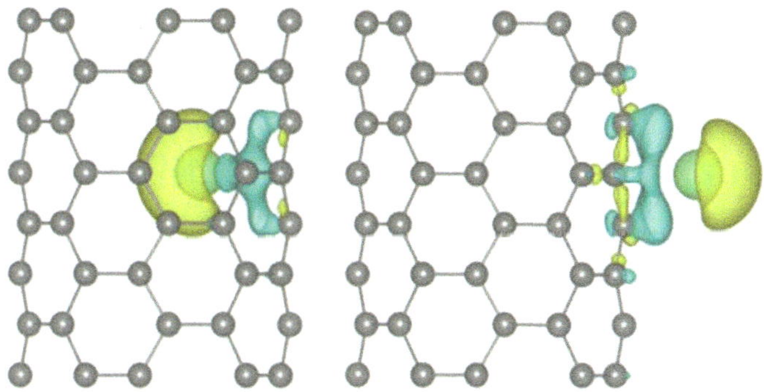

Fig. 5 The geometry of the complex CNT (5, 5) @ Li (Reproduced by permission of the ACS applied nanomaterials)

modeling [96]. According to the results of this investigation, it could be deduced that the internal surface of both considered nanostructures, namely Li-doped and pure is not appropriate for the target applications (Fig. 5).

Further, By taking advantage of the calculations of the ab initio density functional theory (DFT) method, Ganji et al. studied the creation of endoherdral complexes among Si, B, and N-doped C_{60} fullerene and molecular hydrogens, and the calculated capacities were compared with pure fullerene [97]. the energy of binding related to the complex of $nH_2@C_{59}X$ (n = 1 and 2) exhibited that up to 2 molecular hydrogens can create stable complexes with C_{60} nanocage. The attained outcomes revealed that the system of $2H_2@C_{59}Si$ showed an excellent capacity for the storage of molecular hydrogens owing to its ability to the creation of the most stable complex. Figure 6 displays the optimized structure of Si-doped C_{60} fullerene.

In other research, Han et al. carried out an elaborated study to scrutinize in physi-/chemi-sorption characteristics of hydrogen within the walls of (10, 0) SWCNTs by utilizing the DFT method [98]. It was found that the most stable complex was created through physisorption of molecular H_2 exterior the nanotube within a vertical configuration to the axis of the tube. The computation of the Gibbs free energy showed that this energy varied from physisorption to chemisorption by a change in the temperatures.

3.2.2 Heteroatoms Nanostructured Materials

According to the outcomes of different researches, the aimed DOE for storage applications cannot be realized through utilizing carbon-based nanomaterials owing to the undesirable interactions. Importing various heteroatoms into the carbon-based nanoscale materials is a possible approach for improving the interactions. To this end, BNNTs as new tubular structures have been widely investigated. From the discovery of this nanostructure, it has been considered as a favorable alternative for hydrogen

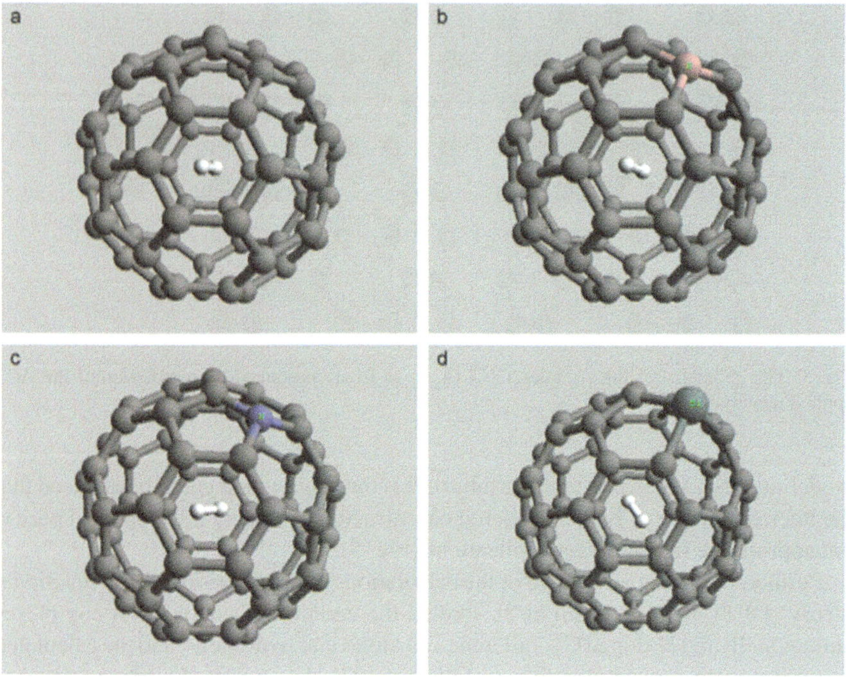

Fig. 6 Optimized geometries of **a** $1H_2@C_{60}$ and $1H_2@X$-doped C_{60} structures with **b** X = B, **c** X = N and **d** X = Si

storage applications [99, 100]. Ma et al. apperceived that the specimens of bamboo-like multiwall BNNT are able to store H_2 to a considerable extent at the temperature of the room [101]. Tang et al. determined that the collapsed structure of BNNTs can reserve even more hydrogen compare to the last investigation [102]. Additionally, theoretical studies confirmed that BNNTs are a more desirable milieu for hydrogen compared to the carbon nanotube. Through meticulous studies of all possible sites of binding of the physisorption of hydrogen on the wall of nanotubes in many structures and configurations, it can be deduced that more efficacious hydrogen binding could be obtained with BNNTs compared to CNTs [103]. The ionic property of the BNNT is the most important, as this enhances the hydrogen binding energy. The simultaneous utilization of potent and heteropolar, bonded porous nanomaterials paves a pathway to discover novel substances for hydrogen storage applications with notable capacity. Bezi et al. researched the formation of endohedral complexes among carbon nitride heterofullerene, namely $C_{44}N_6$, $C_{48}N_{12}$, and $C_{33}N_9$ and hydrogen molecules, through DFT calculations [104]. It was revealed that up to 2 molecular hydrogens could incorporate inside the nanocage and is able to create the complexes with $C_{48}N_{12}$ and C_{12} fullerenes, whilst for other nanostructures only a single hydrogen molecule was presented inside the cages. Further, the achieved results showed that H_2 penetrates inside the C_{60} easier than other considered structure. Ganji et al. carried out

DFT calculations to study the possible creation of $H@C_{60}$ and $H@(BN)_n$–fullerene (n: 24, 36, 60) complexes for hydrogen storage applications [105]. According to this research, $B_{60}N_{60}$ nanocage has a marvelous capability for hydrogen storage applications. Schematic illustration of the optimized structures of (a) $H@B_{24}N_{24}$, (b) $H@B_{36}N_{36}$, (c) $H@B_{60}N_{60}$ and (d) $H@C_{60}$ complexes (Fig. 7).

Regarding the literature, point charges within the surface of the materials can ameliorate the capacity of storage because they improve the energy of binding related to the molecular hydrogens. Diverse theoretical studies have been dedicated to the hydrogen storage performance of silicon carbide (SiC) nanotubes and it was deduced that among 2 energetically stable configurations of Si nanotubes, the form that C and Si atoms possess alternating locations within the walls of nanotubes have plenty of point charges [106, 107]. This occurs due to the great quantity of an electron transfer from Si to C. Therefore, the presence of point charged within the surface of this nanostructure has made them an ideal candidate for application in hydrogen storage. This finding was confirmed through the first principal theoretical research that indicated an improvement of around 20% of the molecular hydrogen binding energy within the silicon carbide (SiC) nanotubes in comparison with pure CNTs. The charges alternation that presents in the wall of SiC nanotubes, create dipole on

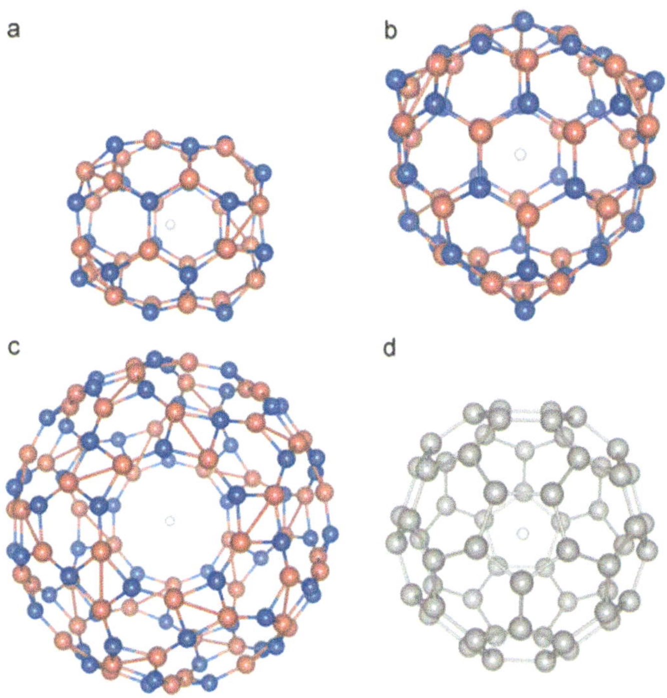

Fig. 7 Schematic illustration of the optimized structures of **a** $H@B_{24}N_{24}$, **b** $H@B_{36}N_{36}$, **c** $H@B_{60}N_{60}$ and **d** $H@C_{60}$ complexes (Reproduced by permission of Phys. E Low-Dimensional Syst. Nanostructures)

the molecular hydrogens, supplying excessive stability. Besides, the simulation of classical Monte-Carlo for bundles of nanotubes exhibit a greater capacity of storage in this SiC nanotube, particularly in low pressures and high temperature [108–110].

3.2.3 Carbon Nanoscrolls

Carbon nanoscroll (CNS) was initially designed and synthesized in 2003 through twisting a nanosheet of graphite and later it was recognized as a fascinating nanostructure for hydrogen materials. The elemental structure of CNS and CNTs are almost similar and the distance of interlayers is alike. The theoretical researches on these nanostructures illuminated a tremendous development of the storage capacity. Although, the distance of interlayers in CNS is not sufficient for hydrogen storage, doping an alkali element can significantly improve the capacity [111–113]. Braga et al. investigated the hydrogen storage capacity of carbon nanoscroll (CNS) by using molecular dynamic (MD) simulation [114]. As described before, CNSs are created from helically wrapping of the layers of graphite. It was revealed that magnificent hydrogen storage capacity within this nanostructure is possible at lower temperatures, whilst by increasing the temperatures, thermal energies remarkably reduced the storage capacity.

3.2.4 Graphene

A new class of nonporous structures has emerged in recent decades and has been widely developed to find an appropriate solution for the hydrogen storage challenge. Their elemental skeleton is consisting of an ultrahigh surface area results in a greater capacity of storage, confirming them as desirable alternative structures compared to CNTs [115, 116]. Besides, carbon-based nanomaterials have favorable stability and sustainability to a broad range of procedures and circumstances and are highly desirable for commercial hydrogen storage applications. In order to enhance the capacity of hydrogen storage, the synthesis of new carbon-based structures containing new features such as proper pores and larger surface areas is indispensable. To this aim, pillared graphene as a cutting-edge three-dimensional nanostructures was suggested. Graphene is constructed from two carbon allotropes, namely, graphene sheets and CNT, and they were combined to form a 3D substance with adjustable pores which is the most significant feature of this structure, as it is essential for hydrogen storage with high efficiency. Tiny pores induce difficulties in the insertion of molecular hydrogen, or will not reserve them. In return, large pores cause vacant spaces inside the nanostructures, where H_2 is merely reserved because of the pressure, same as a vacant tank. Only via an optimal pores size one can achieve the optimum capacities of the target structures. This is the leading criteria in graphene. The diversity in pore size could be obtained via freedom to modify the diameter or length of the tubes simultaneously [117–119]. Khosravi et al. studied the hydrogen adsorption ability of Pt-adsorbed graphene by ab initio vdW computations with the DFT level of theory [120]. It was disclosed that hydrogen could be adsorbed onto the system of Pt-graphene effectively and improve the strength of the substrate. Ganji et al. investigated the adsorption of molecular hydrogens on Si-doped graphene via utilizing DFT computations [121]. The novel advanced Si-decorated graphene could provide a storage capacity of around 15% and could be a promising alternative for hydrogen storage milieus.

3.2.5 Porous Nanotube Networks

As we discussed until here, the proper size of the pore and vast surface area are the most important criteria for improving the capacity of hydrogen storage. For this aim, two classes of substances, namely metal organic framework and nanoporous materials have attracted many attentions [122, 123]. These nanostructures gave essential benefits and important drawbacks. For instance, although MOFs are comprised of a plain cubic framework and possess large pores and high surface area, but these structures are highly sensitive to humid environments. Besides, the infrastructure of carbon nanomaterials has excellent sustainability to an extensive range of processing circumstances, but unfortunately, they cannot provide high surface area compared to MOFs. A feasible strategy for the challenge of hydrogen storage is finding and synthesis of new C-based nanomaterials with large pores and surface areas. A MOF-like plain cubic fundamental skeleton within a formation of graphitic structure could be considered appropriate material. For example, the porous nanotube network (PNN) is a new three-dimensional structure is comprised of interconnected single all nanotubes and are able to create the networks in the form of orthogonal. Because it is well-established that nanotubes can create two-dimensional junctions. PNN is known as the first sample of C-based superstructures consisting of orthogonal-form cubes in the higher scales and nanotubes in the low scales. Theoretical researches have indicated that via selecting the proper dimensions, PNN is capable of surpassing the DOE restrict in both volumetric and gravimetric terms, at lower pressures [124, 125].

4 Conclusion

By glancing over the historical course of the development of industries and the path of industrialization for different countries of the world in recent centuries, it can be simply apperceived that any leap in the industry directly relies on the energy consumption in those countries. Due to a significant reduction in nonrenewable energy resources throughout the world, sustainable development is described as creating a logical balance in preserving the amount of energy resources utilized by humans for current demands while considering the needs of the next generations. Despite the fact that serious dependence on nonrenewable energy resources has created pervasive problems in the course of global advancement, very recently remarkable effective endeavors have been allocated to expand the use of renewable energy resources and the most underlying renewable sources of energy are bioenergy, marine energy, solar energy, hydroenergy, wind energy, and geothermal energy.

Since the future generation of energy must be founded on fundamental infinite resources which render three main benefits, namely reduced consumption of hydrocarbon-based reservoirs, almost zero emission of various GHG_3 and CO_2, and clean and cheap energy for all the inhabitants of the planet, recent exploration has divulged that sing hydrogen as a renewable energy resource can achieve all of the aforesaid goals. In this chapter, hydrogen as an emerging and effective resource of energy has been fully introduced. Furthermore, the current methods and materials used to store hydrogen as an energy resource were discussed. In the next part, the role of nanostructured materials in hydrogen storage applications was thoroughly presented. Until now, a wide spectrum of nanostructures has been synthesized and employed for encapsulation of hydrogen as alternative energy sources and their ability in encapsulation of hydrogen molecules has been investigated through experimental and computational techniques, which confirmed the brilliant capability of nanostructures in the encapsulation of hydrogen as the next generation of renewable energy resources.

References

1. Østergaard PA et al (2020) Sustainable development using renewable energy technology. Renew Energy 146:2430–2437
2. Urbaniec K et al (2017) A holistic approach to sustainable development of energy, water and environment systems. J Clean Prod 155:1–11
3. Nerini FF et al (2018) Mapping synergies and trade-offs between energy and the sustainable development goals and the sustainable development goals. Nat Energy 3:10–15
4. Waas T et al (2010) University research for sustainable development: definition and characteristics explored. J Clean Prod 18:629–636
5. Ukko J et al (2019) Sustainable development: implications and definition for open sustainability. Sustain Dev 27:321–336
6. An Q et al (2018) Dependency network of international oil trade before and after oil price drop. Energy 165:102–1033
7. Wang Q et al (2018) China's dependency on foreign oil will exceed 80% by 2030: developing a novel NMGM-ARIMA to forecast China's foreign oil dependence from two dimensions. Energy 163:151–167
8. Shahbaz M et al (2017) Dynamics of electricity consumption, oil price and economic growth: global perspective. Energy Policy 108:256–270
9. Huang Y et al (2020) The role of forest resources, mineral resources, and oil extraction in economic progress of developing Asian economies. Res Policy 69:101878
10. Johnston JE et al (2019) Impact of upstream oil extraction and environmental public health: a review of the evidence. Sci Total Environ 657:187–199
11. Ramirez IM et al (2017) Contamination by oil crude extraction—refinement and their effects on human health. Environ Pollut 231:415–425
12. Noh SR et al (2019) Hebei Spirit oil spill and its long-term effect on children's asthma symptoms. Environ Pollut 248:286–294
13. Zeng S et al (2017) A review of renewable energy investment in the BRICS countries: history, models, problems and solutions. Renew Sust Energ Rev 74:860–872

14. Shirely R et al (2013) Renewable energy sector development in the Caribbean: current trends and lessons from history. Energy Policy 57:244–252
15. Kalogirou SA (2005) Seawater desalination using renewable energy sources. Prog Energy Combust Sci 31:242–281
16. Panwar NL et al (2011) Role of renewable energy sources in environmental protection: a review. Renew Sust Energ Rev 15:1513–1524
17. Koroneos K et al (2003) Exergy analysis of renewable energy sources. Renew Energy 28:295–310
18. Waheed R et al (2018) Forest, agriculture, renewable energy, and CO_2 emission. J Clean Prod 172:4231–4238
19. Menyah K et al (2010) CO_2 emissions, nuclear energy, renewable energy and economic growth in the US. Energy Policy 38:2911–2915
20. Jaforullah M et al (2015) Does the use of renewable energy sources mitigate CO_2 emissions? A reassessment of the US evidence. Energy Econ 49:711–717
21. Armaroli N et al (2006) The future of energy supply: challenges and opportunities. Angew Chem 46:52–66
22. Narbel PA et al (2014) Estimating the cost of future global energy supply. Renew Sust Energ Rev 34:91–97
23. Nojavan S et al (2017) Application of fuel cell and electrolyzer as hydrogen energy storage system in energy management of electricity energy retailer in the presence of the renewable energy sources and plug-in electric vehicles. Energy Convers Manag 136:404–417
24. Abe JO et al (2019) Hydrogen energy, economy and storage: review and recommendation. Int J Hydrog Energy 44:15072–15086
25. Lagorse J et al (2008) Energy cost analysis of a solar-hydrogen hybrid energy system for stand-alone applications. Int J Hydrog Energy 33:2871–2879
26. Salvi BL et al (2015) Sustainable development of road transportation sector using hydrogen energy system. Renew Sust Energ Rev 51:1132–1155
27. Winter C et al (2009) Hydrogen energy—abundant, efficient, clean: a debate over the energy-system-of-change. Int J Hydrog Energy 34:S1–S52
28. Tian F et al (2005) A hydrogen-rich early earth atmosphere. Science 308:1014–1017
29. Wordsworth R et al (2013) Hydrogen-nitrogen greenhouse warming in earth's early atmosphere. Science 339:64–67
30. Genda H et al (2008) Origin of the ocean on the Earth: early evolution of water D/H in a hydrogen-rich atmosphere. Icarus 194:42–52
31. Seiser R et al (2005) The influence of water on extinction and ignition of hydrogen and methane flames. P Combust Inst 30:407–414
32. Pizza G et al (2008) Dynamics of premixed hydrogen/air flames in microchannels. Combust Flame 152:433–450
33. Lamoureux N et al (2003) Laminar flame velocity determination for H_2–air–He–CO_2 mixtures using the spherical bomb method. Exp Therm 27:385–393
34. Das AK et al (2012) Ignition delay study of moist hydrogen/oxidizer mixtures using a rapid compression machine. Int J Hydrog Energy 37:6901–6911
35. Lei X et al (2020) Chemical effects of hydrogen addition on soot formation in counterflow diffusion flames: dependence on fuel type and oxidizer composition. Combust Flame 213:14–25
36. Zhen HS et al (2012) Effects of hydrogen concentration on the emission and heat transfer of a premixed LPG-hydrogen flame. Int J Hydrog Energy 37:6097–6105
37. Mardani A et al (2010) Effect of hydrogen on hydrogen–methane turbulent non premixed flame under MILD condition. Int J Hydrog Energy 35:11324–11331
38. Choudhuri AR et al (2004) Intermediate radical concentrations in hydrogen–natural gas blended fuel jet flames. Int J Hydrog Energy 29:1293–1302
39. Dorofeev SB et al (1994) Evaluation of the hydrogen explosion hazard. Nucl Eng 148:305–316
40. Kim JS et al (2009) Behavior of buoyancy and momentum controlled hydrogen jets and flames emitted into the quiescent atmosphere. J Loss Prev Process Ind 22:943–949

41. Bauer W et al (1974) Helium and hydrogen re-emission during implantation of molybdenum, vanadium and stainless steel. J Nucl Mater 53:127–133
42. Sofronis P et al (1989) Numerical analysis of hydrogen transport near a blunting crack tip. J Mech Phys Solids 37:317–350
43. Obara S (2019) Energy and exergy flows of a hydrogen supply chain with truck transportation of ammonia or methyl cyclohexane. Energy 174:844–860
44. Cho S et al (2019) Improved hydrogen recovery in microbial electrolysis cells using intermittent energy input. Int J Hydrog Energy 44:2253–2257
45. Thornton AW et al, Materials genome in action: identifying the performance limits of physical hydrogen storage. Chem Mater 29:2844–2854
46. Liao S, Musyoka NM, Mathe M, Ren J, Langmi HW (2016) Current research trends and perspectives on materials-based hydrogen storage solutions: a critical review. Int J Hydrogen Energy 42:289–311
47. Nagpal M, Kakkar R (2018) An evolving energy solution: intermediate hydrogen storage. Int J Hydrogen Energy 43:12168–12188
48. Nakagawa T et al (2014) Physical, structural, and dehydrogenation properties of ammonia borane in ionic liquids. RSC Adv 4:21681–21687
49. Purewal J et al (2012) Improved hydrogen storage and thermal conductivity in high-density MOF-5 composites. J Phys Chem 116:20199–20212
50. Zheng J et al (2011) Development of high pressure gaseous hydrogen storage technologies. Int J Hydrogen Energy 37:1048–1057
51. Helmolt R Von, Eberle U (2007) Fuel cell vehicles: status 2007. J Power Sources 165:833–843
52. Barthelemy H, Weber M, Barbier F (2017) Science direct hydrogen storage: recent improvements and industrial perspectives. Int J Hydrogen Energy 42:7254–7262
53. Wang H et al (2009) High performance of nanoporous carbon in cryogenic hydrogen storage and electrochemical capacitance. Carbon 47:2259–2268
54. Ahluwalia RK, Peng JK (2008) Dynamics of cryogenic hydrogen storage in insulated pressure vessels for automotive applications. Int J Hydrogen Energy 33:4622–4633
55. Felderhoff M et al (2007) Hydrogen storage: the remaining scientific and technological challenges. Phys Chem Chem Phys 9:2643–2653
56. Haberbusch MS (2010) Development of No-Vent™ liquid hydrogen storage system for space applications. Cryogenics 50:541–548
57. Berry GD, Aceves M (1998) Onboard storage alternatives for hydrogen vehicles. Energy Fuels 12:49–55
58. Mori D, Hirose K (2009) Recent challenges of hydrogen storage technologies for fuel cell vehicles. Int J Hydrogen Energy 34:5469–5474
59. Cacciola G et al (1984) Cyclohexane as a liquid phase carrier in hydrogen storage and transport. Int J Hydrogen Energy 9:411–419
60. Zheng J et al (2019) Thermodynamic modelling and optimization of self-evaporation vapor cooled shield for liquid hydrogen storage tank. Energy Convers Manag 184:74–82
61. Chung CA, Ho C (2009) Thermal–fluid behavior of the hydriding and dehydriding processes in a metal hydride hydrogen storage canister. Int J Hydrogen Energy 34:4351–4364
62. Ahluwalia RK, Peng JK (2009) Automotive hydrogen storage system using cryo-adsorption on activated carbon. Int J Hydrogen Energy 34:5476–5487
63. Petitpas GA et al (2014) comparative analysis of the cryo-compression and cryo-adsorption hydrogen storage methods. Int J Hydrogen Energy 39:10564–10584
64. Doroodian A et al (2010) Methylguanidinium borohydride: an ionic-liquid-based hydrogen-storage material. Angew Chem 49:1871–1873
65. Sahler S et al (2014) The role of ionic liquids in hydrogen storage. Chem Eur J 20:8934–8941
66. Fan L et al (2016) Atomically isolated nickel species anchored on graphitized carbon for efficient hydrogen evolution electrocatalysis. Nat Commun 7:10667
67. Rowsell JLC, Yaghi OM (2006) J Am Chem Soc 128:1304
68. Chen B et al (2005) High H_2 adsorption in a microporous metal-organic framework with open metal sites. Angew Chem 44:4745–4749

69. Darvish M et al (2012) Adsorption of hydrogen molecules onto Li-decorated titanium met-car cluster: a first-principles study'. Physica E 46:193–197
70. Litster S, Mclean G (2004) PEM fuel cell electrodes. J Power Sources 130:61–76
71. Mehta V, Cooper JS (2003) Review and analysis of PEM fuel cell design and manufacturing. J Power Sources 114:32–53
72. Wu J et al (2008) A review of PEM fuel cell durability: degradation mechanisms and mitigation strategies. J Power Sources 184:104–119
73. Ozarslan A (2012) Large-scale hydrogen energy storage in salt caverns. Int J Hydrogen Energy 37:14265–14277
74. Michaelis J et al (2014) Evaluation of large-scale hydrogen storage systems in the German energy sector. Fuel cells 14:517–524
75. Johnson TA et al (2011) Performance of a full-scale hydrogen-storage tank based on complex hydrides. Faraday Discuss 15:327–352
76. Klerke A et al (2008) Ammonia for hydrogen storage: challenges and opportunities. J Mater Chem 18:2304–2310
77. Otsuka K et al (2012) Hydrogen storage and production by redox of iron oxide for polymer electrolyte fuel cell vehicles. Int J Hydrogen Energy 37:14265–14277. Int. J. Hydrogen Energy 28:335–342 (2003)
78. Tzamalis G et al (2011) Techno-economic analysis of an autonomous power system integrating hydrogen technology as energy storage medium. Renew Energy 36:118–124
79. Amrouche SO et al (2016) Overview of energy storage in renewable energy systems. Int J Hydrogen Energy 41:20914–20927
80. Conibeer GJ et al (2007) A comparison of PV/electrolyser and photoelectrolytic technologies for use in solar to hydrogen energy storage systems. Int J Hydrogen Energy 32:2703–2711
81. Ganji MD (2009) First-principles simulation of the encapsulation of molecular hydrogen in C120 nanocapsules. Physica E 41:1433–1438
82. Ganji MD et al (2009) Density functional theory calculations of hydrogen molecule adsorption on monolayer molybdenum and tungsten disulfide. Physica E 57:1433–1438
83. Ganji MD et al (2016) Computational design of multi-states monomolecular device using molecular hydrogen and C20 isomers. Phys Solid State 58:1476–1482
84. Ganji MD et al (2015) Theoretical insight into hydrogen adsorption onto graphene: a first-principles B3LYP-D3 study. Phys Chem Chem Phys 17:2504
85. Iijima S (1991) Helical microtubules of graphitic carbon. Nature 354:56–58
86. Dillon AC et al (1997) Storage of hydrogen in single-walled carbon nanotubes. Nature 386:377–437
87. Chen P et al (1999) High H_2 uptake by alkali-doped carbon nanotubes under ambient pressure and moderate temperatures. Science 285:91
88. Yang RT (2000) Hydrogen storage by alkali-doped carbon nanotubes–revisited. Carbon 38:623–641
89. Kajiura H et al (2003) Hydrogen storage capacity of commercially available carbon materials at room temperature. Appl Phys Lett 82:1105
90. Ritschel M et al (2002) Hydrogen storage in different carbon nanostructures. Appl Phys Lett 80:2985
91. Hirscher M, Becher M (2006) J Nanosci Nanotechno 3:3
92. Froudakis GE (2002) Hydrogen interaction with carbon nanotubes: a review of ab initio studies. J Phys Cond Mat 14:453
93. Lochan RC, Head-Gordon M (2006) Computational studies of molecular hydrogen binding affinities: the role of dispersion forces, electrostatics, and orbital interactions. Phys Chem Chem Phys 8:1357–1370
94. Froudakis GE (2001) Why alkali-metal-doped carbon nanotubes possess high hydrogen uptake. Nano Lett 1:531–533
95. Ganji MD, Ahmadian N, Goodarzi M, Khorrami HA (2011) Molecular hydrogen interacting with Si-, S- and P-doped C 60 fullerenes and carbon nanotube. J Comput Theor Nanosci 8:1392–1399

96. Anikina E, Banerjee A, Beskachko V, Ahuja R (2019) Li-functionalized carbon nanotubes for hydrogen storage: importance of size effects Li-functionalized carbon nanotubes for hydrogen storage: importance of size effects. ACS Appl Nano Mater 5:3021–3030
97. Ganji MD (2009) Ab initio investigation of the possibility of formation of endohedral complexes between H2 molecules and B-, N- and Si-doped C60 fullerenes. Phys E Low-Dimens Syst Nanostruct 41:1406–1409
98. Han SS, Lee HM (2004) Adsorption properties of hydrogen on (10, 0) single-walled carbon nanotube through density functional theory. Carbon 42:2169–2177
99. Chopra NG et al (1995) Boron nitride nanotubes. Science 269:966–967
100. Wang P et al (2002) Hydrogen in mechanically prepared nanostructured h-BN: a critical comparison with that in nanostructured graphite. Appl Phys Lett 80:318
101. Ma R et al (2002) Hydrogen uptake in boron nitride nanotubes at room temperature. J Am Chem Soc 124:7672–7673
102. Tang C et al (2002) Catalyzed collapse and enhanced hydrogen storage of BN nanotubes. J Am Chem Soc 124:14550
103. Mpourmpakis G, Froudakis GE (2007) Why boron nitride nanotubes are preferable to carbon nanotubes for hydrogen storage?: an ab initio theoretical study. Catal Today 120:341
104. Javan MB, Ganji MD, Sabet M, Danesh N (2011) Incorporation of hydrogen molecules into carbon nitride heterofullerenes: an Ab initio study. J Comput Theor 8:1–5
105. Ganji MD, Abbaszadeh B, Ahaz B (2011) Hydrogen incorporation into BN fullerene-like nanostructures: a first-principles study. Phys E Low-Dimens Syst Nanostruct 44:290–297
106. Mavrandonakis A et al (2003) From pure carbon to silicon–carbon nanotubes: an Ab-initio study. Nano Lett 3:1481
107. Menon M et al (2004) Structure and stability of SiC nanotubes. Phys Rev B 69:115322
108. Pham-Huu C et al (2001) The first preparation of silicon carbide nanotubes by shape memory synthesis and their catalytic potential. J Catal 200:400–410
109. Sun XH et al (2002) Formation of silicon carbide nanotubes and nanowires via reaction of silicon (from disproportionation of silicon monoxide) with carbon nanotubes. J Am Chem Soc 124:14464
110. Mpourmpakis G et al (2006) Nano Lett 61:581
111. Viculis LM et al (2003) Science 299:1361
112. Braga SF et al (2004) Structure and dynamics of carbon nanoscrolls. Nano Lett 4:881–884
113. Mpourmpakis G et al (2007) Carbon nanoscrolls: a promising material for hydrogen storage. Nano Lett 7:1893–1897
114. Braga SF (2007) Hydrogen storage in carbon nanoscrolls: an atomistic molecular dynamics study. Chem Phys Lett 441:78–82
115. Rowsell JLC et al (2005) Strategies for hydrogen storage in metal-organic frameworks. Angew Chemie Int Ed 44:4670
116. Klontzas E et al (2007) Molecular hydrogen interaction with IRMOF-1: a multiscale theoretical study. J Phys Chem C 111:13635
117. Dimitrakakis GK et al (2008) Pillared graphene: a new 3-D network nanostructure for enhanced hydrogen storage. Nano Lett 8:3166
118. Tylianakis E et al (2010) Li-doped pillared graphene oxide: a graphene-based nanostructured material for hydrogen storage. J Phys Chem Lett 1:2459–2464
119. Klontzas E et al (2008) Improving hydrogen storage capacity of MOF by functionalization of the organic linker with lithium atoms. Nano Lett 8:1572
120. Khosravi A, Fereidoon A, Ahangari MG, Ganji MD, Emami SN (2014) First-principles vdW-DF study on the enhanced hydrogen storage capacity of Pt-adsorbed graphene. J Mol Model 20:2230
121. Ganji MD, Emami SN, Khosravi A, Abbasi M (2015) Si-decorated graphene: a promising media for molecular hydrogen storage. Appl Surf Sci 332:105–111
122. Wang Q et al (1999) Computer simulations of hydrogen adsorption on graphite nanofibers. J Phys Chem 110:281–577

123. Han SS et al (2009) Recent advances on simulation and theory of hydrogen storage in metal–organic frameworks and covalent organic frameworks. Chem Soc Rev 38:1460–1476
124. Zhao J et al (2002) Gas molecule adsorption in carbon nanotubes and nanotube bundles. Nanotechnology 13:195
125. Tylianakis E et al (2011) Porous nanotube network: a novel 3-D nanostructured material with enhanced hydrogen storage capacity. Chem Commun 47:2303–2305

Containers for Thermal Energy Storage

Pramod B. Salunkhe and Jaya Krishna Devanuri

1 Introduction

From several decades, phase change materials (PCMs) are playing a major role in management of short and medium term energy storage applications, namely, thermal energy storage [1–3], building conditioning [4–7], electronic cooling [8, 9], telecom shelters [10], to name a few. A major drawback of the PCMs is their poor thermal conductivity. Most of the PCMs (organic and inorganic) have their thermal conductivity less than 2 W/m-K. To mitigate this problem, few methods have been developed, such as, use of metal matrix, insertion of high thermal conductivity materials, use of fins, encapsulation of PCMs, etc. The latter technique is developing significantly. The recent technological advancement has made it possible to encapsulate the PCMs not only at the micro but also at the nano levels. In recent years, several researchers have focused their attention on micro and nano encapsulated PCMs for the improvement of thermal conductivity of the PCMs and hence their thermal performance of the system [11–13]. Cumplido et al. [14] provided a detailed review on the synthesis of micro and nano encapsulated PCMs. Several researchers have described different types and methods of PCM encapsulation [15–17].

The structural strength of the encapsulated PCM or the container of a PCM is of utmost importance. The leakage of PCM would cause contamination with it's

P. B. Salunkhe (✉)
Department of Aeronautical and Automobile Engineering, Manipal Institute of Technology, Manipal Academy of Higher Education, Manipal, Karnataka 576 104, India
e-mail: pramod.salunkhe@manipal.edu

J. K. Devanuri
Department of Mechanical Engineering, National Institute of Technology, Telangana State, Warangal, India

J. Parameswaranpillai et al. (eds.), *Micro- and Nano-containers for Smart Applications*, Composites Science and Technology,
https://doi.org/10.1007/978-981-16-8146-2_13

surrounding system which may pose hazardous and health related problems, especially in applications such as food storage, blood and drug transportation etc. The encapsulated PCM/container should fulfill the following requirements.

i. should be leak proof.
ii. should accommodate for volume change of PCM during the phase change process.
iii. the container or the shell material should have higher thermal conductivity to improve the heat transfer rate to/from surrounding medium.
iv. should not show any chemical reaction with the PCM or the heat transfer fluid.
v. should be structurally stable at specified temperatures.
vi. should be cheap, easily available and easy to manufacture.

For high temperature application, Gavarrell and Fereres [18] reported use of Borosilicate as an encapsulation material in the temperature range of 300–400 °C. Eight spherical capsules of diameter 20 mm were used. The shell thickness was maintained around 1 mm. The metallic and salt PCM were poured in the capsule keeping radially 4 mm empty space.

Though, there are several applications of PCMs, in the present paper, authors have considered four important applications, namely, thermal energy storage, cooling of electronic equipment, food and drug transportation and solar water and space heating applications.

1.1 Thermal Energy Storage

Guo et al. [19] studied different types of containers, namely, shell-and-tube, encapsulated, direct contact and detachable and sorptive type, for mobile thermal energy storage applications. In shell-and-tube type container, heat transfer fluid passes through tube side, whereas shell side contains the PCM. It was reported that though shell-and-tube type containers are simple to use, the higher heat transfer enhancement can be obtained in encapsulated containers. However, PCM leakage after several charging and discharging may be a concern. Subsequently, Zayed et al. [20] reviewed the effect of different container geometries on the thermal performance of a system. It was reported that cylindrical containers of shell-and-tube type leads to the energy efficiency of more than 70%. The rectangular container gives faster charging and higher storage efficiency which make it one of the best candidates for the container geometry. Fukahori et al. [21] proposed use of cylindrical ceramic containers (Al_2O_3) for the metallic PCMs in high temperature applications. Encapsulation of metallic PCMs is challenging due to their corrosive nature and large volume change during the phase change process. The containers comprised of cup and cap type. The cup had inner diameter of 22.2. mm and inside length of 17.99 mm. Aluminium film was used as a sealing material. The experimental results have showed improved cyclic durability.

Huang et al. [22] investigated the effect of height of rectangular container on the thermal performance of a system. The stainless steel container of 2 mm thickness was tested for four different heights, namely, 12, 24, 36 and 48 cm. The length and width were fixed to 200 mm and 50 mm, respectively. The findings suggested to use lower container height. As compared to container height of 48 cm, the 12 cm height led to 50% reduction in melting time and the average temperature was decreased by 30 °C. Ghafari et al. [23] investigated the melting process of nano-enhanced PCM in a spherical container of 52 mm outer diameter and 2 mm wall thickness. The addition of nanoparticles though increased the thermal conductivity, however, the melting rate was not substantially affected. He et al. [24] demonstrated the benefits of incorporating direct contact type containers as compared to indirect contact type. It was reported that direct contact type containers are simple to use and offer higher heat transfer rates as compared to indirect contact type containers.

Salunkhe et al. [25] performed a detailed review on the effect of shell material, encapsulation geometry and core-to-coating ratio on the thermal performance of a system. Silica was suggested as the best shell material based on its high thermal conductivity, good mechanical strength and leak prof characteristics. A lower core-to-coating ratio reduces the heat storage capacity, whereas, higher core-to-coating ratio reduces the structural strength of the shell material. Chiew et al. [26] used cylindrical container of size $100 \times 20.5\ mm^2$ made of aluminum material. The choice of the container material was based on it's higher thermal conductivity. RT35HC was used as a PCM. The macro-encapsulated container doesn't require provision for the air gap. As compared to conventional PCM thermal energy storage system, better thermal cyclic thermal performance was observed. Zarajabad et al. [27] used bare and finned containers in a domestic refrigerator to reduce the electric consumption. Water was used as a PCM. The 0.8 mm thick coper containers had a size of $2 \times 26 \times 38\ cm^3$. The finned container has three fins. The numerical results showed that the bare container maintained the refrigerator's cabin temperature for 30 min, whereas, finned container maintained it for 68 min. The finned container led to 17.4% reduction in energy consumption per day.

Fang et al. [28] used nanocapsules to improve the heat transfer rate in a thermal energy storage system. N-Tetradecane was used as the PCM (60 wt%) and urea and formaldehyde as the shell material. The average size of the nanocapsule of about 100 nm was obtained by maintaining the stirring rate of 1500 rpm. NaCl was added in the range of 2–5% to improve the encapsulation as well as the thermal stability of the encapsulated PCM. Leng et al. [29] encapsulated the PCM, chloride molten salts with encapsulation material, diatomite for high temperature applications. It was reported that the PCM performance was good up to the PCM concentration of 70%. However, with increase in PCM concentration beyond 70%, the PCM leakage was observed when it was subjected to temperature above it's melting point temperature.

Yu et al. [30] microencapsulated the paraffin using double layer of shell material to prevent the spillage of the PCM and improve the heat transfer rate. The first shell layer was applied by TEOS hydrolysed silica and second layer by spherical silica. It was reported that, the thermal conductivity was improved by 133.3%, PCM leakage was prevented up to 65 °C and heat loss was decreased during the phase change process.

Song et al. [31] encapsulated the PCM, lauric acid using kapok fiber in microtubular shape. The thinner shell and large core PCM resulted in enhanced encapsulation ratio and hence the thermal energy storage capacity. It showed excellent thermal characteristics and high latent heat even after 2000 heating and cooling cycles. Liao et al. [32] encapsulated polyethylene glycol using carbonized graphene aerogel/melamine foam (c-GA/MF). It was reported that c-GA/MF not only provided excellent thermal and structural stability but also retained original shape of the composite PCM. A significant increase in thermal conductivity by about 312% was observed. The weight proportion of c-GA/MF was only about 4.6% which substantially increased the weight of the core part. This helped maintain the high enthalpy of the composite PCM.

McMurray et al. [33] demonstrated the use of SiC for the encapsulation of spherical aluminium beads of 1 mm diameter for high temperature application. The experimental results showed that the composite PCM resulted in good thermal and structural properties. Zou et al. [34] demonstrated the use of eutectic mixture Al-Si/Si-rich as core–shell structure for high temperature application up to 650 °C. The Si-rich material has high thermal conductivity of 116–88 W/m-K. Hence, the overall thermal conductivity of the composite PCM is substantially improved. Based on its corrosion resistance property and high latent heat of fusion even after 1200 cycles, this composite mixture was found to be a promising material.

1.2 Cooling of Electronic Devices

Omari et al. [35] computationally studied the effect of different enclosure geometries, such as, rectangular, rounded rectangular and circular on the heat transfer enhancement for cooling of an electronic device. All these geometries contained the same amount of PCM. The best and the worst performance was observed for the rectangular enclosure shifted vertically upward and the circular enclosure, respectively. The highest temperature difference between the two was about 40 °C. The rounded rectangle has slightly improved the efficiency of the system. Shokouhmand [36] studied the melting of lauric acid in a rectangular container made of transparent plexiglass, which is heated from one of its vertical side. The heat conduction was found to be the dominant mechanism during early stage of the melting process, whereas, convection was dominant in the last stage of the melting process. Yin et al. [37] encapsulated 100 gm of composite PCM (paraffin and porous expanded graphite) in an aluminum container of size 50 mm $\times$ 60 mm $\times$ 100 mm. It was reported that this cooling system led to a heat transfer coefficient of 1.25–1.3 times the traditional cooling system.

Ho et al. [38] used interfacial condensation polymerization to prepare the NEPCM. Eicosane was used as a PCM and formaldehyde as the shell. The core and shell materials were used in the ratio of 1:1. The resulting NEPCM size was around 250–350 nm. The thermal performance of the system was found to be increased at low Re number as compared high Re number. Subsequently, Ghalambaz et al. [39] used NEPCM along with free convection mechanism in a porous medium. Nonadecane was used as a core and polyurethane as the shell. Addition of 5% of NEPCM was found to improve the heat transfer rate up to 47%. Seyf et al. [40] performed numerical studies to improve the performance of a microtube heat sink using NEPCM slurry. Spherical nanoparticles of octadecane was used as a PCM. It was reported that with increase in mass concentration of NEPCM slurry improves the thermal performance, however, at the same time increases the pressure drop.

Sabbah et al. [41] used MEPCM-water slurry for heat transfer enhancement in a microchannel heat sink. Cross-linked polymer and paraffin wax were used as the shell and core part. It was suggested that shell part should be around 15–20% of the total mass of the capsule. It was reported that merely using water at high flow rates, use of MEPCM-water slurry at low flow rates is more beneficial in reducing the heat sink temperature. Though the polymer shell could considerably reduce the heat sink temperature, other materials such as copper could have further improved the thermal performance of the system. Ren et al. [42] used the combinations of microencapsulated PCM (MEPCM) to reduce the temperature of an electronic device. Commercial paraffin and calcium carbonate were used as core and shell materials. It was reported that the encapsulated PCM has good anti-leakage and low volume expansion/shrinkage properties. However, the poor thermal conductivity is still a challenge for the MEPCMs, though it is higher than pure PCM.

Rehman et al. [43] filled the copper foam (with 97% porosity and 35 pores per inch) with different PCMs. The combination of copper foam with PCM substantially improved the thermal conductivity. The highest temperature reduction was found to be 25 °C. Selvaraj [44] encased the graphene nanoplatelets with alumina nanoparticles and processed it further to form the alumina-graphene (core–shell) nanocomposite. The above mixture resulted in increase in heat transfer coefficient by 15–16 times as compared to that of water. Faraji et al. [45] placed the mixture of n-eicosane and copper nanoparticles in a rectangular container of aspect ratio of 0.5 and characteristic length of 0.08 m. All the walls of enclosure were maintained as adiabatic. The addition of nanoparticles enhanced the thermal conductivity and hence the heat transfer rate.

1.3 Food and Drug Storage/Transportation

Food and drug storage/transportation are of prime importance in the present world. The temperature controlled environment is mandate to safeguard the food and drug products. Depending up on the nature of products, the required temperature may be ambient, regulated room temperature (20–25 °C), refrigerated (2–8 °C), cryogenic (as low as −150 °C) [46] etc. The majority of the heat load comes from fluctuating ambient temperature. To regulate the required temperature a sophisticated container is required which can cope up with fluctuations in ambient temperature and interruptions in power supply. Containers with phase change materials (PCMs) can be a viable option to mitigate the above problems.

Highlighting the importance of regulated temperature for food items, Singh et al. [47] presented a thorough review on food packaging materials. Alehosseini and Jafari [48] provided a comprehensive review of micro/nano encapsulated PCMs employed in food industry. Azzouz et al. [49] investigated experimentally the performance of a household refrigerator by considering an eutectic mixture PCM with freezing point of −3 °C. The PCM was encapsulated in a plexiglass rectangular container placed at the back side of the evaporator. The placing of PCM was observed to improve the performance of the refrigerator. Johnston et al. [50] employed a composite nanostructured calcium silicate PCM for the transportation of chilled food. The food products were placed in an encapsulated PCM container made of a paperboard, covered with bubble wrap. The adopted approach could maintain the inside container temperature of 10 °C for a period of 5 h.

Gin and Farid [51] investigated the influence of PCM placed in the internal walls of a freezer. The PCM considered was eutectic composition: water—ammonium chloride with melting point of −15.4 °C. The PCM is placed in a rectangular aluminum container with an internal gap of 10 mm. The freezer with PCM was noted to show better food quality. Salgaonkar et al. [52] suggested natural deep eutectic solvent PCMs to regulate the temperature in refrigerated vehicles. Macro and microencapsulation methods were put forward with stainless steel, aluminum, natural fibre, high density poly ethylene as container materials. Oro et al. [53] performed experimental and numerical investigation of chilly bins for effective storage and transportation of ice cream and coffee. The commercial PCMs considered were RT-2, PT-15 and PT-63. The PCMs were placed inside the aluminum cylinder covered with insulating material at the outer side. The chilly bins with PCM was noted to serve for longer duration. Oro et al. [54] employed Cristopia E-21 with 2% oxyethylmethylcellulose (thickening agent) as PCM package materials to regulate the temperature of ice cream containers. It was observed that the use of PCM improved the quality of ice cream. Leducq et al. [55] employed eutectic PCM mixture of water and sodium chloride with melting point of −21 °C. The PCM was covered with polystyrene material made in the form of rectangular bricks. The ice cream quality was noted to improve by the usage of PCM.

Fioretti et al. [56] studied the performance of a PCM based refrigerated container. PCM used was Rubitherm RT35HC paraffin wax encapsulated in polyethylene panels. Due to the usage of PCM the peak heat load was observed to reduce by 20% and total energy by 4.7%. Arjenaki et al. [57] used tetradecane to regulate the temperature of meat. Tetradecane was placed in a polyethylene package. The usage of the PCM packing was noted to minimize the rise in temperature and preserve the quality of meat. Mousazade et al. [58] employed three eutectic PCMs viz. E-26, E-29, and E-32 in the refrigerated truck. The PCMs were encapsulated in a polyurethane foam core with Aluzinc sheet metals at the exterior surface. The thermal performance of PCMs in the cold box was investigated at different truck speeds. When E-26 was employed the truck was able to travel maximum distance. Xiaofeng and Xuelai [59] performed experimental and numerical investigation by considering n-octanoic acid-myristic acid composite and potassium sorbate-water composite. The study was aimed at improving the economy of logistics and to ensure the quality of food material during transportation. Polyurethane and vacuum insulation board were employed as inner materials and inner and outer surfaces were assembled with galvanized iron plates. The approach of multi temperature zone insulation was observed to be effective for cold storage applications.

One of the important factors to safeguard and improve the shelf life of drugs is it's temperature. The containers used for the storage and transportation of drugs should resist the heat loads and cope up with harsh environmental conditions so that the potency of the drugs is not lost. Huang and Piontek [60] carried out experimental investigation to analyze the performance of PCM based container used for transportation of drugs and food items. PCM employed was water and an alkane mixture with melting point of 5 °C. High-density polyethylene (HDPE) was used as the encapsulating material. The PCM container was able to maintain 4–5 °C of internal temperature for about 80 h. Aniruddha and Devakar [61] presented various PCM based passive and active containers for freeze-sensitive (2–8 °C) vaccines. It was mentioned that the usage of PCM improve the protection and reduce the overall transportation cost.

Dingley et al. [62] investigated the usage of paraffin wax for emergency drug transportation bags. The plastic container walls filled with paraffin wax was compared with the container with hallow walls. The container impregnated with wax provided thermal buffering effect. Du et al. [63] carried out numerical investigation for PCM based portable box for cold chain applications. Five PCMs (RT0, RT2HC, RT3HC, RT4HC, RT5HC and RT8HC) with two insulating materials (polyurethane and vacuum insulated panel) were considered and analysis was performed for five different configurations. The influence of PCM quantity positioned at different walls was examined. RT2HC with 20% at top and each side of the wall with vacuum insulated panel was observed to perform better. Zhao et al. [64] developed a cold box made of nanocomposite eutectic PCM (tetradecane and lauryl alcohol + expanded graphite) to meet the temperature requirements of a drug cold chain logistics. The developed equipment could maintain the required temperature for a longer duration.

1.4 Solar Water and Space Heating

Sharma and Chen [65] reviewed the usage of PCMs for thermal energy storage for solar water heating systems. Salunkhe and Devanuri [66] gave a thorough review on suitable PCMs for solar water and space heating applications. Primary and secondary selection criteria for a PCM were discussed. It was mentioned that the compatibility of container material plays an important role for the durability of PCM based solar water space heating systems. Prakash et al. [67] studied the performance of a rectangular capsules made of galvanized iron impregnated with PCM for solar water heating system. Based on the numerical study it was noticed that the PCM layer at the bottom was effective during off sun-shine hours. Bouadila et al. [68] investigated the usage of two rectangular cavities filled with paraffin. The cavities were placed below the absorber plate. The efficiency of the system was noted to vary between 25–35%.

Kaygusuz [69] employed calcium chloride hexahydrate and sodium sulfate decahydrate in a cylindrical PVC plastic container and observed to be more attractive when compared to rock and water based thermal storage systems. Cabeza et al. [70] used sodium acetate trihydrate—graphite PCM composite as heat storage medium in aluminum bottles. The bottles were placed in the top layer of the storage tank. The usage of PCM module was noted to supply hot water for longer duration. Al-Hinti et al. [71] encapsulated paraffin wax in cylindrical aluminum containers and placed in a storage tank at two levels. The suggested design had an advantage of 13–14 °C higher temperature over the storage tank without PCM.

Porteiro et al. [72] included three different PCMs in a 60 L hot water storage tank. The PCMs employed were Delta Cool-DC58 (hydrated salt), Pure Temp-PT58 (Vegetable based) and Thermusol-HD60 (hydrated salt). Among these PCMs, DC58 and PT58 are macro and HD60 is micro encapsulated in cylindrical shells placed in a cylindrical container made of carbon steel. DC58 and PT58 were noted to have fastest and slowest solidification rates. Kumar and Mylsamy [73] and Manirathnam et al. [74] analyzed the thermal performance of an evacuated tube water heating system by incorporating PCM with and without nano-composites. Kumar and Mylsamy [73] employed paraffin with CeO_2 $-0.5, 1, 2$ mass percentage of nano composite whereas Manirathnam et al. [74] considered paraffin wax with 1 mass percentage of SCi and CuO nanoparticles. The cylindrical containers with nano composite PCM was noted to significantly improve the thermal performance.

Barba and Spiga [75] analytically investigated the discharging process of eutectic mixture of (Ammonium nitrate (37.5%) + Magnesium nitrate hexahydrate (62.5%)) by encapsulating in rectangular, cylindrical and spherical shaped containers. Polyethylene was used as the container material. Based on the study, the solidification rate was noted to be higher for the spherical capsules. Afshan et al. [76] employed HS89 PCM in a spherical capsule made of stainless steel. The encapsulated spherical capsules were arranged as a packed bed in a cylindrical container. With the increase in aspect ratio the stratification was noted to increase but with PCM based storage system the arrangement of PCM capsules was observed to influence more when compared to aspect ratio. Mettawee and Assassa [77] placed paraffin wax below the absorber plate. The absorbed and stored heat was transferred to the pipe which carries water. The useful heat gain was noted to increase with the increase in mass flow rate. Xie et al. [78] reviewed the usage of PCMs in water tanks. Advantages and disadvantages of using various container shapes viz. rectangular, cylindrical, spherical and microencapsulated were discussed. It was mentioned that spherical shaped containers exhibit minimum heat loss. But when issues such as space and production process constraints are considered cylindrical containers are more advantgous.

Tyagi et al. [79] in their review pertaining to the usage of PCMs in solar air heaters mentioned about the usage of PCMs to make the system more effective and efficient. Enibe [80] carried out thermal analysis of a PCM based solar air heater. The PCM employed was paraffin wax encapsulated in a rectangular container. The encapsulated PCM was placed between the absorber plate and the insulation. The overall efficiency was in the range of 7.5–18%. Moradi et al. [81] conducted experiments by placing paraffin wax below the absorber plate. To improve the thermal transport aluminum powder with 0.5% wt paraffin was added. Aluminum plates were placed above and below the PCM. Initially, the numerical methodology was validated with experiments and later optimization was carried out based on thermal efficiency and inlet–outlet temperature difference. El Khadraoui et al. [82] in their experimental study filled the cavity between the absorber plate and insulation with paraffin wax and compared it's performance with solar air heating system without PCM. The efficiencies of the system without and with PCM were observed to be 17% and 33%, respectively.

Karthikeyan and Velraj [83] performed mathematical modelling for paraffin wax encapsulated in plastic spherical containers randomly packed in a storage tank. Inline to this model, Esakkimuthu et al. [84] carried out experiments using cylindrical tank filled with HS 58 (inorganic PCM) spherical capsules. The PCM capsule is covered with high density polyethylene. For the mass flow rate of 200 kg/hr the thermal transport was noted to be uniform with minimum pressure drop. Bouadila et al. [85] conducted experiments on a packed bed absorber arranged with spherical capsules. The spherical capsules were impregnated with SN27 and covered with polyolefin. The spherical capsules were coated black to improve the absorptivity of solar radiation. The energy and exergy efficiencies were noted to vary between 32–45% and 13–25%, respectively. Inline to this work, Arfaoui et al. [86] placed two-packed bed absorber i.e. two rows of spherical capsules above the absorber plate. The energy efficiency for this configuration was noted to be around 47%.

Charvat et al. [87] employed aluminum panels impregnated with Rubitherm RT42 as heat storage medium for solar air heating system. The panels were of rectangular cross section with internal dimensions of 0.44 m × 0.62 m × 1.8 m. The experiments were performed with 100 panels. The numerical model using TRNSYS showed good agreement with the experiments. The usage of PCM was observed to be attractive when compared to sensible heat storage medium. Lin et al. [88] made use of S21-commercial salt hydrate PCM for the thermal energy storage of solar air heating systems. The PCM was encapsulated in plastic bricks. By considering both average heat transfer effectiveness and average charging time the multi objective optimization was performed for the thermal storage system. Based on the study an optimized design was proposed. Saxena et al. [89] experimentally investigated the thermal performance of an air heating system with three different thermal energy storage materials. The materials employed were granular carbon powder, paraffin wax and combination of both. The considered thermal energy storage materials were encapsulated in a cylindrical copper tube and was placed between the glass cover and absorber plate. The combination of paraffin wax and granular carbon powder was observed to attain a thermal efficiency of 78.31%. Further, Saxena et al. [90] extended the study for helical tubes impregnated with granular carbon powder, paraffin wax and combination of both. The paraffin wax—granular carbon composite was noted to attain an efficiency of 79.10%. Jawad et al. [91] proposed solar air heater with aluminum chip and paraffin wax—nanoSiC composite as thermal energy storage media. The design could attain an outlet air temperature of 64.4 °C.

The PCM containers and their details for various applications are provided in Table 1 as shown below.

Table 1 Details about PCM containers for various applications

Sr. No	PCM	Shell/container material	Container/encapsulation size	Container/encapsulation geometry	Application	References
1	Metallic	Al_2O_3	Macro	Cylindrical	Thermal energy Storage	Fukahori et al. [21]
2	RT42	Stainless steel	Macro	Rectangular		Huang et al. [22]
3	RT35HC	Aluminum	Macro	Cylindrical		Chiew et al. [26]
4	N-Tetradecane	Urea and formaldehyde	Nano	Spherical		Fang et al. [28]
5	Al-Si alloy	Silica-rich layer	Macro	Cylindrical		Zou et al. [34]
6	Paraffin	Double layers of Silica	Micro	Spherical		Yu et al. [30]
7	Paraffin with porous expanded graphite	Aluminum	Macro	Rectangular	Electronic cooling	Yin et al. [37]
8	Nonadecane	Polyurethane	Nano	Spherical		Ghalambaz et al. [39]
9	Paraffin	Calcium carbonate	Micro	Spherical		Ren et al. [42]
10	Lauric acid	Plexiglass	Macro	Rectangular		Shokouhmand and Kamkari [36]
11	Eicosane	Formaldehyde	Nano	Spherical		Ho et al. [38]
12	Eutectic	Plexiglass	Macro	Rectangular	Food storage/transportation	Azzouz et al. [49]
13	Nano-structured calcium silicate—Rubitherm RT6 composite	Bubble wrap	Macro	Paper board container		Johnston et al. [50]

(continued)

Table 1 (continued)

Sr. No	PCM	Shell/container material	Container/encapsulation size	Container/encapsulation geometry	Application	References
14	Eutectic composition: water—ammonium chloride	Aluminum	Macro	Rectangular		Gin and Farid [51]
15	Natural deep eutectic solvents	Stainless steel, aluminum, natural fibre, high density polyethylene	Macro/micro	Rectangular		Salgaonkar et al. [52]
16	Commercial PCMs: RT-2, FT-15, PT-63	Aluminum with polystyrene/glass wool insulation	Macro	Cylindrical		Oro et al. [53]
17	Cristopia E-21 with 2% oxyethylmethylcellulose as thickening agent	–	Macro	Rectangular		Oro et al. [54]
18	Eutectic mixture of water and sodium chloride	Polystyrene	Macro	Rectangular		Leducq et al. [55]
19	RT35HC paraffin wax	Polyethylene panel with plastified metal sheet closing layer	Macro	Rectangular		Fioretti et al. [56]
20	Tetradecane	Aluminum	Macro	Rectangular		Arjenaki et al. [57]
21	Eutectic PCMs: E-26, E-29, and E-32	Polyurethane foam core with Aluzinc sheet metals at the exterior surface	Macro	Rectangular		Mousazade et al. [58]

(continued)

Table 1 (continued)

Sr. No	PCM	Shell/container material	Container/encapsulation size	Container/encapsulation geometry	Application	References
22	n-octanoic acid-myristic acid composite, potassium sorbate-water composite	Polyurethane	Macro	Rectangular		Xiaofeng and Xuelai [59]
23	Water and alkane mixture	High-density polyethylene (HDPE)	Macro	Rectangular	Drug storage/transportation	Huang and Piontek [60]
24	Paraffin wax	Plastic	Macro	Rectangular		Dingley et al. [62]
25	RT0, RT2HC, RT3HC, RT4HC, RT5HC and RT8HC	Polyurethane and vacuum insulated panel	Macro	Rectangular		Du et al. [63]
26	Tetra- decane (TD) and lauryl alcohol (LA) – expanded graphite (EG)	Polyethylene with vacuum insulated panel	Macro	Rectangular		Zhao et al. [64]
27	Paraffin wax	Galvanised iron	Macro	Rectangular	Solar water heating	Prakash et al. [67]
28	Calcium chloride hexahydrate, sodium ulphate decahydrate	PVC plastic	Macro	Cylindrical		Kaygusuz [69]
29	Paraffin wax	Copper	Macro	In the gap between the circular pipe and absorber plate		Mettawee and Assassa [77]
30	Sodium acetate trihydrate-graphite compound (90: 10) vol%	Aluminum	Macro	Cylindrical		Cabeza et al. [70]

(continued)

Table 1 (continued)

Sr. No	PCM	Shell/container material	Container/encapsulation size	Container/encapsulation geometry	Application	References
31	Paraffin wax	Aluminum	Macro	Cylindrical		Al-Hinti et al. [71]
32	Paraffin	–	Macro	Rectangular		Bouadila et al. [68]
33	Delta Cool-DC58 (hydrated salt), Pure Temp-PT58 (Vegetable based), Thermusol-HD60 (hydrated salt)	Carbon steel	DC58-Macro, PT58-Macro, HD60-Micro	Cylindrical		Porteiro et al. [72]
34	Eutectic mixture: Ammonium nitrate (37.5%) + Magnesium nitrate hexahydrate (62.5%)	Polyethylene	Macro	Rectangular, cylindrical and spherical		Barba and Spiga [75]
35	HS89	Stainless steel	Macro	Spherical		Afshan et al. [76]
36	Paraffin wax + Sci & CuO Nano composites	Aluminum	Macro	Cuboid		Manirathnam et al. [74]
37	Paraffin + CeO_2 nano composite	Aluminum	Macro	Cylindrical		Kumar and Mylsamy [73]
38	Rubitherm RT55	Stainless steel	Macro	Cylindrical		Carmona et al. [92]
39	Paraffin wax	Galvanized iron	Macro	Rectangular		Shalaby et al. [93]
40	Paraffin wax	–	Macro	Rectangular	Solar space heating	Enibe [80]

(continued)

Table 1 (continued)

Sr. No	PCM	Shell/container material	Container/encapsulation size	Container/encapsulation geometry	Application	References
41	Paraffin wax	Plastic	Macro	Spherical		Karthikeyan and Velraj [83]
42	HS 58 (Inorganic)	High density polyethylene	Macro	Spherical		Esakkimuthu et al. [84]
43	SN27	Polyolefin coated black	Macro	Spherical		Bouadila et al. [85]
44	Rubitherm RT42	Aluminum	Macro	Rectangular		Charvat et al. [87]
45	Paraffin wax	Aluminum	Macro	Rectangular		El Khadraoui et al. [82]
46	AC27 (Calcium chloride hexahydrate with additives)	Polyolefin coated black	Macro	Spherical		Arfaoui et al. [86]
47	Paraffin wax + aluminum powder with 0.5% wt of paraffin	Aluminum	Macro	Rectangular		Moradi et al. [81]
48	S21-Commercial salt hydrate	Plastic	Macro	Rectangular		Lin et al. [88]
49	Granular carbon powder, Paraffin wax	Copper	Macro	Cylindrical		Saxena et al. [89]
50	Paraffin wax + nano-SiC composite	Copper	Macro	Cylindrical		Jawad et al. [91]

2 Conclusions

The present work reviews different containers used for the phase change materials for various applications, namely, thermal energy storage, electronic cooling, food and drug transportation and solar water and space heating. It is reported that the rectangular containers followed by cylindrical one are the most widely used containers due to fast charging of the PCMs. The metallic containers, such as, copper, aluminum, stainless steel are preferred for high heat transfer applications, whereas, polyurethane, high-density polyethylene and vacuum insulated panels are used where heat transfer is to be inhibited such as in food and drug storage /transportation applications. Among metallic containers, aluminum is widely used due to it's high thermal conductivity, good corrosion resistance and substantially lower weight. For encapsulated PCMs silicon is widely used as a shell material due to it's high thermal conductivity and high structural strength.

Future work needs to be done in the improvement/reduction of the heat exchange between the container and the PCM. There is a good scope in this domain by devising a new technology or using the existing technology such as use of fins, metal matrix, a better container material etc.

References

1. Ali S, Deshmukh SP (2019) An overview: applications of thermal energy storage using phase change materials. Mater Today: Proc 26:1231–1237
2. Borri E, Sze JY, Tafone A, Romagnoli A, Li Y, Comodi G (2020) Experimental and numerical characterization of sub-zero phase change materials for cold thermal energy storage. Appl Energy 275(April):115131
3. Tiwari V, Srinivasan P (2020) SWOT analyses of high-temperature phase change materials for thermal energy storage. Mater Today: Proceed
4. Joulin A, Zalewski L, Lassue S, Naji H (2014) Experimental investigation of thermal characteristics of a mortar with or without a micro-encapsulated phase change material. Appl Therm Eng 66(1–2):171–180
5. Lakhdari YA, Chikh S, Campo A (2020) Analysis of the thermal response of a dual phase change material embedded in a multi-layered building envelope. Appl Thermal Eng 179:115502
6. El Ouali A et al (2019) Heat transfer within mortar containing micro-encapsulated PCM: numerical approach. Constr Build Mater 210:422–433
7. Saxena R, Rakshit D, Kaushik SC (2020) Experimental assessment of Phase Change Material (PCM) embedded bricks for passive conditioning in buildings. Renew Energy 149:587–599
8. Tan FL, Tso CP (2004) Cooling of mobile electronic devices using phase change materials. Appl Therm Eng 24(2–3):159–169
9. Behi H, Ghanbarpour M, Behi M (2017) Investigation of PCM-assisted heat pipe for electronic cooling. Appl Therm Eng 127:1132–1142
10. Sundaram AS, Seeniraj RV, Velraj R (2010) An experimental investigation on passive cooling system comprising phase change material and two-phase closed thermosyphon for telecom shelters in tropical and desert regions. Energy Build 42(10):1726–1735
11. Mehryan SAM, Ghalambaz M, Sasani Gargari L, Hajjar A, Sheremet M (2020) Natural convection flow of a suspension containing nano-encapsulated phase change particles in an eccentric annulus. J Energy Storage 28(January):101236

12. Praveen B, Suresh S, Pethurajan V (2019) Heat transfer performance of graphene nano-platelets laden micro-encapsulated PCM with polymer shell for thermal energy storage based heat sink. Appl Thermal Eng 156:237–249
13. Tahmasebi A, Zargartalebi H, Mehryan SAM, Ghalambaz M (2020) Thermal and hydrodynamic behavior of suspensions comprising nano-encapsulated phase change materials in a porous enclosure. Int Commun Heat Mass Transf 116(June):104634
14. Rodríguez-Cumplido F, Pabón-Gelves E, Chejne-Jana F (2019) Recent developments in the synthesis of microencapsulated and nanoencapsulated phase change materials. J Energy Storage 24(April):100821
15. Khadiran T, Hussein MZ, Zainal Z, Rusli R (2015) Encapsulation techniques for organic phase change materials as thermal energy storage medium: A review. Sol Energy Mater Sol Cells 143:78–98
16. Yataganbaba A, Ozkahraman B, Kurtbas I (2017) Worldwide trends on encapsulation of phase change materials: a bibliometric analysis (1990–2015). Appl Energy 185:720–731
17. Giro-Paloma J, Martínez M, Cabeza LF, Fernández AI (2016) Types, methods, techniques, and applications for microencapsulated phase change materials (MPCM): a review. Renew Sustain Energy Rev 53:1059–1075
18. Gimenez-Gavarrell P, Fereres S (2017) Glass encapsulated phase change materials for high temperature thermal energy storage. Renew Energy 107:497–507
19. Guo S et al (2018) Mobilized thermal energy storage: Materials, containers and economic evaluation. Energy Convers Manage 177(June):315–329
20. Zayed ME et al (2020) Recent progress in phase change materials storage containers: Geometries, design considerations and heat transfer improvement methods. J Energy Storage 30:101341
21. Fukahori R, Nomura T, Zhu C, Sheng N, Okinaka N, Akiyama T (2016) Macro-encapsulation of metallic phase change material using cylindrical-type ceramic containers for high-temperature thermal energy storage. Appl Energy 170:324–328
22. Huang S et al (2020) Experimental study on the influence of PCM container height on heat transfer characteristics under constant heat flux condition. Appl Thermal Eng 172:115159
23. Ghafari S, Khorshidi J, Niazi S, Samari F (2020) New correlations for investigating the melting of phase change material loading green synthesized CuO nanosheets in a spherical container: experimental study. J Energy Storage 32(July):101752
24. He S, Wang W, Wei L, Ding J (2020) Heat transfer enhancement and melting behavior of phase change material in a direct-contact thermal energy storage container. J Energy Storage 31:101665
25. Salunkhe PB, Shembekar PS (2012) A review on effect of phase change material encapsulation on the thermal performance of a system. Renew Sustain Energy Rev 16(8)
26. Chiew J, Chin CS, Toh WD, Gao Z, Jia J (2018) Low-temperature macro-encapsulated phase change material based thermal energy storage system without air void space design. Appl Therm Eng 141(June):928–938
27. Ghahramani Zarajabad O, Ahmadi R (2018) Employment of finned PCM container in a household refrigerator as a cold thermal energy storage system. Thermal Sci Eng Progress 7:115–124
28. Fang G, Li H, Yang F, Liu X, Wu S (2009) Preparation and characterization of nano-encapsulated n-tetradecane as phase change material for thermal energy storage. Chem Eng J 153(1–3):217–221
29. Leng G et al (2018) Micro encapsulated & form-stable phase change materials for high temperature thermal energy storage. Appl Energy 217(February):212–220
30. Yu X, Luan J, Chen W, Tao J (2020) Preparation and characterization of paraffin microencapsulated phase change material with double shell for thermal energy storage. Thermochimica Acta 689:178652
31. Song S et al (2019) Natural microtubule encapsulated phase change material with high thermal energy storage capacity. Energy 172:1144–1150

32. Liao H, Chen W, Liu Y, Wang Q (2020) A phase change material encapsulated in a mechanically strong graphene aerogel with high thermal conductivity and excellent shape stability. Compos Sci Technol 189:108010
33. McMurray JW, Jolly BC, Raiman SS, Schumacher AT, Cooley KM, Lara-Curzio E (2020) Ceramic encapsulated metal phase change material for high temperature thermal energy storage. Appl Thermal Eng 170(October):115003
34. Zou Q, Jie J, Shen Z, Han N, Li T (2019) A new concept of Al-Si alloy with core-shell structure as phase change materials for thermal energy storage. Mater Lett 237:193–196
35. El Omari K, Kousksou T, Le Guer Y (2011) Impact of shape of container on natural convection and melting inside enclosures used for passive cooling of electronic devices. Appl Therm Eng 31(14–15):3022–3035
36. Shokouhmand H, Kamkari B (2013) Experimental investigation on melting heat transfer characteristics of lauric acid in a rectangular thermal storage unit. Exp Thermal Fluid Sci 50:201–222
37. Yin H, Gao X, Ding J, Zhang Z, Fang Y (2010) Thermal management of electronic components with thermal adaptation composite material. Appl Energy 87(12):3784–3791
38. Ho CJ, Liu YC, Ghalambaz M, Yan MW (2020) Forced convection heat transfer of nano-encapsulated phase change material (NEPCM) suspension in a mini-channel heatsink. Int J Heat Mass Transfr, 155: 2020
39. Ghalambaz M, Hashem Zadeh SM, Mehryan SAM, Haghparast A, Zargartalebi H (2020) Free convection of a suspension containing nano-encapsulated phase change material in a porous cavity; local thermal non-equilibrium model. Heliyon 6(5):e03823
40. Seyf HR, Zhou Z, Ma HB, Zhang Y (2013) Three dimensional numerical study of heat-transfer enhancement by nano-encapsulated phase change material slurry in microtube heat sinks with tangential impingement. Int J Heat Mass Transf 56(1–2):561–573
41. Sabbah R, Farid MM, Al-Hallaj S (2009) Micro-channel heat sink with slurry of water with micro-encapsulated phase change material: 3D-numerical study. Appl Therm Eng 29(2–3):445–454
42. Ren Q, Guo P, Zhu J (2020) Thermal management of electronic devices using pin-fin based cascade microencapsulated PCM/expanded graphite composite. Int J Heat Mass Transf 149:1–16
43. ur Rehman T, Ali HM, Saieed A, Pao W, Ali M (2018) Copper foam/PCMs based heat sinks: an experimental study for electronic cooling systems. Int J Heat Mass Transf 127:381–393
44. Selvaraj V, Krishnan H (2020) Synthesis of graphene encased alumina and its application as nanofluid for cooling of heat-generating electronic devices. Powder Technol 363:665–675
45. Faraji H et al (2020) Cooling of recent microprocessors by the fusion of nano-enhanced phase change materials. Materi Today: Proceed
46. Sykes C (2018) Time- and temperature-controlled transport: supply chain challenges and solutions. P and T 43(3):154–158
47. Singh S, Gaikwad KK, Lee M, Lee YS (2018) Temperature-regulating materials for advanced food packaging applications: a review. J Food Meas Char 12(1):588–601
48. Alehosseini E, Jafari SM (2019) Micro/nano-encapsulated phase change materials (PCMs) as emerging materials for the food industry. Trends Food Sci Technol 91(October):116–128
49. Azzouz K, Leducq D, Gobin D (2009) Enhancing the performance of household refrigerators with latent heat storage: an experimental investigation. Int J Refrig 32(7):1634–1644
50. Johnston JH, Grindrod JE, Dodds M, Schimitschek K (2008) Composite nano-structured calcium silicate phase change materials for thermal buffering in food packaging. Curr Appl Phys 8(3–4):508–511
51. Gin B, Farid MM (2010) The use of PCM panels to improve storage condition of frozen food. J Food Eng 100(2):372–376
52. Salgaonkar CP, Kulkarni RS, Kulkarni DD (2011) Eutectic: Phase Change Material for Food Storage. Int J Curr Eng Technol 4(4):99–101
53. Oró E, Cabeza LF, Farid MM (2013) Experimental and numerical analysis of a chilly bin incorporating phase change material. Appl Therm Eng 58(1–2):61–67

54. Oró E, De Gracia A, Cabeza LF (2013) Active phase change material package for thermal protection of ice cream containers. Int J Refrig 36(1):102–109
55. Leducq D, Ndoye FT, Alvarez G (2015) Phase change material for the thermal protection of ice cream during storage and transportation. Int J Refrig 52:133–139
56. Fioretti R, Principi P, Copertaro B (2016) A refrigerated container envelope with a PCM (Phase Change Material) layer: experimental and theoretical investigation in a representative town in central Italy. Energy Convers Manage 122:131–141
57. Arjenaki NO, Soltanizadeh N, Hamdami N (2018) Designing an active phase change material package for thermal and qualitative protection of meat. Food Packag Shelf Life 21(July):2019
58. Mousazade A, Rafee R, Valipour MS (2020) Thermal performance of cold panels with phase change materials in a refrigerated truck. Int J Refrig 120:119–126
59. Xiaofeng X, Xuelai Z (2021) Simulation and experimental investigation of a multi-temperature insulation box with phase change materials for cold storage. J Food Eng 292(August):110286
60. Huang L, Piontek U (2017) Improving performance of cold-chain insulated container with phase change material: an experimental investigation, Appl Sci (Switzerland), 7(12):2017
61. Aniruddha C, Devakar D (2019) Episode II: efficient cold chain transport using phase change materials (PCMs). PLUSS Advanced Technologies Pvt. Ltd., Gurgaon, Haryana
62. Dingley J, Thatcher N, Williams D (2019) A study of temperature control in different designs of emergency drug transport bags. Anaesthesia 74(7):868–874
63. Du J, Nie B, Zhang Y, Du Z, Wang L, Ding Y (2020) Cooling performance of a thermal energy storage-based portable box for cold chain applications. J Energy Storage 28(January):101238
64. Zhao Y, Zhang X, Xu X, Zhang S (2020) Development of composite phase change cold storage material and its application in vaccine cold storage equipment. J Energy Storage 30(April):101455
65. Sharma A, Chen C (2009) Solar water heating system with phase change materials. Int Rev Chem Eng (I. RE. CH. E.), 1(4):297–307
66. Salunkhe PB, Jaya Krishna D (2017) Investigations on latent heat storage materials for solar water and space heating applications. J Energy Storage 12:243–260
67. Prakash J, Garg HP, Datta G (1985) A solar water heater with a built-in latent heat storage. Energy Convers Manage 25(1):51–56
68. Bouadila S, Fteïti M, Oueslati MM, Guizani A, Farhat A (2014) Enhancement of latent heat storage in a rectangular cavity: Solar water heater case study. Energy Convers Manage 78:904–912
69. Kaygusuz K (1995) Experimental and theoretical investigation of latent heat storage for water based solar heating systems. Energy Convers Manage 36(5):315–323
70. Cabeza LF, Ibáñez M, Solé C, Roca J, Nogués M (2006) Experimentation with a water tank including a PCM module. Sol Energy Mater Sol Cells 90(9):1273–1282
71. Al-Hinti I, Al-Ghandoor A, Maaly A, Abu Naqeera I, Al-Khateeb Z, Al-Sheikh O (2010) Experimental investigation on the use of water-phase change material storage in conventional solar water heating systems. Energy Conver Manage 51(8):1735–1740
72. Porteiro J, Míguez J, Crespo B, de Lara J, Pousada J (2016) On the behavior of different PCMs in a hot water storage tank against thermal demands. Materials 9(3):213
73. Kumar PM, Mylsamy K (2020) A comprehensive study on thermal storage characteristics of nano-CeO_2 embedded phase change material and its influence on the performance of evacuated tube solar water heater. Renew Energy 162:662–676
74. Manirathnam AS, Dhanush Manikandan MK, Hari Prakash R, Kamesh Kumar B, Deepan Amarnath M (2020) Experimental analysis on solar water heater integrated with Nano composite phase change material (SCi and CuO). Mater Today: Proceed
75. Barba A, Spiga M (2003) Discharge mode for encapsulated PCMs in storage tanks. Sol Energy 74(February):141–148
76. Afshan ME, Selvakumar AS, Velraj R, Rajaraman R (2020) Effect of aspect ratio and dispersed PCM balls on the charging performance of a latent heat thermal storage unit for solar thermal applications. Renew Energy 148:876–888

77. Mettawee EBS, Assassa GMR (2006) Experimental study of a compact PCM solar collector. Energy 31(14):2958–2968
78. Xie L, Tian L, Yang L, Lv Y, Li Q (2017) Review on application of phase change material in water tanks. Adv Mech Eng 9(7):1–13
79. Tyagi VV, Panwar NL, Rahim NA, Kothari R (2012) Review on solar air heating system with and without thermal energy storage system. Renew Sustain Energy Rev 16(4):2289–2303
80. Enibe SO (2003) Thermal analysis of a natural circulation solar air heater with phase change material energy storage, 28(14)
81. Moradi R, Kianifar A, Wongwises S (2017) Optimization of a solar air heater with phase change materials: experimental and numerical study. Exp Thermal Fluid Sci 89(January):41–49
82. El Khadraoui A, Bouadila S, Kooli S, Guizani A, Farhat A (2016) Solar air heater with phase change material: an energy analysis and a comparative study. Appl Therm Eng 107:1057–1064
83. Karthikeyan S, Velraj R (2012) Numerical investigation of packed bed storage unit filled with PCM encapsulated spherical containers—a comparison between various mathematical models. Int J Therm Sci 60:153–160
84. Esakkimuthu S, Hassabou AH, Palaniappan C, Spinnler M, Blumenberg J, Velraj R (2013) Experimental investigation on phase change material based thermal storage system for solar air heating applications. Sol Energy 88:144–153
85. Bouadila S, Kooli S, Lazaar M, Skouri S, Farhat A (2013) Performance of a new solar air heater with packed-bed latent storage energy for nocturnal use. Appl Energy 110:267–275
86. Arfaoui N, Bouadila S, Guizani A (2017) A highly efficient solution of off-sunshine solar air heating using two packed beds of latent storage energy. Sol Energy 155:1243–1253
87. Charvát P, Klimeš L, Ostrý M (2014) Numerical and experimental investigation of a PCM-based thermal storage unit for solar air systems. Energy Build 68(PARTA):488–497
88. Lin W, Ma Z, Ren H, Gschwander S, Wang S (2019) Multi-objective optimisation of thermal energy storage using phase change materials for solar air systems. Renew Energy 130:1116–1129
89. Saxena A, Verma P, Srivastava G, Kishore N (2020) Design and thermal performance evaluation of an air heater with low cost thermal energy storage. Appl Thermal Eng 167(November):114768
90. Saxena A, Agarwal N, Cuce E (2020) Thermal performance evaluation of a solar air heater integrated with helical tubes carrying phase change material. J Energy Storage 30(April):101406
91. Jawad QA, Mahdy AMJ, Khuder AH, Chaichan MT (2020) Improve the performance of a solar air heater by adding aluminum chip, paraffin wax, and nano-SiC. Case Stud Thermal Eng 19:100622
92. Carmona M, Rincón A, Gulfo L (2020) Energy and exergy model with parametric study of a hot water storage tank with PCM for domestic applications and experimental validation for multiple operational scenarios. Energy Convers Manage 222(July):113189
93. Shalaby SM, Kabeel AE, Moharram BM, Fleafl AH (2020) Experimental study of the solar water heater integrated with shell and finned tube latent heat storage system. J Energy Storage 31

Containers Based Drug Delivery for Neuroscience

Dania Alphonse Jose, Krishnapriya, and P. S. Baby Chakrapani

List of Abbreviations

BBB	Blood Brain Barrier
CNS	Central Nervous System
DALY	Years of Life Lost (YLL)
YLD	Years Lived with Disability
CMT	Carrier Mediated Transport
RMT	Receptor Mediated Transport
CDDS	Chemical Drug Delivery System
CSF	Cerebrospinal Fluid
CPP	Cell Penetrating Peptides
FUS	focused ultrasound
Tf	iron binding protein transferrin
TfR	Tf receptor
PEG-PLA	Poly ethylene glycol-poly lactic acid

D. A. Jose · Krishnapriya · P. S. Baby Chakrapani (✉)
Centre for Neuroscience, Cochin University of Science and Technology (CUSAT), Kochi 682022, Kerala, India
e-mail: chakrapani@cusat.ac.in

Department of Biotechnology, Cochin University of Science and Technology (CUSAT), Kochi 682022, Kerala, India

P. S. Baby Chakrapani
Inter University Centre for Nanomaterials and Devices (IUCND), Cochin University of Science and Technology (CUSAT), Kochi 682022, Kerala, India

Centre of Excellence in Advanced Materials, Cochin University of Science and Technology (CUSAT), Kochi 682022, Kerala, India

J. Parameswaranpillai et al. (eds.), *Micro- and Nano-containers for Smart Applications*, Composites Science and Technology,
https://doi.org/10.1007/978-981-16-8146-2_14

1 Introduction

The brain is a delicate and vital organ of the human body; hence it is safeguarded from circulating blood via a barricade. This unique membranous fence made of epithelial cells is known as the Blood–Brain Barrier (BBB). This fence protects the central nervous system (CNS) from the entry of therapeutic products drafted to help in cerebral diseases. The BBB is a dynamic anatomical barrier separating the parenchyma of the cerebellum, sealing the brain from any blood-borne diseases. It prevents the entry of any molecule from the circulatory system into the brain [1]. This aspect of BBB is a boon and a curse. Most CNS disorders remain incurable mainly due to two reasons namely, most brain diseases are complex and the other is the mode of efficient drug delivery hinders drug development. In the past 20 years, there has been significant growth in neuroscience and drug delivery development. Micro and macromolecules are equally studied to treat various brain disorders [2]. However, only small less than 400 Da lipid-soluble molecules can cross the BBB; most macromolecules cannot cross. Hence, this fence created by the BBB stops 95% of potential therapeutic products. Drug delivery methods that can effectively and preferentially deliver drugs to CNS is of great interest.

The existing methods for the treatment of neurological disorders are not well established. However, enthusiastic researchers bring up encouraging developments that put the future in a bright spot. It gives reasonable hope in the treatment of CNS disorders. Active research is required to solve the mystery of drug delivery to the brain and to develop less toxic drug molecules than what is existing in the market presently.

2 Blood Brain Barrier

In 1885, Paul Ehrilch noticed that when given an intravenous injection of trypan blue dye in laboratory animals, all tissues were coloured save the brain tissue. In 1913, a student of Edwin Goldmann injected trypan blue directly into the cerebrospinal fluid and found that it stains the brain tissue but not the rest of the body tissues ending the belief that brain tissues do not have the affinity for such dyes. This experiment proved that the brain was separated from the rest of the body by a barrier. However, the anatomy and nature of this ghost barrier evaded the scientific community for an extended period. In 1933, Reese and Karnovsky's [3] followed by Brightman and Reese [4] provided the final proof of a barrier that came to be known as the Blood–Brain Barrier (BBB).

In the CNS, the capillaries that help in blood circulation lack the minute pores through which solute movement occurs and are sealed by tight junctions. The BBB is composed of brain microvascular endothelial cells, astrocytes and pericytes. Another significant feature of this BBB is that, unlike in other tissues, they lack clefts between cells and do not contain fenestrae and pinocytes; hence diffusion occurs transcellular.

The other functions of BBB include transport of ions, nutrients required for CNS functioning and the removal of toxic material out of the CNS by transport systems.

As mentioned before, BBB is selective, so only small lipophilic molecules can passively cross this barrier [5], making the brain practically impossible to the entry of any other materials. A majority of the brain surface area is composed of the microvessels that constitute the main course in the brain through which chemicals can enter the brain [6]. Due to the proximity of the capillaries in the brain and the anatomy of vessels (they have a smaller diameter and thinner walls than the vessels of other tissues), once a drug crosses the BBB, it can rapidly reach all brain tissue [7].

The blood–brain barrier is an interface between the brain and blood. BBB protects the CNS from blood-borne disease-causing organisms in healthy individuals, and it blocks the diffusion of inflammatory molecules to the brain. Moreover, the endothelial cells break down any solutes crossing the cell membrane. However, with the help of transporters, specific solutes can undergo catalyzed transport. Glucose transport is an example for this method; it is an essential nutrient required for brain functioning but has low lipid solubility (which prevents its entry across BBB). The glucose transporter ensures movement of glucose across this fence [8]. The BBB is disrupted in conditions like multiple sclerosis, Parkinson's disease, Alzheimer's disease, diabetes, seizures, and stroke.

3 Types of Drug Delivery

Drug delivery is the mode of administering a pharmaceutical compound to achieve a therapeutic effect in animals or humans. There are mainly three classifications in drug delivery; controlled rate, slow delivery and targeted drug delivery. A controlled rate drug delivery aims at releasing the drug in a finite dose to the desired site at a fixed time. This method ensures the maximum efficacy of the drug with minimum side effects. On the other hand, slow drug delivery provides prolonged effect. The frequency of drug administration can be reduced and comparatively fewer side effects. The third method is targeted drug delivery or smart drug delivery. As the name states, this method delivers the specific dose of medicine exactly to the targetted area. The CNS is separated from the rest of the body by a series of physiological, biochemical, and metabolic barriers. These barriers include the blood–brain barrier, blood tumour barrier and blood-cerebrospinal fluid barrier. These drug delivery systems are being pursued commonly based on their applicability, ability to cross the various barriers and area of target. As mentioned above the treatment of CNS-related disorders are particularly cumbersome and challenging

4 Brain as a Target for Drug Delivery

There are different type of brain disorders that affects the proper functioning of the CNS. It includes psychiatric disorders, neurodegenerative diseases, neurodevelopmental disorders, are few among them. Recent trends have shown that one in every six people has some form of brain ailment. The cost of treatment amounts to more than a trillion dollars annually. Brain diseases are the major contributor to deaths and disability worldwide. In 2016, the Global Burden of Diseases, Injuries, and Risk Factors Study (GBD) conducted a study. They tried to estimate the degree of CNS diseases and summed up the findings based on disability-adjusted life-years (DALY), which is calculated based on the sum of years of life lost (YLL) and years lived with disability (YLD). They found that neurological diseases are the leading cause of DALY and second in place for causes of death between 1990 and 2016 [9].

Ailments are on the rise due to diet, exercise, lifestyle and genetics. The complexity of the disease is one of the main hurdles for treating these disorders. Most brain-related diseases are not curable yet. The other challenge is to deliver the drug correctly to the brain. There are limited techniques for drug delivery to the brain. Crossing the blood–brain barrier has plagued the medical field for a long time.

5 Strategies for Drug Delivery to the Brain

Several strategies have been experimented with for accurate drug targeting to circumvent the BBB. The methods broadly fall into three categories: techniques that will disrupt the BBB, manipulate the drugs, or alter the process of drug delivery.

Osmotic or biochemical disruption is the commonly used methods to disrupt BBB. Mannitol injection into the neck region can help for temporary disruption of BBB. This temporary disruption is reversible but remains open for approximately 20–30 mins. During this period, the drugs can be delivered to the site of interest. The drug can quickly diffuse through the gaps formed in the BBB. This method also utilizes administering chemotherapeutic agents to treat patients with glioma, germ cell tumour in the brain, and cerebral lymphoma. However, this is not ideal because disrupting the BBB means opening up the CNS for external factors. This procedure makes the CNS stripped naked from its defense mechanism, making it vulnerable. With the disruption of BBB by any means, the brain is open to infections, blood-borne diseases, exposed to toxins, and chemicals circulated. Both the osmotic and biochemical disruption methods increase the penetrating property of the drug through the circulatory system. But this penetration is not selective to the brain alone but the entire body. The loss of BBB can alter the balance of the whole system. It may lead to stress, increase in cranial pressure, exposing therapeutic agents to normal brain tissues [10]. Intra-arterial administration of chemotherapeutic agents has increased survival rates from 11.4 to 17.5 months in patients [11].

Manipulation of BBB using drugs is hence a better method for drug delivery. They include prodrugs, carrier-mediated transport (CMT), Receptor-mediated transport (RMT), Chemical drug delivery systems (CDDS) and using lipophilic analogues of the drug. However, even these measures have their pros and cons. Some have toxic effects, low efficacy, are invasive methods, while others are not sufficiently safe [12].

Altering the drug delivery model includes systemic agents that must penetrate the BBB so that drugs can enter the CNS. Despite advancements in science and technology, drug delivery methods are still inefficient. Many potential drugs cannot penetrate BBB and enter the brain at therapeutic concentrations. Very few compounds can get past the BBB, and these have been indentified as drug delivery modes. One such method is to use lipophilic compounds that can cross the BBB, and encapsulating the drug in the core of a lipophilic compound. Similarly, another attempted drug delivery mode is coupling a therapeutic drug with efflux or transporter pumps [13].

Alternatively, disruption of BBB, which can be classified into physical or non-physical modes is another method. The drug molecules can pass into the brain parenchyma through an opening in the endothelial cells caused by the disruption method. For drug delivery in brain tumours, on applying an osmotic pressure gradient across the BBB, the BBB is disrupted via osmotic shrinkage [14]. The drug carriers used can increase the longevity and half-life in the blood. So the drug clearance via excretion by the liver, spleen and kidney will be less affected making the drug more potent [15].

Below is a detailed elucidation of the various drug delivery methods and their pros and cons.

5.1 Intra-cerebellar Delivery

Intra cerebellar method is a direct approach but invasive at the same time. But it is feasible only in limited disorder treatments and possible only in a few patients. The drug delivers by direct injection and controlled release. Disruption of BBB though promising, is not localized and hence the systemic approach leads to complications like seizures and tumour migration possibilities [11]. However this method can be specifically targeted at a site of interest which can reduce the off-target toxicity in comparison with systemic approach [16].

An alternative, less invasive method is to administer the drug into the cerebrospinal fluid (CSF). CSF is in continuous interaction with the interstitial fluid in the brain. Hence injecting the therapeutic compound in the CSF can aid in crossing the BBB, and we can deliver the drug to the cerebral ventricles. The diffusion pathways in the brain are complex and limited to small compounds. The diffusion coefficient decreases with size; in other words, the smaller the molecule, the better the diffusion [17]. The compound in the CSF will diffuse only at a slower rate which may hinder its penetration property. It is a significant problem, especially in humans, due to its size, compared to other species. Due to this, even promising results observed in rats [18] may not be reproducible in humans. Another problem related to this mode of drug delivery is that CSF clearance occurs at a high rate. In humans, CSF turnover is every 5 or 6 h. Hence, intraventricular injections will show the same result as slow intravenous injections. Brain penetration will be minimal if it's a big molecule like a neurotrophic factor.

5.2 *Colloidal Drug Carriers*

A colloidal drug delivery system is drug vector that can deliver the drug to the target site. Colloidal systems carry nano-sized particles of the range of 1–500 nm. Examples of colloidal drug carriers are liposomes, micellar solutions and nanoparticles. They have low toxicity and are a targeted form of drug delivery [19]. These tiny particles can incorporate the drug and deliver it to the target. The main attraction of colloidal systems is its size and surface morphology. Moreover, we can modulate these properties to attain the needed release speed and biological activity.

Colloidal carriers can be polymeric or lipid materials that encapsulate therapeutic agents, modifying the human body's drug release profile and rate. The release profile, biological activity, and compatibility may depend on the size, surface charge, morphology, and matrix composition. Colloidal systems are administrable for all routes (oral, pulmonary, nasal, and ophthalmic), even for the intravenous route being their size below the smallest vessels (5–6 μm in diameter) [15].

5.2.1 Liposomes

In 1961, Alec D. Bangham introduced liposomes. Liposomes are lipid bilayers encapsulating an aqueous core. Mainly phospholipids are used for liposome preparation, a hydrophilic head and two chains of hydrophobic tails. This property makes liposomes water-soluble in one end and insoluble in another. This lipid bilayer forms a closed structure to protect its hydrophobic tails from the aqueous solution. The water-soluble drug gets trapped in the core, while water-insoluble medicines will incorporate into the phospholipid layer.

Liposomes deliver a drug in various ways to the target site. In some cases, these liposomes attach to the cell membrane, fuse with them and release the drug. In the case of phagocytic cells, liposomes are engulfed, and then lysosomes act on these phospholipid layers breaking it and releasing the drug.

BBB selectively allows the passage of only specific types of molecules. One such property is lipophilicity. Lipophilic products cross the BBB by simple diffusion if the molecule is less than 400 kDa. Initially, a lipophilic analogue of the drug served this purpose. Later it was performed by encapsulation of the therapeutic compound in a lipophilic moiety. Hence, liposomes became the first generation of colloidal drug delivery methods [20]. Not surprisingly, practically all medicinal products available in the market use this lipid-soluble property to cross BBB and deliver the drug. This is not an invasive method and is more favoured because it requires only simple oral administration.

Liposomes are of three different types based on the size and bilayer number. Small and unilamellar ranges of the size 10–50 nm, large-sized and unilamellar of the size 50–100 nm and multilamellar liposomes of the size 20–100 nm.

One example of liposome-mediated drug delivery is the development of dopamine hydrochloride-bearing liposomes. Positively charged small liposomes are sonicated and packed with dopamine hydrochloride. An in-vivo study using Sprague Dawley rats (SD) showed that liposomes are an effective drug delivery method. Drug degradation tendency can be reduced by encapsulating it in a liposome [21]. Similarly, monoclonal antibodies encapsulated in liposomes coated with transferrin receptor or insulin receptors, has been constructed to deliver therapeutic products across BBB. For the treatment of Parkinson's disease using a rat model, monoclonal antibodies coated with liposomes with tyrosine hydrolase were effective [22]. Researchers have also explored modifying the physical structure, drug delivery efficacy and drug leakage in liposome encapsulated drug delivery method [23], and liposome action is extensively studied and experimented [24].

5.2.2 Miscelles

Micelles are another colloidal form of drug delivery obtained by autoaggregation in an aqueous environment. The amphiphobic areas usually obtained from poly ethylene glycol, polycaprolactone and fatty acids. In contrast, amphiphilic portions are usually from poly ethylene glycols. They are amphiphilic block copolymers of size 5–50 nm. The drug gets packed in the core and sealed in the case of micellar drug delivery and it is found to be an advantage of this drug delivery system. These drugs maintain a concentration higher than their solubility with the surrounding water. Hydrogen bonding is formed between the aqueous surrounding and core, hence protecting the core from lysis attack of enzymes. Their chemical compositions, length, and weight can easily be modified to make an ideal drug delivery system. The stability of micelles can be increased by adding functional block copolymer groups that can cross-link [25]. Polymeric micelles have a dimension of approximately 10–100 nm. It can dissolve a high quantity of drugs and prevent drug release and degradation [15]. An example of this is pluronic micelles incorporated with low molecular weight drugs. It facilitates drug delivery to the brain with better pharmacokinetics (due to high stability in plasma) [26].

5.2.3 Nanoparticles

Particles of size 10–1000 nm in diameter are called nanoparticles. When discussing whether nanoparticles can be considered as a container for drug delivery, the first question is whether nanoparticles can cross BBB. First of all, their size (<200 nm), can penetrate leaky BBB, seen commonly in brain tumour conditions.

The drug of interest will be bound as a solid dispersion or adsorbed on the nanoparticle's surface. Hence this can facilitate drug delivery and more excellent retention in such conditions [27]. It is one of the most promising forms of drug delivery. However, due to their small size, nanoparticles need not cross over in a healthy BBB.

An example of this is poly ethylene glycol-poly lactic acid (PEG-PLA) complexed nanoparticles. These are unable to cross the healthy BBB [28]. To cross the leaky BBB, multiple factors like shape, size and properties of the surface influence [29]. Nanoparticles can also carry drug payloads, modify pharmacokinetics and control drug release. Nanoparticles used for drug delivery are generally made of polymers of either natural or artificial origin on which drugs bound as a solution or dispersion. The drugs are generally chemically attached or absorbed on their surface.

Transport across BBB has been shown using animal experiments for compounds such as adriamycin, kyotorphin, loperamide and dalargin. They were attached to a biodegradable polymer (poly(butyl)-cyanoacryl, coated with a surfactant (polysorbate 80)) [30]. The first nanoparticle-based drug delivery was using hexapeptide dalargin to the brain. It was a Leu-enkephalin analogue with opioid activity [31]. With the nanotechnology boom, nanoparticles have been exploited as an intriguing tool to cross the BBB and enhance drug delivery potential. Typically nanoparticle designing is performed in a way; the drug will circulate more in the blood than other traditional drugs. Hence, it maintains a higher concentration at the BBB interface, giving the drug a higher chance to cross the BBB.

Polymeric nanoparticles are efficient for drug delivery in the brain. These encapsulate the drug and protect it from the first-pass metabolism, making it possible to cross the BBB without damaging the BBB [32]. There are many advantages in using nanoparticles that have been discussed. They are primarily used for drug delivery, targeted delivery and their potential to be used both as a diagnostic tool and therapeutic property is being exploited. Primarily, nanotechnologists will focus on improving the stability, ensuring distribution of active compounds, targeting and loading drugs, and, most importantly, improving barrier interaction. The major drawback of nanotechnology and nanoparticles is cytotoxicity. The exact mechanism by which the nanoparticles can transport drug into the brain is still unclear. So the existing hypothesis is focus on particles size, material composition, and structure of the nanoparticle. One of the studies found that nanoparticles mimic the properties of compounds that generally cross the BBB. One example of this is polysorbate-coated nanoparticles. These incline to imitate low-density lipids (LDL), making the nanoparticle possible to cross the BBB without any hindrance [33]. The endocytosis of therapeutic molecules depends on the particle size, surface property and charge of the material. A study showed nanoparticles with an average diameter of 10 nm without any charge can circulate in the blood for a prolonged period. And it showed a low rate of endocytosis by the immune phagocytosis [15]. Few nanotherapeutics like SGT-53 can cross BBB, while drugs like Doxil can disrupt BBB [34].

Many researchers spoke regarding the internalization of nanoparticles into the endothelial cells through BBB [35]. However, recent researches have pointed out that majority of these nanoparticles get trapped in the cell, and only 5% of them successfully crossed the BBB even after 48 h [36]. The immune system of the body considers nanoparticles as foreign objects. So they become opsonized, and the body's defense mechanism tries to remove it by phagocytosis. However, chemical modification in the surface of nanocarrier by polymers like poly ethylene glycol which imparts stability and extend the period of drug in the blood is a promising future method of drug delivery [36].

5.3 Pro Drugs

A prodrug is a medical compound that can transform itself into a pharmacologically active drug on administration [37]. The theory behind prodrugs is the structural changes needed for a drug to produce the desired action. Chemical modifications are done on these drugs to improve their membrane permeability or lipophilicity. Some of the shortcomings of these drugs are difficulty in formulation, fast clearance from blood and poor taste. This modified drug-prodrug with its improved characteristics is better than its parent drug form. It must undergo a biochemical/chemical conversion and change to its active form to become functionally active. In neuroscience, prodrugs are commonly used to deliver a drug to the brain. These prodrugs are disguised as a lipophilic compound and hence can easily cross the BBB.

A well-known example is heroin which is morphine with a diacyl derivative which makes it more lipophilic. Heroine can cross BBB 100 times more efficiently than morphine [38]. Similarly, codeine is also a morphine derivative. Codeine has a methylated hydroxyl group, giving codeine a better permeability (tenfold better) than morphine. The most commonly seen method in prodrug designing is to couple the drug through an ester linkage, which is an easy and reversible procedure. The esterification of carboxylic, amino or hydroxyl groups can increase the lipophilicity of the parent drug. The CNS is rich in esterases and hence these prodrugs can easily be converted back to its parent drug in the CNS. Thus ester-based prodrugs are particularly favoured in neuroscience.

5.4 Microbubbles Coupled with Focused Ultrasound

Small monolipid molecules (gaseous bubbles) that can pass through the BBB due to their lipophilic nature are called microbubbles. They are named so due to their bubble-like appearance. One main drawback of this structure is that it is large and cannot passively diffuse across the BBB unless provided with a supporting force. Recently it has been noted that they can transiently move through the BBB under burst type exposure of energy. This energy is produced from focused ultrasound (FUS). When projected to the BBB, an ultrasound causes interference in the tight junctions, increasing the BBB's permeability, allowing microbubbles to pass through [39]. The detachment of tight junctions on the application of the ultrasound will not cause any neuronal damage [40]. A targeted method as only the area where ultrasound has disrupted the barrier allows for better permeability. Microbubbles combined with focused ultrasound have become an effective drug delivery mode in neuroscience. The site-specificity and focused ultrasound can transiently open BBB reversibly. The microbubbles filled with active therapeutic molecule passes through these gaps. The ultrasound will focus site-specifically, allowing the microbubble to cross the disrupted BBB and precisely deliver the drug to the site.

Focused ultrasound can deliver various therapeutic molecules in diseased models (preclinical stage). Chemotherapeutic compounds, therapeutic peptides, nanoparticles and even monoclonal antibodies can deliver by this method. Over the years, significant advances have been made in the development of microbubbles for human treatment. Some are entering clinical trials to treat patients with neurodegenerative diseases like Alzheimer's disease, Parkinson's disease, dementia and ALS. Preliminary results published for treatment against brain tumours are promising and shows efficacy and no toxicity [39]. Few commercialized microbubbles in the market are Optison® (from GE Healthcare, USA), Definity® (from Lantheus Medical Imaging, USA), and SonoVue® (from Bracco, Italy). Application of FUS has transpired, so the high risk associated with invasive surgeries will reduce. Repeated treatments are possible with microbubbles without complications of surgical methods. The delivery of microbubbles can be now performed by simple means.

Moreover, this removes the risky arterial catheterization used for other treatments. FUS induced BBB opening is irreversible and can be closed within 6 h. The BBB, once closed, was observed to function normally and tends to remain closed [41].

The method is not as easy as it sounds, mainly due to two reasons. First of all, the drug encapsulation should be stable without any leakage. Secondly, the size of the loaded and unloaded bubble should remain the same as this can interfere with the BBB cross over. The ideal drug release method for crossing the BBB is still under debate. High power ultrasound also has the potential for causing thermal coagulation. The cavitation formed due to the collapse of gas bubbles causes bioeffects. The effects range from reopening of the BBB to haemorrhage [42, 43]. Most of the FUS studies performed in healthy animals found that BBB closes by itself. However, these results need not be similar in cases of diseased models and once opened with FUS, BBB might not be able to repair itself. A recent study shows that the kinetics of FUS-mediated BBB opening is different in a model of Alzheimer's disease and control models [44].

5.5 *Chemical Delivery Systems*

This method utilizes the principle of sequential metabolism. Chemical modifications are made to chemical delivery systems to make it an inactive derivative of the original drug but comparable to the original parent molecule. CDS can be used for site-specific targeting and can deliver the drug and activate drugs at their target site or enhanced site delivery. Or through a multistep transformation via chemical or enzymatic mode. In addition to being lipophilic, they exploit the bidirectional property of the BBB.

Once the drug crosses the BBB and enters the brain, the inactive drugs precursors are 'locked' inside the CNS and prevent them from travelling out of the CNS. The CDS, when administered, is a lipophilic and inactive compound making it easy to diffuse through the BBB. The target moiety in the CDS gets oxidized and makes it hydrophilic, which can no longer cross back through the BBB. When entering the brain, the inactive drug compound will be lipophilic, making it possible to cross the BBB. Once it enters the brain, a hydrophilic molecule will not cross out of the BBB, locking the drug inside the brain. The enzymatic release of the active drug can precisely act in the brain in a sustained manner. CDS is used as a carrier to take molecules to the brain and used to retain specific compounds in the brain that have no access to the brain in other conditions. An example for this type of targeted delivery is corticosteroid hormones [45].

5.6 Exosomes

Extracellular vesicles that are of endocytic origin are exosomes. They come from the cells and are present in fluids like saliva, urine and plasma. Exosomes contain siRNA and hence play roles in several physiological processes. The function of exosomes in the brain includes synaptic plasticity, cell communication, neurogenesis and neuronal response.

Exosomes can cross BBB and release their contents, and this property is exploited for drug delivery to treat neurological disorders. One study demonstrated the successful delivery of siRNA to the brain via exosomes in animal models [46]. Exosomes can be used as a target for drug delivery for the brain. Exosomes successfully delivered siRNA to the brain and resulted in specific gene knockdown [46]. Later, similar experiments showed successful delivery of exosomes performed in mice delivered via intranasal injection [47]. Recent studies show that exosomal movement is bidirectional, but there exist no clear evidence for this. One hypothesis of transport of exosomes is said to be a transcellular method through endocytosis. An in vitro model performed exosome transfer based on an adsorptive transcytosis method [48]. The crossing of exosomes is found to be present in both healthy and diseased conditions.

However, exosomal movement across BBB was significantly higher when peripheral lipopolysaccharides were administered [48]. Another study showed that exosomes could cross BBB in both healthy and stroke-like conditions, proving that exosomes function well even in stress conditions, making it an ideal drug delivery method for traeatment of diseases [49]. A study found that exosomes were internalized through endocytosis and then released on the other side of the BBB. This endocytosis might be clathrin dependant because when given chlorpromazine (CPZ), an inhibitor for clathrin-dependent endocytosis, there was decrease in the exosomal migration across the BBB [50]. Based on the payload of the exosomes, the mode by which exosome uptake occurs is different [49]. At the same time, during disease, crossing BBB might vary as the cargo can differ [49].

The capacity of the exosome to cross the BBB makes it a suitable container for drug delivery to CNS. Animal studies use exosomes for the treatment of Parkinson's disease. This study used an exosome-based delivery, and drug accumulation is observed in the target neurons [51]. In another study of morphine addiction, exosomes were used to deliver MOR siRNA into the brain of a mouse, which reduced the MOR in the brain that inhibited morphine addiction-related relapse [52]. Hence exosomes are a promising container-based drug delivery method for many brain disorders, including mental illnesses. This promising line of research needs further investigation for application in neuroscience.

5.7 Cell-Penetrating Peptides

They are peptide sequences that are oligocationic or amphiphilic with a length of 10–30 amino acids. These can move across mammalian plasma membranes. Hence intracellular and transcellular transport of even large molecules is possible through this method. However, the mechanism by which this movement happens is not clear yet. One such theory of action occurs via endocytotic uptake. For drug delivery using CPP it has to be chemically coupled to the drug. Initially, ligands of receptors and transporters were in use for the delivery system.

In contrast, the free endogenous ligands bound competitively. To fix this, researchers were compelled to find other alternatives. One was to use a random ligand from an existing library, and the other was to remodify the current ligands. Redesigning a ligand has its merit in that this design modification increases drug-binding capacity. This method is different from the Trojan horse methodology because CPP cannot transport specifically across the BBB and can lead to peripheral side effects. CPP modifications on the drug enhanced its ability to improve targeted delivery to the brain [53] and high binding affinity [54] capacity [55, 56].

5.8 Receptor-Mediated Transport

In the CNS, the macromolecular movement occurs through an interaction between proteins and the receptors of the endothelial cells. Uptake of macromolecules occurs through endocytosis or transcytosis [57]. It is present in various cells, neurons and endothelial cells. Unidirectional transcytosis includes the transfer of molecules from the apical to the basolateral membrane. All endothelial cells of the brain can perform transcytosis [58]. There are three types of endocytic vesicles in the brain; clathrin-coated vesicles are majorly involved with RMT. Once these vesicles internalize, they move to the common intracellular pathway [59].

Receptor-mediated transfer (RMT) provides large peptides, including enzymes, hormones, growth factors and proteins. Delivery to the brain using RMT is proposed based on creating antibodies and ligands against receptors present on the surface of

CNS endothelial cells. Since the number of receptors on the cell surface is limited, this process is usually a saturable method. This system uses peptide-specific receptors like insulin, leptin, transferrin and mediates the uptake of the corresponding endogenous peptide. The drug (large peptide) is conjugated with the transport vector by done by avidin–biotin technology. Hence can be used only for a relatively small number of molecules. Thus, it can provide a therapeutic effect only for few active compounds. The target receptor generally has low capacity and susceptible to downregulation. That also can interfere with the transport of the endogenous substrate.

A standard receptor-mediated transfer that occurs in the CNS is the transport of iron across BBB. It is by binding the protein transferrin (Tf) to the receptor (TfR). This receptor is a target for multiple studies. The therapeutic molecules are functionalized with anti-TfR antibodies and cross the BBB. A few examples of this type of RMT are MYBE/4C1 antihuman TfR conjugated with the drug doxorubicin used as an anticancer drug [60].

Similar to this anti-insulin receptor (IR), it was developed to transport therapeutic materials for the brain. After showing good internalization, HIRMAb delivers to the brain after intravenous administration. HIRMAb attached to L-idorunidase is used to deliver enzymes in a genetic disorder like Mucopolysaccharidosis. It is a lysosomal storage disease [61].

Molecular Trojan horses are created by joining therapeutic proteins with insulin receptors, including transferrin receptors and monoclonal antibodies for efficient BBB penetration [57, 62–64]. The reported successful RMT formulation with high permeability across BBB includes Angiopep-2 [65]. Angiopep-2 is a peptide ligand of low-density lipoprotein receptor [66, 67]. Similarly, FC5 and FC44 (which are single domain antibodies) are vectors developed for brain delivery [68].

5.9 Carrier Mediated Transport

Carrier-mediated transport (CMT) is a suitable drug delivery system for the transportation of small molecules. They can pass the BBB down a concentration gradient. They are also suitable carriers for even nutrients and hormones. The transporters are substrate selective and can transport materials in both uni and bi-directionally through a mono-, co-, or counter- transport mode. Some drugs found to have an affinity for the BBB due to their structure; L-DOPA is an example. It is a dopamine precursor and is used in treating Parkinson's disease. L-DOPA is transported to the brain via the LAT-1 carrier protein. L-DOPA is a neutral amino acid, and LAT-1 is a carrier protein for a neutral amino acid [69].

For CMT, drugs need modification to increase their affinity for carrier proteins, with a moiety. An example of such modification is the addition of cysteine. Addition of a neutral amino acid to a drug can disguise it as a large neutral amino acid. Hence, this method of drug delivery expects an increased uptake [69]. Another method is to conjugate the drug to an antibody against a receptor of BBB. Such receptors include the human insulin receptor or the transferrin receptor. The drug binds to the receptor in BBB, and transport is facilitated across the BBB, acting as a Trojan horse in molecular transport. These proteins or Trojan horses are widely used as therapeutics [62]. For drug delivery to the CNS, structural similarities between drug molecules and endogenous transporter substrates need further research.

6 Future Prospects

Significant progress has been seen in the area of drug targeting in neuroscience to treat disorders of the CNS. However, most of these finding remain in labs, and until recent year, only a few nanoparticles are under evaluation clinically, but no nanoformulation is under development. The efficacy of these drugs is not yet satisfactory.

After evaluating efficacy and toxicity, moving from the lab (animal models) to the clinical stage (human trials) poses various challenges. The FDA approval for marketing the formulations is required once the trials in humans are successful. The three stages of clinical trials are:

Phase I: Perform on healthy human individuals to standardize dose along with toxicity evaluation.
Phase II: Perform on a larger group of patients who has the illness of interest to evaluate safety and efficacy.
Phase III: Is random, including both control and placebo groups [70].

Over the years, several drug delivery formulations meant to deliver therapeutics to the CNS have been granted approval by the FDA like ANG1005, SGT-53 etc. The transition of these drugs from pre-clinical investigations to being marketed stipulates the feasibility of container-based drug delivery systems. However, fewer formulations in the pipeline of clinical stages show that the extrapolation of lab results to human studies faces many challenges.

The studies described in this chapter prove the feasibility of drug delivery methods that can cross the BBB and successfully treat diseases pertaining to the CNS. However, there are back draws, and many improvements are required for crossing the remaining hurdles. An in-depth scientific evaluation is needed to address the issues faced in drug delivery strategies for neuroscience.

7 Conclusion

The BBB infrastructure has always made drug delivery to the brain a complex issue that has left the scientist puzzled. However, many methods have evolved in the past decade, and delivery methods are no more a challenge with advancements in this field. The above-mentioned novel drug delivery systems help and promises new opportunities in CNS therapys. Various containers like colloidal drug carriers, microbubbles, receptor-mediated, prodrugs etc are being developed with great envision. With safety and efficacy as focal points, and the need for ideal treatment, it requires focused research in developing drug delivery systems for brain-related disorders.

References

1. Banks WA (2016) From blood–brain barrier to blood–brain interface: new opportunities for CNS drug delivery. Nat Rev Drug Discov 15:275–292
2. Goyal D, Shuaib S, Mann S, Goyal B (2017) Rationally Designed Peptides and peptidomimetics as inhibitors of amyloid-β (Aβ) aggregation: potential therapeutics of Alzheimer's disease. ACS Comb Sci 19:55–80
3. Reese TS, Karnovsky MJ (1967) Fine structural localization of a blood-brain barrier to exogenous peroxidase. J Cell Biol 34:207–217
4. Brightman MW, Reese TS (1969) Junctions between intimately apposed cell membranes in the vertebrate brain. J Cell Biol 40:648–677
5. Engelhardt B, Sorokin L (2009) The blood–brain and the blood–cerebrospinal fluid barriers: function and dysfunction. Semin Immunopathol 31:497–511
6. Lund-Andersen H (1979) Transport of glucose from blood to brain. Physiol Rev 59:305–352
7. Saunders NR et al (2014) The rights and wrongs of blood-brain barrier permeability studies: a walk through 100 years of history. Front Neurosci 8
8. Stewart PA, Tuor UI (1994) Blood-eye barriers in the rat: Correlation of ultrastructure with function. J Comp Neurol 340:566–576
9. Feigin VL et al (2019) Global, regional, and national burden of neurological disorders, 1990–2016: a systematic analysis for the Global Burden of Disease Study 2016. Lancet Neurol 18:459–480
10. Rodriguez A, Tatter S, Debinski W (2015) Neurosurgical techniques for disruption of the blood-brain barrier for glioblastoma treatment. Pharmaceutics 7:175–187
11. Gumerlock M, Belshe B, Madsen R, Watts C (1992) Osmotic blood-brain barrier disruption and chemotherapy in the treatment of high grade malignant glioma: patient series and literature review. J Neurooncol 12
12. Mukhopadhyay A, Mukhopadhyay B, Basu SK (1995) Circumvention of multidrug resistance in neoplastic cells through scavenger receptor mediated drug delivery. FEBS Lett 376:95–98
13. Misra A, Ganesh S, Shahiwala A, Shah SP (2003) Drug delivery to the central nervous system: a review. J Pharm Pharm Sci 6:252–273
14. Taylor EM (2002) The impact of efflux transporters in the brain on the development of drugs for CNS disorders. Clin Pharmacokinet 41:81–92
15. Craparo EF, Bondì ML, Pitarresi G, Cavallaro G (2011) Nanoparticulate systems for drug delivery and targeting to the central nervous system. CNS Neurosci Ther 17:670–677
16. Duskey JT et al (2017) Current strategies for the delivery of therapeutic proteins and enzymes to treat brain disorders, Int Rev Neuroboiol 137:1–28
17. Buchwald P, Bodor N (2001) A simple, predictive, structure-based skin permeability model. J Pharm Pharmacol 53:1087–1098
18. Alessandrini A, Namura S, Moskowitz MA, Bonventre JV (1999) MEK1 protein kinase inhibition protects against damage resulting from focal cerebral ischemia. Proc Natl Acad Sci 96:12866–12869
19. Deol P, Khuller G (1997) Lung specific stealth liposomes: stability, biodistribution and toxicity of liposomal antitubercular drugs in mice. Biochim Biophys Acta Gen Subj 1334:161–172
20. Kshirsagar NA, Pandya SK, Kirodian GB, Sanath S (2005) Liposomal drug delivery system from laboratory to clinic. J Postgrad Med 51(Suppl 1):S5-15
21. Shapira I, Budman DR, Bradley T, Gralla R (2009) Evolving lipid-based delivery systems in the management of neoplastic disease. Oncol Rev 3:113–124

22. Uppadhyay AK, Dixit VK (1998) Bioadhesive liposomes bearing levonorgestrel as controlled drug delivery system. Pharmazie 53:421–422
23. Karande SC et al (1995) Successful treatment of antimony-resistant visceral leishmaniasis with liposomal amphotericin B (L-AmpB-LRC) in a child. Trop Doct 25:80–81
24. Pardridge WM (2007) shRNA and siRNA delivery to the brain. Adv Drug Deliv Rev 59:141–152
25. Jain NK, Rana AC, Jain SK (1998) Brain drug delivery system bearing dopamine hydrochloride for effective management of parkinsonism. Drug Dev Ind Pharm 24:671–675
26. Yang S, Zhu J, Lu Y, Liang B, Yang C (1999) Body distribution of camptothecin solid lipid nanoparticles after oral administration. Pharm Res 16:751–757
27. Saraiva C et al (2016) Nanoparticle-mediated brain drug delivery: Overcoming blood–brain barrier to treat neurodegenerative diseases. J Control Release 235:34–47
28. Huang L, Liu Y (2011) In vivo delivery of RNAi with lipid-based nanoparticles. Annu Rev Biomed Eng 13:507–530
29. Tosi G, Costantino L, Ruozi B, Forni F, Vandelli MA (2008) Polymeric nanoparticles for the drug delivery to the central nervous system. Expert Opin Drug Deliv 5:155–174
30. Lee J, Duan W, Mattson MP (2002) Evidence that brain-derived neurotrophic factor is required for basal neurogenesis and mediates, in part, the enhancement of neurogenesis by dietary restriction in the hippocampus of adult mice. J Neurochem 82:1367–1375
31. Wohlfart S, Gelperina S, Kreuter J (2012) Transport of drugs across the blood–brain barrier by nanoparticles. J Control Release 161:264–273
32. Deo MR, Sant VP, Parekh SR, Khopade AJ, Banakar UV (1997) Proliposome-based transdermal delivery of levonorgestrel. J Biomater Appl 12:77–88
33. Tiwari G et al (2012) Drug delivery systems: an updated review. Int J Pharm Investig 2:2
34. Juillerat-Jeanneret L (2008) The targeted delivery of cancer drugs across the blood–brain barrier: chemical modifications of drugs or drug-nanoparticles? Drug Discov Today 13:1099–1106
35. Pirollo KF et al (2016) Safety and efficacy in advanced solid tumors of a targeted nanocomplex carrying the p53 gene used in combination with docetaxel: a phase 1b study. Mol Ther 24:1697–1706
36. Bramini M et al (2014) Imaging approach to mechanistic study of nanoparticle interactions with the blood-brain barrier. ACS Nano 8:4304–4312
37. Rautio J, Laine K, Gynther M, Savolainen J (2008) Prodrug approaches for CNS delivery. AAPS J 10:92–102
38. Oldendorf WH, Hyman S, Braun L, Oldendorf SZ (1972) Blood-brain barrier: penetration of morphine, codeine, heroin, and methadone after carotid injection. Science 80–178:984–986
39. Chen K-T, Wei K-C, Liu H-L (2019) Theranostic strategy of focused ultrasound induced blood-brain barrier opening for CNS disease treatment. Front Pharmacol 10
40. Hynynen K, McDannold N, Sheikov NA, Jolesz FA, Vykhodtseva N (2005) Local and reversible blood–brain barrier disruption by noninvasive focused ultrasound at frequencies suitable for trans-skull sonications. Neuroimage 24:12–20
41. McDannold N, Arvanitis CD, Vykhodtseva N, Livingstone MS (2012) Temporary disruption of the blood-brain barrier by use of ultrasound and microbubbles: safety and efficacy evaluation in Rhesus Macaques. Cancer Res 72:3652–3663
42. Vykhodtseva NI, Hynynen K, Damianou C (1995) Histologic effects of high intensity pulsed ultrasound exposure with subharmonic emission in rabbit brain in vivo. Ultrasound Med Biol 21:969–979
43. Vykhodtseva N, McDannold N, Hynynen K (2008) Progress and problems in the application of focused ultrasound for blood–brain barrier disruption. Ultrasonics 48:279–296
44. Burgess A, Hynynen K (2014) Drug delivery across the blood–brain barrier using focused ultrasound. Expert Opin Drug Deliv 11:711–721
45. Bodor N, Buchwald P (2010) Retrometabolism-based drug design and targeting. In: Burger's medicinal chemistry and drug discovery. Wiley & Sons, Inc. Pharmazie 65(6):395–403

46. Alvarez-Erviti L et al (2011) Delivery of siRNA to the mouse brain by systemic injection of targeted exosomes. Nat Biotechnol 29:341–345
47. Zhuang X et al (2011) Treatment of brain inflammatory diseases by delivering exosome encapsulated anti-inflammatory drugs from the nasal region to the brain. Mol Ther 19:1769–1779
48. Matsumoto J et al (2017) Transmission of α-synuclein-containing erythrocyte-derived extracellular vesicles across the blood-brain barrier via adsorptive mediated transcytosis: another mechanism for initiation and progression of Parkinson's disease? Acta Neuropathol Commun 5:71
49. Chen CC et al (2016) Elucidation of exosome migration across the blood-brain barrier model in vitro. Cell Mol Bioeng 9:509–529
50. Dutta D, Donaldson JG (2012) Search for inhibitors of endocytosis. Cell Logist 2:203–208
51. Haney MJ et al (2015) Exosomes as drug delivery vehicles for Parkinson's disease therapy. J Control Release 207:18–30
52. Liu Y et al (2015) Targeted exosome-mediated delivery of opioid receptor Mu siRNA for the treatment of morphine relapse. Sci Rep 5:17543
53. Qin Y et al (2011) Liposome formulated with TAT-modified cholesterol for improving brain delivery and therapeutic efficacy on brain glioma in animals. Int J Pharm 420:304–312
54. Wei X et al (2015) A D-Peptide ligand of nicotine acetylcholine receptors for brain-targeted drug delivery. Angew Chemie Int Ed. 54:3023–3027
55. Díaz-Perlas C et al (2017) Phage display as a tool to discover blood-brain barrier (BBB)-shuttle peptides: panning against a human BBB cellular model. Biopolymers 108, e22928
56. Thom G et al (2018) Isolation of blood-brain barrier-crossing antibodies from a phage display library by competitive elution and their ability to penetrate the central nervous system. MAbs 10:304–314
57. Xiao G, Gan LS (2013) Receptor-mediated endocytosis and brain delivery of therapeutic biologics. Int J Cell Biol (2013)
58. Pulgar VM (2019) Transcytosis to cross the blood brain barrier, new advancements and challenges. Front Neurosci 13:1–9
59. Rodriguez-Boulan E, Kreitzer G, Müsch A (2005) Organization of vesicular trafficking in epithelia. Nat Rev Mol Cell Biol 6:233–247
60. Gregori M et al (2016) Novel antitransferrin receptor antibodies improve the blood-brain barrier crossing efficacy of immunoliposomes. J Pharm Sci 105:276–283
61. Boado RJ et al (2008) Genetic engineering of a lysosomal enzyme fusion protein for targeted delivery across the human blood-brain barrier. Biotechnol Bioeng 99:475–484
62. Pardridge WM, Boado RJ (2012) Reengineering biopharmaceuticals for targeted delivery across the blood–brain barrier, Methods Enzymol 503:269–292
63. Boado RJ, Hui EK-W, Lu JZ, Pardridge WM (2012) Glycemic control and chronic dosing of rhesus monkeys with a fusion protein of iduronidase and a monoclonal antibody against the human insulin receptor. Drug Metab Dispos 40:2021–2025
64. Pardridge WM (2008) Re-engineering biopharmaceuticals for delivery to brain with molecular Trojan horses. Bioconjug Chem 19:1327–1338
65. Zlokovic BV, Deane R, Sagare AP, Bell RD, Winkler EA (2010) Low-density lipoprotein receptor-related protein-1: a serial clearance homeostatic mechanism controlling Alzheimer's amyloid β-peptide elimination from the brain. J Neurochem 115:1077–1089
66. Kanwar JR, Sriramoju B, Kanwar RK (2012) Neurological disorders and therapeutics targeted to surmount the blood-brain barrier. Int J Nanomedicine 7:3259–3278
67. Demeule M et al (2008) Identification and design of peptides as a new drug delivery system for the brain. J Pharmacol Exp Ther 324:1064–1072
68. van Rooy I et al (2010) Identification of peptide ligands for targeting to the blood-brain barrier. Pharm Res 27:673–682
69. Pardridge WM, Oldendorf WH (1975) Kinetics of blood-brain barrier transport of hexoses. Biochim Biophys Acta Biomembr 382:377–392
70. Valentinuzzi ME, Friedman LM, Furberg CD, DeMets DL (2004) Fundamentals of clinical trials, 3rd edn. Biomed Eng 3:43

CDs: A Potential Candidate for Improving Water Solubility and Stability of Hydrophobic Guest Molecules

Sahar Amiri and Sanam Amiri

1 Introduction

CDs are cyclic oligosaccharides composed of Glucopyranose units whcih α-[1,4] linked Glucopyranose units with a specific cone-shaped structures consisting a hydrophil exterior surface and a hydrophob inert cavity (Fig. 1) [1–3]. Three common CDs are α, β, and γ with of 6, 7, and 8 Glucopyranose units which are crystalline, homogeneous and non-hygroscopic substances produced via starch enzymatic degradation (Table 1) [2–4].

CDs shows low toxicity for administered orally and intravenously so have been used in pharmaceutical applications. Although their absorption is negligible, no side effects on guest molecules have been observed. Formation of the inclusion complex is based on non-covalent interactions without formation of covalent bonds. Formation of reversible bonding such as hydrogen bonds, van der Waals forces between the host and the guest molecules are critical in complex formation. In this mechanism entropy increases via release of water molecules from cavity and enthalpy decreases via formation of new hydrogen bonding [2]. Chemical reactions between CDs and various guest molecules lead to formation intramolecular reversible interactions and form supramolecular structures. The formation of inclusion complexes is able to modify the physical, chemical, and/or biological properties of the guest molecules [2].

Results showed that inclusion complex of guest molecules and CDs are crystalline with two main type; Channel structures and Cage structures. To synthesis of CDs

S. Amiri (✉)
Department of Polymer Engineering, Science and Research Branch, Islamic Azad University, Tehran, Iran
e-mail: S.amiri@srbiau.ac.ir

S. Amiri
Department of Polymer Engineering, Amirkabir University of Tehran, Tehran, Iran

J. Parameswaranpillai et al. (eds.), *Micro- and Nano-containers for Smart Applications*, Composites Science and Technology,
https://doi.org/10.1007/978-981-16-8146-2_15

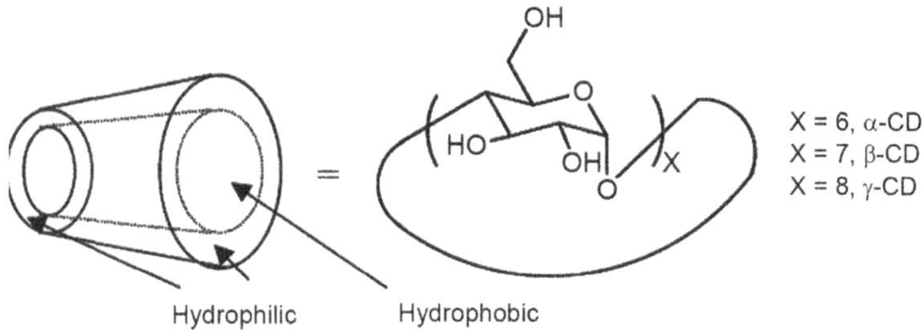

Fig. 1 Schematic diagram of CDs [2]

Table 1 CDs properties [1]

Property	α-CD	β-CD	γ-CD
Number of glucopyranose unit	6	7	8
Molecular weight (g/mol)	972	1135	1297
Solubility in water at 25 °C (% w/v)	14.5	1.85	23.2
Outer diameter (°A)	14.7	15.3	17.5
Cavity diameter (°A)	5.1	6.2	8.1
Height of torus (°A)	7.8	7.8	7.8
Cavity volume (°A^3)	174	262	427
Surface tension (MN/m)	71	71	71
Melting temperature range (°C)	255–260	255–265	240–245
Crystal water content (wt%)	10.2	13–15	8–18
Water molecules in cavity	6	11	17

derivations with different solubility and different functional groups, hydroxyl groups of the CDs can be modified chemically such as Aminations, Esterifications or Etherifications. By modification of CDs, solubility, stability against light or oxygen may improved [5–7].

As one of their important properties, CDs have the capability to form supramolecular interactions with various hydrophob and hydrophil compounds [2, 8]. The most important application of CDs is in drug based compounds, so toxicity and safety profiles of CDs is very important for researchers [9, 10]. Using CDs and their derivatives in various applications is well evidenced by the increasing number of marketed or approved medicinal, skin and cosmetic products containing CDs which potentially broadening the application areas of both CDs and functionalized compound by CDs [5]. The most applicable CDs can be divided to α-CD, β-CD and γ-CD. β-CD is the most common CD in medical and cosmetic applications [9, 10]. Of all the CD derivatives, available, HP-ß-CD is the safest, as it does not permeate the membranes, reduced haemolytic potential, making it suitable for parenteral use as well as for oral applications. γ-CD show low irritation after intramuscular injection, rapid and complete degradation by intestinal enzymes to glucose and the least toxic CD, at least of the three natural CDs, so used widely in industries applications.

1.1 Inclusion Complex Formation

What is considered CDs' amazing property is their structural features which allow for the selective formation of inclusion complexe formation with a different range of molecules. This ability is also known as molecular recognition, or chiral recognition when dealing with enantiomeric compounds. In this process, the guest compounds are partially or fully located inside the hydrophobic cavity of CDs [host] which is based on non-covalent bonding [2, 11–13]. Cavity of CDs showed hydrophobicity, the insertion of a guest molecule in the CD cavity is dependant to amount of water derived out from the cavity by substitution a more non-polar guest molecule which is energetically favorable for both the CD and the guest molecules in an aqueous or non-aqueous solution [11]. When a hydrophobic guest molecule is added to CD solution, water molecules in the CD cavity are substituted for the guest molecules. Complex formation induces structural changes in the CD and changes guest properties [2, 8]. Hydrophobicity of CD cavity provides a suitable microenvironment for formation of newlinkage between molecules and leads to formation of a new complex and induce physicochemical or biological changes in guest molecule's properties, such as enhancing the therapeutic potential, solubility, diffusion and decreasing drug decomposition before they enter tissues [14, 15].

By formation of complex between CDs and various compounds may also increasing solubility and stability of volatile and unstable entrapped molecule against the environmental and chemical trigires such as oxidation, light and heat, limit volatility and sublimation [2, 3, 16, 17]. Scheme 1 present the inclusion complex formation route between water molecules and drug molecules.

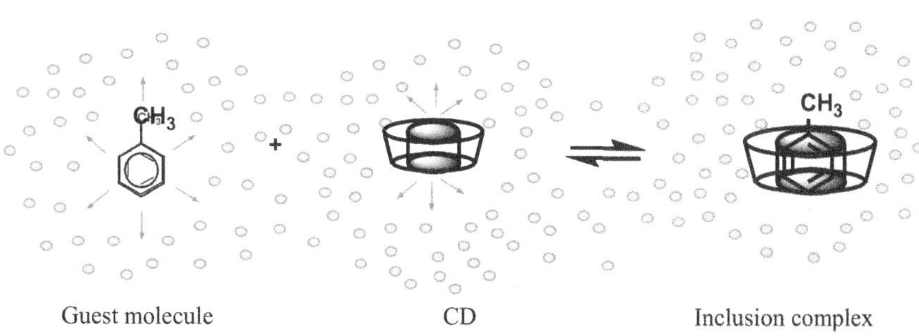

Scheme 1 Schematic presentation of inclusion complex formation (2)

1.2 Inclusion Complex Formation Mechanism

CDs as host molecules can entrap guest molecules in their cavity via structural fitting of intended substance in CD cavity. Guest molecule with larger size than the CD cavity space, can't be fully included in cavity so partially complexes formed [18]. Hydrophobic drug molecules inserted in CDs cavity via enthalpy release which is related to exitance of water due to the insertion of guest molecules of CDs as the polar–apolar interactions [19, 20]. Formation of revesible interactions led to formed a crystalline product. If guest molecule was large molecule such as linear polymer, the arrangement is channel type or column structure and guest molecule can incorporate in CD cavity and make an infinite cylindrical channel [7, 21].

One technique in formation of inclusion complex is co-precipitation method which the guest molecule add to a solution of CD with mixing at various conditions and precipitate has been collected [82, 83]. For guests with low solubility in water, an organic solvent can be used to dissolve the guest but the solvent must be removed by evaporation or other methods. Solvent has important role in inclusion complex formation between CD and guest molecule. By increasing solubility of CD in the solvent, higher molecules are available for formation of new interactions and yield of reaction is increased. Temperature act as a double effect parameter for increasing the conversion and also heat can also destabilize the complex.

2 Applications of CDs

CDs are a family of enzymatically modified with specific molecular structure and shape and alter physical, chemical consisiting a hydrophil outer surface and a lipophil inner cavity. CDs family can formed inclusion complex with a wide range of monomers so can be used in various applications such as medical, chemical, food and packaging industries. Inclusion complex formation may improve aqueous solubility of entrapped compounds and also increase bioavailability of hydrophob and volatile substances, reduce their irritation, prevent their incompatibility and mask their unpleasant odor and taste [2].

Inclusion complex formation and encapsulation may caused to beneficial changes in the properties of the guest molecules and increase shelf time of guest moleccules such as:

- Stabilize volatile compounds, light-sensitive and oxygen-sensitive compounds
- Control of Solubility of guest molecules
- Modification of the chemical and physical properties of guest molecules
- Protection against degradation of substances by UV light, oxidation and chemical reactions
- Masking of unpleasant smell or taste, pigments or the color of guest molecules
- Controlled released of guest molecule in drug delivery system [9].

2.1 Pharmaceuticals Applications of CD

In pharmaceutical industry, inclusion complex formation between hydrophobic or hydrophilic drugs and CDs increased water solubility and bioavailibilty of drug which can act as carrier for targeting drug delivery. Formation of inclusion complex is based on reversible non-covalent bonds [5, 6].

Major challenge in drug delivery is poor bioavailability of drugs which is related to poor solubility and low permeability of drugs. For an efficient drug delivery, the required amount of drug must be delivered to target site in the necessary period of time without undesirable effects on the other cells in the body with both efficiently and precisely. Hydrobic drugs show low solubility in water which restric their application in drug delivery systems, by formation of inclusion solubility of drugs improves [7]. CDs are able to formed reversible interaction with both hydrophob or hydrophil drugs. When the drug molecules are entrapped in CDs cavity or surface, hydrogen bonding or Vander Waals interactions are formed or broken during complexation [22–24].

In drug delivery applications, the sufficient amount of drug must be delivered to target sites in the necessary period of time without undesirable effects on the other cells in the body with both efficiency and precision. CDs are used for formation of inclusion complex with drugs in which the drug molecules are encapsulated and their undesirable properties are eliminated. Then it is delivered to target site with a controlled rate [25]. The most important challenge in drug delivery systems is drugs poor bioavailability which is because of poor solubility and low permeability of drugs. Various methods have been used to overcome this problem such as encapsulation of poor water-soluble drugs in CDs. Entrapment of hydrophobic guest molecules in CDs hydrophobic cavity, improve their aqueous solubility, permeability, dissolution rate and presystemic metabolism [26, 27].

2.2 Drug Delivery Systems Based on CDs Nanocontainers

CDs-based nanocarriers are good candidate for delivery of specific drugs to target cell without other cells damaged such as anticancer drugs. By insertion drug molecules in CD cavity, properties such as drug loading capacity and solubility, toxicity, controlled drug release, protect drugs from environmental conditions may be changed [4, 28, 29]. Supramolecular hydrogels and nanocapsules are formed based on formation of interaction between CDs and various polymers which intermolecular conjugation between drugs and hydrogels is due to a cross linking reaction and used as drug delivery systems [33, 34–37]

2.3 Water Solubility

One of the most impressive characteristic of encapsulated drugs in CDs cavity, is improved aqueous solubility of drugs. Hydrophobic drugs are inserted in the CDs hydrophobic cavity via formation of hydrogen bonding, while hydrophilic side of CD form interactions with the solvent and increase solvability of drugs. In the case of volatile drugs, encapsulation stabilizes guest and may increase its solubility [30, 31]. A wide range of drugs (about half the drugs) are water insoluble or poor water-soluble which limited their bioavailability, efficacy, safety and patient compliance. Formation of interaction between CDs and drugs lead to higher number of hydrogent bonding which improve drugs water solubility, dissolution and permeability with controlled release of the drugs and other encapsulated compounds [2, 32].

2.4 Drug Bioavailability

Most of drugs are water insoluble and show low bioavailability, low efficacy, safety and patient compliance. For overcome this problem, entrapment of drugs in CDs cavity via inclusion complex formation is most applicable and may improve their water solubility, dissolution and permeability.

When bioavailability of drugs improved, drug permeability at the surface of the biological barrier is improved [17].

2.5 Mask Unpleasant Odor, Taste and Side Effects of Drugs

Addition to improve drug water affinity, dissolution rate of drug and absorption rate of drugs, via formation on inclusion complex with CDs, contact time between the drug and the body decreased, limited drug irritation and amount of contact with taste receptors, improved drug safety which is very important in the preparation of oral solutions [33].

2.6 Improve Drug Photostability and Shelf Life

Unstable and volatile drug usage is limited because of their short shelf life and reactions with active agents such as dehydration, hydrolysis, oxidation and photodecomposition. Encapsulated of drug in CDs cavity and due to reversible interaction between drug and active agents, drugs protect against various degradation processes which affected by drug functional group [26]. Spices, flavors and essential oils such as lemon oil are volatile substances at room temperature and can't be used as ordinary applications and became stable at room temperature via formationof inclusion complex by CDs [17, 22]. Formation of inclusion complex postponed drug hydrolysis compare to free drugs and protected it [32, 38, 39].

2.7 Drug Delivery Carriers Based on CDs

Drug delivery carriers based on CDs show wide range application as carrier in pharmaceutical formulations for reach specific dosage of drugs to a targeted site with controlled release rate and efficiency. Novel carriers or capsules will be design via inclusion complex formation between CDs and drugs with controlled release [40]. CDs carriers and capsules are applicable as oral, transdermal or ocular delivery systems.

2.8 Oral, Rectual and Nasal Route Drug Delivery Systems

Oral route drug delivery is limited due to gastrointestinal absorption and drug delivery is either controlled via dissolution, diffusion role, osmotical pressure, density or pH [12]. In this method, CDs-based carriers moderate drug concentration in target site and prevent an un-controlled aglomoration or drug concentration which hydrophobic drug incorporated in CD cavity, solubility of drug improved and delivered to the surface of the ocular barrier [29, 41–43].

One of the challenges for drug delivery systems is delivery of bit and nauseating drugs which is hard for children or infants and elderly. Rectal drugs delivery systems are a good candidate to delivery of bit and nauseating drugs which is hard for children or infants and elderly. This route has some limitation such as drugs with low rectal bioavailability or limited absorbing surface and also for drug which are sensitive to rectum microorganisms. To overcome limitation of rectal drugs delivery systems, specific drugs encapsulated in CDs cavity for masking undesirable taste and bitter taste [34, 44, 45].

Lipophilic drugs and large hydrophilic drugs such as peptides and proteins with poor water solubility didn't deliver via nasal absorption. CD based drug carriers are good for deliver drugs with low oral bioavailability and incitement gastrointestinal via nasal delivery systems. Complexation of CDs with nasal drugs led to higher solubility of drug, improved drug delivery through biological barriers and enhancing their nasal absorption [34, 35].

2.9 *Transdermal Drug Delivery System*

Transdermal systems are novel route to deliver drugs across skin into systemic circulation and can be used for numerous clinical indications. This procedure, which administers the drug through the skin, interacts with any component of skin and skin irritation for local and systemic action without altering the pH of the skin. In this system drug released in blood circulation over a long-desired period of time with a controlled release rate without any harmful effect in the body cell. stratum corneum permeation is critical parameter for the transdermal delivery and drugs must maintain sufficient stability until delivered to target site. CDs and their derivatives are used as transdermal absorption promoters in topical delivery which stabilized drug in transdermal drug delivery for safe therapeutically effects and didn't interact or irritated to skin or change pH of the skin [9, 10, 18, 36, 37].

3 Drug Delivery Applications of Drug-Encapsulated Drugs

3.1 *Improve Water Solubility of Acetamidophenol Suppositories*

Suppositories are solid drug formulations that are softened for rectal administration at body temperature and can be useful in the long-term treatment of diseases. Suppositories have some limitations in the routine use that make them the patient's last resort [33, 38, 39]. Researches indicated that by insertion of lipophil and hydrophil polymers with water solubility led to a suppositories with a slow and controlled rate of release. Polyethylene glycol (PEG) is a polymer with good water solubility and also wide range of molecular weight is available [38]. PEGs of varying molecular weight may be combined to achieve the optimal rate of dissolution of drugs and the extrication of drugs can be influenced by the quality of the base and their concentration [46].

Acetamidophenol (ACP) is a crystalline powder with a $T_m = 170$ °C, is used to alleviate physical ache and eliminate fever. However, this compound shows low water solubility and suppository bases poor dissolution, bioavailability of this compound is limited. Subsequently, to improve the solvability of ACP and to consider the effect of interaction between ACP and β-CD on solubility and release of ACP from suppository bases (PEG and Witepsol), different mass ratios of β-CD: ACP are prepared to form inclusion complexes. Rectal suppositories are prepared with different suppository bases such as PEG and Witepsol H15. The prepared inclusion complexes were then formulated in suppository formulations by fusion method (Table 2). Weight variation, disintegration, melting time, mechanical strength and dissolution test were investigated [47].

Figure 2 and Table 3 shows XRD spectrums of pure ACP, β-CD, P2, P3, P4

Table 2 Formulation of obtained suppositories [47]

Formula	H1	H2	P1	P2	P3	P4	P5
PEG 4000/400 (mass ratio = 6:4)	–	–	1.2/0.8	1.2/0.8	1.2/0.8	1.2/0.8	1.2/0.8
H 15 (gr)	1.4	1.4	–	–	–	–	–
ACP (gr)	0.3	–	0.3	–	–	–	–
ACP: β-CD mass ratio	–	8:1	–	1:1	2:1	4:1	8:1

and P5 in different conditions and their peaks which clearly indicated formation of ACP/β-CD complex.

The most typical structure is when a guest molecule in the host cavity positions, and the molecules interact with each other to form new structures. Peaks obtained at $2\theta = 13, 18, 20, 25$ verify Channel structure and the peaks obtained at $2\theta = 13, 18, 20, 25$ verify cage structure. Results showed that ACP/β-CD is a combination of channel and cage structure. Figure 3 shows the dissolution profile from various suppositorial formulations. Results show that the percentage of accumulative drug release from Witepsol-based suppositories, H1 and PEG-based suppositories is 35.49 and 59.06% respectively, after 4 h. Fatty bases (Hard Fat H15) with the dissolution media are hydrophobic and non-miscible. Their release rate is therefore lower compared to hydrophilic PEG bases [28]. The rate of drug release in formulations P2, P3, P4 and P5 decreases with formation of complex between β-CD and ACP. ACP is insoluble in water but also the obtained complex reduces the diffusion of drugs from polymer base and extends the dissolution time. In the formulation, unreacted molecules induce the initial blast, which follows the gradual release from suppositories [48].

Unlike other formulations, in P2 sample ratio of guest to host is 1:1 specifies the most desirable ratio with the rhe less release after 8 h of 19.6 percent. About P3, P4 and P5, as the quantity of ACP exceeds the quantity of β-CD, the rate of release of drugs increases, indicating less interaction between drugs and CD cavity and higher number of unreacted drug molecules, leading to a higher rate of release. Cumulative release rate in fatty-based suppositories, H1 and H2, is found to be 35.49 and 29.83% after 4 h, respectively, which supports the fact that complex formation by adding β-CD to the medication reduces drug dissolution rate [8]. Results indicate that in samples with mass ratio 1:1 of β-CD and ACP show the most desirable properties with release percent of 19.3% of ACP after 8 h [47].

3.2 *Ciprofloxacin/CDs Nanocontainers*

Fluoroquinolones (FLQ) are synthetic antibiotics that serve as an important class of antibiotics and antimicrobials which act against a wide range of Gram-positive and Gram-negative bacteria. The limitation of such compounds is side-effects for humen [49]. FLQs have different form (Scheme 2) but bioavailability of ciprofloxacins is limited in the aqueous fluids due to their poor solvability and low release rate.

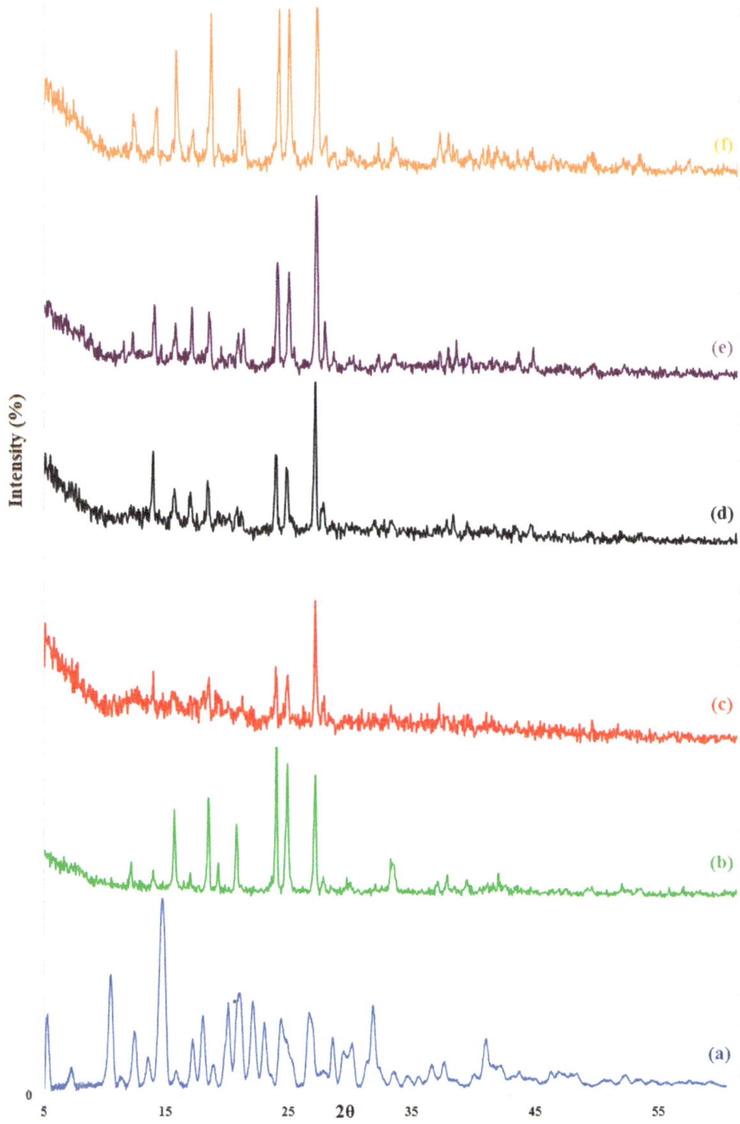

Fig. 2 XRD spectrums: **a** β-CD, **b** ACP, **c** P2 containing ACP/β-CD (1:1), **d** P3 containing ACP/β-CD (2:1), **e** P4 containing ACP/β-CD (4:1), **f** P5 containing ACP/β-CD (8:1)

Table 3 Peaks found in XRD in ACP/β-CD complex [47]

Compound	Peak (2θ)
2θ = 12.05, 15–16, 18, 20, 23.4, 24.3, 26, 32.45, 40.8	ACP
2θ = 5–7, 8, 10, 12, 13.4, 16, 17, 18, 19, 21, 22, 24, 27, 32, 34.8	β-CD
2θ = 5 7, 8, 12, 13, 14, 15, 17, 18, 23, 24, 26, 32, 35	P2
2θ = 5–8, 10, 12, 13, 15, 16, 17, 18, 20, 23–24, 26, 31	P3
2θ = 5–8, 11, 13, 14, 15, 16.65, 17, 20–21, 23, 24, 26, 27, 32	P4
2θ = 5–8, 12, 13, 15–16, 7, 18, 20, 23, 24, 26–27, 32.7, 36	P5

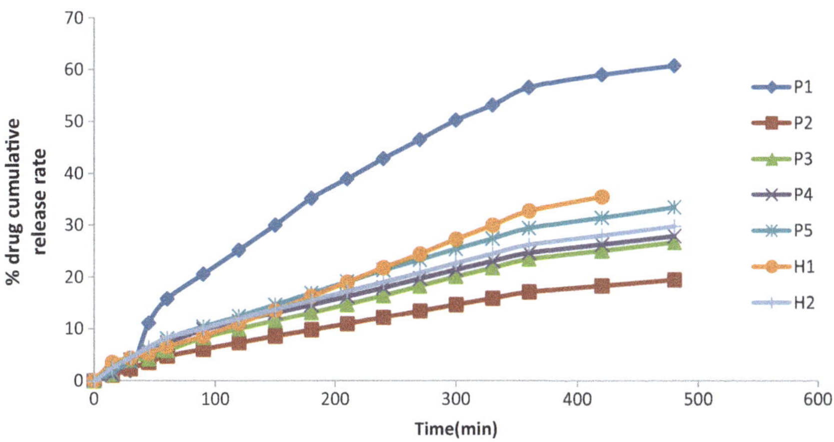

Fig. 3 In vitro release of ACP and ACP: β-CD from PEG and Witepsol based suppositories

Supramolecular containers of ciprofloxacin/CDs were prepared to overcome this critical problem by means of complex formation between ciprofloxacin and CDs (α-CD and β-CD) under different conditions.

FTIR, SEM, DSC and XRD have shown the existence of ciprofloxacin tanks. When ciprofloxacin is encapsulated with CDs via inclusion complexes, the slow release of the drug from the container results in an effective delivery system for hydrophobic drugs which ensures extended delivery [50, 51]. Structures obtained were characterized after the new bindings between CDs and ciprofloxacin [50]. ^{1}H-NMR is one of the most important and precise methods that confirmed hydrophobic guest penetration in the cavity of CDs [2, 40–42, 52]. ^{1}H-NMR spectra signals have been assigned to their respective monomers and it can show the inclusion of complexes of ciprofloxacin as guest with α-CD and β-CD as host molecule (Figs. 4 and 5). with formation of complex between these two compounds, internal protons

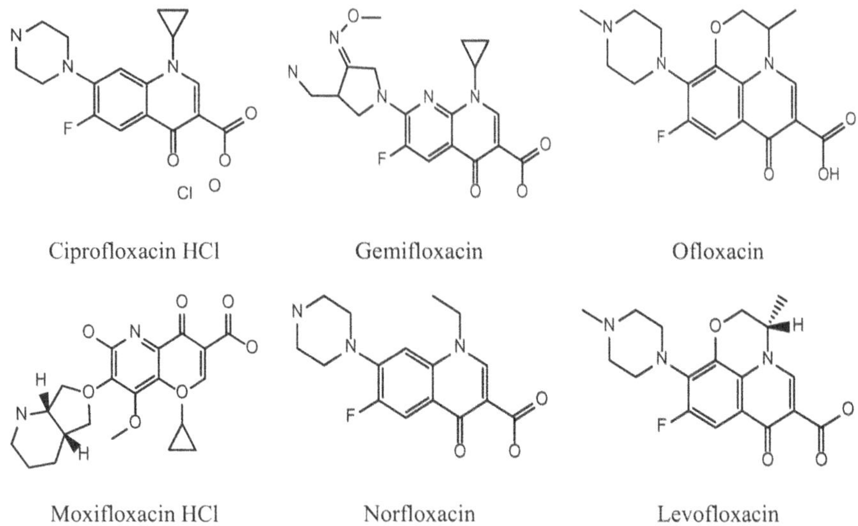

Scheme 2 Chemical formulas and structures of fluoroquinolones

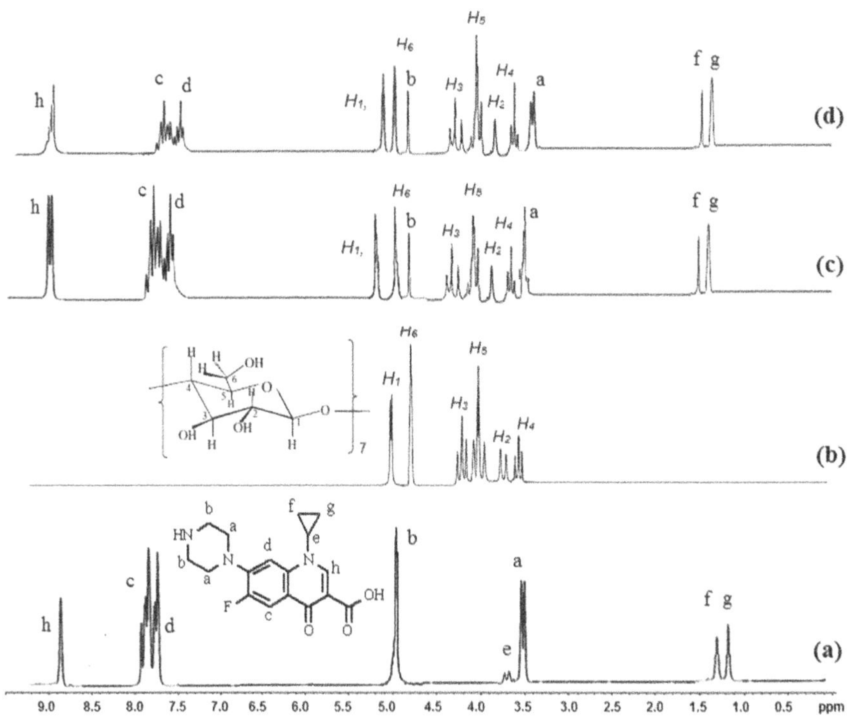

Fig. 4 ^{1}H-NMR of **a** Ciprofloxacin, **b** β-CD, **c** β-CD/Ciprofloxacin under sonic energy and **d** β-CD/Ciprofloxacin at room temperature

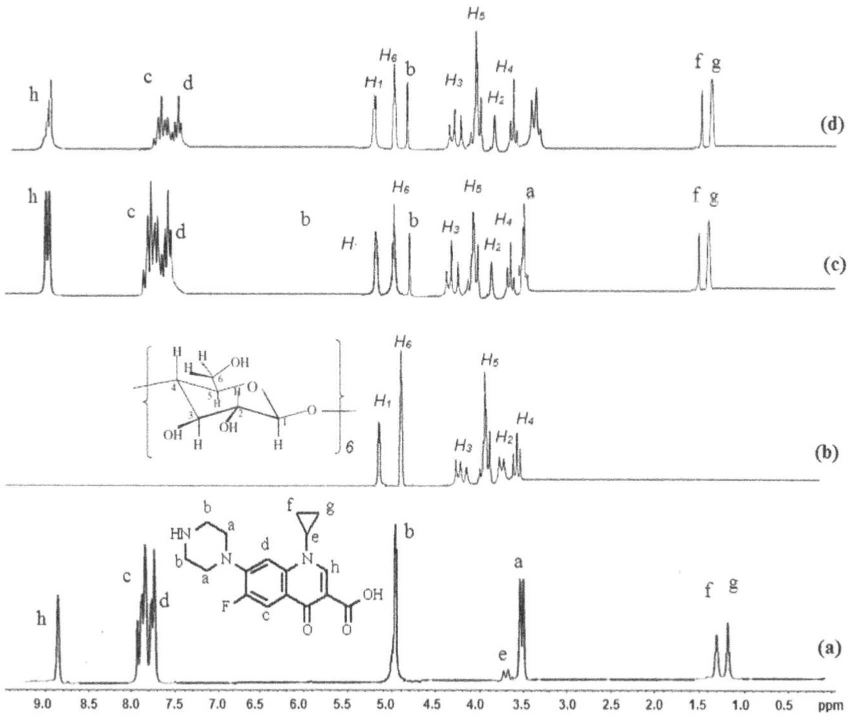

Fig. 5 ^{1}H-NMR of **a** Ciprofloxacin, **b** α-CD, **c** α-CD/Ciprofloxacin under sonic energy and **d** α-CD/Ciprofloxacin at room temperature

(H-3, H-5 and H-6) of CDs substantially shifted compared to external protons (H-2 and H-4) when cavitated in CDs [40–42].

Ciprofloxacin molecules are deeply penetrated into the β-CD cavity, so that internal protons of CDs participated in the interaction that makes the interaction site the secondary alcoholic group at H-3. This shift thus proves the formation of the inclusion complex between CDs and ciprofloxacin, indicating FTIR inclusion complex results [40–44, 52].

UV–visible spectroscopy was used for determiniation of CDs/ciprofloxacin solubility ($\lambda_{max, cipro} = 280$ nm) which CDs didn't show charastric peaks at the desired λ_{max}. For preparation of standard drug solution, the ration of 1:1 of ethanol/water was used. The obtained calibration curve which is showed in Fig. 6a is linear. Figure 6 shows the increase concentration of released ciprofloxacin in the presence of higher amount of CD in complex due to the entrapment of guest product in the cavity of CDs and also solvability and bioavailability of guest moleculesshift to higher amounts [2].

Ciprofloxacins bioavailability is restricted due to their poor solvability and low release rate in water based solutions. Obtained complexes between CDs and ciprofloxacin, increased its solvability significantly (Fig. 7). Through complexation, hydrogen bonding formed between ciprofloxacin/CDs which increased release of

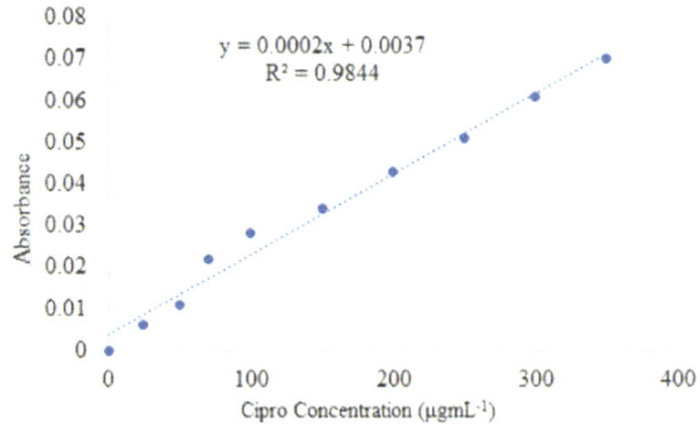

(a)

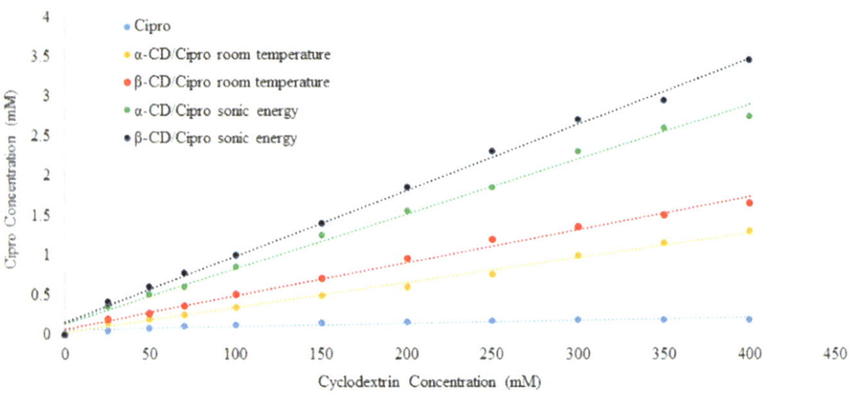

(b)

Fig. 6 **a** Calibration curve and **b** relationships between ciprofloxacin and CD concentrations

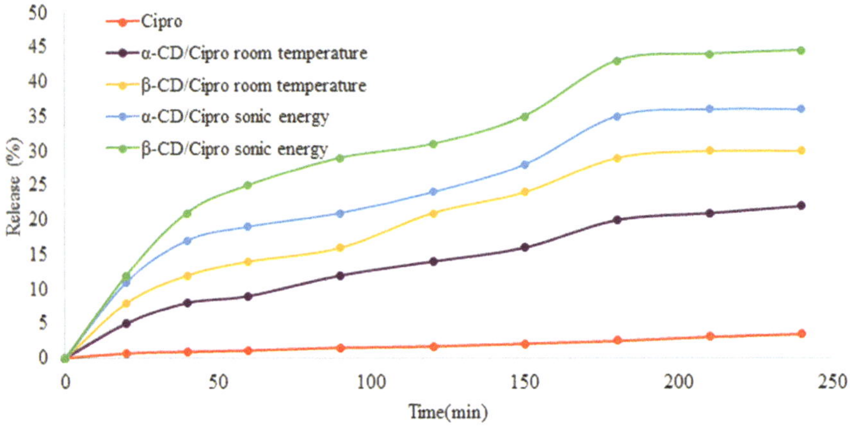

Fig. 7 Release profiles of ciprofloxacin and its inclusion complex with CDs at various conditions at pH = 7.2

release of ciprofloxacin from ciprofloxacin/CDs nano containers and improved solubility of hydrophobic ciprofloxacin. However, addition of CD resulted in controlled diffusion. Therefore, prepared nanofibers showed release activity with a slow and continuous drug release. Under sonic energy, sufficient energy is provided for existing water molecules from CDs cavity to replace with the guest molecules in CDs cavity [34, 45, 50].

3.3 *Ibuprofen/CDs Nanocontainers*

Ibuprofen (IBU) is a nonsteroidal mitigating medicine with a wide range applications, such as treating torment, fever, irritation, difficult menstrual periods, headaches, and rheumatoid joint inflammation [18, 35–37]. Poor dissolvability and low release rate of IBU in the watery liquids limited IBU application and bioavailability. To defeat the impediment of IBU and improving its watery dissolvability, supramolecular nanocontainers of IBU/CD (CD) were readied by means of development of incorporation complex (IC) among IBU and CDs (α-CD and β-CD) at different conditions. Summary of IBU in CDs pit changed its properties, for example, improvement of its solvency, stage dissolvability and in vitro disintegration. It additionally made it a potential contender to create controlled medication conveyance frameworks [37]. Obtained IBU nanocontainers were portrayed with XRD which is one of the valuable techniques for the location of development of incorporation complex among CDs and visitor particles by means of appearance of a couple of new pinnacles and moving of specific pinnacles. The structure of CDs incorporation edifices are separated to channel structur and enclosure type. By consolidation of IBU visitor particles in CDs hole in the structure of channel, trademark tops are seen as Table 4. In the pen type, every hole is closed off on the two sides by neighboring CDs and trademark crests are seen (Table 4) [37]. The XRD results asserted that the edifices shaped for incorporation are crystalline. Results indicated to be under sonic vitality allowed the arrangement of crystalline structure in which CD particles are stacked along an IBU

Table 4 XRD peaks of α-CD, β-CD, IBU and their inclusion complexes [37]

Compound	Peak (2θ)
IBU	12, 18, 20, 21, 22, 24, 25
α-CD	10, 14, 15, 16, 18, 19, 21, 23, 24, 25, 26, 27
β-CD	5, 8, 12, 15, 16, 16, 17, 18, 20, 21, 22, 23, 24, 25, 27, 28, 30, 32
α-CD/IBU sonic energy	11, 12, 14–16, 17, 19–21, 23
α-CD/IBU room temperature	12, 13, 15, 18, 21
β-CD/IBU sonic energy	12, 12, 14, 16, 17, 19, 21, 23–25, 27
β-CD/IBU room temperature	10, 12, 15, 18–20, 21, 25

pivot to form a pipe. At room temperature, precious stones acquired are a mixture of the form of confines and the channel-style gems [37].

^{1}H-NMR pictures of the host and geust molecules at different conditions were display in Fig. 8. All signs of the were allocated to their comparative pure compounds and the IBU/α-CD and IBU/β-CD consideration peaks. Due to hydrophobicity of IBU inside protons (H-3, H-5 and H-6) of CDs altogether moved contrasted with outside protons (H-2 and H-4). Inside hydrogens of CDs partook in the connection because of the profound infiltration of the IBU particle into the β-CD and the site of assembly could be the optional alcohol gathering at H-3 [5–8, 24, 53–55]. Hence, this move gives the sign of the development of the incorporation complex among CDs and IBU, which showed FTIR after effects of consideration edifices. Disappering hydroxyl groups of IBU at 9.1 ppm, indicated the arrangement of hydrogen holding among IBU and CDs [37, 40–42, 50, 52].

The IBU content was measured at a wavelength of 264 nm using UV–visible spectroscopy. IBU was solved in ethanol and diluted by 50% v/v ethanol to desire concentrations. CDs have n no absorption at the $\lambda_{max} = 264$ nm which is the IBU absorption $ÿ_{max}$. IBU's calibration curve is a direct line [37]. Figure 9 indicated amount of IBU directly dependant to CD concentration [37, 56].

The release rate of IBU and IBU nanoconatiners were determined to evaluate the IBU release profiles at pH = 7.2 (Fig. 10) [37].

3.4 Increase Solvability and Bioavailability of Silver Sulfadiazine (SSD)

Because of the improvement of hypertrophic scars (HTSs), new strategies for presenting consume care tranquilize conveyance frameworks have pulled in developing interest [57, 58]. Diverse factors, for example, Heat, power, synthetics, grinding or radiation cause consume wounds to skin [59, 60]. Be that as it may, attractive properties of a fitting consume treatment, for example, low water solvency, bioavailability and biocompatibility can once in a while be found in normal prescriptions. The most usually utilized prescriptions in consume treatment are antimicrobial balms, for example, SSD, mafenide, silver nitrate, and povidone-iodine which help lessen the danger of disease and anti-infection agents, for example, oxacillin, mezlocillin, and gentamicin help treat a contamination [61–63]. SSD is a successful antimicrobial agent against various bacterial diseases and outside tainting from consume damage. Be that as it may, it is halfway solvent in watery frameworks and along these lines, its dissolvability, bioavailability, in vitro disintegration and lethal potential to microorganisms have been constrained. In spite of the way that SSD hinders the development of microbes and diminishes the danger of microorganisms spreading to encompassing skin, some reactions were watched [19, 64]. Because of this restriction, solvency of SSD were improved by means of development of incorporation complex among CDs and SSD and got nanocontainers were portrayed [65].

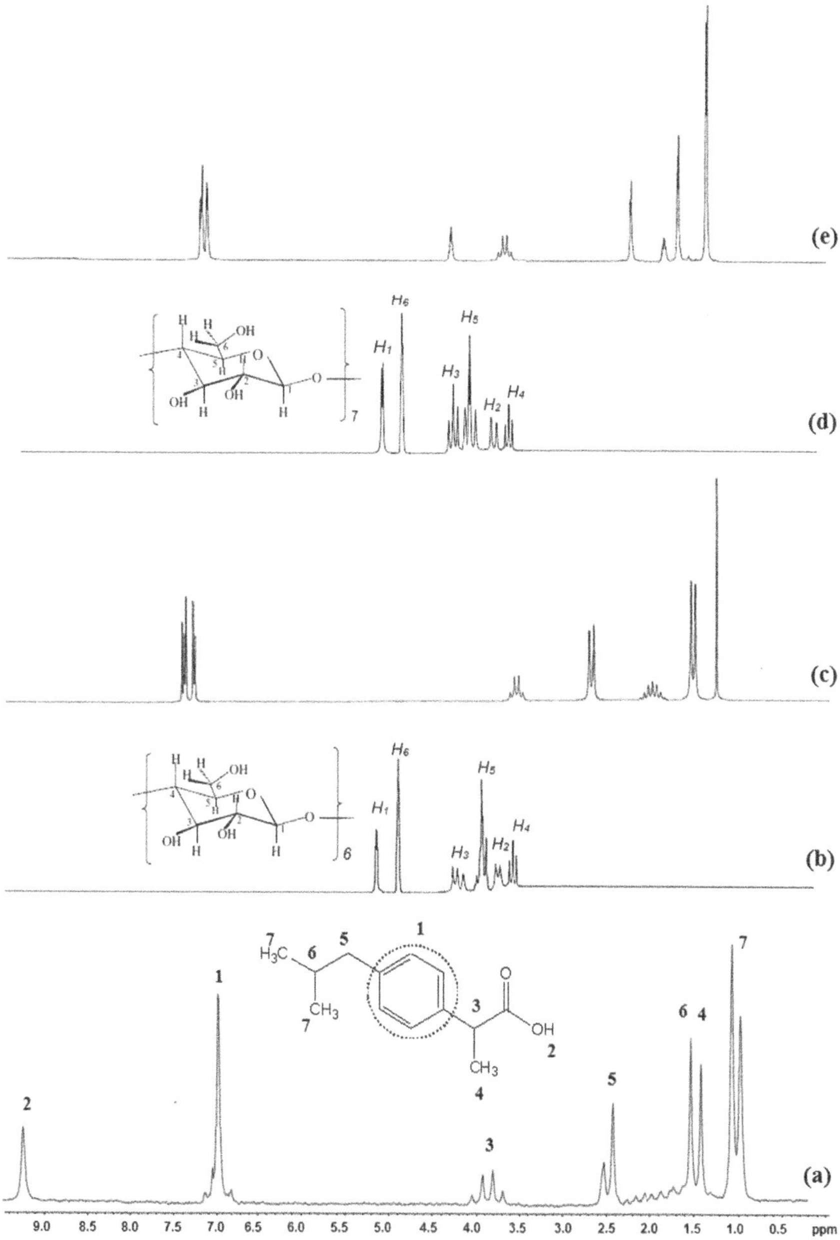

Fig. 8 ^{1}H-NMR of **a** IBU, **b** α-CD, **c** α-CD/IBU at room temperature, **d** β-CD and **e** β-CD/IBU under sonic energy

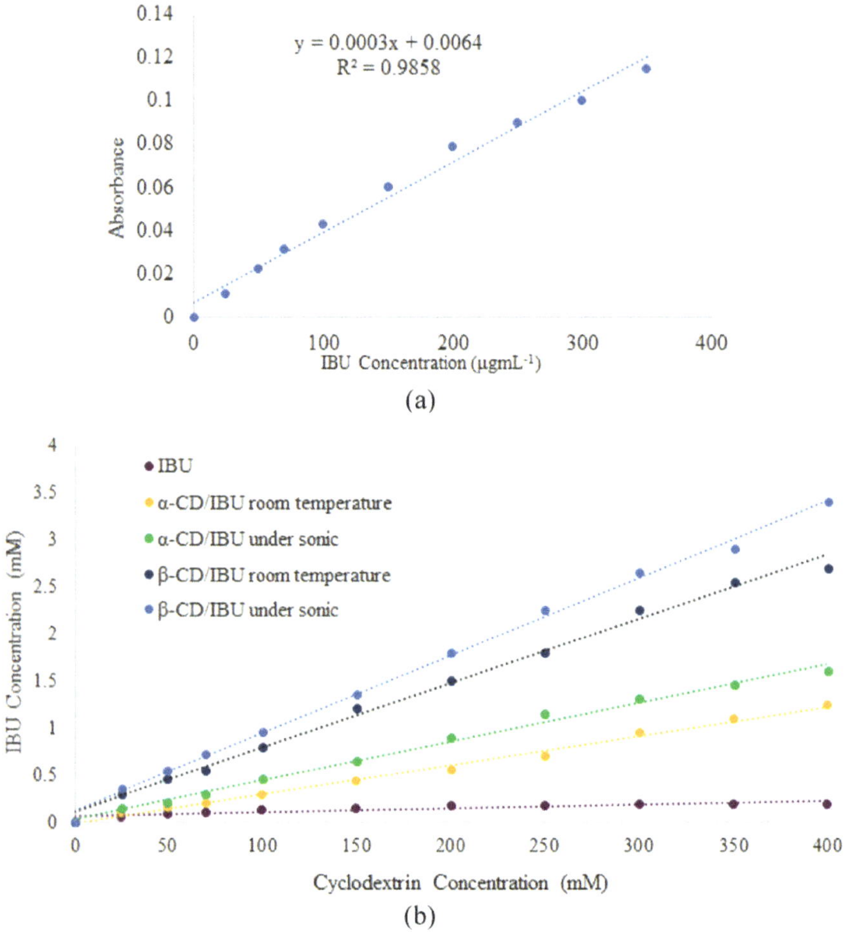

Fig. 9 **a** Calibration curve and **b** Relationships between IBU and CD concentrations

CDs derivatives did not demonstrate any peaks at the $\lambda_{max,\,SSD} = 263$ nm (Figs. 11 and 12) and solvability of complexes is shown in Table 5. Solubility result indicated that that β-CD/SSD complexes obtained under sonic energy have had the greatest effect on SSD solubility which is related to new interaction between two components [65].

Pure SSD demonstrates poor water solvency and arrangement of consideration complex altogether improved the dissolvability of SSD which is an aftereffect of the development of reversible hydrogen securities by means of incorporation complex arrangement of CD and SSD and hence, its water dissolvability increased [65].

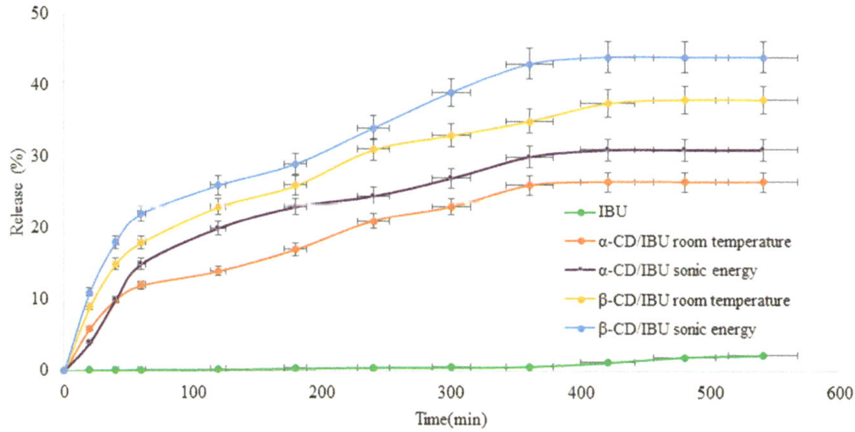

Fig. 10 Release profiles of IBU and its inclusion complex with CDs at various conditions at pH = 7.2

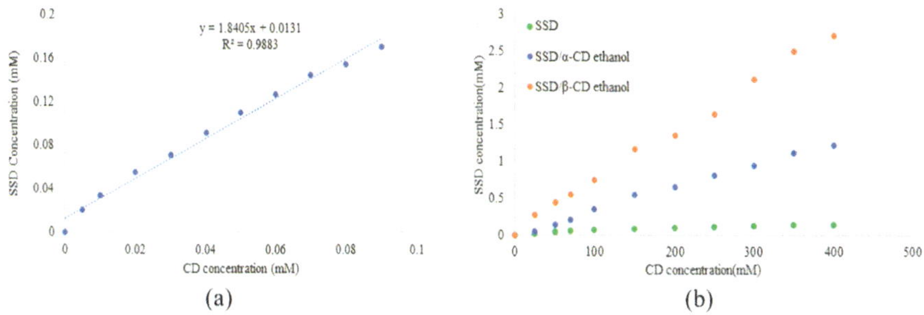

Fig. 11 **a** Calibration curve and **b** relationships between SSD and CD concentrations for SSD/CDs inclusion complex in ethanol at room temperature

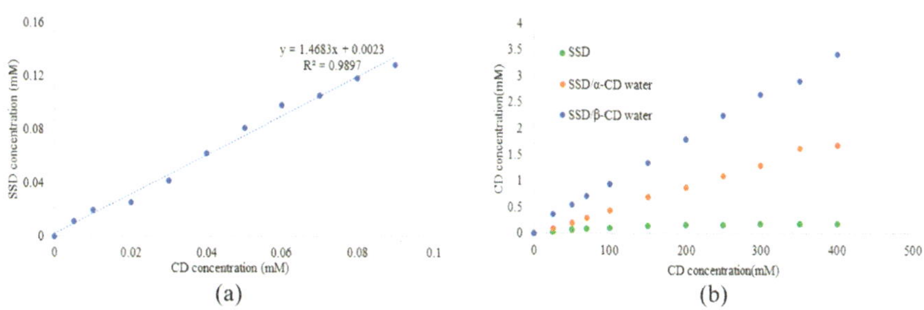

Fig. 12 **a** Calibration curve and **b** relationships between SSD and CD concentrations for SSD/CDs inclusion complex in water at room temperature

Table 5 Water solubility of SSD and its complexes at various conditions [65]

Sample	Solubility (μg/ml)
SSD	3.5
SSD/β-CD in ultrasonic bath in water	201.41
SSD/β-CD at room temperature in ethanol	285.57
SSD/α-CD in ultrasonic bath in ethanol	186.05
SSD/α-CD at room temperature in water	207.29

4 Drug Safety and Stability

A few medications may cause aggravation and effectsly affect body so little dose of them can be utilized and accordingly, their utilization is restricted. In spite of the fact that, the utilization of insecure medications is restricted because of their short timeframe of realistic usability and corruption caused substance responses, for example, drying out, hydrolysis, oxidation and photodecomposition with dynamic operators. To conquer this issue, medications can start responses with CDs, exemplify in CD hole and subsequently, the connection among medication and dynamic operators and bioconversion at the ingestion site are restrained, so sedate security improves. Embodiment procedure may cover the medication and ensure it against different corruption forms which rely upon the nature and impact of the included utilitarian gathering on medication strength and vehicle. For instance, if there should arise an occurrence of trimeprazine and promethazine, photostability and time span of usability expanded by incorporation complex arrangement among CD and medication [27, 30, 31]. Likewise, development of incorporation complex ensured the medication and delayed medication hydrolysis contrasted with free medications [65].

4.1 Improving Water Solubility and UV-Stability of Ofloxacin

Fluoroquinolones act as antimicrobial agent to prevent certain microbial infections with some side effects [66, 67]. Ofloxacin (OFL) act against a wide variety of bacterial putridity which limited due to its poor water solvability and photostability limits its uses, so it is a challenge to enhance its water compatibility and strength. Complex formation between α-CD and OFL has been synthesized for the production of nanocontainers with desirable water solubility and stability under various environmental parametres. CD/OFL complexes formed under sonic energy increased solvability of OFL in ethanol (Table 6) [68].

Characterization of the complexes obtained, incorporation of OFL in the study of α-CD and thermal analysis confirmed formation of new interaction via reversible bonding between host and guest (Fig. 13). Guest molecule was introduced into the cavity of the host molecule by existence of water from cavity and insertion guest

Table 6 Each reaction Content in formation of inclusion complex between α-CD and OFL [68]

Paper	α-CD	OFL	PEG 6000 (20% w/w)	Sonic energy	Room temperature
C-1	■	■	–	■	–
C-2	■	■	–	–	■
C-3	■	■	■	■	–
C-4	■	■	■	–	■

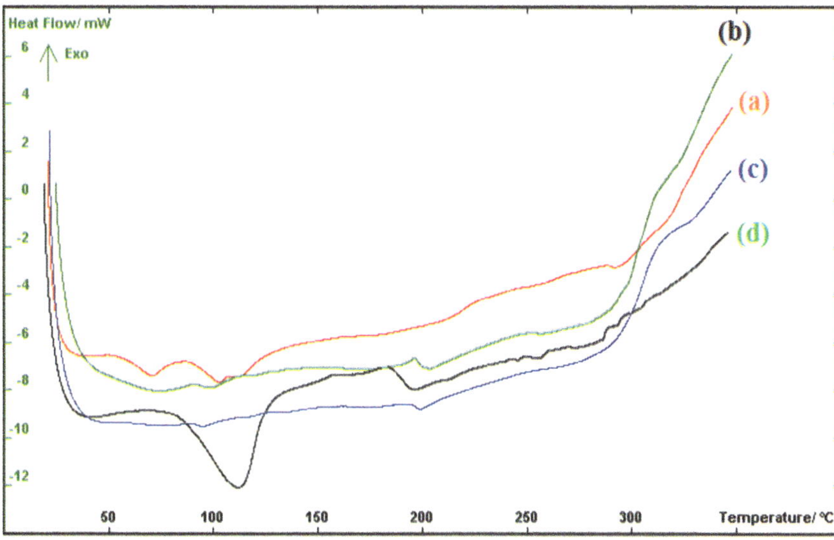

Fig. 13 DSC images of **a** α-CD, **b** OFL, **c** C-1 and **d** C-2

molecule in appropriated position, so the T_m peak of the guest molecule disappeared. The α-CD thermogram displayed two typical endothermic peaks, one at 90 °C attributable to CD water loss and the other at 310 °C due to thermal CD decomposition (Fig. 13a).

DSC spectrum of OFL showed two peaks which are related to T_m and T_d of OFL and were seen at 136 and 198 °C, respectively (Fig. 13b). OFL melting temperature once disappeared due to OFL incorporation in α-CD cavity (Fig. 13c, d), and once when PEG is used in CD/OFL complexes, both OFL-related peaks have disappeared which can be due to complete OFL injection in α-CD cavity. PEG with hydroxyl groups led to higher number of reversible hydrogen bonding interaction which are able to entrapped higher number of OFL molecules [37, 50, 65, 68].

Drugs water solvability is an essential criteria for using as medical application and low soluble drugs must be modified to reach an acceptable range of solubility. OFL is characterized by poor water solubility (0.09 mg/ml), and this problem limits its application, thus improving its solubility by encapsulating it in α-CD [34, 37, 45, 50, 65, 68].

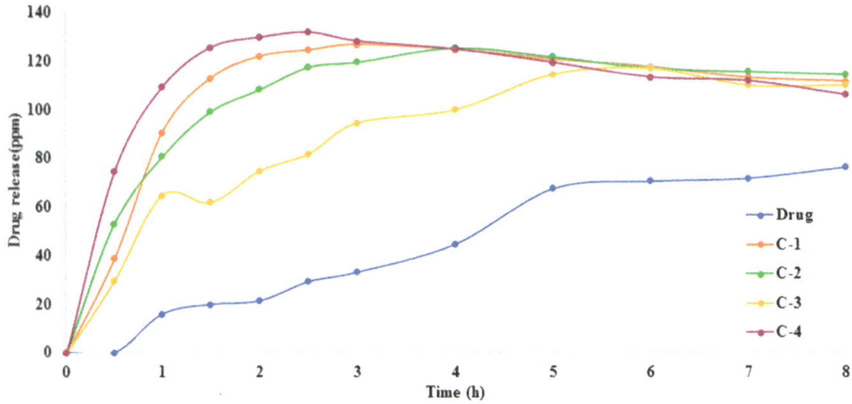

Fig. 14 Dissolution profile of OFL and its inclusion complexes at various conditions

Controlled-release systems are designed to improve medication efficacy and temporal presence of drug in the body. OFL entrapped in α-CD, shielding medicines from physiological degradation and removal, increasing enduring admission, enhancing control in drug product manufacturing, and expanding patent security and enhancing its bioavailability. Figure 14 showed OFL's time-programmed discharge behavior from its complexes and showed an initial burst release of OFL is controlled after 3 h.

Controlled releas of OFL from containers with a slow and contionous rate due to formation of reversible interactions between OFL and CD cavity. When higher number oof hydrogen bonding formed, higher amount of OFL entrapped and so releas rate improved [34, 37, 45, 50, 65, 68].

5 CDs Based Carriers

Phytochemicals play a key role in people's medical uses and phytonutrients tend to be highly useful products and can solve health problems. Different chemical functionlas can be used for the decrease side effects of severe infectious diseases [69–71].

Phytochemicals show anti-oxidant, anti-bacterial and fungicidal activities but their poor solubility in water and their bioavailability restricted their application [20, 72, 73]. Encapsulation of Phytochemicals with CDs protect them from hydrolysis and oxidation and release in a sustained pattern by specific and controlled rate, so high concentrations of drug didn't present in target site [2, 73–75]. Encapsulation enhances bioavailability and water solubility of immobilized phytochemicals in CD cavity and make them chemically stable. The core material can also be protected against light, heat and oxidation [76].

5.1 Cinnamon Essential Oil Nanocapsules with Controlled Release and High Solubility

Different groups of phytochemicals are useful in medicinal services and can be utilized for anticipation of interminable dynamic infections and utilized broadly in sustenance and dietary enhancement businesses that have additionally prompted solid nourishment with great tasting, for example, phenolic mixes. Phenolic Compounds can go about as free radicals to kill and hinder or stifle dynamic cancer-causing agents [2–4]. Cinnamon basic oil (CEO) is a phytochemical and a standout amongst the most famous flavors and nourishment added substances with a wide scope of uses [2–4]. Bioavailability, dissolvability and strength of phenolic mixes decide the assimilation of CEO mixtures in body and CEO demonstrates a low water solvency. On account of low aqueous solvability of cinnamon oil, its skin aggravation, poor bioavailability, unfavorably susceptible impacts and its severe taste, this compound must be encapsulated in good containers [4–6].

Cinnamon basic oil (CEO) nanocontainers were readied by means of incorporation complex (IC) development among CDs and CEO at two. Obtained carriers were described by XRD, SEM, DSC and FTIR (Table 7) [45]. UV spectroscopy was utilized to decide the substance of CEO at a $\lambda_{max} = 360$ nm. The standard CEO arrangement was set up in a 50% v/v ethanol in water. The adjustment bend of the technique was straight [45]. Water solvability of of complexes are shown in Table 8 which shows that formation of new reversible hydrogen bonding between host and guest under sonic energy had the highest effect on solubility of CEO [54].

Table 7 Solubility of CEO and its complexes at various conditions [45]

Sample	Solubility (μg/ml)
CEO	0.015
CEO/β-CD under sonic energy	290.37
CEO/α-CD under sonic energy	227.24
CEO/β-CD at room temperature	263.08
CEO/α-CD at room temperature	140.57

Table 8 Solubility of CUR and its complexes at various conditions [34]

Compound	Solubility (μg/ml)
CUR	0.195
α-CD/CUR at room temperature	382.43
β-CD/CUR at room temperature	369.78
α-CD/CUR under sonic energy	553.17
β-CD/CUR under sonic energy	586.75

6 Improve Bioavilability of Curcumin via Encapsulation in CDs

Curcumin (CUR) has pulled in wide consideration and been broadly utilized due to its anticancer and antioxidant properties against a few pathogens, which is a vital substance comprises of an extensive gathering of synthetic substances that are normally happened and created by plants and has a few medical advantages for people [21, 77, 78]. CUR has different pharmacological applications as against disease and calming however its short disposal half-life, unsteadiness, insolubility in water, moderate take-up by the cells and fast discharge, limit its bioavailability [79–81]. For this reason, supramolecular nanocontainers of CDs/CUR were set up by incorporation complex arrangement among CDs and other guest molecules. Based on the outcomes got it is reasoned that consideration complexation of CUR with CDs, gives high solvency and moderate arrival of CUR [34]. Water solvability of other complexes are shown in Table 8. which showed that by applying sonic energy led to higher efficiency of encapsulation, higher hydrogen bondings and let to improvr dissolution rate of CUR [34].

7 Transdermal Delivery Microneedle Patches Containing Insulin-CD

Diabetes is a chronic disorder of hyperglycemia and a serious clinical syndrome that can stay quietly undetected for a long time to come. Insulin is a hormone that controls carbohydrates and fat metabolism in the body and has been secreted by pancreatic β-cells to monitor blood sugar levels [1]. To efficient delivery of insulin, microneedle patches (MNPs) are designed to deliver insulin [5, 6]. MNs are applied to the surface of the skin and pierce the epidermis painlessly, producing microscopic aqueous pores that disperse drugs through the dermal microcirculation [7, 12]. Some MNPs are made of disposable biodegradable and biocompatible materials. MNs made up of biodegradable materials have higher product consumption charges and no possible biohazardous waste after use [11]. The polymer microneedle technology attracted many research interests, due to all the above described advantages. Due to its good availability, biodegradability, biocompatibility, antimicrobial properties and low toxicity, chitosan has recently been used as a drug carrier for insulin. In long-term therapeutic systems the tendency of insulin sites to accumulate and form fibrils. Complexation of CDs represents a specific and efficient technique to improve stability of insulin against temperature. Dissolving polymer microneedles based on CD and chitosan microneedles can dissolve into the skin after insertion and release encapsulated insulin as they dissolve in a controlled manner. The drug release results ascertained that the release was due to diffusion and degradation of polymeric carriers and release of insulin. Phosphate-buffered saline (pH = 5) was used as a release medium because it closely resembles extracellular fluids and plasma. Drug release from bio-based polymeric microneedles caused by formation and breakage of hydrogen bonding which formd between insulin and carriers, then diffused in the polymer matrix and then release in the target medium (Figs. 15, 16, 17 and 18) [48].

The rate of cumulative release gradually increased within a period of 7 h. It was observed that about 80% of the drug encapsulated in the microneedle array was released within a period of 7 h. In this study, it is observed prolonging drug release rates for 24 h without high initial burst release at microneedles (M_1) (CS/I) and (M_2)

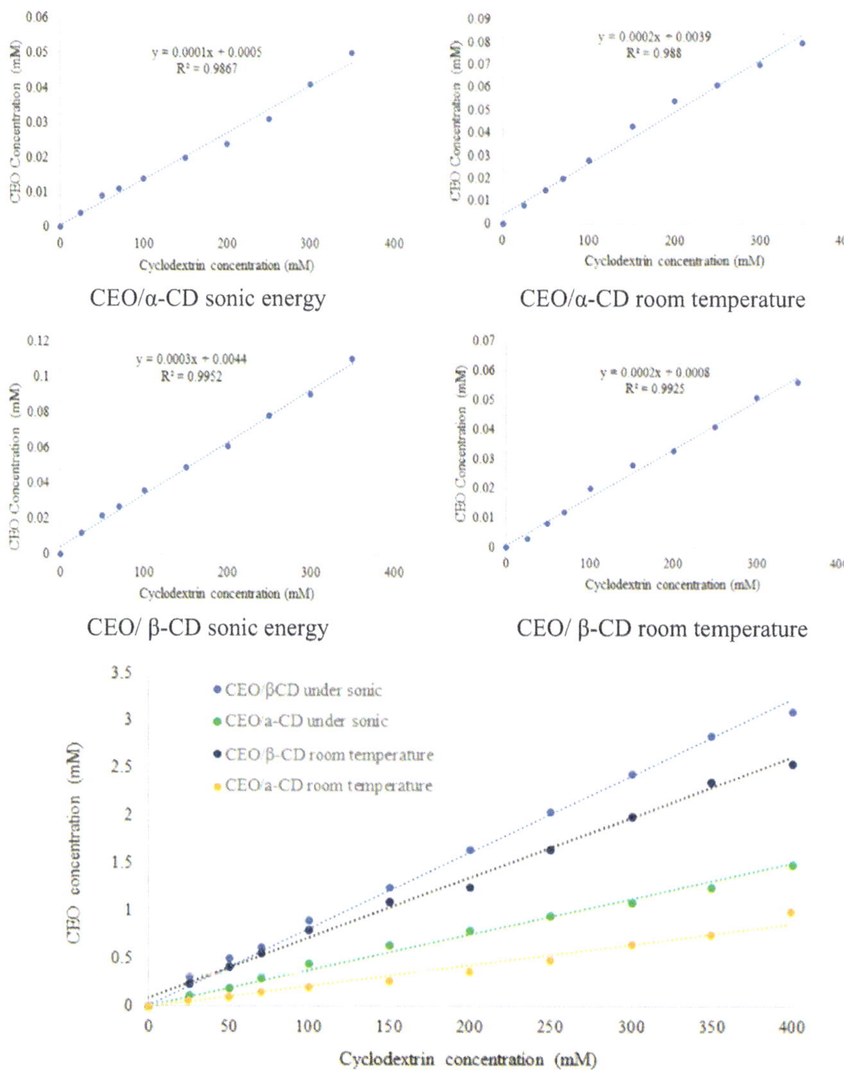

Fig. 15 Relationships between CEO and CD concentrations

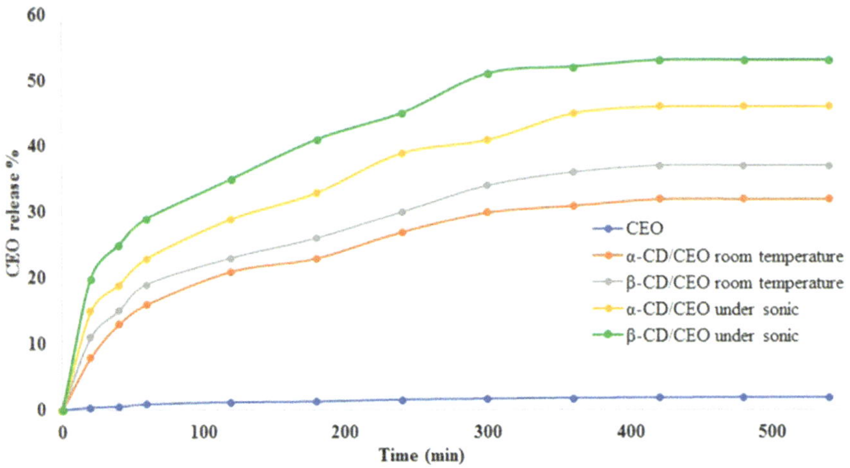

Fig. 16 Release profiles of CEO and its inclusion complex with CDs at various conditions at pH = 7.4

(CS/Insulin/β-CD 0.5-I) (Fig. 19). Initial slight burst release suggests that insulin density is low at the surface of the microneedles which is related to equal solvability of the polymer and the drug in the medium and distribution was homoguoneous among the microneedles [49]. Figure 19, microneedle (M_3) (CS/Insulin/β-CD 0.75) shows the cumulative release amount in the microneedles improved with the increasing concentration of β-CD. Indeed, by increasing the concentration of β-CD in formulation (microneedle (M_3) (CS/Insulin/β-CD 0.75), the more hydroxyl groups are prone to form reversible hydrogen bonds and it will be able to encapsulated more insulin. Moreover, release profile at microneedle (M_3) showed that the drug was released with initial burst and became sustained later with the maximum drug released around a period of 24 h beyond which low further release was observed. The process of burst release (within the first 7 h) was mainly triggered by surface-bound or adsorbed drug desorption. And the second step was a relatively slow release up to 24 h, which attributed erosion or degradation to drug diffusion through the micro-particle matrix and micro-particle matrix [48].

Further release can also be explained by the fact that maximum drug is encapsulated within the microneedles and released only after the degradation of the polymer [50]. As a result, MMWCS, β-CD and insulin successively formed complexes as micro-carriers, and insulin release regulated against formulation factor was predicted. The MMWCS and β-CD are suggested as a successful insulin delivery candidate [39].

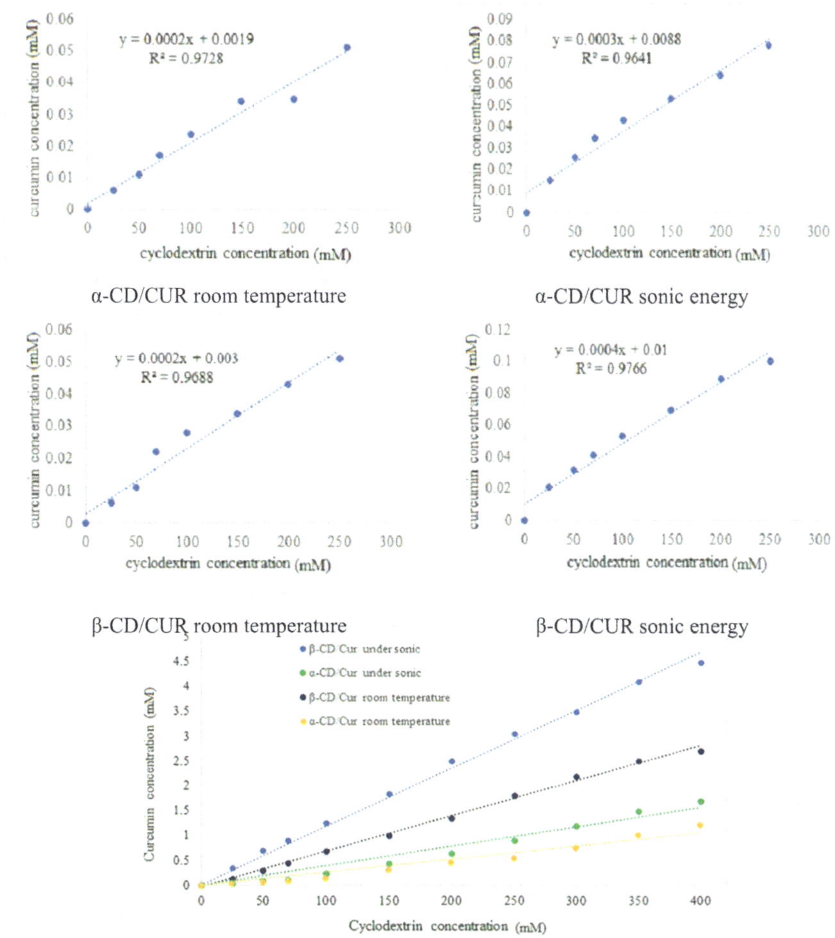

Fig. 17 Relationships between CUR and CD concentrations

8 Summary

CDs can form inclusion complex with guest molecules for numerous purposes such as stabilization of volatile compounds, improvement of bioavailability and water solubility of poor soluble compounds, encapsulation and target drug delivery, controlled drug release, masking the undesirable properties of guest molecules and improving stability of guest molecules etc.

Major challenge in drug delivery is limited water solubility of hydrophobic drugs which restricted bioavailability and permeability of drugs. Inclusion complexes formation between drugs and CDs may improve drugs solvability, safety and biocompatibilty of low soluble drugs by increasing the drugs solubility. In this review, some

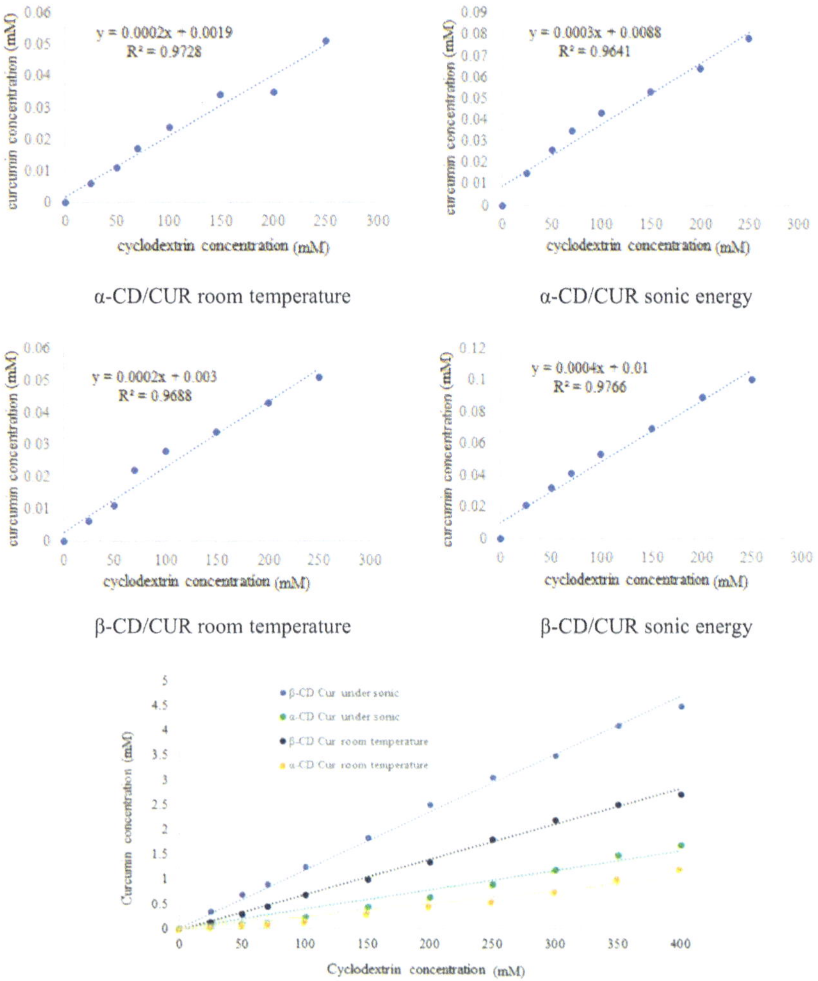

Fig. 18 Release profiles of CUR and its inclusion complex with CDs at various conditions at pH = 7.4

novel applications of CDs in formation of inclusion complexes with hydrophobic drugs and phytochemicals are discussed. Also, experimental results showed these polymeric microneedles based on CDs, may be a suitable carrier for control released the insulin for transdermal delivery.

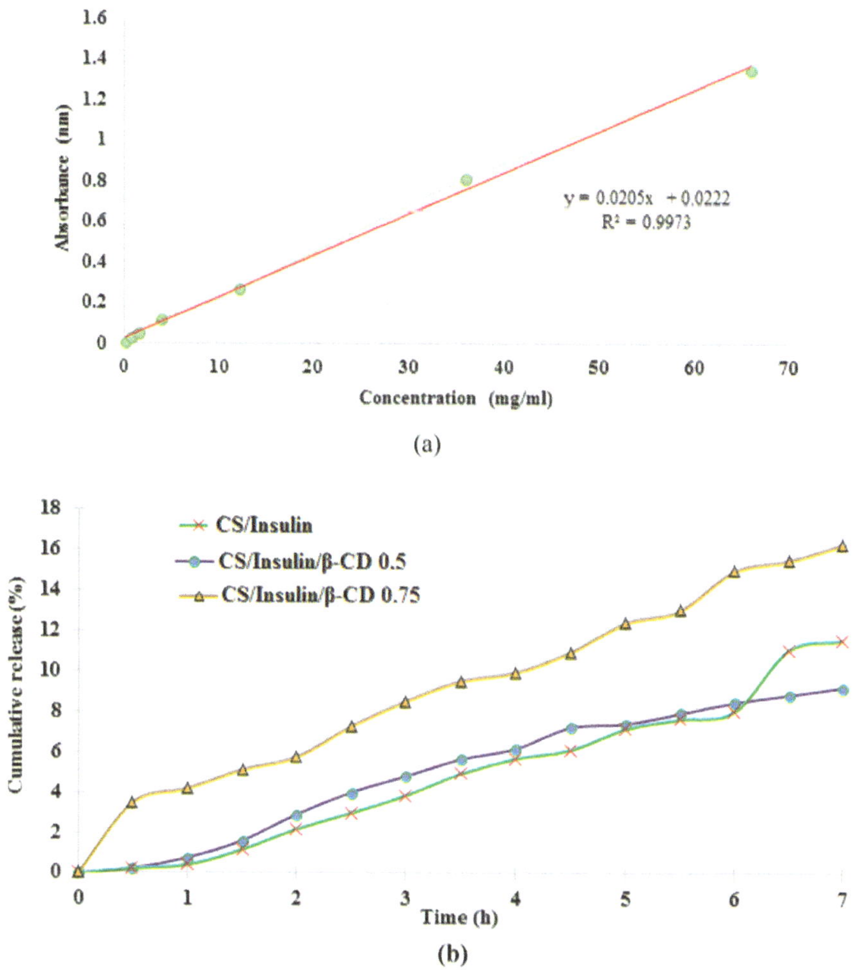

Fig. 19 **a** Calibration curve and **b** drug release profile of Insulin from microneedles

References

1. Szetjli J (2009) Chem Rev 98:1743–1753
2. Amiri S, Amiri S (2017) CDs, properties and industrial applications. Wiley
3. Hedges RA (1998) Chem Rev 98:2035–2044
4. Jansook P, Ritthidej GC, Ueda H, Stefansson E, Loftsson T (2010) J Pharm Pharm Sci 13:336–350
5. Villiers A (1891) Seances Acad Sci 112:536–538
6. Szetjli J (1998) Chem Rev 98:1743–1753
7. Szetjli J (1989) TIBTRCH 7:171–174
8. Szejtli J (1982) CDs and their inclusion complexes. Akadémiai Kiadó, Budapest, Hungary
9. Tavassoli-Kafrani E, Goli SAH, Fathi M (2017) Food Bioprocess Technol. https://doi.org/10.1007/s11947-017-2026-9

10. Benavides S, Cortés P, Parada J, Franco W (2016) Food Chem 204:77–83
11. Hirayama F, Uekama K (1987) Methods of investigating and preparing inclusion compounds. In: Duchêne D (ed) CDs and their industrial uses. Paris, Editions de Santé, pp 131–172
12. Chen G, Jiang M (2011) Chem Soc Rev 40:2254–2266
13. Nimse SB, Kim T (2013) Chem Soc Rev 42:366–386
14. Semsarzadeh MA, Amiri S (2012) Silicon 4:151–156
15. Semsarzadeh MA, Amiri S (2013) Bull Mater Sci 36:989–996
16. Eastburn SD, Tao BY (1994) Biotechnol Adv 12:325–339
17. Buchwald P (2002) J Phys Chem B 106:6864–6870
18. Wang Q, Sanming L, Che X, Fan X, Lib C (2010) J Pharm 5:188–193
19. Percival SL, Bowler PG, Russell D (2005) J Hosp Infect 60:1–7
20. Yadav VR, Suresh S, Devi K, Yadav S (2009) Pharm Sci Tech 10:752–762
21. Atares L, De Jesús C, Talens P, Chiralt A (2010) J Food Eng 99:384–391
22. Loftsson T, Duchene D (2007) Int J Pharm 329:1–11
23. Hakkarainen B, Fujita K, Immel S, Kenne L, Sandstrom C (2005) Carbohydr Res 340:1539–1545
24. Irie T, Uekama K (1997) J Pharm Sci 86:147–162
25. Aggarwal S, Singh PN, Mishra B (2002) Pharmazie 57:191–193
26. Archontaki HA, Vertzoni MV, Athanassioum Malaki MH (2002) J Pharm Biomed Anal 28:761–769
27. Arima H, Yunomae K, Miyake K, Irie T, Hirayama F, Uekama K (2001) J Pharm Sci 90:690–701
28. Pandey S, Varshney HM, Gupta MM (2013) RJCES 1:19–25
29. Rajendiran N, Mohandoss T, Thulasidhasan J (2016) Phys Chem Liq 54:193–212
30. Choi HG, Lee BJ, Han JH, Lee MK, Park KM, Yong CS (2001) Drug Dev Ind Pharm 27:857–862
31. Chutimaworapan S, Ritthidej GC, Yonemochi E, Oguchi T, Yamamoto K (2001) Drug Dev Ind Pharm 26:1141–1150
32. Dalmora ME, Dalmora SL, Oliveira AG (2001) Int J Pharm 222:45–55
33. Baviskara P, Bedsea A, Sadiqueb S, Kundea V, Jaiswal S (2013) Int J Pharm Sci Rev Res 13:70–76
34. Amiri S, Rahimi A (2017) Int J Polym Mater Po. https://doi.org/10.1080/00914037.2018.1482467
35. Chen L, Qifeng D, Liu C, Chen J, Song L, Chen X (2012) Front Med 6:195–203
36. Farmer S, Anderson P, Burns P, Velagaleti R (2002) Technology 32:28–42
37. Masoumi S, Amiri S, Bahrami SH (2017) J Ind Text. https://doi.org/10.1177/1528083718764910
38. David SJ (2016) Pharmaceutics dosage form and design, 2nd edn, chapter 4, pp 4–7
39. Jahromi MM, Ghaemi H, Tafti MA, Arabzadeh A, Afsharypuor S (2015) J Nat Pharm Prod 10:1–3
40. Amiri S, Rahimi A (2014) J Polym Res. https://doi.org/10.1007/s10965-014-0566-5
41. Amiri S, Rahimi A (2014) J Polym Res. https://doi.org/10.1007/s10965-014-0624-z
42. Amiri S, Rahimi A (2015) J Polym Res. https://doi.org/10.1007/s10965-015-0699-1
43. Patel R, Patel M (2010) Drug Discov Ther 4:442–452
44. Melle-Franco M, Strutyński K, Borges F, Brett CMA, Manuela EPJ (2017) J Environ Sci Health Part A. https://doi.org/10.1080/10934529.2016.1258864
45. Amiri S, Rahimi A (2018) J Ind Text https://doi.org/10.1177/1528083718764911
46. El-Majri MA, El-Basir MM (2016) J Res Pharm 7:87–90
47. Amiri S, Askari F (2018) Pharm Innov. https://doi.org/10.1007/s12247-018-9328-y
48. Gowthamarajan K, Kulkarni TG, Nenkateswaran G, Samanta MK (2002) Indian J Pharm Sci 1:525–528
49. Sousa J, Alves G, Oliveira P, Fortuna A, Falcão A (2017) Eur J Pharm Sci 97:30–37
50. Masoumi S, Amiri S, Bahrami SHJ (2018) TEXT 1–10. https://doi.org/10.1080/00405000.2017.1398625
51. Sahoo S, Chkaborti CK, Mishrab SC, Nath U (2011) Int J Pharm Pharm Sci 3:165–170
52. Semsarzadeh MA, Amiri S (2013) Bull Mater Sci 36:989–996

53. Gould S, Scott RC (2005) Food Chem Toxicol 43:1451–1459
54. Szente L, Szejtli J, Kis GL (1998) J Pharm Sci 87:778–781
55. Liu L, Guo QX (2002) J Incl Phenom Macroc Chem. 42:1–14
56. Higuchi T, Connors A (1965) Phase-solubility techniques. Wiley-Interscience, New York
57. Aziz Z, Abu SF, Chong NJ (2012) Burns 38:307–318
58. Atiyeh BS, Costagliola M, Hayek SN, Dibo SA (2007) Burns 33:139–148
59. Szegedi A, Popov M, Yonchev K, Makk J, Mihalya J, Shestakov PJ (2014) Mater Chem B 2:6283–6292
60. Ebrahimi A, Kalantar Motamedi MH (2012) Trauma Mon 17:255–258
61. Jodar Karin SP, Balc Victor M, Chaud Marco V, Tubino M, Valquiria MH, Yoshida M, Oliveira JR, Marta MDC (2015) J Pharm Sci 104:2241–2254
62. Ratnaparakhi MP (2012) Int J Pharm Pharm Sci 4:305–316
63. Venkataraman M, Nagarsenker M (2012) AAPS Pharm Sci Technol 14:254–264
64. Ping L, Xiangmin X, Longlong W, Binjie L, Zhao Y (2015) Med Chem Commun 6:2204–2208
65. Amiri S, Nalbandi B (2018) J Inorg Organomet Polym. https://doi.org/10.1007/s10904-018-0809-8
66. Zhanel GG, Ennis K, Vercaigne L, Walkty A, Gin AS, Embil J, Smith H, Hoban DJA (2002) Drugs 62:13–59
67. Neu HC (1989) Am J Med 87(Suppl. 6C):2S-9S
68. Amiri S, Mehrizi M (2020) J Text Inst 111. https://doi.org/10.1080/00405000.2019.1657614
69. Dillard CJ, German JB (2000) J Sci Food Agric 80:1744–1756
70. Sharma G, Singh RP, Chan DCF, Agarwal R (2003) Anticancer Res 23:2649–2655
71. Singh RP, Sharma G, Sivanandhan D, Agarwal C, Agarwal R (2003) Biomark Prev 12:933–939
72. Awad AB, Williams H, Fink CS (2003) J Nutr Biochem 14:111–119
73. Marques HMC (2010) Flavour Fragr J 25:313–326
74. Munin A, Edwards-Lévy F (2011) Pharmaceutics 3:793–829
75. Fang Z, Bhandari B (2010) Trends Food Sci Technol 21:510–523
76. Chandrasekar V (2010) Optimizing the microwave-assisted extraction of phenolic antioxidants from apple pomace and microencapsulation in CDs. MSc Thesis, Purdue University Graduate School
77. Arancibia M, Giménez B, López-Caballero ME, Gómez-Guillén MC, Montero P (2014) LWT Food Sci Technol 59:989–995
78. El-Baroty GS, Abd El-Baky HH, Farag RS, Saleh MA (2010) Afr J Biochem Res 4:167–174
79. Pereva S, Sarafska T, Bogdanova S, Spassov TJ (2016) Drug Deliv Sci Technol. https://doi.org/10.1016/j.jddst.2016.04.006
80. Hussein K, Turk M, Wahl MA (2007) Pharm Res 24:585–592
81. Tozuka Y, Fujito TY, Moribe K, Yamamato K (2006) J Incl Phenom Macrocycl Chem 56:33–37

Containers for Encapsulation of Aroma/Flavour for Food Applications

Lucía M. Quintero-Borregales, Silvia Goyanes, and Lucía Famá

1 Introduction

Food is one of the three most essential needs for all living beings, as the main source of nutrients and energy to develop and grow. It is necessary to ingest a certain amount every day as fuel to cover the daily nutritional requirements in order to strengthen the body and the mind [1]. Nevertheless, when people think about food, the perceptions that recreate or recognise aromas and flavours related to lived experiences occur. Aromas and flavours are responsible for the taste and smell that will inspire the desire to ingest food but also, they are the decisive factors for a consumer choosing on the quantity and frequency to eat the product. A good aroma and flavour experience can leave people craving more. These reasons lead producers to carry out different kinds of studies on food products before presenting them to the consumers, undertaking them through sensorial tests where all aromas and flavours dissolved in the food are not only perceived on the tongue as sweet, sour, bitter, salty or umami, but also by smell [2].

Several foods are rich in vitamins, carbohydrates, enzymes, protein and/or minerals, among others. However, time and type of processing until its final consumption might jeopardise their aromas and flavours due to their exposure to light, heat, oxygen and temperatures [2], ruining the final product. A successful consumer experience is the main target for the food industry to innovate and find solutions to the problems affecting their commodities. Framed in the context of aroma and flavour loss

L. M. Quintero-Borregales · S. Goyanes (✉) · L. Famá (✉)
Universidad de Buenos Aires, Facultad de Ciencias Exactas y Naturales, Departamento de Física, Laboratorio de Polímeros y Materiales Compuestos, Buenos Aires, Argentina,
CONICET-Universidad de Buenos Aires, Instituto de Física de Buenos Aires-CONICET (IFIBA), Buenos Aires, Argentina
e-mail: goyanes@df.uba.ar

L. Famá
e-mail: lfama@df.uba.ar

J. Parameswaranpillai et al. (eds.), *Micro- and Nano-containers for Smart Applications*, Composites Science and Technology,
https://doi.org/10.1007/978-981-16-8146-2_16

during food processing and storage, scientists have focused on investigating different methodologies to avoid aroma and flavour degradation by developing improved techniques for protecting them [3, 4]. These methods take into consideration the individual properties of each compound to come up with new techniques and shells' materials to reach the goal, meeting the constantly changing customers' demands and trends, not only with flavours but also adding nutritional content [5, 6].

Encapsulation has brought the attention towards describing a process to protect active compounds, entrapping them as particles and enclosing them within a shell, protecting the active compound against evaporation, reaction, or migration in food, as to mask undesired aromas and flavours [7].

When designing an encapsulation process, its purpose and final use must be primarily considered. Since the encapsulation can be used to enhance or mask a bioactive until its final release. For example, in the case of chewing gum, the aim is to improve the flavour through the encapsulation of the active compound. A successful process seeks to enhance the flavour by releasing it during a longer chewing time. In the case of nutraceuticals, it is sought not only to mask the flavour but also to protect the bioactive compound from the digestive process until it reaches the intestine where it is meant to degrade and be bioabsorbed by the cells of the intestinal walls. A recent challenge has been taking microencapsulation to nano scale [8]. Moreover, this is also why multilayer encapsulations have had a great outcome. This type of encapsulation allows the bioactive agent to be transported through different environments where each layer can be specially designed to overcome it.

This chapter focuses on the exposition and discussion of the process of encapsulating aroma and flavour through different techniques in order to either mask or prevent their degradation overtime to conserve food richness and tastefulness until their consumption. The relevance and the effect of aromas in food products, as well as the significance of encapsulation techniques (spray drying and electrohydrodynamic micro or nanoencapsulation) in industrial processes will be discussed, considering the researches in the area. Additionally, the advantages and the issues of the addition of aromas and flavours in food, their further advances in the market and the tendencies in encapsulation technology will be exposed. Particularly, the benefits of cyclodextrins as encapsulating matrices will be presented.

2 Aromas and Flavours Generalities

Aroma is usually a mixture of volatile and odorous mostly lipophilic and hydrophilic organic compounds that have a low molecular weight (<1000 Da) [9], high saturated vapour pressure of water and are usually found in low concentrations (ppb or ppm) [10]. It is typically released before and after the product has been ingested, depending on the features and physical state of the core. They can also be classified as hydrocarbons, alcohols, aldehydes, ketones, esters, acids, sulphides, among others. Table 1 exhibits the molecular structures of some of the most commonly employed aromas utilised in the food industry [2, 10]. These structures are widely employed in the food

Table 1 Molecular structures of some of the most commonly employed aromas in the food industry

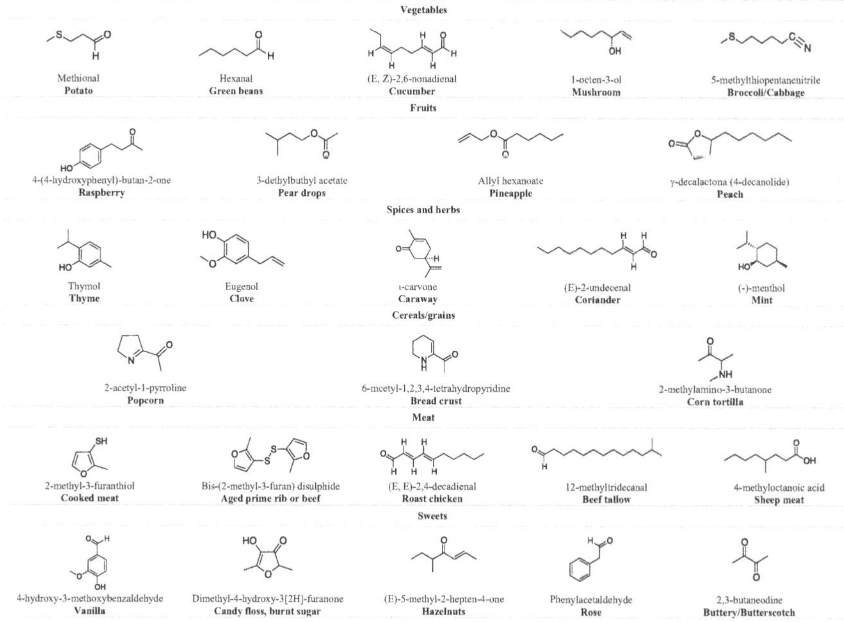

industry to simulate vegetables (potato, mushroom, green beans, cucumber, broccoli and others), fruits (raspberries, pineapple), spices and herbs (thyme, clove), cereals and grains (popcorn, bread crust, corn tortilla), meat (cooked meat, aged prime rib, roast chicken, beef tallow, sheep meat) and sweets (vanilla, candy floss, hazelnuts, rose and butterscotch) [11].

In particular, Thomas et al. considered the effect of aeration as an aroma perception increaser and potential compensator of the perceived aroma loss of low-fat dairy products by implementing a heating experiment on an in vitro model built to study aroma release at oral processing temperatures [12]. Sáenz-Navajas et al. studied the red wine mouthfeel by investigating the possible intermodal effect of aroma and establishing relationships between sensory dimensions and chemical measurements [13]. On another hand, Bertelsen et al. evaluated the effects of different aromas among young adults. Vanilla aroma resulted the most promising with regards to the improvement of sweetness and women were more susceptible to the "sweet" effect. Regarding the intensity of the sweet taste, it was only found in the 22–25 years old young adults but not in the 19–21 years old ones [14].

On the other hand, flavour is a combination of taste, smell, texture and mouthfeel as main stimuli factors that contribute to flavour perception. It is also one of the most important and distinctive traits in consumers' acceptability of food products, since the new generations prefer to ingest visually attractive and flavoured food. However, not only is the flavour determinant but also the odour/aroma have particular importance in the acceptance of food as well, being particularly responsible for its typical perception, reaching the nose directly before consumed (orthonasal) or when it is in the mouth (retronasal) [15].

Kang et al. observed the potential harms of alkenylbenzenes and aldehydes compounds (found in tobacco) in youth from inhaling flavours of e-cigarettes by studying the DNA adduct formation, which could possibly lead to carcinogenesis, conducting a structural similarity analysis to the chemicals without DNA adduct information [4]. Shi et al. investigated the effect of the starches-based coatings on the quality of ripened dry-cured hams during long-term storage by evaluating the changes in different volatile flavours compounds through multivariate statistical analysis, finding that 2-nonanone, nonanal, amyl alcohol and 2-heptanone could be employed as markers to distinguish between the coated and uncoated hams [3]. On another hand, Chai et al. evaluated the effects of enzymatic hydrolysis of tallow and addition of sodium chloride (NaCl) on the formation of 3-monochloropropane-1,2-diol esters and sensory characteristics of beef flavours finding that the enzymatic hydrolysis conditions had a significant effect on the formation of the compound during the flavour preparation [15]. McKay et al. tested a complex mixture of five compounds (guaiacol, o-cresol, 4-ethylphenol, 2-iso-butyl-3-methoxypyrazine, and 2,4,6-trichloroanisole) in a partially de-aromatised red wine matrix by a trained panel, finding the wines with the complex mixtures elicit an olfactory response that single compounds on their own did not show [16].

Figure 1 exhibits an example of aromas and flavours wheel of cross-products for training analysis panels. These aromas can be classified as flowery, fruity, spicy, roasty, smoky, microbiological, earthy and chemical, and are highly important to the food industry.

2.1 Perception of Aromas and Flavours

When imagining a meal, the senses of taste and smell are activated allowing to have a multisensory experience. In this context, an increasing number of researches focus on analysing and demonstrating how all of the human senses play a significantly important role in the perception of aromas and flavours through sensorial studies, analytical techniques and/or quantification methods [16, 17].

Determining which of thousands of volatile compounds contribute to the aroma profile of the food has been one of the biggest challenges for food chemists. Decades of research have concluded on the need for human senses, especially the nose as a detector for sensory analysis, to work along with gas chromatography–olfactometry (GC-O) and accurate quantification methods to successfully identify specific compounds [15–18].

Several studies for sensory analysis such as the qualitative methods allow to determine the properties of people's food preferences. This evaluation is important as an interdisciplinary science to measure, analyse and interpret stimuli reactions perceived through the human basic senses (hearing, sight, smell, taste and touch)

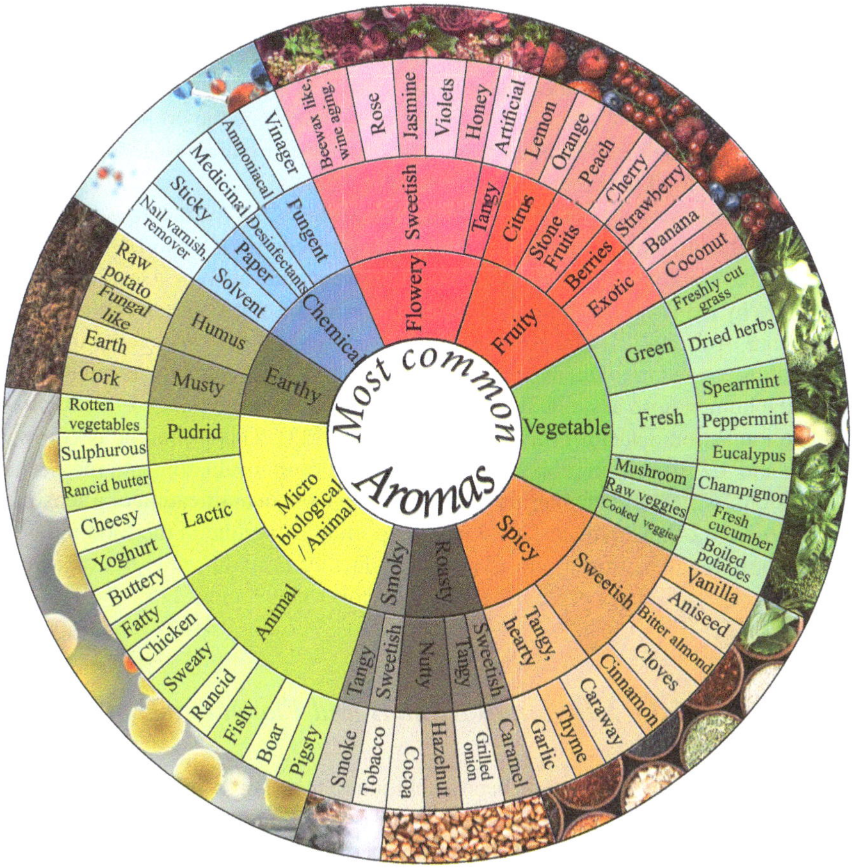

Fig. 1 Commonly employed aromas in the food industry

[19]. It can provide useful information to understand which attributes should be present and which ones should be reduced or eliminated [20].

The sensorial analysis can be divided into three kinds of tests: discriminative (used to determine whether or not a difference exists among samples), descriptive (employed to identify sensory characteristics that are important in a product and give information on the degree or intensity of those characteristics) and affective or hedonic (to measure how much a panellist likes a product sample based on its sensory characteristics) [2].

The perception of aromas and flavours is translated from the information gathered by multiple sensory afferents, including gustatory, olfactory and somatosensory fibres. Even when their perception comes from the gathering of multiple sensory afferents, it is possible to identify the different contributing inputs. Thus many researchers have aimed to understand the sensorial studies into the food industry

[21, 22]. These sensorial studies can be carried on either with trained or untrained panellists but always in accordance with international reference standards [20].

Some other quantitative methods such as statistical analysis and gas chromatography–olfactometry (GC-O) are also employed to determine the properties of food. The plotting of such results is essential for scientists to collect real, reproducible information on food products across the world [19].

Furthermore, sensory aspects are an excruciating factor for the consumer food selection. The olfactory food profile is determined by comparing it to 10–20 aromas, for which the employment of GC-O has become an essential analytical technique used nowadays by many food producers to determine which isolated aroma contributes to the flavour of food [22]. The compounds are analysed by gas chromatography and assessed by trained individuals, who describe the aroma and estimate its intensity by using the human nose as a GC detector in parallel to a second detector [2], such as a flame ionization detector or a mass spectrometer [23]. This technique becomes a precise and descriptive approach to characterize stimuli, evaluating and measuring impressions, which also enables the comprehension and quantification of a sensorial characteristic [24]. Moreover, GC-O is an important technique to the food industry due to its implementation to elucidate the molecular basis of varietal aromas and flavours to optimise food processing and to discover new aromas [25].

2.2 Issues of the Addition of Aromas and Flavours in Food

Encapsulation has for a long time been a solution to incorporate bioactives and lipids in water-based products. The selection of the best encapsulation process for a specific aim is a challenge that must be taken into a strict consideration since the efficiency of aroma and flavour encapsulation is a delicate balance between core–shell stability. Furthermore, the shell's material should ideally cover a number of characteristics to achieve its most relevant goal, protect the active material until its release and increase its content in the final product [26]. Moreover, issues like the water solubility, physical state, aroma and flavour retention, and profile and cost of the technology scaling-up are often faced in the food industry when incorporating these active materials.

It is well known that there are substances such as many bioactive lipids and minerals wanted to be incorporated into food products. A low water solubility or oil solubility can limit their incorporation as they cannot be dispersed into water-based products (fortified waters, soft drinks, nutritional beverages, sauces, desserts and yogurts) [26, 27]. A key factor in this is that, for the food industry, all aromas and flavours need to meet "food grade qualification". These food grade active materials can be mostly found as liquids and solids and only some as gases at room temperature. In some cases, the typical physical state of an active material may not be appropriate for its incorporation into a food product, having to modify its physical properties in order to make its incorporation possible [26].

It is also important to take into account and somehow assure the chemical, biochemical and physical stability of the incorporating compound in the final product.

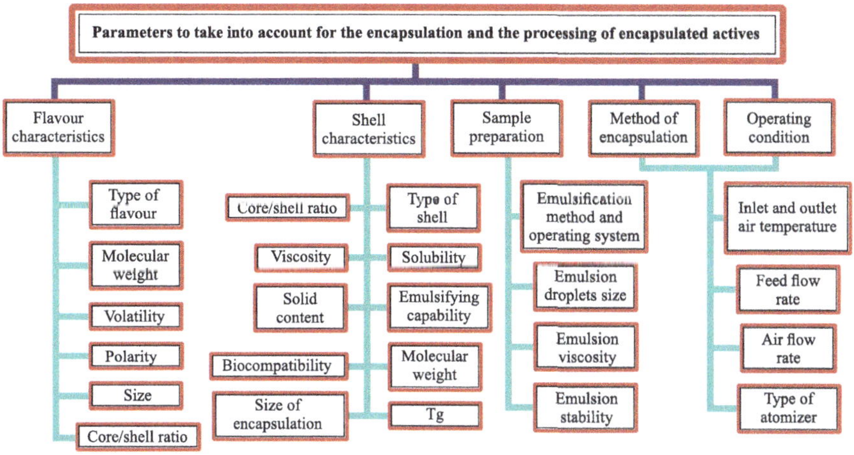

Fig. 2 Parameters to take into account for the encapsulation and the processing of encapsulated actives

For example, the stability of the encapsulation of nutraceuticals until their final destination, avoiding its early liberation causing the release of their undesired characteristics. This can also lead to their manufacture, storage and transportation being difficult [26, 28]. In the encapsulation process, the parameters to take into account for the encapsulation and the processing of encapsulated actives are shown in Fig. 2, such as the characteristics of the flavour and shell, the sample preparation, the methods of encapsulation and the operating condition. The most important flavour characteristics are the type of flavour, molecular weight, volatility, polarity, size and concentration, as well as the shell characteristics like the type, concentration, solubility, viscosity, solid content, emulsifying capability, biocompatibility, molecular weight, size of encapsulation and T_g.

During the sample preparation, factors like the emulsification method and the operating system and the emulsion droplet size, viscosity and stability need to be considered. The encapsulation method and operation condition require special attention on the inlet and outlet temperature, feed flow and air rate and the type of atomiser.

On the other hand, many active ingredients incorporated into a food can be limited by their characteristic and undesirable odours or flavours, such as saltiness, sweetness, sourness, bitterness and astringency. It is essential to find a proper strategy to encapsulate them in order to avoid this problem without risking their quality. For example, omega 3, 6 and 9 [29]. Furthermore, a company's growth is essential to ensure their products can outlast current trends and climb its way into meeting the market's demand. Towards this goal, scaling up is needed for expanding and adapting in order to successfully reach a greater number of people [30]. Leading to a higher production either with the same or less resource, always looking out for the same or higher output at a lower cost.

Additionally, the cost of scaling up depends on the size of the encapsulation needed. At a micro scale, a low-cost technique such as spray-drying can be used. While for a nano scale some modifications or a different technique are needed, such as electrospraying, which tends to have a more complicated process than spray-drying.

3 Micro and Nanoencapsulation of Aromas and Flavours

Aromas and flavours are typically found in liquid and solid state but when these compounds are in the gaseous state they can be perceived by the olfactory epithelium and can generate a stimulus called either odour if the product emanates a bad smell and aroma when the emanated smell is pleasant [15]. These compounds can be incorporated into the food in their processing or may be formed during their cooking. On the other hand, flavours tend to be expensive. This issue led to food manufacturers concern about the preservative methodology for these additives. In this context, encapsulation has been one of the most widely studied techniques in the last decades for this matter. Since aroma and flavour have high importance in the food industry, the ultimate goal of encapsulation is to control their release and to improve stability during the processing and consumption of the final product. For this purpose, aroma is usually encapsulated to expedite its handling during the production process conferring it more stability, textural effect, facilitating its incorporation to the commodity and its controlled release in the food industry [10]. Moreover, the healthy food tendency promotes the inclusion of nutraceuticals, such as fish oil, among others in different foods [31]. However, the inclusion of these nutritional supplements tends to be a problem due to their bad flavour and odour [32], which are widely inhibited by encapsulating them for their final release in the intestine.

As for the industry and as a global definition, encapsulation is a process capable of enclosing a particle of a material (solid, liquid or gas) or a mix of materials, called core within another material, denominated shell. The shell is also defined by many researchers as a thin layer of coating with a range of sizes and many morphologies as a result of the different matrices (encapsulated core within the shell) [3]. The encapsulation technique is implemented by different industries (food, chemical, pharmaceutical, etc.) and has become important due to increasing the active material life. In addition, it can protect it from any factor that can cause its degradation (oxygen, light, heat, water) and/or to mask undesirable odours until its later release. The encapsulation release of aroma or flavour to a specific location can be triggered by an appropriate temperature, mechanical pressure, moisture, pH, and enzymatic release [32]. Different methodologies have been employed to carry out encapsulation and will be covered in this chapter.

In general, the encapsulation techniques are shown in Fig. 3. (a) shows a core, also known as active compound and (b-e) can be classified as monocore (a core within its shell), multilayer (more than one shell overlapped), polycore (more than one core within its shell) and amorphous (a core with an undefined morphology), respectively [32].

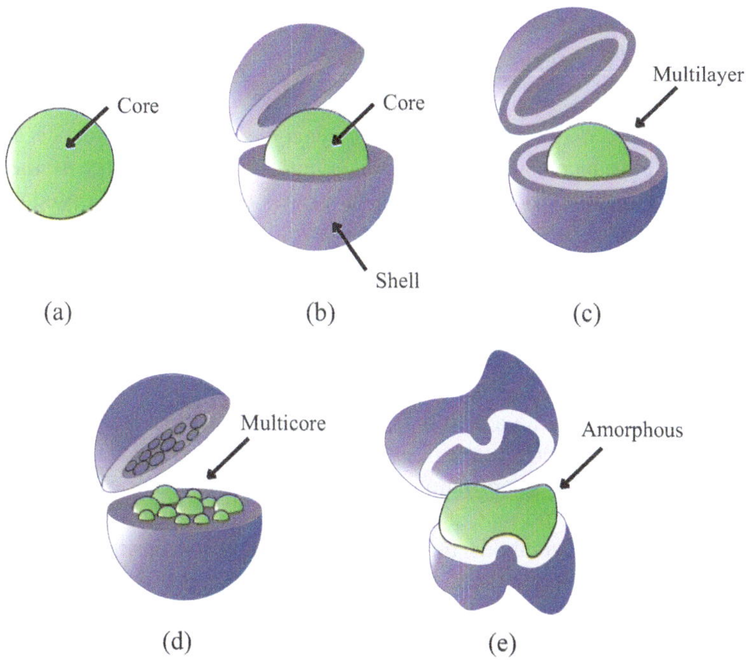

Fig. 3 Different morphologies of encapsulation, **a** core, **b** monolayer, **c** multilayer, **d** multicore and **e** amorphous

The matrix functionality is critically determined by the chemical composition and the properties and physical states of the material used in the shell. Ideally, it should have desired characteristics like the ease of handling throughout the encapsulation process and ability to effectively disperse or emulsify the encapsulated components and to stabilise the composite structure. It should not react with the active core but it should seal and protect it during the capsules manufacturing and storage. Additionally, it should properly swell, have a high solubility and dispersibility under certain conditions for controlled release, as well as be inexpensive raw material and food-grade approved. Mostly, a mix of components is employed to achieve the goal. The selection of the right shell is determining in the encapsulation process due to its dependence on the nature of the core and the desired barrier properties [33].

The most common encapsulation materials are polysaccharides (including cyclodextrin and alginates) and in recent years there has been a growing trend for the use of proteins. Furthermore, all shell materials have been classified as water-soluble polymers, water-insoluble polymers, water-soluble non polymeric materials, water-insoluble non polymeric material [34–36].

3.1 Importance of Aroma and Flavour Encapsulation in Food Processes

The encapsulation of active compounds has become a common practice and has gained great importance in the last few decades as an approach to improve the retention, controlled release, and preservation of aromas and flavours. Consequently, a broad range of inclusion methods has been developed to produce pastes, powder (most common), capsules, granules and emulsion droplets, as types of encapsulation to cover the desired forms for all different aromas and flavours' final use. Powders are the easiest to handle and most preferred form of encapsulation. It is easily employed in a variety of food products such as bread, cakes, candies, chocolates, instant desserts, milk powder, instant drinks, extruded snacks, etc. However, due to the properties of the active ingredients and the coating material, there is a different inclusion method suitable for every type of product [34]. Table 2 exhibits some examples of recent encapsulation studies that employed matrices with different core/shell materials, including CDs that are the most used in the food industry.

The most common methods for producing powders encapsulation are: spray drying, used for room temperature flavours and aqueous releases like the microencapsulation of orange aroma [35]; extrusion, which applies higher temperatures to release the flavours [36, 37]; melt extrusion, which is the best inclusion method for oxygen sensitive flavours compounds because of the shell's properties it can provide, allowing a high control of the final product feature [38]; molecular inclusion with β-cyclodextrin (β-CD) that is employed for the preservation of polyphenols obtained from different natural resources [37, 39, 40]; aroma and flavour emulsions, which are implemented in ice creams and beverages [41]; micro and nanoemulsion, the best method for transparent and low viscosity isotropic dispersions by the draw-on of water and oil phase stabilisation [42]; and micro and nanoencapsulation, which work great for mixing the aromas and flavours compounds to help extending their release [34].

In the food industry, many factors such as temperature, pH, among others, need special consideration during processing in order to avoid aroma and flavour degradation. Furthermore, this is the main reason why some of them need to be added in different stages of the process. Depending on desired final product aromas and flavours may need to be added at the final stage in order to reduce their thermal loss [43]. Figure 4 illustrates the main advantages of encapsulation that allow it to enhance flavours through the protection of the active compound, its controlled delivery, preservation/masking and handling. The controlled delivery can take place through the diffusion (core/shell characteristics), melting, fracture, swelling, erosion and degradation of the capsule [44].

A primary reason for encapsulation is the protection of the active ingredients from their loss of functionality or their degradation before they achieve their desired effect. It aims to preserve the stability of the compounds during processing, storage and consumption but one of the problems during this process is the interaction between substances that can damage the shell's life [45].

Table 2 Examples of encapsulation using different core/shell materials

Core	Carrier/shell's material	Method of encapsulation	Results	References
Orange	Maltodextrin (DE 15) and gum arabic	Spray-drying	Significant differences between the powders collected in aroma retention	[61]
Orange essential oil (OEO)	Zein and β-CD	Molecular inclusion	The OEO delayed the microbiological spoilage of cakes from 30 to 150 days, suggesting its utilization as a natural antifungal in bakery products	[95]
Black pepper essential oil	Lactoferrin and sodium alginate	Complex coacervation	Obtained capsules capable of transporting and resisting oral and gastric conditions tested in vitro	[62]
D-limoneno	β-CD	Molecular inclusion	D-limonene showed relevant clinical antibacterial activity	[98]
Limonene, nonanal, 2-methylpyrazine, benzaldehyde, geraniol, and ethyl benzoate	Oil-in-water octanolwater, water–air, and octanol–air	Emulsion	Correlations between the aroma compound release rates and the octanol–water, water–air, and octanol–air partition coefficients	[63]
Vitamins A and E	Sodium caseinate (SC), tween 80 (TW80) and maltodextrin (MD)	Extrusion and spraydried	Thermally stable microcapsules (up to 170 °C)	[64]
Anthocyanins extracted from grape skin	Sodium alginate	Emulsification/ internal gelation followed by spray/freeze-drying techniques	Microencapsulation greatly enhanced light and thermal stabilities of anthocyanins, spray-dried microcapsules resulted the most stable	[65]

(continued)

Table 2 (continued)

Core	Carrier/shell's material	Method of encapsulation	Results	References
Ascorbic acid (AA)	Maltodextrin (DE 10–15) and gum arabic	Melt extrusion	Maltodextrin, maltodextrin-gum arabic and maltodextrins-trehalose successfully encapsulated AA powders	[66]
Phenolic compounds in blackberry	β-CD, chitosan, xanthan and hydrogel	Lyophilization	High antioxidant activity in microcapsules coated with β-cyclodextrin	[71]
Grape skin extract	Gum Arabic, β-CD, and hydroxypropyl-β-cyclodextrin	Molecular inclusion	Inclusion complexes powder of phenolic compounds from grape skin	[72]
Caffeic acid, trans-ferulic, acid and p-coumaric acid	β-CD	Molecular inclusion	Inclusion complexes in the hydrophobic cavity of β-cyclodextrin by non-covalent bonds with different physicochemical properties	[69]
Tea polyphenols	Hydroxypropyl-β-CD	Freeze-drying and molecular inclusion	Improved antioxidant activity and bioavailability of the inclusion complexes	[103]
Amino acid based ionic liquids	β-CD	Co-precipitation and molecular inclusion	Successful formation of AAILs and β-CD	[111]
Glimepiride	β-CD	Solvent evaporation and kneading	The complexes prepared with both methods showed better drug release as compared with the pure drug	[110]
Coriander essential oil	Cyclodextrin nanosponges	Gelation	Synthetized cyclodextrin nanosponges are stable at temperatures over 200 °C and capable to providing a controlled release of antimicrobial active agent	[101]

(continued)

Table 2 (continued)

Core	Carrier/shell's material	Method of encapsulation	Results	References
Cinnamon oil	Cyclodextrin nanosponges	Molecular inclusion	Achieved a better controlled release by its encapsulation, allowing cinnamon essential oil to be effective at lower concentration	[94]

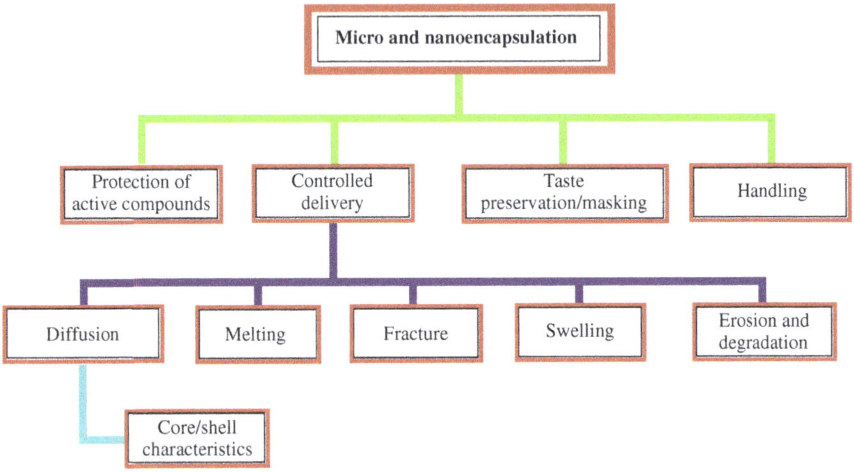

Fig. 4 Micro and nanoencapsulation purposes

Many producers around the globe introduced vitamins, minerals, polyphenols [46] among other functional compounds due to the health benefits they bring, finding difficulties in keeping them stable towards many factors like their high sensitivity to the environment (oxygen, light, etc.) in their processing and gastrointestinal conditions. As a solution to keep the bioactive compound functionality through its delivering process, encapsulation has been employed as an approach because it can avoid or retard the oxidation or hydrolysis process, effectively preventing the degradation until the active compound has reached the desired location delivering a better colour, flavour and texture [47].

Furthermore, bioactive compounds like antioxidants, minerals, vitamins, phytosterols [48] and living cells as probiotics [49], can be successfully delivered at a desired place and time using the encapsulation technique [50]. By applying this process, the sensory and functional benefits can be achieved through a proper delay of the release of the active compounds. In fact, they can be classified into two different types, delayed and sustained. The delayed-release refers to a mechanism that allows the retard deliverance of active compounds to the desired location. Some examples are currently implemented in probiotic bacteria to protect them from gastric acidity and be finally discharged in the small intestine: the flavours of the ready-to -eat meals are liberated by microwave heating [51] and the activation of encapsulated sodium bicarbonate by baking [52]. On the other hand, sustained release is a mechanism designed to maintain a constant concentration of actives compounds at the desired site. The most commonly found example encompasses encapsulated flavours and sweeteners for chewing gum, reducing the rate of delivery of the flavour throughout a long time of chewing [53].

Despite the application of the encapsulation process through the delayed and sustained type of deliveries, some factors can affect the flavour release rate mechanism, as the flavour molecules migrate over a certain period of time. Such as a higher molecular weight of the carrier material minimises de aroma diffusion and lowers retention of flavour compounds. This possibly occurs due to lower degrees of interactions between the aroma and the carrier material when an amorphous coating material is in a glassy state [54]. The glassy to rubbery state transformation is described by the glass transition temperature (T_g). When the coating material is at a lower temperature than the T_g, it remains in its glassy state, but at the T_g, its viscosity reduces and its molecular mobility increases showing poor barrier characteristics [34].

Nevertheless, amorphous materials with higher molecular weight have a lower residual opening and a low porosity, tending to improve protection against flavour release at temperatures above the T_g, whilst amorphous materials with lower molecular weight have better storage stability [55]. This leads to the morphology being dependent on the properties of the shell and the active compound. Most of the aromas and flavours are hydrophobic in their natural state and do not mix well with hydrophilic shells.

On the other hand, the process of delivering the active compound can take place through diffusion. It is a two steps encapsulation release mechanism that is induced by the solubility of the main compound and its mobility through the shell's material walls. It is provoked by the vapor pressure of the volatile compound on both sides of the encapsulation material and depends principally on the active core diffusion from the interior to the exterior of the matrix and, posteriorly, on the separation of the volatile compound to the adjacent food components (away from the matrix surface) [34, 43].

Additionally, melting is another mechanism of delivery. It is mostly applied in the food industry to keep the aromas and flavours protected while cooking. It involves the need for heat or moisture to release the aroma compound by thermally melting the fat in the coating material or by dissolving the matrix with moisture in a water-soluble coating material liberating the active at the desired cooking stage. Additionally, the fracture mechanism consists of the rupture of the encapsulation walls by external physical forces (e.g. by chewing it), once the core's material breaks the aroma and flavours are quickly released [34].

Similarly, swelling takes place when the encapsulating material is maintained in a thermodynamically complementary medium. That will be controllably absorbed (with different solvents mediums) by the shell material allowing the aroma and flavour to be diffused through it towards its surroundings [34]. The erosion mechanism of delivery consists of the erosion, degradation, dissolution and diffusion processes taking all place at once, towards the liberation of the aroma and flavour active compound from the encapsulation matrix. In this process, the erosion and degradation decrease the thickness of the matrix walls, thus the diffusion of the active compound through the wall can be increased. This affects the shape of the matrix depending on if the erosion only takes place on the delivery surface (heterogeneous erosion) or if it also affects the active compound within (homogeneous erosion) [43].

As mentioned before, one of the reasons to encapsulate the aromas and flavours is that many active compounds employed in the food industry are undesired and should be masked until they pass the taste buds on the tongue. Some examples of the active ingredients are the added vitamins, polyphenols and minerals to improve the nutritional properties of the final products. Also, is it to mention the case of nutraceuticals, in which encapsulation is sought to resist the different stages during the digestion process and be delivered to the intestine. Increasing its bioavailability to be degraded and finally absorbed by the cells of the intestine walls [8, 56].

Another factor of the micro and nanoencapsulation techniques is how they can be implemented to provide easier handling of the active ingredient in the food processing, as well as, simplify handling by converting a liquid into a solid, giving the matrix a uniform position and promoting the simple mixing [43].

As previously mentioned, the cyclodextrins (CDs) compounds are *par excellence* used in the industry as encapsulants, which is why they are so attractive to work with.

3.2 Aromas and Flavours Encapsulation with Cyclodextrins

Cyclodextrins are natural, odourless, tasteless [57] and cone-shaped cyclic oligosaccharides that are attained from enzymatically modified starch. They have a hydrophilic surface due to their hydroxyl external groups, and hydrophobic interior constituted of glucose units. This characteristic has made them one of the most remarkable macrocyclic molecules with a major interest for scientists. Due to its small diameter and its restricted cone shape and side rim depth of ~8 Å, it allows the encapsulation of external molecules within [58]. Figure 5 exhibits the three types of CDs: α, β and γ-CD that mostly differ on the number of glucose units and their internal diameter. The α-CDs are shaped by 6 units of glucopyranose, have an internal cavity of ~5 Å, are mostly employed to form inclusion complexes with benzene derivatives and are about 8 times more soluble in water than β-CD but have lower solubility than γ-CD at room temperature; the β-CDs are the most common and extensively used in the food industry, they are ~7 Å wide and contain 7 glucopyranose units and form complexes with aromatics or heterocycles, and the γ-CDs have 8 units of glucopyranose, their internal cavity is ~9 Å and they are mostly employed to form complexes with macrocycles or steroids [59].

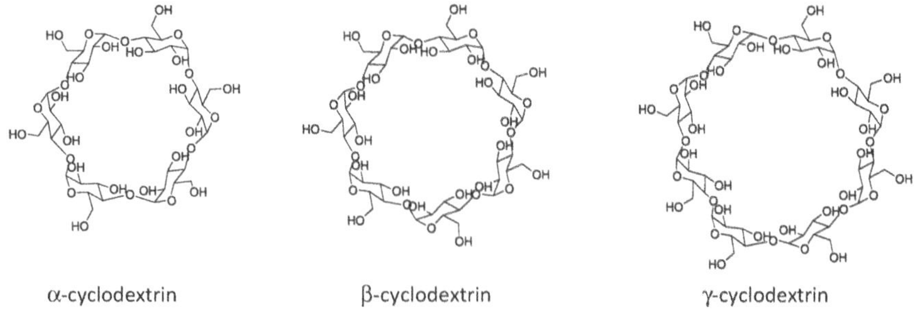

Fig. 5 Types of cyclodextrins

CDs are widely employed in supramolecular chemistry because of their high biocompatibility, complex forming capacities, high availability in multiple sizes, high reactivity from binding a range of molecular guests in water and their low cost [44]. Additionally, being called cage molecules, they have the ability to form encapsulation complexes with other molecules through host–guest interactions, which make them very useful for practical application, including pharmacy, medicine, foods, cosmetics, toiletries, catalysis and chromatography, among others [58]. Figure 6 (a) illustrates the primary and secondary face of a CD, (b) the formation of an inclusion complex and (c) their stoichiometries that can be often formed with 1 host-1 guest, forming a 1:1 complex, but one or two molecules of each can also react to form complexes with 2 hosts-1 guest (2:1), 1 host-2 guests (1:2) or 2 hosts-2 guests (2:2) [60].

The main application of CDs is to protect unstable compounds. They are employed as encapsulation systems to enhance solubility, protection, controlled release, improve organoleptic behaviour and mask off flavours [61] by encapsulating them in their inner lipophilic cavity via hydrophobic interactions (molecular inclusion). Many studies have been reported on the improvement of the aromas and flavour characteristics with CDs [62, 63].

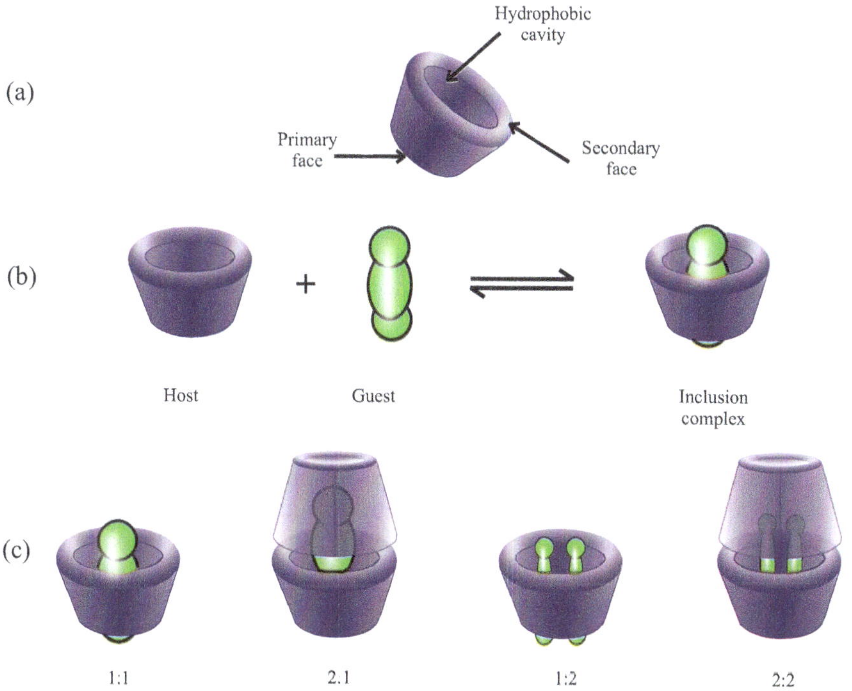

Fig. 6 **a** cyclodextrin, **b** formation of an inclusion complex, **c** stoichiometries of the cyclodextrins complexes

In the last decades, the growth of the food industry has increased research into adding value to already known products and new developments. One of the improvements in the food field has been the encapsulation of thymol as a stable dry powder formulation of CDs with high inclusion efficiency, for its implementation as an ingredient at an industrial scale. This was acquired by solubilizing it in aqueous solutions finding the Kc (formation constant) value to increase with the pH of the media until neutral pH, forming stable solid complexes with economic and technological advantages for the food industry [64]. The effect of β-CDs was also studied to minimise microbiological risk as a possible stabiliser for nisin (an antimicrobial peptide extensively used as a food preservative on cooked pork meat). It offers a new simple method in the food industry that could provide a better antibacterial effect by delaying its degradation and improving its stability when surrounded by trypsin [65, 66]. Some others procedures have been reported for the antibacterial effect of d-limonene complexed with β-CD [67], antibacterial nanoparticles embedded in electrospun CD nanofibers [68], to improve peptide affinity and stability by complexing to CDs [69], the insertion of coriander essential oil in CD nanosponges and for the synthesis of many inclusion complexes [70] for aromas and flavours in the food industry [71, 72].

The encapsulation of CD complexes can be achieved by countless methods depending on the properties of the molecule and on selected CD (Table 2) [72]. These methods are physical blending, kneading, co-precipitation, solvent evaporation, neutralisation precipitation, milling, spray-drying, lyophilization/freeze drying, microwave irradiation and supercritical antisolvent. The physical blending method consists of creating a solid blend of flavours and CDs. It is prepared by machine-driven trituration in a mortar or sieve to get a final product with a wanted size to be then stored under controlled conditions. Kneading technique is the simplest and lowest cost method to prepare the inclusion complexes. The aroma or flavour is added into a barely hydrated CD paste to be kneaded for some time and dried to be taken into the wanted particle size [73]. Co-precipitation method comprises the addition of aroma or flavour to a CD solution to co-precipitate into the formation of a complex. This process involves keeping the system under controlled stirring and protected from light [74]. Solvent evaporation technique encompasses getting molecular dispersion of the aroma or flavour by dissolving it along with the CDs in different but mutually miscible solvents in order to achieve the inclusion complex after the solvent's evaporation [75]. Neutralisation precipitation method centres on dissolving the aroma or flavour in an alkaline solution to be mixed with an aqueous solution of CDs to then be neutralised with an acid solution until obtaining the inclusion complex as a white precipitate [76]. In milling technique, after being physically blended, the mix passes through an oscillatory or ball milling with different balls sizes until obtaining the desired particle size. This is a low-cost technique since it does not include the employment of solvents [77]. Spray-drying allows to obtaining dry powder products from the compound in solution. It is mostly used as a preservation method by eliminating the water content making the product's storage easier [78]. Lyophilization/freeze drying consists of freezing the aroma/flavour complex solution in order to expose it to a drying process afterward to eliminate the solvents at low pressure, obtaining a porous, amorphous and highly compatible aroma/flavour-CD complex powder [79]. The microwave irradiation method is a quick process with a high a yield of the product that involves microwaving the aroma/flavour-CD complex in a regular microwave oven after being dissolved in a blend of water and organic solvents to be filtered and dried afterward [80]. Supercritical antisolvent is a solvent-free, non-flammable, inexpensive and fast technique that implements carbon dioxide as a new complexation medium to prepare the inclusion complex in a solid state [80].

The inclusion complexes can sometimes be analysed spectroscopically by studying the changes in the spectra intensity (increases/decreases) and shifts of the peaks that can occur when the active compound is protected by the formation of the complex.

A recent study by Day et al. (2020) remarks on the importance of the quantification of the hygroscopic response of the complex formed with β-CD to predict particle deposition patterns within the pulmonary system with the aim of delivering therapeutics to a target location and optimizing drug efficacy by analysing the morphology-dependant resonances arising in Raman spectra, dynamic vapour sorption and bulk solution refractive index and density measurements [81]. However, since the inclusion complexes are mostly based on the desired stoichiometry, most quantification methods may not provide accurate information on the stability of the complex [82].

On the other hand, CDs can also form non-inclusion complexes by building hydrogen bonds with other molecules through the hydroxyl groups on their external surface. They can develop as non-cyclic oligosaccharides and polysaccharides to yield lipophilic complexes insoluble in water. These are only able to solubilize in water by the formation of micro and nano size aggregates complexes, similar to micelles that are held together by hydrogen bonds and hydrophobic forces [58].

3.3 *Micro and Nanoencapsulation Techniques*

Up to the present time, the usual encapsulation micro and nano techniques have been successfully used for the physical or mechanical encapsulation of aromas and flavours active compounds. Particularly, micro and nanoencapsulation physical techniques [83], like spray-drying, extrusion, electrohydrodynamic (electrospinning and electrospraying), have offered numerous benefits over orthodox methods [43].

Microencapsulation technique consists of generating the encapsulation particles of solids, droplets of liquid or a combination of both within micro sized capsules with a thin shell or coating material. Mostly employing gelatine, lipids, starch, cellulosic proteins (proteins, polysaccharides, wax or polyacrylic acids) as shells [84]. It shows many advantages over the unencapsulated compounds since it protects the core material from its surroundings, providing it with flavour-masking, controlled release rates, reduced toxicity and protection against oxidation and moisture [85].

The techniques mainly differ on the size of the particles that are able to produce, generating capsules in the micro and nano scales respectively [86]. Even though

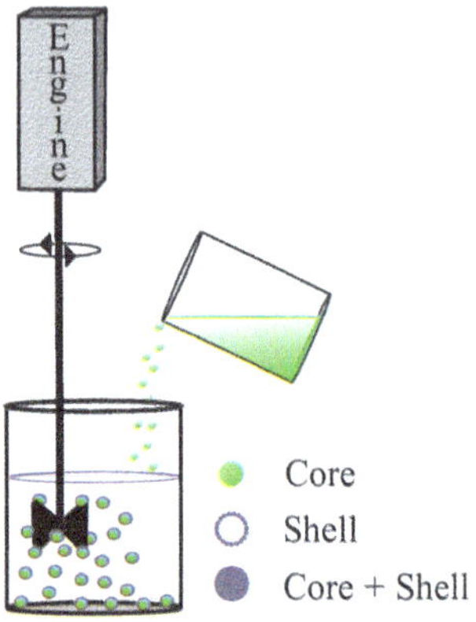

Fig. 7 Coacervation method process

microencapsulation has been the industry preferred technique for long, nanoencapsulation has become important for the food industry for its high efficiency, offering a wide range of benefits on storage stability and control release compared to microencapsulation [86]. Moreover, nanoencapsulation is an emerging and promising technology that allows for the minimisation of dosage and the addition of volatile compounds in the food industry. It confers protection and availability of the active component [87] and improves its storage stability with the utilization of nano size (up to 200 nm) encapsulation [88].

Micro and nanoencapsulation techniques can be achieved by a handful of methods as coacervation, liposomes and emulsification methods, among others. The coacervation method, illustrated in Fig. 7, refers to a colloidal phenomenon that comprises liquid–liquid phase separation of a single blend of two oppositely charged polymers induced by the modification of the pH, ionic strength, temperature and solubility under controlled conditions [89].

The coacervation method can occur by simple or complex coacervation, depending on the number of employed polymers. A single coacervation utilises a single polymer to obtain coacervates from the dehydration caused by adding a coacervating agent (a salt) to the colloidal solution and complex coacervation while the complex coacervation involves a blend of two or more polymers types for the encapsulation shell [90]. An interesting experiment employs the simple coacervation method for the preparation of semi-solid food. It disperses sodium-containing soy protein and gum Arabic coacervates into food gels to enhance food saltiness. It is capable of reducing the sodium content by 30% without risking flavour and texture qualities [91].

Micro and nanoencapsulation by complex coacervation have been widely employed in the food industry for its simplicity, low cost, reproducibility and high encapsulation efficiency. A recent study on the coacervation method made use of the interaction of oppositely charged polyelectrolytes in the aqueous form to obtain a high encapsulation efficiency (up to 99%) at a lower processing cost. It emphasises that the utilization of food-grade shell materials synthesized at ambient temperature makes the coacervation an appropriate choice in the food industry [33] for its highly effective and complex but achievable scaling-up.

Additional research has developed a novel industrially scalable microencapsulation process. It consists of the employment of an "in situ" complex coacervation method to retain up to 80% of the total d-limonene over 72 days of storage in sealed vials at room temperature by spray drying. This collapses multiple process steps into one to obtain dry complex coacervate microcapsules of d-limonene using alginate and gelatine as wall materials [8]. Micro and nanoencapsulation can also take place through the liposomes method composed of phospholipids which have both a hydrophilic and a hydrophobic ending group and have been investigated for drug delivery, cosmetics, pharmaceuticals and food industry. It has shown a huge potential for encapsulating aromas and flavours through its ability to entrap lipophilic compounds within the lipid bilayer of the hydrophobic region and the hydrophilic compound within an aqueous medium in the hydrophilic region [92]. It also has high bioavailability, biocompatibility and great permeability. On the other hand, this method sets its limitations on higher scaling-up costs compared to the coacervation method, encapsulation with poorer physical and chemical stability and the need of extra complex treatment. However, some food industries and researchers take into consideration these limitations to employ this method [93, 94]

Emulsification or ionic gelation method is a mild, simple and solvent-free encapsulation technique with great potential to protect compounds against the effect of negative environmental conditions [95]. This technique focuses on the interactions between two charged polymers (a positive and a negative), producing a hydrophobic active components emulsion in a polymer to be dripped into a calcium solution and crosslinked with polyelectrolytes by multivalent ions without the implementation of high temperatures or toxic crosslinking agents. This method is mostly applied by the extrusion technique to fabricate spherical gel microparticles [39, 96]. A recent review focuses on the study of the gelation of proteins in aqueous solution induced by heating or cooling as the principal mean to give desirable texture to food products, taking into consideration the gel stiffness and the microstructure of the gelled mixtures [97].

In addition, nanoencapsulation also exhibits a lower tendency for generating particle aggregation creating more stable encapsulation systems, as in the case of emulsification encapsulation. Moreover, it has the characteristic of having a thinner and weaker particle light scattering, which does not modify the food colour characteristics and makes it suitable for clear or transparent food products and even beverages. Furthermore, this technique allows the active compounds to be more accessible and available for its absorption than microencapsulated compounds [34].

It is important to mention that, even when nanoencapsulation looks superior, both techniques are very promising and highly suitable for aromas and flavour retention in the food industry, especially on functional foods. These processes are achievable through the application of methodologies like spray drying and electrohydrodynamic techniques. Their use needs to be founded on the knowledge about the properties of the core to be encapsulated, the encapsulating material and the suitability of the delivery system [86].

3.3.1 Spray-Drying

Spray-drying is the most commonly employed technique for the encapsulation of active compounds and the production of flavour powders due to it being a rapid, continuous, low cost, reproducible and scalable process for producing dry powders from a liquid solution [98]. It consists of four phases, starting with the preparation of a liquid solution (core and shell material dissolved on a water-based stable emulsion or suspension), being fed into a drying chamber or gas medium through an atomiser or high pressure spray nozzle entailing the contact of the solution with a stream of hot air (phase 2). Then, the formation of spray droplets that will lose moisture in the drying chamber (phase 3) and lead to a solid particle, which is subsequently removed from the air stream and collected into a receiver (phase 4). Final dried powder particles are obtained in only a few milliseconds (Fig. 8). The atomiser and nozzle opening size of this encapsulation method can be modified to achieve micro or nano scale products [83].

Some factors should be considered for micro and nanoencapsulation using spray-drying technique to be successful [99]. Among others, the nature of the active compound to be encapsulated is relevant, highlighting the importance of the utilization of only water-based dispersion with a highly proven stability over time. It is also important to use the appropriate shell materials (mostly hydrophilic polysaccharides, proteins and active hydrophobic or hydrophilic molecules). The nozzle feeding rate and viscosity of the solution should be checked and adjusted looking to avoid total or partial blockage that would affect the atomisation. It could increase the shell thickness and eventually affect the bioavailability of the active compound

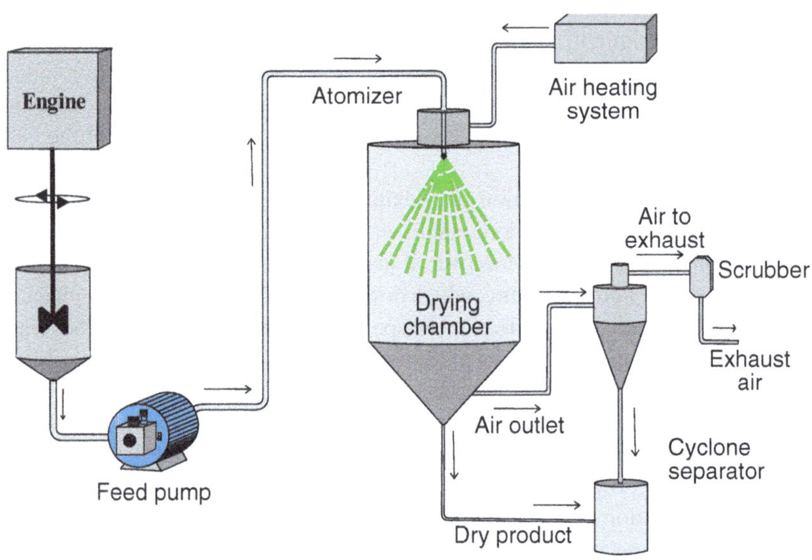

Fig. 8 Spray-drying method process

for its diffusion. Moreover, the temperature of the drying medium should be kept under 100 °C otherwise, at high temperatures, some active compounds could be lost by evaporation [8, 34, 89].

3.3.2 Electrohydrodynamic

Electrohydrodynamic micro and nano techniques are simple, economic and scalable methods for encapsulating aromas and flavours [100]. They consist of generating liquid droplets on the needle tip through the application of a large electrical potential difference, deforming it into a Taylor cone to be ejected from it. They are called *electro* due to the electric energy and *hydrodynamic* for fluid dynamics applied to liquids [34]. These techniques are widely employed in the food industry for their potential industrial and scientific applications obtained by using non thermal methods like electrospinning and electrospraying to encapsulate heat sensitive flavours avoiding thermal degradation [101, 102]. Even when these techniques are based on the same non-thermal principle and prevent the agglomeration complication, they mainly differ in the concentration of the polymeric solution to be employed. Electrospinning jet from the spinneret stabilises and lengthens high concentrated polymers generating continuous micro or nanosized fibres while the electro-spraying jet destabilises to produce fine droplets or particles with low concentration polymers transforming them into micro or nanosized particles. Additionally, some materials like many proteins cannot be electrospun by themselves, thus needing a plasticiser, but can be electrosprayed with no extra care [103].

Electrospinning

Electrospinning is a simple, extremely flexible and low-cost technique capable of producing continuous fibres and nonwoven mats with diameters from 2 nm to micro size with a high surface area using different polymers possibly producible on large scale. This technique allows full control over the diameter, surface morphology and pore size by adjusting many parameters of the solution during the electrospinning process such as viscosity, concentration, electrical conductivity, temperature, humidity and atmospheric pressure [103, 104]. A high voltage potential difference produces the formation of the fibres in the spinneret or grounded electrode whilst the solvent of the polymeric solution evaporates on its trajectory from the needle. Among many other researches, this technology has been employed to encapsulate curcumin within electrospun protein fibres to generate bioactive coatings for food packaging, showing a promising release behaviour for packaging of high water content [105], and for the development of bilayer films with antioxidant and antimicrobial activity from biodegradable polymers by encapsulating rosemary polyphenols in the inner layer. It showed antioxidant activity in food [106]. Figure 9 illustrates the technique process with different grounded fixed and rotating electrodes.

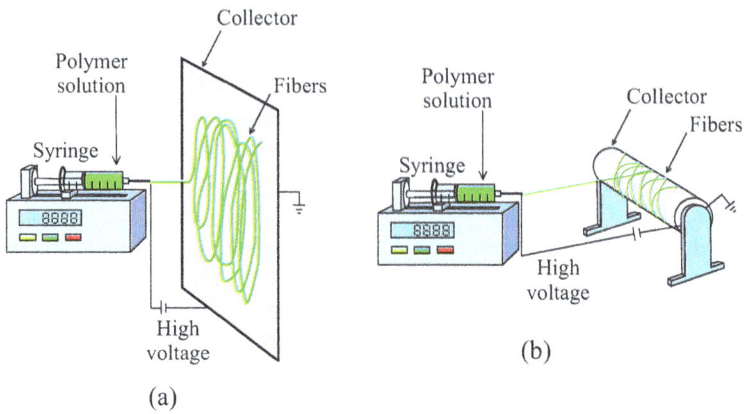

Fig. 9 Electrospinning technique process with: **a** fixed electrode and **b** rotating electrode

Electrospraying

Electrospraying is a variation of the electrospinning technique, also called electrohydrodynamic atomisation, and it is employed for micro and nanomaterials due to the flexibility of the technique that allows having control over the composition, structures, sizes, morphologies and shapes of the final product. In this technique, the liquid is broken into droplets by its atomisation under the effect of a high voltage potential difference, thus the solution in the needle is deformed into a Taylor cone to consequently be ejected from it towards the collector on an electrically grounded target whilst the solvent evaporates through the ejecting process (Fig. 10). Factors like the diameter of the droplet are dependent on the voltage, flow rate, polymer, solvent and concentrations employed [107], and better and smoother surface morphology can be achieved by using a closed chamber with a nitrogen flow. This technique was used to optimise the encapsulation of *mentha longifolia L.* essential oil into Balangu (*Lallemantia royleana*) seed gum, increasing their utility as flavouring and bioactive agents in foods and beverages [108].

Fig. 10 Electrospraying technique process

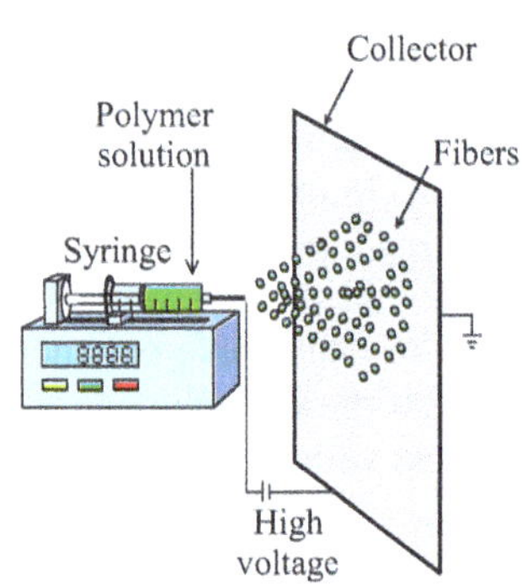

3.3.3 Advantages of Electrohydrodynamic Techniques over Spray Drying

For decades, spray drying has been widely employed to produce near-spherical particles. It consists of transforming a liquid solution into powder through a rather simple process, generating droplets through an atomiser to interact with a hot gas stream that quickly evaporates the solvent from the droplet to form an individual particle [109]. This technique has a main limitation, pumping the liquid solution to produce atomised droplets into the drying chamber. This inconvenience is due to the required temperature of the gas stream to successfully evaporate the solvent risking the active compound by a possible thermal degradation [108, 110]. Therefore, the development of electrohydrodynamic techniques as a novel particle engineering technique is booming in many fields such as food and drug industry, surging as an improvement to spray drying. Thus, it allows the formation of tiny droplets from a jet of conductive liquid under the stimulus of an electric field difference that with a simple set up can produce micro and nano sized particles within a reduced size distribution. Besides, the applicability of these techniques expands to ambient conditions that could minimise the risk of heat-induced degradation and withhold the possibility of their use as atomiser in spray drying technique [111]. Some other advantages of the electroyhydrodynamic techniques are their simplicity and cost-effectiveness for preparing functional fibres with porous structures with a wide surface area to volume ratio, as well as their high encapsulation efficiency [83]. Moreover, in the case of the proteins, the low feed flow rate generally perceived as a drawback of the technique makes the global shell tension to which proteins are exposed to relatively low [111].

4 Aromas and Flavours Encapsulation Market and Tendencies

The global encapsulation of aromas and flavour is a competitive market, which makes manufacturers centre on the adoption of new technologies to expand, including agreements, mergers/acquisitions and new product launches to reinforce the food industry. It was already worth USD 14.6 billion by 2019 with a value projection over USD 17.4 billion by 2027 [112], only taking into account 35 countries as exhibited in Fig. 11. The US and Europe are leading the industrial share with 30% and 26% of the overall industry since 2017, respectively [113].

The fast-growing market can be attributed to the increasing population along with their rising needs. The potential applications of encapsulation for odour and flavour masking, controlled release, improved flavours, fortified foods, the extension of shelf life and the precision in delivering nutrients, which are among the principal features boosting the food industry development (including beverages). These characteristics have been reported as the main stimulus for technological developments partnering with research and development investments [114, 115]. They are only able to reach

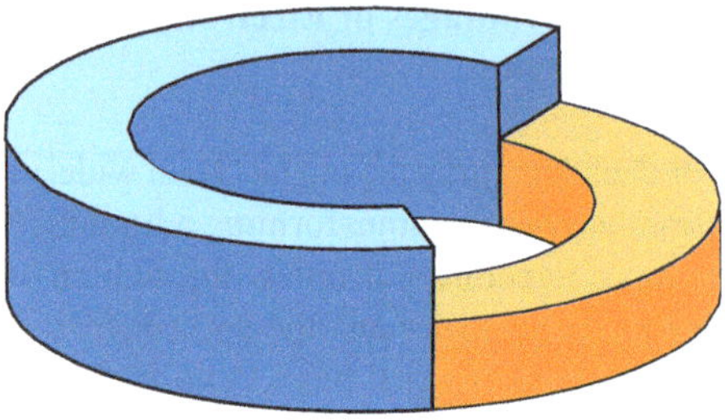

Fig. 11 Aroma and flavour's market value in 2019 and 2027

quality and compete in major markets focusing on new products launches and innovations to cope with market trends led by the rising consumers' population's spending on flavoured food.

Trends are fast-moving transformations that not only change consumers' lifestyle but also influence their diet and eating habits. Busier lifestyles leave less time for meal preparations causing a shift in the tendency from natural and nutritional food towards fast and ready to eat food. Changes in the consumers' diet made them aware of the lack of nutrients and vitamins. Moreover, the consumers' health awareness leads to expanding the need for encapsulated minerals and vitamins motivating the industrial growth into developing an essential oils and natural extracts market, which is thought to increase 4.7% by 2024. In addition, flavour blends currently hold the highest market revenue share of 40% and are thought to keep increasing [115–117]. Food processing dominates the market, leading the physical techniques with yearly revenue over USD 5.5 billion, chemical and physiochemical techniques holding over 13% of the global market, whilst microencapsulation process accounts for about 40% of the overall market and the increasing trend for nanocapsules is thought to drive up 4.6% its demand by 2027. Expanding its utilization to processed meat, fish and increasing the ready-to-eat food shelf-life [113, 118].

5 Regulations

Consumers constantly changing tastes, raising needs and interests for natural products and how the current aromas and flavours affect their health, make the food industry reorganize around these new demands. This has created a concern of reliable and consistent regulatory standards across the global food industry, having to assure their products quality and safety of flavours employed to increase the delight of food and beverages consumption [119].

As established by the regulation No. 1334/2008 of the European Parliament and of the Council of December 2008 indicates that "*Flavouring are used to improve or modify the odour and/or taste of foods for the benefit of the consumer and flavourings should not be used in a way as to mislead the consumer about issues related to, amongst other things, the nature, freshness, quality of ingredients used; the naturalness of a product"... "The approval of flavourings should also take into account other factors relevant to the matter under consideration including societal, economic, traditional, ethical and environmental factors, the precautionary principle and the feasibility of controls"* [120].

All flavour manufactures have to follow certain regulations established by the main regulatory bodies like the Food and Drug Administration (FDA) for the United States and the European Food Safety Authority (EFSA) for Europe. As well as, some other governmental and industries such as the Joint FAO/WHO Expert Committee on Food Additives (JEFCA), the Flavours and Extract Manufacturers Association (FEMA), the Food Chemical Codex (FCC), and the Council of Europe's Index System. Most countries follow the US or Europe regulations to avoid having different ranks of classification that can cause to further confusion for all involved. For example, even when the US and European regulations are similar, they slightly differ on the cataloging of a natural flavour. While the FDA defines it as "an essential oil, oleoresin, essence or extractive, protein hydrolysate, distillate, or any product of roasting, heating or enzymolysis", the EFSA establishes the "source material must be vegetable, animal or microbiological. Must be produced by a traditional food preparation process" [60, 121].

Another example of these discrepancies is the case of the molecule of ethyl vanillin that provides the well-known and legendary vanilla flavour. This organic compound is typically synthetized, which in the US it would be explicit on the product's label as "artificial vanilla flavour" while in Europe, it would be implicit with the absence of the "natural" word on the label, only saying "vanilla flavouring" [121, 122].

6 Conclusions

Aromas and flavours have proven to be among the most important active ingredients in the food industry. They are two of the organoleptic properties that most influence the expectations of consumers and are generally associated with the quality and freshness of the product. The main interest of aromas and flavours is that they remain invariant until they reach the consumer's palate. In the case of nutraceuticals, which have been in vogue all over the world in recent years due to their great benefits, the main objective is that they reach the intestine and are released inside it slowly. A common problem with flavours and nutraceuticals is that they must retain all their properties in the final product in order to reach their consumption, thus they should not be altered with processing.

To achieve this, different encapsulation methods have been investigated, being spray drying, extrusion, molecular inclusion with β-CD, emulsion, and electrohydrodynamic the most promising. In this sense, proper capsule design has been shown to be critical. In particular, polysaccharides and cyclodextrins are some of the most widely used shell materials for protection and enhancement. However, even when cyclodextrins have a great ability to form stable complexes and mask aromas and flavours, their quantification is still a promising field of investigation.

Due to the very promising response, the application of the encapsulation of aromas, flavours and nutraceuticals in the food industry has shown fast growth. Due to the demanding and changing market, the food industry had to collaborate with scientists to offer better quality and healthier products to customers. Particularly, it promoted further development and research on techniques such as spray-drying, electrospray or electrospinning to achieve efficiency, increase stability and have better control of the aroma and flavour release. These techniques have been investigated for a long time; however, they still have much more to offer and there is a great expectation to improve them.

Acknowledgements The authors want to thank Universidad de Buenos Aires (UBACYT 2018 20020170100381BA), ANPCyT (PICT 2019–04509, PICT 2017–2362 and PICT Startup 2016–4639).

References

1. Gupta P (2020) Food addiction and intermittent fasting. J Addict Res 4(1). https://doi.org/10.33140/jar.04.01.03
2. Simat T (2017) Panel training on odour and aroma perception for sensory analysis. DLG Expert Rep 1–24 [Online]. https://www.dlg.org/en/food/topics/dlg-expert-reports/sensory-technology/dlg-expert-report-1-2017/. Accessed 13 Aug 2020
3. Shi Y, Li X, Huang A (2020) Multivariate analysis approach for assessing coated dry-cured ham flavor quality during long-term storage. J Food Sci Technol. https://doi.org/10.1007/s13197-020-04579-z
4. Kang J, Valerio L (2020) Investigating DNA adduct formation by flavor chemicals and tobacco by products in electronic nicotine delivery system (ENDS) using in silico approaches. Toxicol Appl Pharmacol 398:115026. https://doi.org/10.1016/j.taap.2020.115026
5. Aheto JH et al (2020) Multi-sensor integration approach based on hyperspectral imaging and electronic nose for quantitation of fat and peroxide value of pork meat. Anal Bioanal Chem 412(5):1169–1179. https://doi.org/10.1007/s00216-019-02345-5
6. Liu D et al (2020) Characterization of Jinhua ham aroma profiles in specific to aging time by gas chromatography-ion mobility spectrometry (GC-IMS). Meat Sci 168. https://doi.org/10.1016/j.meatsci.2020.108178
7. Ogrodowska D et al (2020) Pumpkin oil addition and encapsulation process as methods to improve oxidative stability of fish oil. Lwt 124. https://doi.org/10.1016/j.lwt.2020.109142
8. Assadpour E, Jafari SM (2019) Advances in spray-drying encapsulation of food bioactive ingredients: from microcapsules to nanocapsules. Annu Rev Food Sci Technol 10(1):103–131. https://doi.org/10.1146/annurev-food-032818-121641
9. Liberto E et al (2020) Chapter 1 Headspace sampling: an 'evergreen' method in constant evolution to characterize food flavors through their volatile fraction. In: Advanced gas chromatography in food analysis. The Royal Society of Chemistry, pp 1–37
10. Buettner A, Schieberle P (2000) Influence of mastication on the concentrations of aroma volatiles—some aspects of flavour release and flavour perception. Food Chem 71(3):347–354. https://doi.org/10.1016/S0308-8146(00)00199-0

11. Letelier L et al (2020) Southern species from the biodiversity hotspot of central chile: a source of color, aroma, and metabolites for global agriculture and food industry in a scenario of climate change. Front Plant Sci 11:1–16. https://doi.org/10.3389/fpls.2020.01002
12. Thomas CF et al (2020) Design of in vitro model to study oral aroma release: experimental study and numeric simulation of heat transfer in a foamed dairy matrix. J Food Eng 278. https://doi.org/10.1016/j.jfoodeng.2020.109940
13. Sáenz-Navajas MP et al (2020) Effect of aroma perception on taste and mouthfeel dimensions of red wines: correlation of sensory and chemical measurements. Food Res Int 131. https://doi.org/10.1016/j.foodres.2019.108945.
14. Bertelsen AS et al (2020) Individual differences in sweetness ratings and cross-modal aroma-taste interactions. Foods 9(2):9–15. https://doi.org/10.3390/foods9020146
15. Du W et al (2020) Key aroma compounds in Chinese fried food of youtiao. Flavour Fragr J 35(1):88–98. https://doi.org/10.1002/ffj.3539
16. Spackman C, Lahne J (2019) Sensory labor: considering the work of taste in the food system. Food Cult Soc 22(2):142–151. https://doi.org/10.1080/15528014.2019.1573039
17. McKay M et al (2020) Investigation of olfactory interactions of low levels of five off-flavour causing compounds in a red wine matrix. Food Res Int 128:108878. https://doi.org/10.1016/j.foodres.2019.108878
18. Wang S, Chen H, Sun B (2020) Recent progress in food flavor analysis using gas chromatography–ion mobility spectrometry (GC–IMS). Food Chem 315(2019):126158. https://doi.org/10.1016/j.foodchem.2019.126158
19. Lotfabadi SV, Mortazavi SA, Yeganehzad S (2020) Study on the release and sensory perception of encapsulated d-limonene flavor in crystal rock candy using the time–intensity analysis and HS-GC/MS spectrometry. Food Sci Nutr 8(2):933–941. https://doi.org/10.1002/fsn3.1372
20. Chambers E (2019) Analysis of sensory properties in foods: a special issue. 8(8)
21. Margeta P et al (2019) Importance of sensory evaluation in assessment of egg quality. Poljoprivreda 25(1):56–63. https://doi.org/10.18047/poljo.25.1.8
22. Natrella G et al (2020) Short communication: sensory characteristics and volatile organic compound profile of high-moisture mozzarella made by traditional and direct acidification technology. J Dairy Sci 103(3):2089–2097. https://doi.org/10.3168/jds.2019-17059
23. Barbará JA et al (2020) Volatile profile and aroma potential of tropical syrah wines elaborated in different maturation and maceration times using comprehensive two-dimensional gas chromatography and olfactometry. Food Chem 308:125552. https://doi.org/10.1016/j.foodchem.2019.125552
24. Tranchida PQ (2019) Advanced gas chromatography in food analysis. Royal Society of Chemistry
25. Nollet L (2019) Food aroma evolution: during food processing, cooking and aging, 1st edn. CRC Press
26. Nada A et al (2019) Fabrication and bioevaluation of a medicated electrospun mat based on azido-cellulose acetate via click chemistry. Cellulose 26(18):9721–9736. https://doi.org/10.1007/s10570-019-02739-9
27. Mcclements DJ (2015) Nanoparticle- and microparticle-based delivery systems. CRC Press, pp 1–550
28. Rezaei A, Fathi M, Jafari SM (2019) Nanoencapsulation of hydrophobic and low-soluble food bioactive compounds within different nanocarriers. Food Hydrocoll 88:146–162. https://doi.org/10.1016/j.foodhyd.2018.10.003
29. Dinu V et al (2019) Mucin immobilization in calcium alginate: a possible mucus mimetic tool for evaluating mucoadhesion and retention of flavour. Int J Biol Macromol 138:831–836. https://doi.org/10.1016/j.ijbiomac.2019.07.148
30. Selmer I et al (2019) Encapsulation of fish oil in protein aerogel micro-particles. J Food Eng 260:1–11. https://doi.org/10.1016/j.jfoodeng.2019.04.016
31. Shilomboleni H, De Plaen R (2019) Scaling up research-for-development innovations in food and agricultural systems. Dev Pract 29(6):723–734. https://doi.org/10.1080/09614524.2019.1590531

32. Jamshidi A et al (2020) Advantages of techniques to fortify food products with the benefits of fish oil. Food Res Int 137:109353. https://doi.org/10.1016/j.foodres.2020.109353
33. Walia N et al (2019) Methods for nanoemulsion and nanoencapsulation of food bioactives. Environ Chem Lett 17(4):1471–1483. https://doi.org/10.1007/s10311-019-00886-w
34. Ribba L et al (2014) Electrospun nanofibrous mats: from vascular repair to osteointegration. J Biomed Nanotechnol 10:3508–3535. https://doi.org/10/1166/jbn.2014.2046
35. Saifullah M et al (2019) Micro and nano encapsulation, retention and controlled release of flavor and aroma compounds: a critical review. Trends Food Sci Technol 86:230–251. https://doi.org/10.1016/j.tifs.2019.02.030
36. Zhang R et al (2020) Microencapsulation of anthocyanins extracted from grape skin by emulsification/internal gelation followed by spray/freeze-drying techniques: characterization, stability and bioaccessibility, vol 123. Elsevier Ltd
37. Grasso S (2020) Extruded snacks from industrial by-products: a review. Trends Food Sci Technol 99:284–294. https://doi.org/10.1016/j.tifs.2020.03.012
38. Jedlińska A et al (2019) Industry-scale spray-drying microencapsulation of orange aroma. Int Agrophys 33(3):397–405. https://doi.org/10.31545/intagr/110857
39. Simsek T et al (2019) Combined computational and experimental study on the inclusion complexes of β-cyclodextrin with selected food phenolic compounds. Struct Chem 30(4):1395–1406. https://doi.org/10.1007/s11224-019-01347-4
40. Da Rosa CG et al (2014) Encapsulation of the phenolic compounds of the blackberry (Rubus fruticosus). LWT Food Sci Technol 58(2):527–533. https://doi.org/10.1016/j.lwt.2014.03.042
41. Chang D et al (2019) Ascorbic acid encapsulation in a glassy carbohydrate matrix via hot melt extrusion: preparation and characterization. Food Sci Technol 39(3):660–666. https://doi.org/10.1590/fst.02918
42. Kuck LS, Noreña CPZ (2019) Application of gum Arabic, β-cyclodextrin, and hydroxypropyl-β-cyclodextrin to microencapsulation by molecular inclusion of grape skin extract (Vitis labrusca var. Isabel). J Food Process Preserv 43(2):1–7. https://doi.org/10.1111/jfpp.13874
43. Dasgupta N, Ranjan S, Gandhi M (2019) Nanoemulsions in food: market demand. Environ Chem Lett. https://doi.org/10.1007/s10311-019008562
44. Hashidzume A, Yamaguchi H, Harada A (2019) Cyclodextrin-based rotaxanes: from rotaxanes to polyrotaxanes and further to functional materials. Eur J Org Chem 2019(21):3344–3357. https://doi.org/10.1002/ejoc.201900090
45. Tamaru S et al (2019) High correlation between octanol-air partition coefficient and aroma release rate from O/W emulsions under non-equilibrium. Food Res Int 116:883–887. https://doi.org/10.1016/j.foodres.2018.09.024
46. Sonawane SH, Bhanvase BA, Sivakumar M (2020) Encapsulation of active molecules and their delivery system. Elsevier Inc., pp 1–3782020. https://doi.org/10.1016/C2018-0-05369-4
47. Dutta S et al (2020) Nanoencapsulation of green tea polyphenols. In: Nanoengineering in the beverage industry, vol 20. Elsevier Inc., pp 229–261. https://doi.org/10.1016/b978-0-12-816677-2.00008-9
48. Popović DA et al (2019) Encapsulation technologies for polyphenol-loaded microparticles in food industry. In: Green food processing technique. Elsevier Inc., pp 335–367. https://doi.org/10.1016/b978-0-12-815353-6.00012-4
49. Mujica-Alvares J, Barra PA (2020) Encapsulation of vitamins A and E as spray-dried. Molecules 25(6):1357
50. Mohammadi M et al (2020) Phytosterols as the core or stabilizing agent in different nanocarriers. Trends Food Sci Technol 101:73–88. https://doi.org/10.1016/j.tifs.2020.05.004
51. Estevinho BN (2020) Nanocarriers loaded with nutraceuticals and bioactive ingredients (vitamins and minerals). In: Nanotechnology in the beverage industry. Elsevier Inc., pp 373–412. https://doi.org/10.1016/b978-0-12-819941-1.00013-4
52. Cravotto G, Binello A (2019) Effect of emerging processing methods on the food quality. Eff Emerg Process Methods Food Qual 147–161. https://doi.org/10.1007/978-3-030-18191-8
53. Dordoni R et al (2019) Enrichment of whole wheat cocoa biscuits with encapsulated grape skin extract. Int J Food Sci 2019:9161840. https://doi.org/10.1155/2019/9161840

54. Hinderink EBA et al (2019) Dynamic flavor release from chewing gum: mechanisms of release. Food Res Int 116:717–723. https://doi.org/10.1016/j.foodres.2018.09.002
55. Cheng J et al (2020) Development and characterization of Al-based amorphous coating. Jom 72(2):745–753. https://doi.org/10.1007/s11837-019-03966-y
56. Li R et al (2019) Glass transition, structural relaxation and stability of spray-dried amorphous food solids: a review. Dry Technol 37(3):287–300. https://doi.org/10.1080/07373937.2018.1459680
57. Aree T (2019) Understanding structures and thermodynamics of β-cyclodextrin encapsulation of chlorogenic, caffeic and quinic acids: implications for enriching antioxidant capacity and masking bitterness in coffee. Food Chem 293:550–560. https://doi.org/10.1016/j.foodchem.2019.04.084
58. Tian B et al (2020) The application and prospects of cyclodextrin inclusion complexes and polymers in the food industry: a review. Polym Int. https://doi.org/10.1002/pi.5992
59. Crini G et al (2018) Fundamentals and applications of cyclodextrins. In: Fourmentin S, Crini G, Lichtfouse E (eds) Cyclodextrin fundamentals, reactivity and analysis. Environmental chemistry for a sustainable world, vol 16. Springer, Cham. https://doi.org/10.1007/978-3-319-76159-6_1
60. Ahuja K et al (2018) Encapsulated flavors and fragrances market size by product (flavor blends, fragrance blends, essential oils & natural extracts, aroma chemicals), by technology (physical process [atomization {spray drying, spray chilling, spray disk}, extrusion, fluid bed technique], chemical & physicochemical process), by encapsulation process (micro encapsulation, nano encapsulation, hybrid technology, macro encapsulation), by end-use (food & beverages [beverage mixes, pressed tablets, chewing gum, bake mixes], toiletries & cleaners) industry analysis report, regional outlook, growth potential, price trend, competitive market share & forecast, 2018–2024 [Online]. https://www.gminsights.com/methodology/detail/encapsulated-flavors-and-fragrances-market. Accessed 24 June 2020
61. Zhang YM, Liu YH, Liu Y (2020) Cyclodextrin-based multistimuli-responsive supramolecular assemblies and their biological functions. Adv Mater 32(3):1–19. https://doi.org/10.1002/adma.201806158
62. Pinto Heckert Bastos L et al (2020) Encapsulation of the black pepper (Piper nigrum L.) essential oil by lactoferrin-sodium alginate complex coacervates: Structural characterization and simulated gastrointestinal conditions. Food Chem 126345. https://doi.org/10.1016/j.foodchem.2020.126345
63. Simionato I et al (2019) Encapsulation of cinnamon oil in cyclodextrin nanosponges and their potential use for antimicrobial food packaging. Food Chem Toxicol 132:110647. https://doi.org/10.1016/j.fct.2019.110647
64. Kringel DH et al (2020) Free and encapsulated orange essential oil into a β-cyclodextrin inclusion complex and zein to delay fungal spoilage in cakes. J Food Process Preserv 44(5):1–10. https://doi.org/10.1111/jfpp.14411
65. Rodríguez-López MI et al (2019) Thorough characterization and stability of HP-β-cyclodextrin thymol inclusion complexes prepared by microwave technology: a required approach to a successful application in food industry. J Sci Food Agric 99(3):1322–1333. https://doi.org/10.1002/jsfa.9307
66. Kfoury M et al (2019) Encapsulation in cyclodextrins to widen the applications of essential oils. Environ Chem Lett 17(1):129–143. https://doi.org/10.1007/s10311-018-0783-y
67. Li J et al (2019) Protective effect of β-cyclodextrin on stability of nisin and corresponding interactions involved. Carbohydr Polym 223:115115. https://doi.org/10.1016/j.carbpol.2019.115115
68. Costa MDS et al (2019) Comparative analysis of the antibacterial and drug-modulatory effect of D-limonene alone and complexed with β-cyclodextrin. Eur J Pharm Sci 128:158–161. https://doi.org/10.1016/j.ejps.2018.11.036
69. Celebioglu A et al (2019) One-step green synthesis of antibacterial silver nanoparticles embedded in electrospun cyclodextrin nanofibers. Carbohydr Polym 207:471–479. https://doi.org/10.1016/j.carbpol.2018.12.008

70. Cesari A et al (2020) Improvement of peptide affinity and stability by complexing to cyclodextrin-grafted ammonium chitosan. Polymers (Basel) 12(2):1–18. https://doi.org/10.3390/polym12020474
71. Li Y et al (2020) Ferulic acid-β-cyclodextrin inclusion complexes: application on the preservation of hairtail (Trichiurus lepturus). Int J Food Prop 23(1):282–296. https://doi.org/10.1080/10942912.2020.1722161
72. Zhong Y et al (2020) Inclusion complexes of tea polyphenols with HP-β-cyclodextrin: preparation, characterization, molecular docking, and antioxidant activity. J Food Sci 85(4):1105–1113. https://doi.org/10.1111/1750-3841.15083
73. Ramos ÓL et al (2019) Advances in processing technologies for bio-based nanosystems in food, 1st edn. CRC Press, pp 1–396. https://doi.org/10/1201/9781315177328
74. Adhikari L et al (2019) Binary complexes of glimepiride with β-cyclodextrin for improved solubility and drug delivery. Indian Drugs 56(3):54–60
75. Banjare MK et al (2020) Inclusion complexation of novel synthesis amino acid based ionic liquids with β-cyclodextrin. J Mol Liq 299:112204. https://doi.org/10.1016/j.molliq.2019.112204
76. Sousa S et al (2019) Poly(lactic acid)/Cellulose films produced from composite spheres prepared by emulsion-solvent evaporation method. Polymers (Basel) 11(1):1–19. https://doi.org/10.3390/polym11010066
77. Wang H et al (2019) Inclusion complexes of lycopene and β-cyclodextrin: preparation, characterization, stability and antioxidant activity. Antioxidants 8(8). https://doi.org/10.3390/antiox8080314
78. Oliva E et al (2020) New lipidyl-cyclodextrins obtained by ring opening of methyl oleate epoxide using ball milling. Biomolecules 10(2):1–18. https://doi.org/10.3390/biom10020339
79. Cui L et al (2020) Combining solid dispersion-based spray drying with cyclodextrin to improve the functionality and mitigate the beany odor of pea protein isolate. Carbohydr Polym 245:116546. https://doi.org/10.1016/j.carbpol.2020.116546
80. Fonseca L et al (2019) Characterization of inclusion complex of croton zehntneri essential oil and β-cyclodextrin prepared by spray drying and freeze drying. Rev Virtual Quim 11(2):529–542. https://doi.org/10.21577/1984-6835.20190040
81. Vilela A (2018) Introductory chapter: generation of aromas and flavours. Gener Aromas Flavours 1–8. https://doi.org/10.5772/intechopen.81630
82. Sabisch M (2020) The complex regulatory landscape for natural flavor ingredients [Online]. https://www.sigmaaldrich.com/technical-documents/articles/white-papers/flavors-and-fragrances/natural-flavor-ingredients-regulations.html. Accessed 18 Aug 2020
83. Day CPF et al (2020) Quantifying the hygroscopic properties of cyclodextrin containing aerosol for drug delivery to the lungs. Phys Chem Chem Phys 20(22):11327–21133. https://doi.org/10.1039/d0cp01385d
84. Benson G (2019) Development of quantification methods and evaluation of drugs photostability in the presence of cyclodextrin polymer and human serum albumin protein. PhD dissertation, Leicester School of Pharmacy, De Montfort Univ., Leicester
85. Das S et al (2020) Microwave-assisted β-cyclodextrin/chrysin inclusion complexation: an economical and green strategy for enhanced hemocompatibility and chemosensitivity in vitro. J Mol Liq 310:113257. https://doi.org/10.1016/j.molliq.2020.113257
86. Lopez-Polo J et al (2020) Humectability and physical properties of hydroxypropyl methylcellulose coatings with liposome-cellulose nanofibers: food application. Carbohydr Polym 231:115702. https://doi.org/10.1016/j.carbpol.2019.115702
87. Selvamuthukumaran M (2019) Handbook on spray drying applications for food industries, 1st edn. CRC Press, pp 1–362. https://doi.org/10.1201/9780429055133
88. Dhakal SP, He J (2020) Microencapsulation of vitamins in food applications to prevent losses in processing and storage: a review. Food Res Int 137:109326. https://doi.org/10.1016/j.foodres.2020.109326
89. Grand D, Goll H (2020) Microencapsulation process and product. US 10561621 B2

90. Medeiros AKOC et al (2019) Nanoencapsulation improved water solubility and color stability of carotenoids extracted from Cantaloupe melon (Cucumis melo L.), vol 270. https://doi.org/10.1016/j.foodchem.2018.07.099
91. Suyanto A et al (2019) Nano-emulsion and nano-encapsulation of fruit flavor: review. IOP Conf Ser Earth Environ Sci 292(1). https://doi.org/10.1088/1755-1315/292/1/012025
92. Ozkan G et al (2019) A review of microencapsulation methods for food antioxidants: principles, advantages, drawbacks and applications. Food Chem 272:494–506. https://doi.org/10.1016/j.foodchem.2018.07.205
93. Rodrigues do Amaral P, Lopes Andrade P, Costa de Conto L (2019) Microencapsulation and its uses in food science and technology: a review. Microencapsul Process Technol Ind Appl 1–18. https://doi.org/10.5772/intechopen.81997
94. Li Y et al (2020) Salt reduction in semi-solid food gel via inhomogeneous distribution of sodium-containing coacervate: effect of gum Arabic. Food Hydrocoll 106102. https://doi.org/10.1016/j.foodhyd.2020.106102
95. Timilsena YP et al (2019) Complex coacervation: principles, mechanisms and applications in microencapsulation. Int J Biol Macromol 121:1276–1286. https://doi.org/10.1016/j.ijbiomac.2018.10.144
96. El-Kader AA, Hashish HA (2019) Encapsulation techniques of food bioproduct. Egypt J Chem. https://doi.org/10.21608/ejchem.2019.16269.1993
97. Liu W et al (2019) Structural stability of liposome-stabilized oil-in-water pickering emulsions and their fate during: in vitro digestion. Food Funct 10(11):7262–7274. https://doi.org/10.1039/c9fo00967a
98. Lee BB, Bhandari BR, Ching SH, Howes T (2019) Improving hydrophilic barriers of encapsulated compounds in Ca-alginate microgel particles through a new ionotropic gelation method for double emulsion droplets. Food Biophys 14(4):365–382. https://doi.org/10.1007/s11483-019-09586-y
99. Afzaal M et al (2020) Encapsulation of bifidobacterium bifidum by internal gelation method to access the viability in cheddar cheese and under simulated gastrointestinal conditions. Food Sci Nutr 1–9. https://doi.org/10.1002/fsn3.1562
100. Nicolai T (2019) Gelation of food protein-protein mixtures. Adv Colloid Interface Sci 270:147–164. https://doi.org/10.1016/j.cis.2019.06.006
101. Bratovcic A, Suljagic J (2019) Micro- and nano-encapsulation in food industry. Croat J Food Sci Technol 11(1):113–121. https://doi.org/10.17508/cjfst.2019.11.1.17
102. López-Córdoba A et al (2018) Electrospinning and electrospraying technologies and their potential application in the food industry. In: Rai R, Bai JA (eds) Nanotechnology applications in the food industry. CRC Press, Taylor & Francis Group. ISBN: 978-1-49-878483-2
103. Beyerinck R et al (2019) Spray drying processes for forming solid amorphous dispersions of drugs and polymers. US 10383941 B2
104. Tabarestani HS, Jafari SM (2019) Production of food bioactive-loaded nanofibers by electrospinning. Elsevier Inc., pp 1–25
105. Huang X et al (2019) Hierarchical electrospun nanofibers treated by solvent vapor annealing as air filtration mat for high-efficiency PM2.5 capture. Sci China Mater 62(3):423–436. https://doi.org/10.1007/s40843-018-9320-4
106. Estevez-Areco S et al (2018) Release kinetics of rosemary (Rosmarinus officinalis) polyphenols from polyvinyl alcohol (PVA) electrospun nanofibers in several food simulants. Food Packag Shelf Life 18:42–80. https://doi.org/10/1016/j.fpsl.2018.08.006
107. Chen L et al (2019) A comprehensive review of electrospinning block copolymers. Soft Matter 15(12):2490–2510. https://doi.org/10.1039/c8sm02484g
108. Senthil Muthu Kumar T et al (2019) A comprehensive review of electrospun nanofibers: food and packaging perspective. Compos Part B Eng 175:107074. https://doi.org/10.1016/j.compositesb.2019.107074
109. Alehosseini A et al (2019) Electrospun curcumin-loaded protein nanofiber mats as active/bioactive coatings for food packaging applications. Food Hydrocoll 87:758–771. https://doi.org/10.1016/j.foodhyd.2018.08.056

110. Estevez-Areco S et al (2020) Active bilayer films based on cassava starch incorporating ZnO nanorods and PVA electrospun mats containing rosemary extract. Food Hydrocoll 108:106054. https://doi.org/10.1016/j.foodhyd.2020.106054
111. Wang P et al (2020) Electrospraying technique and its recent application advances for biological macromolecule encapsulation of food bioactive substances. Food Rev Int 1–23. https://doi.org/10.1080/87559129.2020.1738455
112. Flores M, Toldrá F (2020) Chemistry, safety, and regulatory considerations in the use of nitrite and nitrate from natural origin in meat products. Meat Sci 171:108272. https://doi.org/10.1016/j.meatsci.2020.108272
113. Rezaeinia H et al (2019) Electrohydrodynamic atomization of balangu (Lallemantia royleana) seed gum for the fast-release of mentha longifolia L. essential oil: characterization of nanocapsules and modeling the kinetics of release. Food Hydrocoll 93:374–385. https://doi.org/10.1016/j.foodhyd.2019.02.018
114. O'Sullivan JJ et al (2019) Atomisation technologies used in spray drying in the dairy industry: a review. J Food Eng 243:57–69. https://doi.org/10.1016/j.jfoodeng.2018.08.027
115. Di Pretoro A, Manenti F (2020) Spray drying. In: Springer briefs in applied sciences and technology. Springer, pp 65–74. https://doi.org/10.1007/978-3-030-34572-3_7
116. Batens M et al (2020) Feasibility of electrospraying fully aqueous bovine serum albumin solutions. Eur J Pharm Biopharm 147:102–110. https://doi.org/10.1016/j.ejpb.2019.12.011
117. Wood L (2019) Encapsulated flavors and fragrances market—global industry analysis, trends, market size, and forecasts up to 2024—ResearchAndMarkets.com [Online]. https://www.businesswire.com/news/home/20190325005770/en/Encapsulated-Flavors-Fragrances-Market---Global-Industry. Accessed 24 June 2020
118. Infinium-Global-Research (2019) Encapsulated flavors and fragrances Market (technology—physical process, physicochemical process, chemical process, extrusion, fluid bed, and others; Encapsulation process—hybrid-encapsulation, micro-encapsulation, nano-encapsulation, and macro-encaps [Online]. https://www.infiniumglobalresearch.com/consumer-goods-packaging/global-encapsulated-flavors-fragrances-market. Accessed 24 June 2020
119. Zion-Market-Research (2018) Global encapsulated flavors and fagrances market set for rapid growth, to reach USD 8,122.74 million by 2024 [Online]. https://www.zionmarketresearch.com/news/encapsulated-flavors-fragrances-market. Accessed 24 June 2020
120. Hegde IA (2018) Encapsulated flavors and fragrances market worth over $8 billion by 2024: Global Market Insights, Inc. [Online]. https://www.globenewswire.com/news-release/2018/07/04/1533218/0/en/Encapsulated-Flavors-and-Fragrances-Market-worth-over-8-billion-by-2024-Global-Market-Insights-Inc.html. Accessed 24 June 2020
121. Zion J, John MR (2018) Global encapsulated flavors and fragrances market worth over USD 8,122.74 million by 2024: Zion Market Research [Online]. https://www.globenewswire.com/news-release/2018/10/09/1618625/0/en/Global-Encapsulated-Flavors-and-Fragrances-Market-Worth-Over-USD-8-122-74-Million-By-2024-Zion-Market-Research.html. Accessed 24 June 2020
122. Hallagan JB, Hall RL, Drake J (2020) The GRAS provision—the FEMA GRAS program and the safety and regulation of flavors in the United States. Food Chem Toxicol 138:111236. https://doi.org/10.1016/j.fct.2020.111236

Incorporation of Novel Nanocontainers into Corrosion Protective Coatings on Metals to Induce Self-healing: A Multistimuli Approach

George Kordas

1 Introduction

Corrosion is a slow destruction process of metals after electrochemical reaction with their environment. Corrosion controls the lifespan of airplanes, automobiles and ships. Protection offered by chromic compounds was proved to be carcinogenic and its use was banned in 2007 [1, 2]. In recent years, active anti-corrosion protection systems have been developed promising not only friendly to the environment protection technology but also "self-healing" properties after corrosion. This technology prolongs the life of the metal components. Major airplane, automobile and ship companies today use epoxy resins developed by themselves, like BOING, or DAIMLER, FIAT and EADS uses epoxy coatings developed by independent companies, like Mankiewicz Gebr. & Co., Chemetal GmbH, VARNISH Srl, Re-Turn AS, etc. to protect airplanes, ships and automobiles. Recently, nano-containers (e.g. CeMo) were incorporated into these coatings charged with corrosion inhibitors (e.g. 8-hydroxyquinoline, 8-HQ) for self-healing, water and chlorine nanotraps for slow down the diffusion of water and chlorine into these coatings, CuO microcontainers loaded with SeaNine™ or Bromosphaerol (BR) for antifouling protection. All these technologies were developed in the framework of the MULTIPROTECT and MUST European programs [3, 4]. This way, not only self-healing was induced into the coatings but also "multifunctionality", to maximize the financial benefit of companies applying this technology. This new technology has led to protective coatings that exceed the properties of chromium(VI) based coatings.

G. Kordas (✉)
Sol-Gel Laboratory, Institute of Nanoscience and Nanotechnology, NCSR Demokritos, Aghia Paraskevi Attikis, Greece

Peter the Great St. Petersburg Polytechnic University, St. Petersburg, Russian Federation

J. Parameswaranpillai et al. (eds.), *Micro- and Nano-containers for Smart Applications*, Composites Science and Technology,
https://doi.org/10.1007/978-981-16-8146-2_17

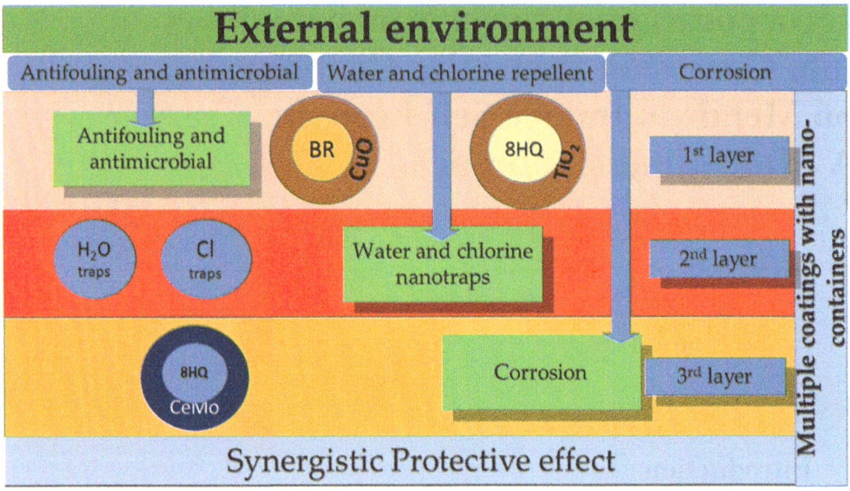

Fig. 1 Multilevel protection of metals via nano/micro-containers (BR Bromosphaerol, 8HQ 8 Hydroxyquinoline, CeMo Cerium/molybdenum Oxide, CuO Copper oxide)

For the first time in literature, our group developed and introduced the "multiple level protection" technology, where the protective systems use "smart" nanocontainers not only reacting satisfactorily to external influences, but also respond to changes in their internal structure. The epoxy coatings were developed by our group or given by commercial companies like Mankiewicz Gebr. & Co., Chemetal GmbH, VARNISH Srl, Wilckens Farben GmbH and Re-Turn AS. The tests were done in the lab and also in the field. Figure 1 describes schematically this revolutionary method of self-healing based on the multifunctional protection approach. This technology is described in aggregated in this publication that is the subject of many recent publications in the literature [5–36]. Figure 1 summarized the active ingredients incorporated into the protective system that respond to different types of mechanisms of destruction due to external environment:

1. The first level of protection will be consisted by CuO nanocontainers enriched with Bromosphaerol or SeaNine™. The antimicrobial action occurs via TiO_2 nanocontainers.
2. The second level of protection will be provided by the incorporation of water and chlorine nanotraps.
3. The third level of protection will be based on the encapsulation of inorganic anticorrosion substances (corrosion inhibitors) in different types of micro/nanocontainers (e.g. CeMo(8-HQ)) that will, upon request, suppress corrosion.

2 Experimental Procedures

2.1 Characterization Methods

The morphology of the coatings and nanocontainers were determined by Scanning Electron Microscopy (SEM) using a PHILIPS Quanta Inspect (FEI Company) microscope with W (tungsten) filament 25 kV equipped with EDAX GENESIS (Ametek Process and Analytical Instruments). The size and zeta potential of the nanocontainers was measured by Dynamic Light Scattering technique (DLS) using a Malvern Instruments Zetasizer DTS 1060. The pore size of the nanocontainers was done using a volumetric static sorption apparatus BET (Autosorb-1 MP, Quantachrome Instruments). The nanocontainers were filled with chemicals (inhibitors, antifoulants, etc.) using a home-made vacuum apparatus. The corrosion resistance was evaluated using the electrochemical impedance spectroscopy (EIS) apparatus of Solartron (ModuLab XM MTS—Materials Test System). The analysis and data acquisition were done with the ModuLab software. The experiments were done at room temperature inside a Faraday cage at the open circuit potential in an electrochemical cell from Plexiglas filled with 0.5 M NaCl aqueous solution with a saturated calomel reference electrode (SCE), a counter electrode made from a rectangular platinum plate (~1 * 1 cm^2) and coated metallic substrate as working electrode (area of about 3.15 cm^2). The FRA frequency was ranged between 100 kHz and 5 MHz with a 10 mV rms voltage. The self-healing effect was determined using a SVET and SIET commercial system from Applicable Electronics, measured by the ASET program (Science wares). More details are given in the literature [18]. The encapsulation capacity of chemicals into the nanocontainers was assessed by thermogravimetric analysis (TGA) via a Perkin Elmer (Pyris Diamand S II) apparatus (heating rate = 10 °C min^{-1} in air). The release of the inhibitor from nanocontainers was determined by a HITACHI model 100-40 spectrophotometer apparatus connected with a stirrer and an automatic sampler. This instrument was developed by Microtechniki Analytical Instruments. All measurements were done at room temperature. For the synthesis of coatings and nanocontainers, deionized water was used produced by Millipore Milli-Q apparatus. The homogenization of the suspensions was carried out in an ultrasonic processor (Elma Sonic Bath, S.30H).

2.2 Synthesis of the Nanocontainers

The preparation of nano/micro-containers requires three stages. The basic principle is the production of the nucleus where its size becomes the basis of the size of the nano/micro-container. A study should be carried out that the various parameters determine its size. The nucleus can be a polymer or silica. On this nucleus is built a shell with the sol–gel method which will be the nano/micro-container. The composition of the shell can be organic or inorganic. Organic shell is built on silica. Inorganic is created on polystyrene. The silica nucleus is removed with a dilute HF solution. The removal of polystyrene is done with slow calcination. In inorganic containers should be done thorough control of the structure of the surface and porosity. Also, the containers must be loaded with chemicals that will be triggered to leave the containers when an external need arises such as corrosion or cancer. The main reason we incorporate different molecules in the containers is the avoidance of incorporation into the matrix of the bind them into the epoxy resin and will not be able to operate autonomously. This is why the porosity of the containers must be examined with BET techniques to make optimum distribution. The containers are chemically filled under special conditions. In organic containers, we change the pH inside the shell so as to create a charge on the internal surface positive or negative while at the same pH the chemical compounds are charged in reverse negatively or positive so that the Coulomb attraction forces to pull the molecules into the inside the shell. In the end, the loads are neutralized and the molecules can come out of the containers with an external stimulus like cancer. Such examples were shown in the literature [37]. The loading of the inorganic containers requires a special vacuum device in which the containers are placed inside in a glass pot. The vacuum breaks with chemical input in a liquid form from a funnel. The destruction of the vacuum leads to the encapsulation of the chemicals into the containers. Once the nano/micro-containers are produced with the appropriate cargo, one has to determine their action, like anticorrosion, antibacterial, antifouling, etc. This is demonstrated in the following discussion.

2.3 *Synthesis and Characterization of the Nano/Micro-containers*

2.3.1 Chemicals

All the chemicals were analytical grade reagents. Cerium Acetylacetonate (III) (Ce (ACAC) 3, Aldrich), Polyvinyl Pyrrolidone (PVP, average molecular weight: 55,000, Aldrich), potassium hypersulfate (KPS, Aldrich), titanium tetraisopropoxide (TTIP Aldrich), iron nitrate [$Fe(NO_3)_3$ * $9H_2O$, Aldrich], sodium dodecyl sulphate (SDS, Aldrich), TTIP (Aldrich), PVP, (average molecular weight: 55,000, Aldrich), AMPA (Aldrich), and sodium chloride (Riedel de Haen) were used without further cleaning. Styrene (styrene, Aldrich) were double discharged under reduced pressure before use. 8-Hydroxyquinoline (8-Hydroxyquinoline, 8-HQ) reagents, P-toluene (P-toluene sulfonic acid, P-TSA), 1-H-benzotriazol-4-Sulphuric acid (1-D-benzotriazole-4-sulfonic acid, 1-BSA), 2-mercaptobenzothiazol (2-mercaprobenzothiazole, 2-MB) and 5-amino-1, 3.4-theiodiazol-2-thiol (5-amino-1, 3.4-thiadiazole-2-thiol, 5-ATDT) were used without further cleaning.

Table 1 The conditions used for the manufacture of polystyrene nanospheres at 80 °C

	Methodology I	Methodology II	Methodology III	Methodology IV
Material	Quantity (g)	Quantity (g)	Quantity (g)	Quantity (g)
Styrene	3.62	4.53	9.06	4.53
KPS (a)/AMPA (b)	0.3 (a)	0.65 (a)	1.3 (a)	0.6521 (b)
SDS	0.09	0.21	0.42	
Water	250	460	900	400
Nanosphere size (nm) determined by SEM	150 ± 10	86 ± 10	115 ± 10	300 ± 10

2.3.2 Polystyrene Nanospheres

The method of emulsion polymerization was used to manufacture polystyrene nanospheres. To this end, the reaction was achieved at a 500 cm^3 reactor under the conditions set out in Table 1. To eliminate the effect of oxygen, the solution was purified with argon gas before the begin of the polymerization process. The polymerization lasted 12 h. The polymer solution was centrifuged, the supernatant was discarded and the sediment rinsed with distilled water.

2.3.3 CeMo Nanocontainers

Cerium/Molybdenum oxide (CeMo) nanocontainers were manufactured using negatively charged polystyrene cores. This negatively charged colloidal solution in water mixed with aqueous solutions of $Ce(ACAC)_3$ and molybdenum sodium in the presence of PVP. Subsequently, cerium/molybdenum oxide nanocontainers were obtained by calcination at high temperatures of the coated polystyrene in presence of air. The conditions are given in Table 2. It followed an ageing of the solutions, centrifugation and rinsing of the coated nanospheres with distilled water. For the production of cerium/molybdenum oxide nanocontainers, the polystyrene cores were removed with sintering. Initially, the coated nanosphere were placed in a stirrer and

Table 2 The conditions used to manufacture coated polystyrene nanospheres

	Methodology X
Material	Quantity (g)
Polystyrene	~9
PVP	10.0
$Ce(acac)_3$	5.0
Sodium molybdenum	0.5
Water	1000
Nanosphere size (nm) determined by SEM	145 ± 10

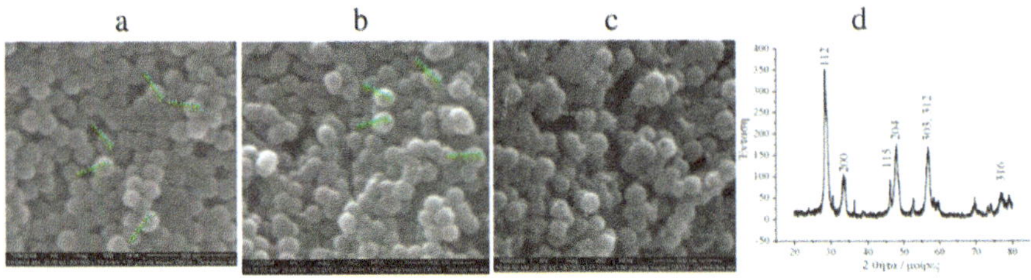

Fig. 2 **a** SEM of polystyrene nanospheres. **b** SEM of polystyrene/CeMo/Au nanospheres (Table 2). **c** SEM of CeMo nanocontainers after sintering at 550 °C. **d** XRD spectrum of CeMo oxide nanocontainers

dried up in glass jar, first at room temperature and then at 60 °C. Then, the composite material was then sintered at 600 °C for 4 h. The heating rate was 10 °C min^{-1}.

Figure 2a depicts the polystyrene nanosphere synthesized as given in Table 2. We observed a homogeneous distribution of their size with an average diameter of about 110 ± 10 nm. Due to necessary gold plating, the diameter is increased by 15 nm. Figure 2b gives the polystyrene nanospheres with cerium/molybdenum and gold coating exhibiting a diameter of 145 ± 10 nm.

After sintering of coated nanospheres, nanocontainers with a diameter of 145 ± 10 nm (Fig. 2c) are obtained. Figure 2d depicts the XRD range of nanocontainers formed at 550 °C. The fact that the peaks are very narrow suggests that the nanocontainers are crystalline. The position of the peaks in the diagram indicates that the nanocontainers consist of cerium molybdenum oxide (33-0330) type (Library: JCPDS-ICDD 1997 International Center for Diffraction Data). More specifically, (112) corresponds to 2-Theta: 28,587, (200) in 2-Theta: 34,196, (115) in 2-Theta: 46,134, (204) 2-Theta: 46,969, (303) and (312) in 2-Theta: 57,875, and (316) in 2-Theta: 76,012.

Figure 3 shows the pore size distribution of the CeMo hollow nanospheres. One can perceive from this figure that four distinctive narrow sizes are observed at 33, 36, 43 and 50 Å.

Fig. 3 Pore size distribution of CeMo

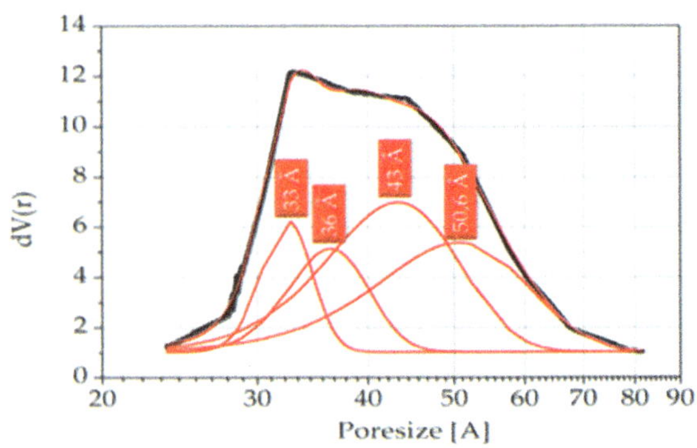

2.3.4 CuO Nanocontainers

The hollow CuO of nanosphere were produced by heating a solution that contained:

- Copper acetate $Cu(OAc)_2$ where the OAc—is acetate ($CH_3CO_2^-$) and has a dark green crystalline color, while the aqueous copper acetate $Cu(OAc)_2 * H_2O$ has more bluish crystalline color.
- Hydrazine was used it as a reducing agent.
- 2-propanol or otherwise isopropanol with a molecular formula C_3H_8O was used as a solvent.

The experiment was done using 2 g of aqueous copper acetate were added under stirring to 100 ml of 2-propanol, which were placed in a triplex flask. The solution was heated by reflux at 81 °C and remained at this temperature for approximately 20 min. Then 2.5 ml of hydrazine aqueous solution was added at dropwise. The originally dark green solution gradually became reddish brown, stayed for about 10 extra minutes while stirring and then if precipitation of solid was observed, separate the liquid part to keep the sediment. After drying in a fume hood, it was centrifuged four times to remove the whole part of the solution, which had not reacted (liquid green) and washed away with absolute ethanol.

Important factors play a role in shaping the size and structure of spheres are:

- The temperature, where the experiment executed. The lower the temperature, the larger spheres.
- The more time the solution remains at a given temperature while stirring, the morphology of the spheres and their size changes.
- The amount of water to which hydrazine will be dissolved yields to increasing the average diameter of the spheres. For example, adding 10 ml of water or 20 ml water leads to spheres of a diameter of 100–300 nm or of 600–700 nm, respectively

Figure 4a shows the SEM of the containers. Their size therefore varies in approximately from 100 to 250 nm with an average diameter of 150 nm.

Figure 4b shows the TGA of the nanocontainers. The TGA curve depicts the loss and the increase of mass. In the diagram below the first fall of the curve (50–180 °C) is due to the loss of mass due to the rejected water and corresponds to 480 μg. The

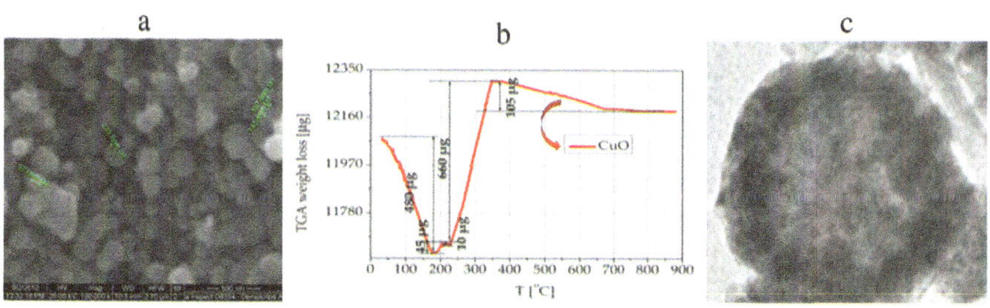

Fig. 4 **a** SEM of CuO nanospheres. **b** TGA curve for CuO nanospheres. **c** TEM of CuO microcontainers

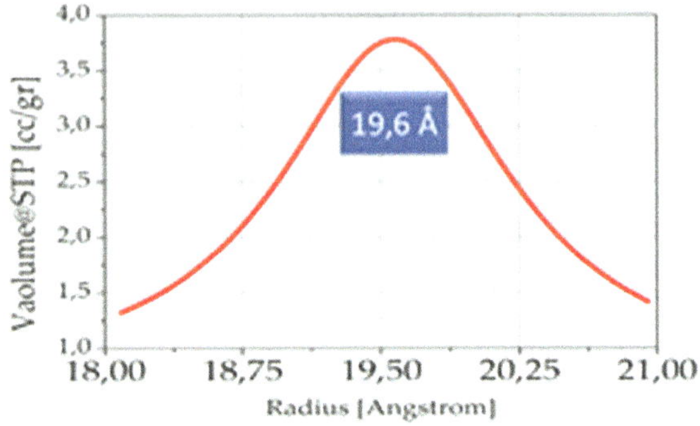

Fig. 5 Distribution of pores

small increase of 45 μg that follows (180–210 °C) is due to the oxidation of Cu_2O ($Cu_2O + 1/2O_2 = 2CuO$), followed by a small drop of 10 mg (210–240 °C), due to the elimination of H_2O molecules that has been absorbed by the product. Then a large increase in mass (240–355 °C) is observed, which is 660 mg due to the oxidation of Cu_2O from the O_2 in the air. In fact, during this period, the product loses mass but at the same time increases, but because the increase is greater than the loss, the final curve rises. Finally, the curve has a drop in the interval of 355–700 °C which is however small compared to the previous loss (105 μg) and above 700 °C, CuO remains with a mass of 12,190 μg, therefore our sample has increased weight of 110 μg. Figure 4c shows the TEM of the CuO containers. One can perceive the hollow nature of the containers from this micrograph. The BET method can measure the specific surface of a sample and study its porosity. Figure 5 shows us the distribution of pores in our sample. As we observe the majority of resources has a radius of ~19.6 Å. There are large pores with 2.5, 3, 4 nm diameter but their concentration is far less.

2.3.5 Water- and Chlorin-Nanotraps

The PMAA has the capacity to absorb water by converting carboxylic groups to the corresponding sodium salts. For the production of nanotraps, 2 g of dry microspheres of PMAA were used in 100 ml of deionized water with ultrasonic stirring. A homogeneous suspension achieved where we added sodium hydroxide (0.40 gr, 10.0 mmol, 0.1 M final concentration). The milky solution became quick transparent. After, we centrifuge the solution at 10,000 rpm for 4 min to give the product form as a hydrogel. Then, we dissolved it again in ethanol and centrifuged at 6000 rpm for 4 min twice. Then it was scraped in a vacuum at 90 °C where it gave the completely dry form of the spherical water traps. Figure 11 shows the water nanotraps produced as described in this section. When 35 gr water are added into a bigger containing 1 gr of nanocontainers, the water is completely absorbed forming a paste. The paste reverses to the nanocontainers of Fig. 6a when water is removed via drying.

The question is how the water absorption capacity depends on the composition parameters. These water traps will be incorporated into epoxy coatings and their

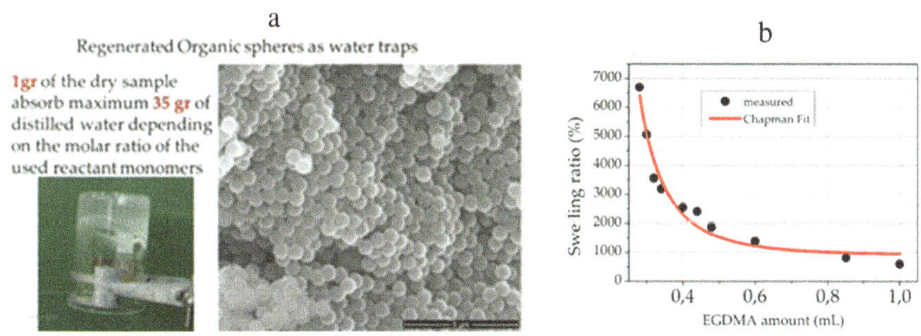

Fig. 6 **a** Water traps adsorbing 35 gr water per 1 gr. **b** Swelling ratio as a function of EGDMA amount

swelling may cause damage to the consistency of the coating. Figure 6b shows the swelling ratio dependency from EDGMA concentration. The dependency follows the Chapman model (y = A * (1−exp(−b * x))^C, A = 922,76,294, b = 5,70,392, and C = −8.56282). The role of traps in coatings to slow water mobility in epoxy coatings was studied theoretically and experimentally [26]. In epoxy coatings incorporated 5% water nanotraps and found to significantly slow the water diffusion rate in the coatings. The diffusion coefficient depends on the concentration of water traps and the morphology of the coatings. In triple epoxy resin coatings that the traps were embedded in the middle layer, the water diffusion coefficient is three times lower than for the pure epoxy coating [26]. Another attempt was made with the production of chlorine nanotraps to slow the diffusion of chlorine molecules in epoxy resins. Their production is first done with the synthesis of silicon dioxide nanospheres with the method Stober [38, 39], secondly by modifying them with 3-glysidoxy-propyl-trimethoxy-silane and finally their modification with diethyl-triamine. Figure 7 shows the structure of the chlorine nanotraps.

2.3.6 Loading of Inhibitors

Nanocontainers were manufactured to be used as filling and releasing corrosion inhibitors. The most studied corrosion inhibitors are: 8-hydroxyquinoline (8-Hydroxyquinoline, 8-HQ), p-toluolomsulphuric acid (P-toluene sulfonic acid, P-TSA), 1-H-benzotriazole-4-Sulphuric acid (1-H-benzotriazole-4-sulfonic acid, 1-BSA), 1-O-benzotriazole (1-H-benzotriazole, 1-UK), 2-mercaptobenzothiazol (2-mercaprobenzothiazole, 2-MB) and 5-amino-1, 3.4-theiodiazol-2-thiol (5-amino-1, 3.4-thiadiazol-2-thiol, 5-ATDT). The extend of filling of nanocontainers with corrosion inhibitors can be determined by the TGA technique by comparing the weight of the nanocontainers filled with 5-ATDT or 2-GB or 8-HQ or 1-BSA with the empty nanocontainers of the same quantity. From the comparison between the two measurements we take the weight of the inhibitors that filled the nanocontainers. This is attempted to show with the Fig. 8 which shows the two measurements of

Conditions used to modify the spheres with the reagent DETA

Reagents		MB (g mol-1)	Density (g ml-1)	Quantity
SiO_2+GPTMS				
monomer	Dithyleno-Triamine	103.17	0.955	0.5ml
	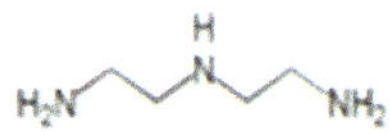 [Bis(2-aminoethyl)amine]			

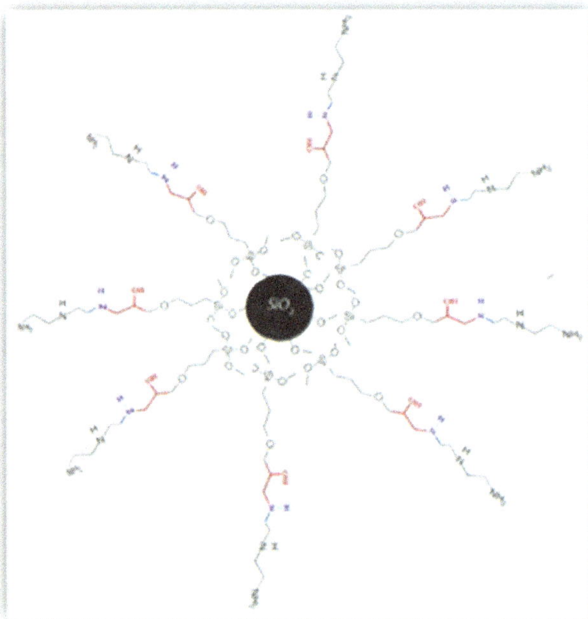

Fig. 7 Chlorine nanotraps

TGA one of a CeMo nanocontainer filled with 8-Hydroxyquinoline (b) and other 8-Hydroxyquinoline (a). Figure 8c is derived from the overlap of measurements a and b where the TGA displacement of the CeMo measurement (c) of the nanocontainer filled with 8-HQ from 8-HQ (a) indicates the weight of 8-HQ; that was trapped in the nanocontainers.

Table 3 summarizes the results of loading we achieved in the nanocontainers with various inhibitors.

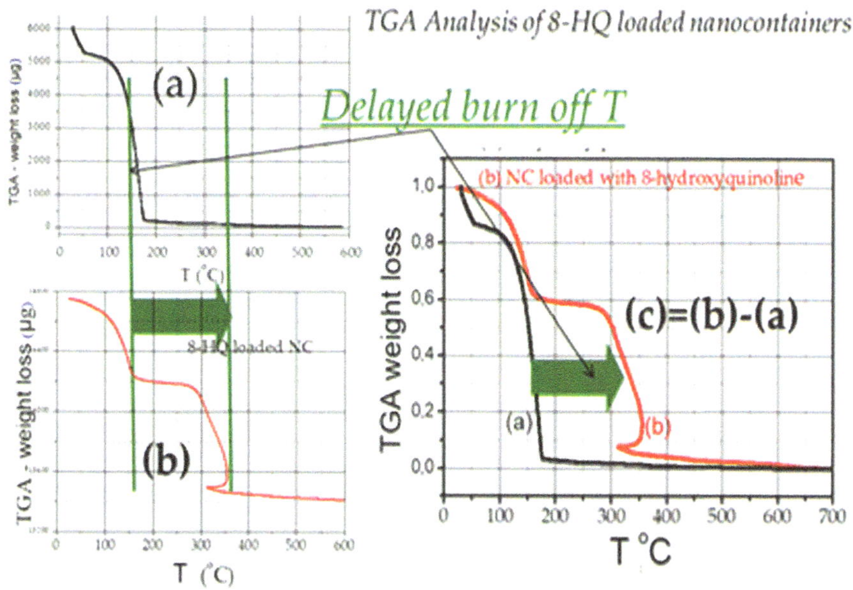

Fig. 8 TGA of 8-HQ (**a**), CeMo (8-HQ) (**b**) and composite TGA (**c**)

Table 3 .Loading achieved in the nanocontainers with various inhibitors

Nanocontainer	Inhibitor	Loading (%)
Titanium oxide	8HQ (8-hydroxyquinoline)	5.3–7.1
Titanium oxide	MBT (2-mercaptobenzothiazole)	5.1–22
Cerium titanium oxide	MBT (2-mercaptobenzothiazole)	19.5–25.3
Cerium molybdate	MBT (2-mercaptobenzothiazole)	12.8–13.4

2.3.7 Release Studies of Inhibitors from the Nanocontainers

In this chapter we will deal with the kinetic release of corrosion inhibitors from CeMo nanocontainers. Figures 9 the release of 8-HQ compound from cerium/molybdenum oxide nanocontainers, respectively. These diagrams show the % permeability versus the time of insertion. At the beginning of time the permeability is 100% because the curves are normalized and release has not happened, yet. As time progresses, the compound is released from the nanocontainers and the instrument begins to absorb the specific wavelength used. This is confirmed by the reduction in % permeability. After the interval of 1500 min, the permeability of the water mixture with 8-HQ reaches 76.1%. The simulations showed that experimental values can be reproduced when with three mechanisms (three-time constants) releasing inhibitors from nanocontainers.

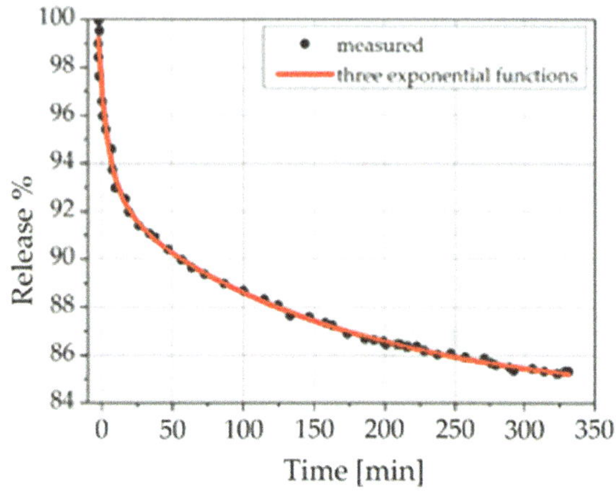

Fig. 9 The release of 8-HQ compound from cerium/molybdenum oxide nanocontainers

2.4 *Antibacterial and Antifouling*

There are applications in planes, ships and cars that require a reduction in the charging of microorganisms. Especially in planes, their insides are charged with microbes after many hours of flight. The steering wheel of cars is also charged with germs which must automatically be killed to avoid infections in drivers and passengers. The possibility of killing microbes from the nanocontainers was examined in the following study.

2.4.1 Assessment of Cell Viability in CeMo Nanocontainers

Figures 10 show the photocatalytic activity of CeMo hollow nanocontainers with respect to the drastic decrease of *E. coli* concentration after illumination with blue-black light on pure culture of *E. coli,* respectively. The system was illuminated with light in the range 350–400 nm and sufficient amount of energy for the nanospheres to

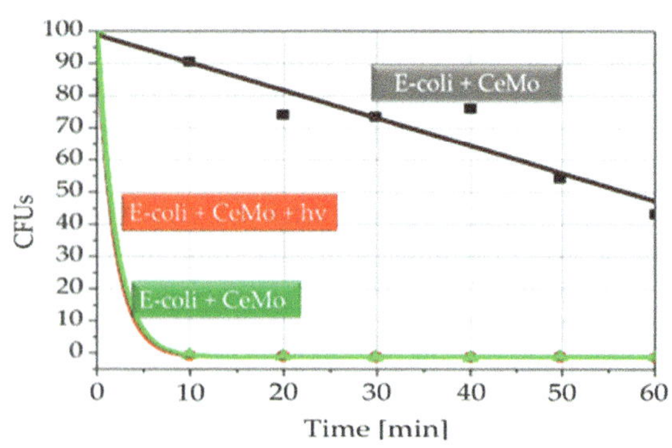

Fig. 10 CeMo hollow nanocontainers photocatalytic activity on cell viability

act as photocatalysts. One very important result is that the CeMo hollow nanospheres exhibit antibacterial activity on *E. coli* in the absence of illumination, namely it is a "zero-light" catalyst. One can think various applications of these nanocontainers.

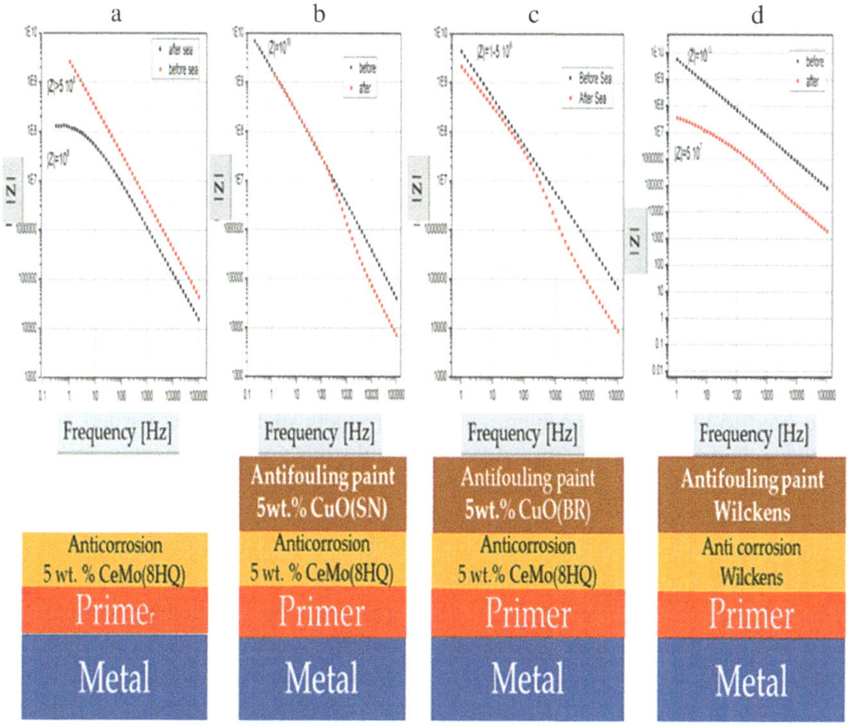

Fig. 11 **a** Shows the FRA of the paint consisting of the primer and anticorrosion paint with 5 wt% CeMo(8HQ) exposed to Mikrolimano sea waters for three months. |Z| is better than to 10^{10} Ω in the sample without exposure to the sea water and |Z| is 10^8 Ω after exposure of the sample to the sea water. **b** Shows the FRA of the primer, the anticorrosion layer with 5 wt% CeMo (8-HQ) and the antifouling top coat with 5 wt% CuO (SN). |Z| is better than 10^{10} Ω in sample unexposed to the sea water and |Z| remains the same after exposure of the sample to the sea water for three months. **c** Shows the FRA of the paint consisting of the primer, anticorrosion paint with 5 wt% CeMo(8HQ) and the antifouling top coat with 5 wt% CuO (SN). |Z| is better than 10^{10} Ω in sample unexposed to the sea water and |Z| is slightly less than 10^{10} Ω after exposure of the sample to the sea water for three months. **d** Shows the FRA of the paint consisting of the primer, Wilckens anticorrosion paint and the Wilckens antifouling paint. |Z| is better than 10^{10} Ω in sample unexposed to the sea water and |Z| drops down to 5 10^7 Ω after exposure of the sample to the sea water for three months

2.5 Antifouling

Antifouling is a process in which a coating is applied to the skeletons of ships in order to avoid the accumulation of organisms on them. Pollution developed in water-immersed structures or ships is divided into two species [40]. The chemical which means that corrosive processes are taking place, which imply that the structures develop a covering of rust and other corrosive products. Organic, where it is done gradually but is much more evident and mainly in a marine environment than in fresh water.

When the surface of a metal is exposed to the atmosphere or to an electrolyte that is automatically ion conductor, then the interface of the metal and the electrolyte becomes load-carrying reactions (electrons) which result in corrosion of the metal. Corrosion of a metal is a destructive attack through electrochemical reaction with the environment.

Previously, the vessels suffered from the development of various organisms that clung to the ship's hull, resulting in its slow during the voyage and its ability to sail against the wind. Both of these decreased its profitability [41, 42]. Several methods were used to resolve this problem. Natural products such as wax, tar and asphalt, wooden enclosures with additional fat or tar and studs positioned next to each other so that these heads create a metal casing. They also tested non-metallic casings such as rubber, vulcanite and cork but were very soon abandoned due to their high cost and difficult application. Copper was originally used as a nail mounted on zinc, nickel, steel and other materials but was soon abandoned, as well as the bronze casing with iron nails due to their corrosion. In 1950 approximately the first organo-metallic dyes with tin, arsenic, mercury and others appeared and after research resulted in tributyltin compounds (TBT; $(C_4H_9)3Sn$), which became particularly known due to its excellent potency and adaptability, but due to its toxicity particularly to oysters, snails and shellfish in general was rejected as well. After research has been found that copper, which has corrosion properties, is safer for the specific use without causing contamination of the environment despite the fact that it has not been ascertained 100%.

In today's era, anti-fouling paints consist of toxic copper, organic ingredients and SeaNine™ that act against the development of organisms. They are applied to the making of ships and to static structures, which are submerged in water, such as docks, bridges, nets, buoys and others. The impact of the development of the organisms on ships is the same as those of older (lower speed, higher fuel consumption and more difficult handling), while in terms of static structures, they become less stable and there may be structural imperfections, which are not easy to distinguish. Usually copper is used as biocidal dye (CuO) and zinc (ZnO). However, because the Zn is less weak as a biocidal product, its use is combined with copper as a booster, resulting in an increase in the toxicity of the final mixture and the transmission of flexibility. Some organisms are still resistant to inorganic Cu and Zn, so it is possible to enrich the anti-fouling paint in order to improve its properties by adding other ingredients that play the role of supporting biocidal products containing Cu, Zn or other [42].

In the epoch of nanotechnology, 2 wt% of CeO_{2-x} nanorods were incorporated into paints to resolve biofouling [43]. The know-how was verified both in the laboratory and in the field. Another group developed vanadium pentoxide nanowires unified in ship paints as anti-biofouling agents [44]. A number of marine organisms like sharks, mussels, and crabs do not exhibit Today, scientists try to imitate these surfaces to solve the problem [45]. In the field of biocides, bromosphaerol was isolated from red algae of the genus Laurencia occurring in the Sea near Corfu. This metabolite was used in a recent work where bromosphaerol was incorporated in CuO and ZnO nanocontainers exhibiting excellent antifouling properties [34].

CeMo nanocontainers have been synthesized by our sol–gel laboratory loaded with 8-hydroxyquinoline (8-HQ) as described in previous chapter. Furthermore, CuO nanocontainers were produced as mentioned above and filled with bromosphaerol. Another set of CuO nanocontainers were filled with SeaNine™. We obtained commercial paints from Wilckens and Re-Turn without any additives and incorporated our nanocontainers and tested the new resulting paints in the Sea water of Mikrolimano (near of Piraeus harbor) and Singapore. The test in Mikrolimano harbor close to Piraeus harbor were done placing the specimens in a tube where on one end a propeller was set moving the water inside the tube with a speed simulating the motion of ships with 14 knots. After three months, the samples were taken out for further laboratory tests. The experiment in Singapore were static placing the samples in a platform hanging for more than 35 months without significant fouling [33–35]. Figure 11a–d summarizes the results of this study.

The measurements show that in the specimen (A) the protection falls significantly after the sample exposure for three months at sea at the port of Mikrolimano. Samples (B) and (C) with coatings that include antifouling paint with 5 wt% CuO (SN) or CuO (BR) offer very good corrosion protection, aside the excellent antifouling protection [32, 33, 36]. In contrast, the sample (D) is worse in protection against corrosion than samples (B) and (C) using nanotechnology.

As far as antifouling protection is concerned, technology has given us very good results in the Wilckens and Re-Turn paints [32, 33, 36]. The results were presented in a publication that the space here does not allow to be presented, again [36]. The technology we developed will persuade the industry to implement it if we prove it works on painted ships for a long time. The warranty given by the manufacturer is a cruising time for a year with 14 knots with a one-month cruise pause a year. Our workshop painted two ships one traveled to the Adriatic Sea between Ancona and Patras painted with the paints of Wilckens and the other in North Sea painted with the paints of Re-Turn. The salinity of sea waters is different and the test is of increased fidelity [46, 47].

Figure 12 shows the two ships we used. The companies left us to paint only one part of the ship with the composition of Fig. 11c and because there are no great possibilities of mass production of nanocontainers we were only able to paint a small fraction of them. The one called Sea Anemos was painted with Wilckens's paints and with our nanocontainers and the other called Berge Arzew painted with Re-Turn paints and sailed in the North Sea for a year. The boats came out after a year from the sea and the results were very impressive. The conclusion is that the fraction of ships

Fig. 12 Two ships painted to a fraction with our technology using commercial paints traveling in two sea with different salinity. Our paints last for a year without any fouling, compared to the commercial paints

painted with our nanotechnology was in excellent condition compared to the paints of the companies as in Fig. 12. An example of high importance for nanotechnology that seems to give solutions to many problems. Here we have a reduction of copper which in commercial paints comes in 40–76%. Reducing copper with own paints is a very good trade result. Because the foul-up is a product that comes from the sea, technology gives us a sensible solution to the problem and will not require extra costs to approve its use in addition to the ecological solution it offers. This is a technological revolution in the field of shipping and nanotechnology. Figure 12 Ships painted with nanotechnology one year traveling in the sea. The autopsy showed that the part that was painted with nanotechnology is in better condition than paints with commercial paints.

It is obvious that the paint industry for corrosion resistant paints and antifouling paints for ships is competitive. There are many companies that have invested large amount of money and developed high-tech products that offer satisfactory solutions. Sigma made the product SYLADVACE™ 700 based on pure silyl acrylate binder. This product demonstrates self-polishing by reducing the thickness of the paint by moving the hull. The AkzoNobel extended the Intersleek™ 900 product to a normal, slippery, low friction surface on which the organisms grow very difficult. The sea Paint Tsugoku, LTD. achieves an exceptionally smooth surface with good self-healing performance. HEMPEL developed a self-polishing anti-fouling paint called Clobic™ NCT. HEMPEL also has the HEMPASIL™ 87500 with great texture in solid. JOTUN has the SeaQuantumPlus™ S to the market for faster boats.

NIPPON launched A-LF-SeaTM incorporating this into polishing copper silyl acrylate antifouling, long-term low friction performance reward is guaranteed and offers a 10% reduction in fuel consumption.

We do not expect to penetrate this area with commercial success because we do not have the funds to refute these technologies of large companies in the sector. All we can do is to propose a new solution based on nanotechnology that perhaps the big company can approach as an alternative to the huge problem of the time. What we offer as new is that new technology will not need licenses to use in different countries because bromosphaerol is a natural product. The CuO nanocontainers anyway is used in an amount of 5 wt% instead of 40–76% of current technology. Because there is a reduction in copper, this will reduce the cost and the copper pollutants in the sea. With them all, we believe that nanotechnology is making a new technological revolution.

3 Corrosion

Our laboratory has been working for the last twenty years to develop the technology of self-healing of corrosion in metals via hollow inorganic nanocontainers [5–36]. There are many publications showing the nanocontainers, the resins we developed and the metals we used [5–36]. Here we will stand in the technology we show with Fig. 13 which incorporates all nano-containers like water traps, chlorine traps, CeMo (8-HQ) nanocontainers and offers multiple properties to protected metals. The coatings on AA2024-T3 metals were exposed to NaCl (aq) 0.5 M for 1, 124 and 288 h. The impedance spectra were measured and displayed in Fig. 13. High frequency spectra

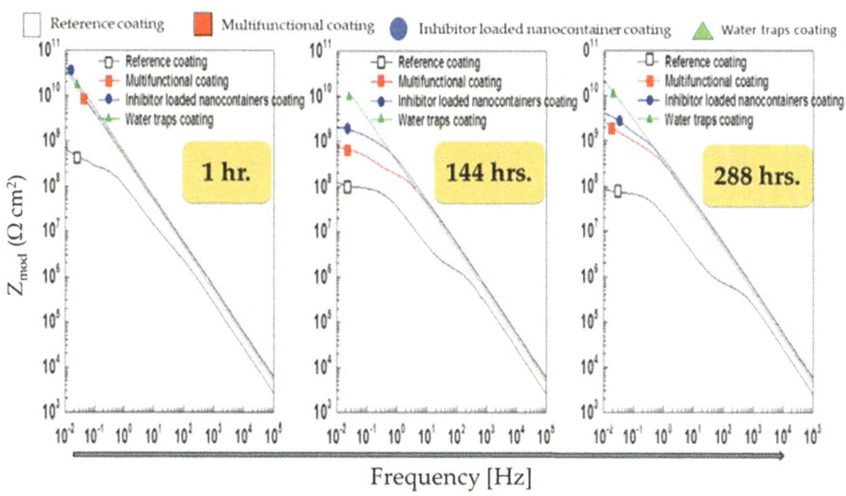

Fig. 13 FRA of multifunctional coating including inhibitor loaded nanocontainers and nanotraps of water and chlorine exposed to 0.5 M NaCl(aq) for 1, 144 and 288 h

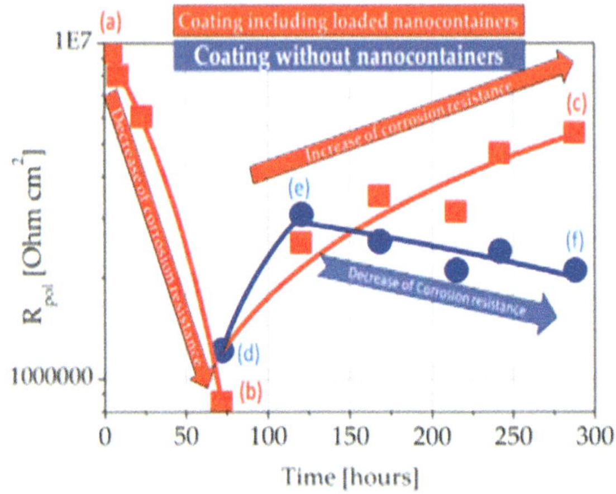

Fig. 14 R_{pol} versus the time of exposure in 0.5 M NaCl (aq) in epoxy coating with ((a)–(b)–(c)-route) and without ((d)–(e)–(f)-route) CeMo (8-HQ) on Mg ZK10 substrate

can be used to understand the development of the protective barrier properties of the coating. Reducing resistance |Z| corresponds to the extent of corrosion. We notice that the protective coatings containing the nanotraps of water and chlorine together with the CeMo (8-HQ) nanocontainers remains unchanged from their original value even in the test of NaCl (aq) 0.5 M 288 h. These results of the FRA revealed that the incorporation of nano- or micro-particles into coatings is of extreme importance for the substantial improvement of the corrosion properties of protective coatings, without damaging the barrier properties and preventing the occurrence of corrosion phenomena. More discussion is done in the publication [32].

Figure 14 shows the R_{pol} versus the time of exposure in 0.5 M NaCl (aq) in epoxy coating with CeMo (8-HQ) on Mg ZK10 substrate. One can perceive from this figure the R_{pol} decrease with the time first from (a) to (b) point and then increases with the time from (b) to (c) time point in coatings with CeMo (8-HQ). The increase of R_{pol} with the time from (b) to (c) is attributed to self-healing of the sample due to the action of 8-HQ from the nanocontainers. On the contrary, we notice a temporary increase from time point (d) to (e) and then we observe a decrease of R_{pol} from time point e to f in samples in coatings without CeMo (8-HQ), due to corrosion.

Important information on self-healing one can get with the SVET technique. Our team made several publications with the teams of UAVR Aveiro and IST Lisbon Portugal [14, 20, 26].

The result of Fig. 14 is of one of many measurements the self-healing phenomenon in our samples. These measurements have been explained in our recent literature extensively [14, 20, 26].

4 Conclusions

The work describes the production of various inorganic nanocontainers and collects the pioneering work done in the Sol–Gel Laboratory of the NCSR Demokritos. This expertise is of technological interest and may in future lead to many other applications, e.g. fertilizers, climatized cloths, nanofluids, etc. that didn't develop during the time we were involved at nanotechnology and our nanocontainers can make significant contribution. This summary of production of nanocontainers can help others to avoid mistakes and make these advances by earning time with the suggestions of this work and published work [5–36]. We have given some examples of successful applications of nanocontainers and future work will show the boundaries of this technology. Hope this work will to inspire young scientists. Our view is that there are many other applications that will cause the development of new smart materials.

Acknowledgements The work is conducted under generous funding from European Union and General Secretary of Research and Technology (GSRT) since 1997. Two EU programs were given the one FP6 MULTIPROTECT and the second FP7 MUST. The GSRT funded the research under MARIPAINTS 1294. We thank our sponsors, numerous scientific collaborators and all the engineers of the participating industries (DAIMLER, FIAT, EADS, Liebherr, Mankiewicz Gebr. & Co., Chemetal GmbH, SIKA, VARNISH Srl, Re-Turn AS and Wilckens Farben GMBH) for the inspiring interactions they are so many not space enough here to list them all but their names can be found in the public domain. The collaboration with the industry was important to determine the compatibility of our nanocontainers with their paints, a task important for commercial applications.

Supported by the Ministry of Education and Science of the Russian Federation on application No. 2020-220-08-6558 for state support of scientific research conducted under the guidance of leading scientists in Russian educational institutions of higher education, scientific institutions, and state scientific centers of the Russian Federation.

References

1. Costa M, Klein CB (2006) Toxicity and carcinogenicity of chromium compounds in humans. Crit Rev Toxicol 36(2):155–163
2. The European Parliament and the Council of the European Union (2003) Directive 2002/95/EC of the European Parliament and of the Council. Off J Eur Union L 37/19
3. https://cordis.europa.eu/project/rcn/74839/factsheet/en
4. http://www.must.risk-technologies.com/
5. Kartsonakis IA, Liatsi P, Danilidis I, Kordas G (2008) Synthesis characterization, and antibacterial action of hollow ceria nanospheres with/without a conductive polymer coating. J Am Ceram Soc 91(2):372–378
6. Kartsonakis IA, Liatsi P, Danilidis I, Bouzarelu D, Kordas G (2008) Synthesis, characterization and antibacterial action of hollow titania spheres. J Phys Chem Solids 69(1):214–221
7. Tapeinos C, Kartsonakis I, Liatsi P, Kordas G (2008) Synthesis and characterization of magnetic nanocontainers. J Am Ceram Soc 91(4):1052–1056
8. Kartsonakis IA, Danilidis IA, Kordas G (2008) Encapsulation of the corrosion inhibitor 8-hydroxyquinoline into ceria nanocontainers. J Sol Gel Sci Technol 48(1–2):2431

9. Tsirimpis A, Danilidis I, Liatsi P, Kordas G (2010) Effect of sweep rate and electrolyte concentration on the potentiodynamic deposition of intrinsically conductive polymer coatings onto Al2024-T3 substrates. Prog Org Coat 67:389–397
10. Kartsonakis IA, Kordas G (2010) Synthesis and characterization of cerium molybdate nanocontainers and their inhibitor complexes. J Am Ceram Soc 93(1):65–73
11. Kartsonakis IA, Danilidis IL, Pappas GS, Kordas G (2010) Encapsulation and release of corrosion inhibitors into titania nanocontainers. J Nanosci Nanotechnol 10(1–9):5912–5920
12. Mekeridis E, Kartsonakis IA, Pappas G, Kordas G (2011) Release studies of corrosion inhibitors from cerium titanium oxide nanocontainers. J Nanopart Res 13(2):541–554
13. Balaskas AC, Kartsonakis IA, Kordas G, Cabral AM, Morais PJ (2011) Influence of the doping agent on the corrosion protection properties of polypyrrole grown on aluminium alloy 2024–T3. Prog Org Coat 71:181–187
14. Balaskas AC, Kartsonakis IA, Snihirova D, Montemor MF, Kordas G (2011) Improving the corrosion protection properties of organically modified silicate–epoxy coatings by incorporation of organic and inorganic inhibitors. Prog Org Coat 72:653–662
15. Balaskas AC, Kartsonakis IA, Tziveleka LA, Kordas G (2012) Improvement of anti-corrosive properties of epoxy-coated AA 2024–T3 with TiO2 nanocontainers loaded with 8-hydroxyquinoline. Prog Org Coat 74(3):418–426
16. Kartsonakis IA, Balaskas AC, George KG (2011) Influence of cerium molybdate containers on the corrosion performance of epoxy coated aluminium alloys 2024–T3. Corros Sci 53:3771–3779
17. Balaskas AC, Kartsonakis IA, Bilalis P, Karatzas A, Kordas G (2011) Epoxy coatings containing nanocontainers loaded with corrosion inhibitors for corrosion. In: Kartsonakis IA, Balaskas AC, Koumoulos EP, Charitidis CA, Kordas G (eds) Protection of AA2024-T3. Meeting Abstracts. pp 1656–1656
18. Kartsonakis IA, Balaskas AC, Koumoulos EP, Charitidis CA, Kordas G (2012) Incorporation of ceramic nanocontainers into epoxy coatings for the corrosion protection of hot dip galvanized steel. Corros Sci 57:50–53
19. Kartsonakis IA, Koumoulos EP, Balaskas AC, Pappas GS, Charitidis CA, Kordas G (2012) Hybrid organic–inorganic multilayer coatings including nanocontainers for corrosion protection of metal alloys. Corros Sci 57:56–66
20. Montemor MF, Snihirova DV, Taryba MG, Lamaka SV, Kartsonakis IA, Balaskas AC, Kordas G, Tedim J, Kuznetsova A, Zheludkevich ML, Ferreira MGS (2012) Evaluation of self-healing ability in protective coatings modified with combinations of layered double hydroxides and cerium molybdate nanocontainers filled with corrosion inhibitors. Electrochim Acta 60:31–40
21. Kartsonakis IA, Balaskas AC, Kordas G (2013) Influence of TiO_2 nanocontainers on hybrid organic-inorganic coatings for corrosion protection of magnesium alloy. Int J Struct Integr 4(1):127–142
22. Balaskas AC, Kartsonakis IA, Kordas G, Cabral AM, Morais PJ (2011) Influence of the doping agent on the corrosion protection properties of polypyrrole grown on aluminium alloy 2024–T3. Prog Org Coat 71(2):181–187
23. Kordas G, Balaskas AC, Kartsonakis IA, Efthimiadou EK (2013) A Raman study of 8-Hydroxyquinoline release from loaded TiO2 nanocontainer. Int J Struct Integr 4(1):121–126
24. Karatzas A, Bilalis P, Kartsonakis IA, Kordas G (2012) Reversible spherical organic water microtraps. J Non-Cryst Solids 358(2):443–445
25. Mekeridis ED, Kartsonakis IA, Kordas G (2012) Multilayer organic–inorganic coating incorporating TiO_2 nanocontainers loaded with inhibitors for corrosion protection of AA2024-T3. Prog Org Coat 73(2–3):142–148
26. Montemor MF, Snihirova DV, Taryba MG, Lamaka SV, Kartsonakis IA, Balaskas AC, Kordas G, Tedim J, Kuznetsova A, Zheludkevich ML, Ferreira MGS (2012) Evaluation of self-healing ability in protective coatings modified with combinations of layered double hydroxides and cerium molibdate nanocontainers filled with corrosion inhibitors. Electrochim Acta 60:31–40
27. Kartsonakis IA, Balaskas AC, Kordas G (2011) Influence of cerium molybdate containers on the corrosion performance of epoxy coated aluminium alloys 2024–T3. Corros Sci 53(11):3771–3779

28. Balaskas AC, Kartsonakis IA, Snihirova D, Montemor MF, Kordas G (2011) Improving the corrosion protection properties of organically modified silicate–epoxy coatings by incorporation of organic and inorganic inhibitors. Prog Org Coat 72(4):653–662
29. Kartsonakis IA, Balaskas AC, Koumoulos EP, Charitidis CA, Kordas G (2012) Incorporation of ceramic nanocontainers into epoxy coatings for the corrosion protection of hot dip galvanized steel. Corros Sci 57:30–41
30. Krzak M, Tabor Z, Nowak P, Warszyński P, Karatzas A, Kartsonakis IA, Kordas G (2012) Water diffusion in polymer coatings containing water-trapping particles. Part 2. Experimental verification of the mathematical model. Prog Org Coat 75(3):207–214
31. Kartsonakis IA, Balaskas AC, Koumoulos EP, Charitidis CA, Kordas G (2013) ORMOSIL-epoxy coatings with ceramic containers for corrosion protection of magnesium alloys ZK10. Prog Org Coat 76(2–3):459–470
32. Kartsonakis IA, Athanasopoulou E, Snihirova D, Kordas G (2014) Multifunctional epoxy coatings combining a mixture of traps and inhibitor loaded nanocontainers for corrosion protection of AA2024-T3. Corros Sci 85:147–159
33. Kordas G (2019) Nanotechnology to improve the biofouling and corrosion performance of marine paints: from lab experiments to real tests in sea. Int J Phys Res Appl 2:053–057
34. Kordas G (2019) Novel antifouling and self-healing eco-friendly coatings for maritime applications enhancing the performance of commercial maritime paints. IntechOpen 1–9
35. Kordas G (2019) Incorporation of spherical shaped CuO@SiO_2 light micro-traps into CuCoMnOx spinels to enhance solar absorbance. J Am Ceram Soc. https://doi.org/10.1111/jace.16851
36. Kordas G (2019) CuO (Bromosphaerol) and CeMo (8 Hydroxyquinoline) microcontainers incorporated into commercial marine paints. J Am Ceram Soc (in press)
37. Kordas G (2019) Nanotechnology in cancer treatment as a Trojan Horse: from the bench to preclinical studies. In Swain SK, Jawaid M (eds) Nanostructured polymer composite for biomedical applications. Micro and nano technologies, pp 323–365
38. Ibrahim IAM, Zikry AAF, Sharaf MA (2010) Preparation of spherical silica nanoparticles: Stober silica. J Am Sci 6:11
39. Kundu D, Nandy S, Naskar MK (2014) Synthesis of mesoporous Stober silica nanoparticles: the effect of secondary and tertiary alkanolamines. J Sol-Gel Sci Technol
40. Davenport J, Oxford P, Awobamise A (1986) A study of the effectiveness of antifouling electrodes in the marine environment. Wat Res 20(6):703–708
41. Singh N, Turner A (2009) Leaching of copper and zinc from spent antifouling paint particles. Environ Pollut 157:371–376
42. Holmes L, Turner A (2009) Leaching of hydrophobic Cu and Zn from discarded marine antifouling paint residues: evidence for trans chelation of metal parathions. Environ Pollut 157:3440–3444
43. Herget K, Hubach P, Push S, Deglmann P, Götz H, Gorelik TE et al (2017) Haloperoxidase mimicry by CeO_{2-x} nanorods combats biofouling. Adv Mater 29:1603823
44. Natalio F, André R, Hartog AF, Stoll B, Jochum KP, Wever R et al (2012) Vanadium pentoxide nanoparticles mimic vanadium haloperoxidases and thwart biofilm formation. Nat Nanotechnol 7:530
45. Magin CM, Cooper SP, Brennan AB (2010) Non-toxic antifouling strategies. Mater Today 13:36
46. Lipizer M, Partescano E, Rabitti A, Giorgetti A, Crise A (2014) Qualified temperature, salinity and dissolved oxygen climatologies in a changing Adriatic Sea. Ocean Sci 10:771
47. Jones JE, Howarth MJ (1995) Salinity models of the southern North Sea. Cont Shelf Res 15:705